Mulching in Agroecosystems

Kashif Akhtar · Muhammad Arif ·
Muhammad Riaz · Haiyan Wang
Editors

Mulching in Agroecosystems

Plants, Soil & Environment

Springer

Editors
Kashif Akhtar
College of Life Science
and Technology
Guangxi University
Nanning, China

Muhammad Riaz
Environmental Sciences
and Engineering
Government College University
Faisalabad, Pakistan

Muhammad Arif
University of Agriculture
Peshawar, Pakistan

Haiyan Wang
College of Agriculture
and Biotechnology
Zhejiang University
Hangzhou, China

ISBN 978-981-19-6409-1 ISBN 978-981-19-6410-7 (eBook)
https://doi.org/10.1007/978-981-19-6410-7

This Springer imprint is published by the registered company Springer Nature Singapore Pte Ltd.
The registered company address is: 152 Beach Road, #21-01/04 Gateway East, Singapore 189721, Singapore

Contents

Soil Section

Effect of Mulches on Mineral Fertilizer (N, P & K) Management and Fertilizer Use Efficiency

Amir Aziz, Noor-us-Saba, Mukkram Ali Tahir, Qura-Tul-Ain, Adeel Ahmad, Ameer Hamza, Humaira Ramzan, and Bilal Ahmad Khan

Abstract Global crop productions are limited due to limited water and nutrient availability. To enhance the availability of nutrients different fertilizers are applied to the soil. But due to extreme temperatures or irregularity in the moisture levels, the efficiency of these fertilizers become reduced. Soil mulching (organic or inorganic) reduces evaporation, control soil temperature and enhances nutrient use efficiency, thereby affects crop yield and production. This chapter highlighted different aspects of mulching including, ways mineral fertilizer loss, strategies to manage mineral fertilizer and fertilizer use efficiency. Additionally, this chapter highlighted the effect of mulch material on mineral fertilizer (N, P & K) management and fertilizer use efficiency. Ways of improving fertilizer use efficiency for some important fertilizers are also discussed in detail.

1 Introduction

Mulch word derived from the German word "molsch" means "easy to decay". It is widely used since ancient times for vegetable production (Lightfoot, 1994). It is defined as on soil surface covered by the spreading of various material to minimizing the population of weed and soil moisture losses and to enhance the yield of the crop (Nalayini, 2007; Kader et al., 2019). Mulches improve soil infiltration, minimize

A. Aziz (✉) · Noor-us-Saba · M. A. Tahir · Qura-Tul-Ain · H. Ramzan
Department of Soil and Environmental Sciences, College of Agriculture, University of Sargodha, Sargodha, Pakistan
e-mail: amirazizuos@gmail.com

A. Ahmad
Institute of Soil and Environmental Science, University of Agriculture Faisalabad, Faisalabad, Pakistan

A. Hamza
School of Environment Sciences and Engineering, China University of Geosciences, Wuhan, China

B. A. Khan
Depaertment of Agronomy, College of Agriculture, University of Sargodha, Sargodha, Pakistan

K. Akhtar et al. (eds.), *Mulching in Agroecosystems*,
https://doi.org/10.1007/978-981-19-6410-7_1

water runoff, control weeds population and also control evapotranspiration. Mulching shows other important environmental effects like minimum nutrient losses, reduce soil compaction and soil erosion, improve soil physical condition and also affect the regulation of temperature and plant roots in soil (Lamont, 2005; Ngouajio & McGiffen, 2004).

To enhance crop productivity and save soil from environmental degradation mulching of field is an appropriate agronomic technique (Giller et al., 2009; Knowler & Bradshaw, 2007). The things that can be used for mulching include polythylene sheets, straw of wheat and barley, corn stems, stones, sand, pebbles, geotextiles of biological nature, vegetative remaining as well as trample stones (Mo et al., 2016; Prosdocimi et al., 2016). There are several benefits of mulching, that's why the status of this technique in the agriculture sector is good. One of its topmost benefits is a decrease in evaporation which in turn improves water use efficiency as well as crop yield. (Awe et al., 2015; Jemai et al., 2013; Li et al., 2015). Secondly, the soil temperature is also maintained which is useful for better germination of seeds and in the initial stages of development also helps in the growth of roots (Zhang et al., 2015; Siczek et al., 2015). Its third benefit includes the change in the microbial biomass as well as soil fertility, which consequently improves soil health and crop productivity (Zhang et al., 2011; Balota et al., 2014; Qiu et al., 2014; An et al., 2015; Huoet al., 2017; Munoz et al., 2017). Fourthly, the organic C of soil is maintained by mulching and nutrient cycles are regulated, consequently, crop yield is increased (Bationo et al., 2007; Liu et al., 2009; Naab et al., 2015; Wang et al. 2016). The metabolism of soil is also improved as mulching improves the enzymatic activities of soil (Elfstrand et al., 2007). Weeds can also be controlled effectively by mulching (Campiglia et al., 2015; Jabran et al., 2015; Nawaz et al., 2017; Splawski et al., 2016). Moreover, in the rainy season mulches prevent soil from runoff and enhance the water penetration capacity (Adekalu et al., 2007; Jordan et al., 2010; Smets et al., 2008). It improves soil aggregation as well as the structure of the soil (Luna et al., 2016). In summers weeds and soil-borne pathogens can also be checked by using black polyethylene mulches. Amongst all the above-mentioned reimbursement of mulching the most vital is soil moisture conservation change in soil temperature (Bu et al., 2013; Li et al., 2004; Montenegro et al., 2013; Zhu et al., 2015). On the other hand, the factors that affect the mulches include the time of mulching, duration of mulching; the material used for mulching, tillage practices and either the mulch is applied in furrows or ridges (Edwards et al., 2000; Li et al., 2004).

2 Mineral Fertilizers and Problems Occurring with Mineral Fertilizers

Mineral fertilizer is materials either manufactured or natural, containing nutrients that are essential for plant growth and development. Mostly mineral fertilizer term is used as synthetic, artificial or chemical fertilizer. Nitrogen, phosphorus and potassium these three nutrients that have to be applied in large quantities to plants. Sulphur, magnesium and calcium are also required in considerable amounts. All these nutrients are major constituents of plant components such as proteins, chlorophyll, nucleic acid etc. All these components are essential for some process like enzyme action, maintenance of internal pressure and energy transfer. Almost seven nutrients are required to plants in small quantities are called micronutrients. The deficiency of any one nutrient can compromise plant growth and development. The relationship is very strong between fertilizer consumption level and agricultural productivity. After various inputs in agriculture after irrigation, contribute to an increase in agriculture production (Pilbeam, 1996).

Soil nutrients can failures to replenish in many countries. This problem can be solved through efficient and balanced use of plant nutrients and through improving the soil management practices. Some plant nutrients requirements can be fulfilled by the application of organic material that is available on the farm or in the community. This material is insufficient to replenish the nutrients of the plant removed from the soil. Plant obtained most of the nutrients from organic manure, soil reserves or recently added fertilizers. Plants uptake of nitrogen (50–70%), phosphate (15%) and potash (50–60%) nutrients during the application season (Pilbeam, 1996). Nitrogen uptake by plant and proportion in soil are varied widely in response to differences in evaporation and rainfall. Under control conditions, almost 50–70% applied nitrogen can be uptake by the plant and in practically, nitrogen losses can be much greater. Nitrogenous fertilizer can be lost by erosion, gaseous emission or leaching. All these processes can vary widely and depending on the environment and agricultural system. Similarly, denitrification and ammonium volatilization also varies and depends on form of N fertilizer used, agro-ecosystem, environmental condition and crop management. Sometime problems arise after the large application of nitrogen fertilizer like ammonium volatilization and denitrification in sugar cane, cotton and rice crop. Farmers are unconcerned about the excessive application of nitrogen fertilizer when economic situation is good, but environmental problems are arises. Many approaches are available now days to control the nitrogen losses by ammonium volatilization and denitrification (Peoples et al., 1995). By the application of mulch former can control the loss of nitrogen. On the other hand, by improving management practices reduced gaseous loss of nitrogen about 14 kg/ha (Matson et al., 1998). By the loss of phosphate and potash from soil system also decrease the crop yield and which represent financial loss to farmer. It may also cause environmental risk, in that soil lost by wind and water erosion to stream, lakes and rivers. Phosphate has both effects; direct and indirect. Increase phosphate availability show positive effect on quality and quantity of crop. Through indirect interaction phosphate increases crop production by adding nitrogen and potassium and also show positive effect on soil organic matter, biological nitrogen fixation, soil erosion control, water holding capacity and other physio-chemical properties of soil (Baanante, 1998).

3 Reasons for Low Fertilizer Use Efficiencies

Soil pH is the major factor of nutrient use efficiency in the soil system. This is for two reasons. Firstly, extremes of soil pH can decrease crop growth and development. For example, at low pH, the toxicity of manganese and aluminium can restrict the crop growth and yield and at high pH of soil deficiencies of micronutrients limits the growth and yield. Secondly, in the soil system, pH markedly affects the chemistry of phosphorus and ultimately its effect on phosphorus adsorption by the interaction that precipitation of phosphorous into solid forms in the soil system (Willett & Higgins, 1978).

The amount and type of clay present in the soil system are strongly affected nutrients availability and phosphorous adsorption. Soils having high clay content retain more nutrients (especially phosphorous) strongly. In high sandy soil, nutrients (P) do not retain and leaching from the soil system. Most phosphorous uptake by the plant from the soil system through the diffusion process (higher concentration in soil solution and low concentration at root surface). On the other hand, drought can severely decrease phosphorous use efficiency. Flooding of soil can also reduce the oxygen status in the soil system and ultimately reduced the nutrient use efficiency by plants (Willett & Higgins, 1978).

4 Ways to Improve Fertilizer Use Efficiencies

The soils with elevated Phosphorus retaining capacity due to adsorption reactions, placing the Phosphorus as band placement is the safest managing practice for soluble Phosphorus fertilizers as this lessens the extent of soil fertilizer contact as well as restricts powerful adsorption. Additionally, broadcast Phosphorus is superlative for sparingly soluble fertilizers (Chien et al., 2009).

Phosphorus fertilizers including MAP, DAP and TSP have analogous Phosphorus use efficiency in many soils, provided there are no further confines to crop development (deficiency of nitrogen will favour MAP, DAP over TSP, or deficiency of calcium will favour TSP). Rock phosphates and struvite are less soluble which give an alkaline reaction in the soil that will usually be less efficient than soluble Phosphorus sources except for acidic soils, or soils susceptible to Phosphorus leaching. Soil acidification enhances the usage efficiency of Phosphorus acidifying fertilizers will have benefits over that alkaline in reaction. Of the various additives and microbial inoculants asserted to enhance usage efficiency of Phosphorus, not a bit have been shown up to date to regularly deliver substantial advantages (Chien et al., 2009).

5 Importance of Mulching

In the top 30 cm of soil, most of the water is available. So, upper soil areas are required to be remained moist to encourage root growth as well as to deliver sufficient water for the plant. Mulching performs a substantial role in conserving soil water. Additionally, mulches inhibit weeds and retain a narrow array of temperature in the soil. Therefore, soil structure, soil moisture as well as optimum fertilizer levels will improve onion production. Inorganic mulches (such as black or white polythene) or organic mulch are a logical expense and preserve soil moisture. Using residues of plants or synthetic materials as a mulch material is a well-recognized procedure for enhancing the effectiveness of several horticultural crops (Mukherjee et al., 2004).

Mulches can conserve soil moisture and reduce evaporation ultimately reducing the irrigation requirements. The mulch materials act as barriers against raindrops beating action and irrigation water which carry spores of diseases. These spores are attached to plant shoots and foliage. Mulches provide nutrition to many organisms (beneficial) which competes against pathogenic spores or inhibit the pathogens by releasing chemicals. In this way, mulching reduces the chances of disease occurring. Mulches also an important part of integrated pest management (IPM) (Chalker-Scott, 2007).

Heavy metals are very harmful to both humans as well as animal's health. Mulching material is a good source for the removal of these harmful metals from the soil system (Chalker-Scott, 2007). For the removal of heavy metals, leaves of pine, poplar and eucalyptus are mostly used (Salim & El-Halawa, 2002). In forest areas, compost and woodchips are used that convert copper metal into a form that is not/less toxic for plant growth and development (Kiikkila et al., 2002).

Mulches deal with different pathogens by decreasing the stress level on plants. Plants get resistant to weed attack. In this way, there will be no use of any type of herbicides and insecticides. Decline the use of these chemicals leads to favour of beneficial organisms in soil and environment and also non-use of these chemicals leads to favour of farmers in sense of no money is used to purchase such chemicals (Chalker-Scott, 2007).

Whenever people use fertilizer, mulches and synthetic chemicals, they estimate the benefits outcomes and cost from investment. Compare to synthetic material, mulches are not so costly in term of crop growth and soil health. By the use of mulches, there will be no cost of purchasing pesticide and other weeds control methods. For rehabilitation of damaged lands, we can use local wood debris to enhance crop growth and development and increase farmer income. Timber and peat harvest residues are locally available mulch material that is economical and enhance crop growth (Kader et al., 2019).

6 Importance of Mulching on Fertilizer Dynamics

Mulches enhance the nutrient status of soil, conserve soil moisture, control soil temperature, weeds control in crop, control erosion losses, and remove the residual effect of heavy metals, fertilizers and pesticide. On the other hand, organic mulch influences the properties of soil and also affect soil health and fertility. Mulch materials also increased the availability of organic carbon, potassium and phosphorus that enhance crop yield and growth. Mulch materials also affect soil pH, nutrient availability and soil salinity (Kar & Kumar, 2007).

6.1 Soil Fertility Improvement

The organic mulches show beneficial impacts on soil health in term of improving nutrient levels. However, material type, climate conditions and characteristics of soil determine the decrease, increase, or no effect on soil nutrients. The application of organic mulches is more beneficial. It is because organic mulches decomposed in the soil system and providing the plant nutrients (described in Fig. 1). Different organic mulches (straw, bark, green manure and wood chips) provide plant nutrients as compared to inorganic mulch materials (Ansari et al., 2001). Mulch materials containing high nitrogen content increase crop yield and production. On the other hand, mulches having low nitrogen content (straw, sawdust and bark) also increase plant nutrients and soil fertility (Chalker-Scott, 2007).

6.2 Lowering the Soil pH

A few mulches are acidifying the soil. However, there is no scientific evidence about soil acidification by the application of mulching. The bark of wood chips and some trees are the main source of acidification. Organic mulch soils are more acidic as compared to inorganic mulch or bare soil. In a nursery, the application of woody mulch can produce phenolic acid due to woody materials decomposition. However, in field conditions, there is a very acidifying effect by the application of woody materials. Some researchers found that, in acidic soil, there is no acidifying effect. But in alkali soil, organic mulch material shows a positive effect for lowering the soil pH (Chalker-Scott, 2007).

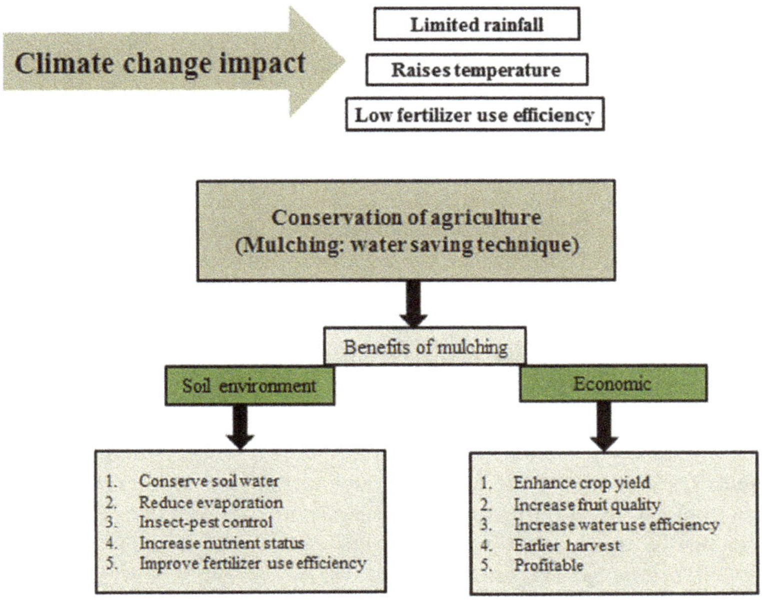

Fig. 1 Effect of mulching on fertilizer use efficiency

7 Mulching and Fertilizer Use Efficiencies

By the application of organic mulching, soil organic matter increased. Ultimately, nutrient availability increased and enhances the biological as well as physical properties of soil (Kader et al., 2019; Wei et al., 2009). Management practices (tillage, irrigation, straw mulching and fertilization) have a significant effect on soil enzymes and total soil organic matter (Ali et al., 2018; Lefèvre et al., 2014; Smith et al., 2011). Application of fertilizers (organic or inorganic), crop rotation and tillage practices improve the soil microorganisms and soil ecosystem. To check the nutrient availability, fertilization practices are an important indicator in soil and are linked with crop production and fertility of the soil. Soil enzymes activities are also increased by the added nitrogen and carbon into soil (Akhtar et al., 2018). Soil enzymes activities have also involved in the availability of nutrients and nutrient cycling. Long term organic mulches application increased soil organic matter and soil enzyme activities. Researchers found that by increasing soil temperature, enzymes sensitivity also increased (Bowles et al., 2014; Stone et al., 2012; Zhou et al., 2013). Atmospheric temperature increases due to global climate warming and ultimately negative impact on crop growth and development. So, a scientist had new innovative approaches for improving efficiency and yield of the crop with minimum impact on ecosystem and environment. For this innovation, direct seeding of wheat after the harvesting of rice and eliminated the burning of residue thus stabilized soil organic matter and reduced environmental pollution (Anand et al., 2016).

8 Conclusion and Remarks

Application of mulching can suppress weed population, conserve soil moisture by reducing evaporation and enhance nutrient status in soil. Different mulching materials significantly impact crop growth, yield and quantity. However, it can be concluded from the literature that, mulches are a good and cheap source to conserve soil moisture, reduce weed populations and control soil temperature. Therefore, in drought/water deficit conditions, the water requirement of the crop could be compensated by properly managed mulching strategies. Moreover, integrated use of mulch with partial rootzone drying is an efficient technique to enhance crop yield and development. All type of mulches improves soil quality concerning crop growth and yield. Besides, mulching also improves fertilizer use efficiency by reducing their losses. Mulching not only creates a hindrance against the volatilization of fertilizers but also improves the moisture and nutrient status of the soil. Though mulching shows positive effects on yield, water use efficiency and nutrient use efficiency. In the future attention could be focused to manage mineral fertilizer (N, P & K) and nutrient use efficiency by using organic and inorganic mulches in combination, as organic mulches will enhance the organic matter of soil as well as nutrient use efficiency.

References

Adekalu, K. O., Olorunfemi, I. A., & Osunbitan, J. A. (2007). Grass mulching effect on infiltration, surface runoff and soil loss of three agricultural soils in Nigeria. *Bioresource Technology, 98*, 912–917.

Akhtar, K., Wang, W., Ren, G., Khan, A., Feng, Y., & Yang, G. (2018). Changes in soil enzymes, soil properties, and maize crop productivity under wheat straw mulching in Guanzhong China. *Soil and Tillage Research, 182*, 94–102.

Ali, S., Xu, Y., Jia, Q., Ma, X., Ahmad, I., Adnan, M., & Jia, Z. (2018). Interactive effects of plastic film mulching with supplemental irrigation on winter wheat photosynthesis, chlorophyll fluorescence and yield under simulated precipitation conditions. *Agricultural Water Management, 207*, 1–14.

An, T., Schaeffer, S., Li, S., Fu, S., Pei, J., Li, H., Zhuang, J., Radosevich, M., & Wang, J. (2015). Carbon fluxes from plants to soil and dynamics of microbial immobilization under plastic film mulching and fertilizer application using 13C pulse-labeling. *Soil Biology & Biochemistry, 80*, 53–61.

Anand, K., Kumari, B., & Mallick, M. A. (2016). Phosphate solubilizing microbes: An effective and alternative approach as biofertilizers. *Journal of Pharmacy & Pharmaceutical Sciences, 8*(2), 37.

Ansari, R., Marcar, N. E., Khanzada, A. N., Shirazi, M. U., Crawford, D. F. (2001). Mulch application improves survival but not growth of Acacia ampliceps Maslin, Acacia niloticaL. and Conocarpuslancifolius L. on a saline site in southern Pakistan. *International Journal of Review, 3*, 158–163.

Awe, G. O., Reichert, J. M., Timm, L. C., & Wendroth, O. (2015). Temporal processes of soil water status in a sugarcane field under residue management. *Plant and Soil, 387*, 395–411.

Balota, E. L., Yada, I. F., Amaral, H., Nakatani, A. S., Dick, R. P., & Coyne, M. S. (2014). Longterm land use influences soil microbial biomass P and S, phosphatase and arylsulfatase activities, and S mineralization in a Brazilian Oxisol. *Land Degradation and Development, 25*, 397–406.

Bannante, C. A. (1998). Economic evaluation of the use of phosphate fertilizers as a capital investment. In A. E. Johnston & J. K. Syers (Eds.), *Nutrient management for sustainable crop protection in Asia*. CAB International.

Bationo, A., Kihara, J., Vanlauwe, B., Waswa, B., & Kimetu, J. (2007). Soil organic carbon dynamics, functions and management in West African agro-ecosystems. *Agricultural Systems, 94*, 13–25.

Bowles, T. M., Acosta-Martínez, V., Calderón, F., Jackson, L. E. (2014). Soil enzyme activities, microbial communities, and carbon and nitrogen availability in organic agroecosystems across an intensively-managed agricultural landscape. *Soil Biology and Biochemistry, 68*, 252–262.

Bu, L., Liu, J., Zhu, L., Luo, S., Chen, X., Li, S., Lee Hill, R., & Zhao, Y. (2013). The effects of mulching on maize growth, yield and water use in a semi-arid region. *Agricultural Water Management, 123*, 71–78.

Campiglia, E., Radicetti, E., & Mancinelli, R. (2015). Cover crops and mulches influence weed management and weed flora composition in strip-tilled tomato (*Solanum lycopersicum*). *Weed Research, 55*, 416–425.

Chalker-Scott L. Impact of mulches on landscape plants and the environment—A review. *Journal of Environmental Horticulture, 25*(4), 239–249.

Chien, S. H., Prochnow, L. I., & Cantarella, A. H. (2009). Recent developments of fertilizer production and use to improve nutrient efficiency and minimize environmental impacts. *Advances in Agronomy, 102*, 267–322.

Edwards, L., Burney, J. R., Richter, G., & MacRae, A. H. (2000). Evaluation of compost and straw mulching on soil-loss characteristics in erosion plots of potatoes in Prince Edward Island Canada. *Agriculture Ecosystems and Environment, 81*, 217–222.

Elfstrand, S., Bath, B., Mårtensson, A. (2007). Influence of various forms of green manure amendment on soil microbial community composition, enzyme activity and nutrient levels in leek. *Applied Soil Ecology, 36*, 70–82.

Giller, K. E., Witter, E., Corbeels, M., & Tittonell, P. (2009). Conservation agriculture and smallholder farming in Africa: The heretics' view. *Field Crops Research, 114*, 23–34.

Huo, L., Pang, H., Zhao, Y., Wang, J., Lu, C., & Li, Y. (2017). Buried straw layer plus plastic mulching improves soil organic carbon fractions in an arid saline soil from Northwest China. *Soil and Tillage Research, 165*, 286–293.

Jabran, K., Mahajan, G., Sardana, V., & Chauhan, B. S. (2015). Allelopathy for weed control in agricultural systems. *Crop Protection, 72*, 57–65.

Jemai, I., Aissa, N. B, Guirat, S. B., Ben-Hammouda, M., Gallali, T. (2013). Impact of three and seven years of no-tillage on the soil water storage, in the plant root zone, under a dry subhumid Tunisian climate. *Soil and Tillage Research, 126*, 26–33.

Jordan, A., Zavala, L. M., & Gil, J. (2010). Effects of mulching on soil physical properties and runoff under semi-arid conditions in southern Spain. *CATENA, 81*, 77–85.

Kader, M. A., Singha, A., Begum, M. A., Jewel, A., Khan, F. H., & Khan, N. I. (2019). Mulching as water-saving technique in dryland agriculture. *Bulletin of the National Research Centre, 43*(1), 1–6.

Kar, G., & Kumar, A. (2007). Effects of irrigation and straw mulch on water use and tuber yield of potato in eastern India. *Agricultural Water Management, 94*(1–3), 109–116.

Kiikkila, O., Derome, J., Brugger, T., Uhlig, C., & Fritze, H. (2002). Copper mobility and toxicity of soil percolation water to bacteria in metal polluted forest soil. *Journal of Plant Soil, 238*, 273–280.

Knowler, D., & Bradshaw, B. (2007). Farmers' adoption of conservation agriculture: A review and synthesis of recent research. *Food Policy, 32*, 25–48.

Lamont, W. J. (2005). Plastics: Modifying the microclimate for the production of vegetable crops. *Horticulture Technology, 15*, 477–481.

Lefèvre, V., Capitaine, M., Peigné, J., Roger-Estrade, J. (2014). Farmers and agronomists design new biological agricultural practices for organic cropping systems in France. *Agronomy for Sustainable Development, 34*(3), 623–632.

Li, F., Wang, J., Xu, J., & Xu, H. (2004). Productivity and soil response to plastic film mulching durations for spring wheat on entisols in the semiarid Loess Plateau of China. *Soil and Tillage Research, 78*, 9–20.

Li, X., Jin, M., Huang, J., & Yuan, J. (2015). The soil–water flow system beneath a cotton field in arid North West China, serviced by mulched drip irrigation using brackish water. *Hydrgeology Journal, 23*, 35–46.

Lightfoot, D. R. (1994). Morphology and ecology of lithic-mulch agriculture. *Geographical Review, 25*, 172–185.

Liu, C. A., Jin, S. L., Zhou, L. M., Jia, Y., Li, F. M., Xiong, Y. C., & Li, X. G. (2009). Effects of plastic film mulch and tillage on maize productivity and soil parameters. *European Journal of Agronomy, 31*, 241–249.

Luna, L., Miralles, I., Andrenelli, M. C., Gispert, M., Pellegrini, S., Vignozzi, N., Solé-Benet, A. (2016). Restoration techniques affect soil organic carbon, glomalin and aggregate stability in degraded soils of a semiarid Mediterranean region. *Catena, 143*, 256–264.

Matson, P. A., Naylor, R., Ortiz-Monasterio, I. (1998). Integration of environmental, agronomic, and economic aspects of fertilizer management, *280*, 112–114.

Mo, F., Wang, J., Xiong, Y., Nguluu, S. N., & Li, F. (2016). Ridge-furrow mulching system in semiarid Kenya: A promising solution to improve soil water availability and maize productivity. *European Journal of Agronomy, 80*, 124–136.

Montenegro, A. A., Abrantes, J. R. C. B., de Lima, J. L. M. P., Singh, V. P., & Santos, T. E. M. (2013). Impact of mulching on soil and water dynamics under intermittent simulated rainfall. *CATENA, 109*, 139–149.

Mukherjee, S., Paliwal, R., & Pareek, S. (2004). Effect of water regime, mulch and kaolin on growth and yield of ber (*Ziziphus mauritiana* Lamk). *The Journal of Horticultural Science and Biotechnology, 79*(6), 991–994.

Munoz, K., Buchmann, C., Meyer, M., Schmidt-Heydt, M., Steinmetz, Z., Diehl, D., Thiele-Bruhn, S., & Schaumann, G. E. (2017). Physicochemical and microbial soil quality indicators as affected by the agricultural management system in strawberry cultivation using straw or black polyethylene mulching. *Applied Soil Ecology, 113*, 36–44.

Naab, J. B., Mahama, G. Y., Koo, J., Jones, J. W., & Boote, K. J. (2015). Nitrogen and phosphorus fertilization with crop residue retention enhances crop productivity, soil organic carbon, and total soil nitrogen concentrations in sandy-loam soils in Ghana. *Nutrient Cycling in Agroecosystems, 102*, 33–43.

Nalayini, P. (2007). Poly-mulching a case study to increase cotton productivity. Senior scientist, Central Institute for Cotton Research, Regional Station, Coimbatore.

Nawaz, A., Farooq, M., Lal, R., Rehman, A., Hussain, T., & Nadeem, A. (2017). Influence of sesbania brown manuring and rice residue mulch on soil health, weeds and system productivity of conservation rice-wheat systems. *Land Degradation and Development, 28*, 1078–1090.

Ngouajio, M., & McGiffen, M. E. (2004). Sustainable vegetable production: Effects of cropping systems on weed and insect population dynamics. *Acta Horticulture, 638*, 77–83.

Peoples, M. B., Freney, J. R., & Mosier, A. R. (1995). Minimizing gaseous losses of nitrogen. In P. E. Bacon (Ed.), *Nitrogen fertilization in the environment*. Marcel Dekker, Inc.

Pilbeam, C. J. (1996). Effect of climate on the recovery in crop and soil of 15N labeled fertilizer applied to wheat. *Fertilizer Research, 45*, 209–220.

Prosdocimi, M., Jordan, A., Tarolli, P., Keesstra, S., Novara, A., & Cerda, A. (2016). The immediate effectiveness of barley straw mulch in reducing soil erodibility and surface runoff generation in Mediterranean vineyards. *Science of the Total Environment, 547*, 323–330.

Qiu, Y., Wang, Y., & Xie, Z. (2014). Long-term gravel—Sand mulch affects soil physicochemical properties, microbial biomass and enzyme activities in the semi-arid Loess Plateau of North-western China. *Acta Agricultrae Scandinavica, Section B. Soil & Plant Science, 64*, 294–303.

Salim, R., & El-Halawa, R. A. (2020). Efficiency of dry plant leaves (mulch) for removal of lead, cadmium and copper from aqueous solutions. *Process Safety and Environmental Protection, 80*(5), 270–276.

Siczek, A., Horn, R., Lipiec, J., Usowicz, B., & Łukowski, M. (2015). Effects of soil deformation and surface mulching on soil physical properties and soybean response related to weather conditions. *Soil and Tillage Research, 153*, 175–184.

Smets, T., Poesen, J., & Knapen, A. (2008). Spatial scale effects on the effectiveness of organic mulches in reducing soil erosion by water. *Earth-Science Reviews, 89*, 1–12.

Smith, A. N., Reberg-Horton, S. C., Place, G. T., Meijer, A. D., Arellano, C., & Mueller, J. P. (2011). Rolled rye mulch for weed suppression in organic no-tillage soybeans. *Weed Science, 59*(2), 224–231.

Splawski, C. E., Regnier, E. E., Harrison, S. K., Bennett, M. A., & Metzger, J. D. (2016). Weed suppression in pumpkin by mulches composed of organic municipal waste materials. *HortScience, 51*, 720–726.

Stone, M. M., Weiss, M. S., Goodale, C. L., Adams, M. B., Fernandez, I. J., German, D. P., & Allison, S. D. (2012). Temperature sensitivity of soil enzyme kinetics under N-fertilization in two temperate forests. *Global Change Biology, 18*(3), 1173–1184.

Wang, Y. P, Li, X. G, Fu, T., Wang, L., Turner, N. C., Siddique, K. H. M., & Li, F. M. (2016). Multi-site assessment of the effects of plastic-film mulch on the soil organic carbon balance in semiarid areas of China. *Agriculture,* 42–51.

Wei, W., Chen, L., Fu, B., Lü, Y., & Gong, J. (2009). Responses of water erosion to rainfall extremes and vegetation types in a loess semiarid hilly area, NW China. *Hydrological Processes: An International Journal, 23*(12), 1780–1791.

Willett, I. R., & Higgins, M. L. (1978). Phosphate sorption by reduced and reoxidized rice soils. *Soil Research, 16*(3), 319–326.

Zhang, S., Li, P., Yang, X., Wang, Z., & Chen, X. (2011). Effects of tillage and plastic mulch on soil water, growth and yield of spring-sown maize. *Soil and Tillage Research, 112*, 92–97.

Zhang, F., Li, M., Qi, J., Li, F., & Sun, G. (2015). Plastic film mulching increases soil respiration in ridge-furrow maize management. *Arid Land Research and Management, 29*, 432–453.

Zhou, Z., Jiang, L., Du, E., Hu, H., Li, Y., Chen, D., & Fang, J. (2013). Temperature and substrate availability regulate soil respiration in the tropical mountain rainforests, Hainan Island China. *Journal of Plant Ecology, 6*(5), 325–334.

Zhu, L., Liu, J., Luo, S., Bu, L., Chen, X., & Li, S. (2015). Soil mulching can mitigate soil water deficiency impacts on rainfed maize production in semiarid environments. *Journal of Integrative Agriculture, 14*, 58–66.

Effects of Mulching on Soil Biota and Biological Indicators of Soil Quality

Noor-ul-Ain, Ali Aslam, and Fasih Ullah Haider

Abstract The concept of soil health has gained importance recently, recognizing the soil as a living entity. In the recent scenario of urbanization and excessive land use, agricultural land is subjected to degradation and desertification. For sustainable agriculture production and ecological interactions, there is a dire need for management strategies to improve soil health and quality. Mulching is among the important conservation strategies to enhance soil health by improving soil biota, organic contents, and soil aggregation. In this chapter, we encompassed the different categories of living entities dwelling in soil and their key activities to enhance ecological relations of soil. Based on the literature study, mulches are proved to be very efficient in improving soil biota, soil moisture retention, maintaining the soil temperature, nutrient dynamics, decrease in severity of soil contaminants, suppression of weeds, and control in insects pests. The addition of mulch in the soil fluctuate a number of indicators of soil biota, which account for soil health. Species diversity, microbial biomass, soil respiration, organic content, and enzymatic respiration mainly determine quality status of soil biota, which are influenced by mulches. We have also given the overview of indices of species diversity, i.e., richness, evenness, and phylogenetic indices altered by the introduction of mulches in soil and thus modify the ratio of pests predators. Moreover, based on field conditions, crop and mulch type, and environment-specific application of mulch can become more productive for soil conservation, plant growth and soil biota.

Noor-ul-Ain (✉)
Center for Genomics and Biotechnology, Fujian Provincial Key Laboratory of Haixia Applied Plant Systems Biology, Fujian Agriculture and Forestry University, Fuzhou, Fujian 350002, China
e-mail: noorulainali001@gmail.com

A. Aslam
Faculty of Agriculture and Veterinay Sciences, Superior University, Lahore, Pakistan

F. U. Haider
College of Resources and Environmental Sciences, Gansu Agricultural University, Lanzhou 730070, China

Gansu Provincial Key Lab of Arid-Land Crop Science, Gansu Agricultural University, Lanzhou 730070, China

K. Akhtar et al. (eds.), *Mulching in Agroecosystems*,
https://doi.org/10.1007/978-981-19-6410-7_2

1 Introduction

1.1 Importance of Soil Biota for Soil Health

Soil biota acts as one indicator of soil health, enabling the soil to function as a living system in an ecosystem for sustainable productivity. Biological activity in the soil mainly occurs in the 30 cm top layer of soil (Serrano et al., 2009). This layer of soil comprises of less than 0.5% biological components and 10% of organic matter generally. Biological/Living components of soil are the soil organisms inhabiting and being decomposed in the soil, for example, protozoa, earthworms, microarthropods, enchytraeid, arthropods, termites, fungi, algae, bacteria, and soil flora (Roger-Estrade et al., 2010). Regardless of being small in size, soil biota acts as a key player in nutrient cycling and accelerate the decomposition process of organic residues. Many protozoans and insects dwelling in soil favour mechanical mixing, enhancing the physical structure while microbes largely contribute in nutrient dynamics in the soil. In simple words, the energy cycle in soil ecosystem is driven by microbes mediated decomposition of deceased plants parts, animals, and organic matter. Thus, these organic constituents are possibly converted into biomass or subjected to the mineralization process yielding CO_2, H_2O, mineral nitrogen, phosphorous etc. (Curtin et al., 2012). The mineralized nutrients are consumed by microbivores such as protozoa and nematodes (Bloem et al., 1997). Likewise, waste material and synthetic organic compounds transformation and degradation are also mediated by soil microbes (Stenberg, 1998).

1.2 Soil Biota Components

Soil is a big reservoir of living entities which are interacting in the diverse system for stabilized soil ecology. Soil components based upon the size (length and diameter) are categorized as macrofauna, mesofauna, and microfauna (Huera-Lucero et al., 2020).

Macrofauna, as their name, indicates a class of large-sized organisms with their diameter ranging from 2 to 20 cm. For example, earthworms, gastropods, isopods, myriapods, some araneidae, and the majority of insects. This category includes invertebrates mostly and is regarded as soil engineers in terms of mechanical manipulation of the soil (Cabrera et al., 2011). Other important members of this category include beetles, ants, snails, slugs, centipedes, millipedes, and, enquitraeid worms.

Intermediate sized organisms having body size between 1 and 2 mm is known as mesofauna. They are also known as microarthropods, belong to an invertebrate group. Their examples include nematodes, rotifers, tardigrades, small araneidae, pseudoscorpions, opiliones, enchytraeids, insect larvae, small isopods, and myriapods (Scheu et al., 2005). However, the key members are of this class are mites and springtails, which constitute a large portion of this group. A square meter of land encompasses thousands of species belonging to this group. In the forest ecosystem, they form the key reservoir and significantly affect the decomposition process (Maillard et al., 2019). They serve as a connecting bridge between macro fauna and micro fauna in the terrestrial ecosystems and act as key players among soil decomposers, and are mainly involved in nutrients fluxes and transformation of leaf litter and organic matter. Owing to a regulatory role in nutrients fluctuation and fluxes, they are regarded as webmaster of the ecosystem (Dervash et al., 2018). Organisms whose bodies size ranges between 20 and 200 μm come under the category of micro fauna. Major representatives of this class are protozoa, fungi, and bacteria. Nevertheless, the upper limit of this class also includes small mites, nematodes, rotifers, tardigrades, and copepod crustaceans. Like a predator, they feed on fungi and bacteria; their pathogenicity activity makes them a bio-control agent and also influences microbial biomass significantly in the soil.

2 Mulching and Soil Health Management

Mulch is defined as a covering of soil by the use of organic or inorganic material to improve plant performance by retaining soil moisture, maintaining soil temperature, reducing weed growth, inhibiting erosion, increasing fertility, and nutrient balance, and avoiding diseases and pests (Robichaud & Ashmun, 2013). Based on materials used for soil covering, mulches are generally classified into two broad types, i.e., inorganic and organic (Table 1). Nevertheless, usage of mulch is strictly dependent on its properties to affect soil characters, decomposing ability, resilience, and most importantly, their ease of access. Mulching increases the roughness of land surface, thereby reducing transportation, controlled flow of water that keeps the soil and water intact (Foltz & Wagenbrenner, 2010; Montenegro et al., 2013; Prats et al., 2016).

2.1 Organic Mulch

Plant and animal materials, ground-covers and compost are common organic mulches and are successfully being employed in agricultural farming (Montenegro et al., 2013). Organic mulches come from plants and animals sources and have reportedly been shown to improve soil health effectively (Adekalu et al., 2007; Teame et al., 2017). Organic mulches are further divided into living and non-living. Common examples of living mulches are C. mucunoides (leguminous), cowpea (field crop) and bracharia (grass) whereas common nonliving mulches include wheat, palm, and rice debris from the plant sources and cow, pig, poultry, horse, and goat faeces from the animal source are examples of nonliving mulches (Abrantes et al., 2018; Akhtar et al., 2018; Gholami et al., 2013). To nurture soil health and to promote sustainability, plant residues are used as mulches as well (Berglund et al., 2006).

Table 1 Describes the role of organic and inorganic mulches on soil moisture and temperature, indirectly affecting soil biota and eventually on crop yield

Mulch type	Crop	Year of cultivation	Impact on soil and biota	Impact on crop yield	Reference
Organic					
Rice straw mulch	Groundnut	2000–2001	Improved the soil temperature and water holding capacity of soil	Improved the crop yield by 28.73%	Ramakrishna et al. (2006)
Maize straw	Soybean	1990	Improved the soil respiration and microbial biomass by 52 and 26% respectively. Also enhanced the enzymatic activity in soil	–	Rabary et al. (2008)
Wheat straw mulch	Maize	2005	Improved the soil organic matter and soil moisture content by 50.87 and 17% respectively and decreased the soil bulk density by 4.9%	Improved the plant height, biological yield, and grain yield by 5.1, 8.08, and 18.1% respectively	Pervaiz, Muhammad and Khuram (2009)
Bahiagrass straw mulch	Sorghum	2006–2007	Significantly enhanced the cation exchange capacity, soil organic carbon, total nitrogen, and available phosphorous contents in soil		Obalum et al. (2011)

(continued)

Table 1 (continued)

Mulch type	Crop	Year of cultivation	Impact on soil and biota	Impact on crop yield	Reference
Wheat straw mulch	Soybean	2008	Minimized the soil compaction and significantly improved the soil bulk density and nodule formation in soil by 22.5 and 93.8% respectively	Significantly enhanced the protein content and yield of plant	Siczek and Lipiec (2011a)
Wheat straw mulch	Watermelon	2010	Improved the temperature and moisture % of soil	Enhanced the crop yield by 10.96%	Parmar et al. (2013)
Rice straw mulch	–	2012	Regulates the soil temperature and soil moisture and significantly minimized the soil erosion and runoff by 21 and 51%, respectively	–	Montenegro et al. (2013)
Rice straw mulch	Rice	2014–2015	Significantly improved the soil moisture, temperature and soil microbial carbon in soil	Improved the root and shoot biomass of plant and enhanced the crop yield by 40%	Dossou-Yovo et al. (2016)
Inorganic					
Saw-dust	Calla lily	1996	Improved the soil moisture and temperature	Significantly improved the tuber yield by 90%	Wright and Burge (2000)
Polythene plastic	Groundnut	2000–2001	Improved the soil temperature and water holding capacity of soil	Improved the crop yield by 40.87%	Ramakrishna et al. (2006)

(continued)

Table 1 (continued)

Mulch type	Crop	Year of cultivation	Impact on soil and biota	Impact on crop yield	Reference
Saw dust	Strawberry	2010	Improved the soil moisture contents and maintain the C:N ratio in soil	Enhanced the crop yield by 79%	Kumar and Dey, (2011)
Polythene plastic	Watermelon	2010	Improved the temperature and moisture % of soil	Enhanced the crop yield by 25.62%	Parmar et al. (2013)
Gravel-sand	Watermelon	2006–2008	Significantly decreased the soil evaporation and improves the soil temperature and water use efficiency	Enhanced the crop yield by 21.9%	Wang et al. (2011)
Polythene plastic	Watermelon	2015	Improved the soil temperature	Enhanced the crop yield by 25%	Maughan and Drost (2016)

2.2 Inorganic Mulch

Inorganic mulches are a cost-effective mean as a soil management strategy popular in low-income countries and are mostly used for persistent agriculture (Ngosong et al., 2016). Recently, the use of plastic mulch has gained hype due to its ability to reduce moisture loss, especially in drought-stricken areas (Li et al., 2004; Yu et al., 2018). Nevertheless, merit and demerits vary with types of mulch being used; for instance, it was reported that organic mulch positively impacts root systems and nodule formation while the negative effect of plastic mulch was seen (Dukare et al., 2017). It is desirable to understand the properties and their potential effect on soil flora and fauna before application at the farmer's level. Plastics, gravels, soils, and carpets are listed in inorganic mulches; their utility varies with purpose, i.e., from protecting soil to erosion, extreme weather, weed control, and moisture retention as well (Ingman et al., 2015). The area under plastic mulch cultivation has been exponentially stretched all over the world, considering water scarcity and their possible ability to cope with less water supply. China has stepped forward as the top most consumer of plastic mulch (Ingman et al., 2015). The use of plastic mulch offers an effective way to curtail water loss, also being an economic strategy for saving water (Table 1). The use of plastic mulch is accompanied by a number of protective roles, i.e., avoid soil erosion, suppress diseases, pests, and weed proliferation, save plants from heat, cold, flooding and drought with improvement in food quality and production (Espí et al., 2006). According to (Zhou et al., 2009), with the use of plastic mulch, production of maize, and yield was boosted, and it served as the best strategy to save water and to regulate the temperature in dry land areas. As a result of plastic mulch, water use efficiency was enhanced, which led to a loss in subsoil water, thereby increasing crop transpiration rate and increasing yield in relation to traditional water application methods. The use of plastic mulch and its intended benefits is dependent on several factors, including type and quality of mulch being used, the surface under coverage, weather conditions, soil type their, and interactive effects. Different colours of plastic mulch showed differential effect which is associated with water conditions and the intended objective for which they are being used, and it offers great challenge (Ashrafuzzaman et al., 2011; Ocharo, Korir & Gweyi-Onyango, 2017).

The impact of plastic mulch on soil biodiversity should ascertain to positively regulate the soil properties and to improve sustainable agriculture food production. For that reason, idiosyncratic responses are pieces of evidence that could support the proliferation of several disease and insect's suppression of others (Torres Bojórquez et al., 2017). It has reportedly been mentioned that plastic mulch supports the arthropod and omnivorous population and their species diversity and regulated the fungi and bacterial community (Addison et al., 2013; Farmer et al., 2017; Qin et al., 2017). Plastic mulch offers higher protection in summer squash against watermelon mosaic virus, and their use was beneficial to protect some cultivars against viruses (Boyhan et al., 2000; Walters, 2003). A number of studies noted no effect of plastic mulch at all or decline in number and density of carabid beetle, springtails, earthworms, parasitic, and predatory organisms and the soil food-web structure (Addison et al., 2013; Stirling & Eden, 2008; Tuovinen et al., 2006).

The massive use of plastic films also includes impact environment, soil and a series of other effects affecting agriculture developments (Gao et al., 2019). The negative effect of a decline in the abundance of soil invertebrate, suppression of microbial growth, and bacterial composition was reported (Schirmel et al., 2018). Plastic mulch regulates biotic and abiotic factors, which lead to the extensive effect of soil microbe's activity and their performance. Hence, the use of plastic mulch, in the long run, offers great danger or drastic effect to soil biota of arable lands and leads to severer consequences to functional ecosystems (Steinmetz et al., 2016). Above all, it increases soil pollution due to their poor breakdown ability, which offers another threat to our environment (Liu et al., 2014; Vox et al., 2016; Wang et al., 2016). In addition, the uses of plastic mulch could cause a negative impact, which needs more in-depth and detailed experiments to be conducted in different regions and ecological zones of the world in order to get a good understanding of their use and its effect on soil microbes.

3 Mulches for Enhancing Biological Activities in Soil

Biological activities in soil are inevitable in order to sustain soil health, plant development and to transform nutrients into acceptable form. Fungi, bacteria, and algae constitute living flora of the soil, while living fauna is formed by protozoa, earthworms, termites, nematodes, arthropods, microarthropods, enchytraeid, etc. (Lal, 1988; Roger-Estrade et al., 2010). Bacteria, fungi and other micro life forms in the soil are involved in nutrient reservoir remodelling and their storage, intact soil particles, decomposition of organic matter, and perform a crucial role in nitrogen and carbon cycle and many other processes that are important to plants. Specifically, biological activities of soil microflora and fauna occur at the highest rate in the rhizosphere near the root zone of plants (Potthoff et al., 2005; Waid, 1999). Detail overview of the importance and metabolic activity was described in detail by Waid (1999).

Greater nutrient and organic matter content, better soil texture, and porosity as well infiltration is directly proportional to biological activities (Lal, 1988; Paz-Ferreiro & Fu, 2016). Among the factors that regulate biological actives include are fertilization, tillage operation, crop rotation, soil amendments, type, and characters of soil, soil composition and crop (Bonilla et al., 2012). Mulches enhance biological activities has become a globally known event (Lal, 1988).

An increase in microbial activity in soil is directly linked to enhanced activities of enzymes such as urease, dehydrogenase, urease, and β-glucosidase. Besides several other beneficial effects, the application of organic mulches stimulates and facilitate the soil flora and fauna effectively (Lal, 1988). In the wake of beneficial effects of soil mulches on soil health, it is suggested to leave the crop remains on the soil in order to maintain physical properties of soil along with increase biological performance (Kahlon et al., 2013; Kashif et al., 2020).

Mulches with a low C/N ratio are regarded as high-quality mulches and are better in strengthening the diversity, growth and population of soil life forms. For example, it was reported that the earthworm population was affected by the use of mulch and its quality. Compared to low-quality mulch, the earthworm population was denser, with high-quality mulch having high C/N ratio application (Tian et al., 1997). The effect of mulch is more pronounced and noticeable in the topsoil layer just after the application of mulch (Yang et al., 2003).

Cultural practices are important to stimulate and facilitate the microorganism in the soils and are also beneficial for sustainable food production and reduction of the incidence of disease and pests (Abawi & Widmer, 2000). Organic mulches are preferably applied as a desire to encourage a favourable microbial community, which results in enzymatic stimulation, thus improving crop performance.

4 Indices of Soil Biota

4.1 Soil Respiration

Soil respiration is the flux of CO_2 released by soil organisms, i.e., plant roots, microorganisms, and soil animals (earthworms or nematodes), as a result of their biological activity (Fu et al., 2020). Recent studies highlighted that soil respiration is an indicator of the soil's ability to support plant growth and soil microbes (Wang et al., 2018). Soil respiration is an indication of the level of microbial development, plant debris, soil organic matter (SOM), and decomposition (Luo et al., 2001; Zheng et al., 2009). In addition, soil respiration in the soil is needed to preserve the soil quality, nutrients transformation (which may be utilized by plants), and as a result, improve the plant growth. Reduction in soil respiration rates means that the soil has little or no SOM or aerobic microbial activity (Raich & Tufekcioglu, 2000; Striegl & Wickland, 2001). It may also indicate that soil properties, i.e., aeration, available nitrogen, and temperature, which has a significant effect on soil respiration, are restricting the SOM decomposition and biological activity (Yinkun et al., 2013). Furthermore, nutrients are not absorbed from SOM to meet the nutrients requirement of plants and soil species due to a reduction in soil respiration (Ren et al., 2018). This has an effect on plant root respiration, which can contribute to plant death (Li et al., 2016). In flooded or saturated soils, incomplete mineralization of SOM occurs regularly, resulting in the formation of compounds such as alcohol and methane, which are classified as toxic to plant roots. Sulfur volatilization and denitrification are normal in such prevailing conditions leading to causes pollution of greenhouse gases and acid deposition (Fu et al., 2020; Li et al., 2016). Recent studies highlighted that agricultural practice such as the incorporation of organic mulches in soil tends to improve the SOM, which usually enhances soil respiration (Liu et al., 2016). It was noted that combination of conventional tillage with decomposing inoculants and wheat (*Triticum aestivum* L.) straw as a mulch significantly enhanced the soil respiration in the hilly region of the South-West region of China by 1.01–5.58 μmol $(m^2\ s)^{-1}$ as compared to sole conventional tillage having no mulch and decomposing inoculants (Sun et al., 2019). In addition, it was also observed that mulching of straw significantly enhanced the soil respiration in soil from 13.7 to 68.7%. In another study (Zhang et al., 2015a, 2015b) reported that soil respiration was significantly improved by the addition of mulches in soil, i.e., 4089 μmol $(m^2\ s)^{-1}$ as compared to soil having no mulches (806.78 μmol $(m^2\ s)^{-1}$). The incorporation of organic straw as mulch generally expected to influence the release of CO_2 from the soil (Shaohui & Jingyu, 1997). In the dryland region of the Loess Plateau, it was observed that the utilization of straw mulching as an amendment significantly enhanced the soil respiration of winter wheat farmland (Guan et al., 2011). In another study (Zhang et al., 2005) observed that the rate of soil respiration tends to improve significantly with the increase in the decomposition rate of straw in the farmland. In a recent study, it was also observed the application of straw as mulch enhanced the soil respiration at the early phase of incubation and then decreased gradually during the later phase of incubation (Fu et al., 2020). Furthermore, fast and rapid microbial community proliferation may have resulted in the allocation of more substrates to their proliferation and development than to respiration, trends in lowering the soil respiration (Lee et al., 2012). Soil amended with straw mulches has a higher rate of soil respiration due to a higher concentration of SOC and organic carbon (Wang et al., 2018). Organic mulches, i.e., rhizodeposition, plant litter, and straw, serves as a substrate for soil microbial population to mineralize into CO_2 (Whitaker et al., 2014). Hence it was concluded that soil microbial population composition was not the only significant determinant of soil respiration, and SOC and potential carbon mineralization played a prominent role in deciding the variation of soil respiration (Zhang et al., 2005). However, in certain studies, it was noted that soil respiration rate under mulching and conventional tillage practices failed to achieve a consensus due to variation in climate, soil, and cropping pattern (Guan et al., 2011), which also highlighted that further deep analysis based research needs to be done to explore the potential effects of mulches on soil respiration with respect to topography, soil, climate, and cropping patterns.

4.2 *Enzymatic Activity*

Soil enzymes catalyze the decomposition of plant residues; play a vital role in nutrients cycling and the release of plant-available nutrients (Siczek & Lipiec, 2011a; Burns et al., 2013; Jabran, 2019). The materials on which soil enzymes act are known as substrate, i.e., plant litter and straw (Acosta-Martinez et al., 1999; Downer et al., 2001). Living and dead microbes, plant roots, and residues and soil organisms are all sources of soil enzymes (Akhtar et al., 2018). Enzymes that have been stabilized in the soil matrix aggregate or form complexes with organic matter (humus), clay, and humus-clay complexes but are no longer connected to viable cells (Rabary et al., 2008; Richter et al., 2011). Enzymatic activities in soil are affected due to variation in soil temperature and pH (Pavan Fernandes et al., 2005). For the recycling of macro-compounds, i.e., pectin, chitin, hemicelluloses, lignin, cellulose, etc., microbial enzymatic activity is needed (Jabran, 2019). Some enzymes, i.e., glucosidases and hydrolases, are only active in the degradation of organic matter, while others are involved in nutrients mineralization, i.e., sulfates, phosphatases, ureases, and amidases (Akhtar et al., 2018). However, there is no clear evidence available that linked enzymatic activity to nutrients supply or crop production, with the exception of phosphatase activity (Burns et al., 2013). Since nutrients mineralization to plant-available forms is achieved with the aid of enzymatic activity, the correlation may be indirect (Wang et al., 2018). In recent years it was highlighted that application of mulches in soil significantly enhanced the enzymatic activities of soil. It was noted that the application of mulches significantly enhanced the beta-glucosidase, amidase, FDA hydrolysis, sulfatase, urease, and phosphatase that improves the nutrients cycling in soil (Zhang et al., 2019; Zheng et al., 2006). In another study, it was recorded that application of straw mulches in soil significantly improved the beta glucosidase, and phosphatase activity in soil by 59.48 and 65.68%, respectively, higher as compared to natural fallow land having no mulches (Rabary et al., 2008). Similarly, (Zhang et al., 2015a, 2015b) observed that the application of wheat straw as mulch in maize (*Zea mays* L.) field significantly improves the enzymatic activities of urease, invertase, protease, and dehydrogenase in soil. Likewise, (Siczek & Lipiec, 2011a) reported that the application of wheat straw as mulch in soybean (*Glycine max* L.) field improved the enzymatic activity of nitrogenase in soil. Application of maple (*Acer palmatum* L.) leaves as mulch in ryegrass (*Lolium perenne* L.) field improves the enzymatic activity of beta-glucosidase and dehydrogenase in soil (Acosta-Martinez et al., 1999; Hai-Ming et al., 2014) reported that incorporation of ryegrass as mulch in rice (*Oryza sativa* L.) field improves the enzymatic activities of phosphatase, arylsulfatase, arylamidase, and beta-glucosidase in soil. Correspondingly, (López et al., 2014) observed that the application of almond (*Prunus dulcis* L.) shells as mulch in avocado (*Persea americana* L.) field significantly enhanced the enzymatic activity of protease, dehydrogenase, and phosphomonoesterase in soil. Yet, to explore the potential effects of mulches on soil enzymatic activities, further deep research is required.

4.3 Organic Carbon and Total Nitrogen

Organic mulches decompose over time, adding organic matter, and nutrients to the soil that are becoming a part of the soil (Huang et al., 2008; Sainju et al., 2008). The key benefit of organic mulches is that they maximize the amount of soil organic carbon (SOC) in soil (Li et al., 2010; Monaco et al., 2008). Soil organic carbon and total nitrogen play a vital role in the soil fertility and quality, as it potentially affects the soil chemical, biological, and physical properties that can affect the agro-ecosystems and crop production (Ge et al., 2015; Peng et al., 2018).

Therefore, conservation of a satisfactory proportion of SOC and total nitrogen in the soil is essential for sustainable agro-ecosystem and crop production (Fig. 1). Recent studies highlighted that application of mulches on a long-term basis in agricultural soil can significantly improve the level of SOC and total nitrogen in the soil. (Hosseini Bai et al., 2014), observed that application of forest residues as mulch improves the water retention in soil and can significantly improve the SOC and total nitrogen in soil by 78.85 and 81.08% higher than soils having no addition of mulch. Similarly, (Bajoriene et al., 2013), conducted four (2008–2011) years study and reported that the application of peat as a mulch in soil significantly improves the SOC in soil by 53.43, 47.28, 54.25, and 46.86%, respectively, higher with respect to years than soil having no mulch. Correspondingly, synthetic and inorganic mulches may not have a proportion of organic matter in it, yet they can enhance the properties of soil by maintaining optimum soil water conservation that significantly increases the soil organic residues decomposition (Hosseini Bai et al., 2014; Al-Bayati & Hamdoon, 2019). Combined application of gravel rocks and date palm (*Phoenix dactylifera* L.) as mulch in a country like Saudi Arabia significantly improves the total nitrogen in soil by 65.9% higher as compared to soil having no mulch (Alharbi, 2017). In another study (Shahadat Hossen et al., 2017), reported that application of organic mulch, i.e., saw-dust improves the SOC and total nitrogen in soil by 63.64 and 68.46%, respectively higher than control, while in the same experiment, it was recorded that application of black polythene sheet as mulch enhances the SOC and total nitrogen in soil by 45.95 and 45.33% higher than soil having no mulch. Similarly, in China, (Peng et al., 2018) reported that application of maize straw potentially improved the SOC and total nitrogen in soil by 14.51 and 7.52% higher than soil having no mulch, while application of plastic mulch enhanced the SOC and total nitrogen in soil by 2.86 and 5.13% higher than soil having no mulch. Likewise, incorporation of maize straw mulch significantly improves the SOC and total nitrogen of soil by 68.70 and 46.06% higher as compared to control (Yang et al., 2020). Still, further research analysis needs to done to explore the potential effects of mulches on SOC and total nitrogen contents in soil.

4.4 Microbial Biomass Carbon (MBC)

One of the sensitive indicators of microbial activity and soil organic matter is microbial biomass carbon (MBC) (Wick et al., 1998). A number of soil management properties greatly influence the content of soil organic carbon and microbial biomass carbon. Measuring and interpreting microbial biomass gives an indication of soil health, quality, sustainability, and aggregation (Smith et al., 1994). The soil covering techniques, for example, mulching, already discussed above, plays a pivotal role in modulating soil temperature and moisture, thus modifying soil ecology. This is

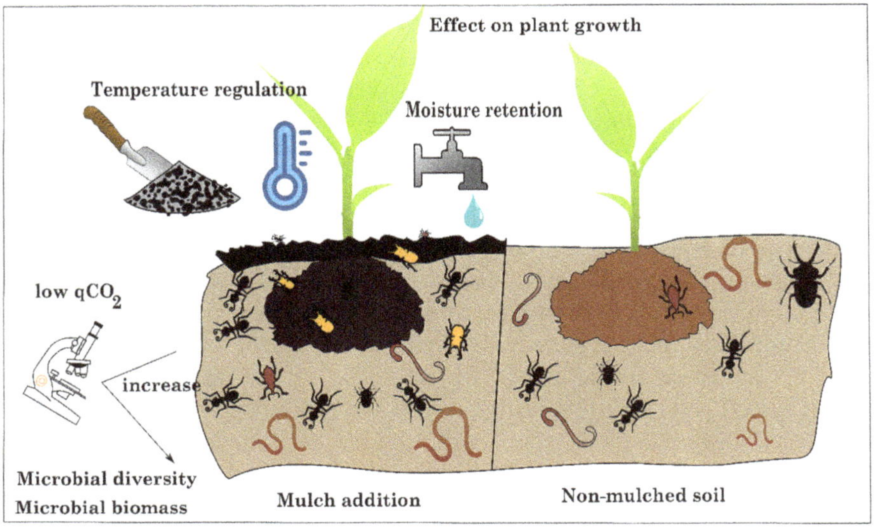

Fig. 1 This figure illustrates the changes in soil fauna due to the addition of mulch in the soil. Microbial diversity and biomass are enhanced due to the regulation of temperature and moisture in the soil

directly related to soil quality in terms of nutrient dynamics and nutrients availability which is accomplished by microbes in the soil (Fig. 1). Microbial biomass has been used as an index for the evaluation of different management practices by determining the labile fractions of water organic carbon (Duda et al., 2003; Liang et al., 1996). A lot of research experiments regarding soil mulching technique and soil biota has been a topic of interest for past three decades especially (Tu et al., 2006) studied the effects of different kind of mulches on soil microbial biomass in organic tomato (*Solanum lycopersicum*) farming systems. Different kind of organic mulches (cotton gin trash (CGT), animal manure (AM) and rye/vetch green manure (RV)) were incorporated in the soil with synthetic fertilizer (SF) as conventional control. Results of two consecutive years revealed the higher microbial biomass in the organically treated soil as compared to conventionally managed soils. CGT addition significantly increased soil MBC and activity around 103–151 and 88–170%. Owning to increase in MBC, mineralizable N was increased by 182–285% cotton gin mulched soil as compared to SF. Straw mulch also added microbial biomass and N availability by 42 and 30%, respectively. Mulch mediated increase in microbial biomass and organic carbon was also reported in different studies using cattle manure compost, saw-dust compost, and rice husk compost (Chowdhury et al., 2000), wheat straw and farmyard manure (Goyal et al., 1999), dairy shed effluent (Zaman et al., 2006), and municipal solid waste compost and cow manure (García-Gil et al., 2000; Peacock et al., 2001). As this is very obvious from the previous literature that frequent tillage and cultivation may lead to a loss in microbial population and thus nutrient status in the soil (Follett & Schimel, 1989) in their experiment compared microbial biomass and nitrogen dynamics in no-tilled, mulched, and ploughed soil. A treatment of native sod and wheat-fallow rotation was established in replicates, which led a decrease in total nitrogen contents by 73, 68, and 50% of native sod in the no-till, mulched, and ploughed treatments, respectively. Likewise, microbial biomass was also declined in corresponding tillage treatments by 57, 52, and 36%. They concluded that respiration was directly proportional to microbial biomass. Anyhow, nitrogen mineralization was not dependent on microbial biomass. Rate of respiration per unit of mineralized N increased in ploughed; stubble mulched soil, no-tilled soil, and sod treatments, giving an indication of lowered C availability with increased tillage. However, managing the soil by incorporating mulch might re-establish microbial communities and improve nutrient dynamics. Incorporation of gravel sand in the soil is a water conservation strategy that is usually practiced in the semi-arid soils of China. 3 different sampling intervals were selected after 16 years of continuous addition of gravel sand in the soil. Results showed that sampling after 11 years of mulch in the soil increased microbial biomass and nitrogen in the soil but there was a dramatic decrease in both parameters after 16 years of sampling.

Certain experiments also suggest that different mulch materials affect differently host crops resulting in altered patterns of nutrient release in the soil (Li et al., 2013; Ngouajio et al., 2007), enhancing microbial biomass and rapid breakdown of non-microbial organic substances in soil. A study conducted by Rabary et al. (2008) revealed that among different mulch and fertilizer type treatments living mulch of kikuyu bean/soybean outperformed in terms of soil microbial biomass and soil respiration as compared to *Desmodium uncinatum* living mulch and control. Some mulches pose inhibitory effect on the soil biota and microbial biomass for example, *Desmodium uncinatum*; this might be due to the fact that *D. uncinatum* contains toxic levels of tannins (Skerman et al., 1988) and roots exudates consisting of isoflavones (Tsanuo et al., 2003). Certain studies also confirm the inhibitory role of organic materials added to the soil as mulches (Hättenschwiler & Vitousek, 2000; Nsabimana et al., 2004). In another experiment, a spring wheat field was mulched with plastic film for various periods of time to unravel its effects on soil microbial biomass and soil fertility. Film mulching in the soil promoted MBC by an increase in temperature with a decrease in soil organic carbon. This study also provided the significance of the time span during which mulch remains in the soil (Li et al., 2004). However, it is suggested to optimize the time and type of mulch before selecting of any mulch according to the objectives of mulching.

4.5 Species Diversity

Owing to ecotoxicity caused by agro-chemicals, mulches nowadays have gained much attention as a strategy to improve soil quality and health. The addition of organic mulches in soil considerably enhances species diversity and population, eventually affecting nutrient status (Bandopadhyay et al., 2018). Typically, biological diversity is regarded as the variability between the living species from different ecological multiplexes in which they are present, termed as species richness and abundances in a spatio-temporal manner. A number of experiments have been conducted to study the impact of mulches on soil biota community size and diversity. Before going into details of the studies related to changes in species abundances and diversity caused by the addition of mulch in soil, the following are some basic concepts regarding mathematical calculations of indices for determining species diversity.

4.5.1 Measuring Species Diversity Indices

Biodiversity indices give an estimate of biological variability at quantitative levels in comparison to different biological entities belonging to diverse components (Pavoine & Ricotta, 2019). Following are the are details to distinguish.

Species Richness

It is the measure of the total number of species that are found in certain biological samples. Richness is directly proportional to the number of species, no matters the abundance or count of a particular species. For example, Margalef's diversity index and Menhinick's diversity index can opt for computational analysis.

Evenness

Evenness gives the idea of smoothness of distribution of a number of individuals belonging to different species in the community. Evenness can be determined by the most commonly used indices i.e. Shannon-Wiener diversity index, Simpson's index, and seldomly used ones include Pielou index, Hill numbers, and Brillouin index.

Taxonomic Indices

These indices are based on the taxonomic distances between any two organisms in the community chosen randomly in one sample. The distances are usually depicted as the length of the branches in phylogenetics.

In the samples which are spatially distributed following three types of indices can be taken into contribution.

Alpha Diversity

Alpha diversity is considered when the number of taxa (normally species) within a specific area, community, or ecosystem is counted.

Beta Diversity

Beta diversity counts for the diversity of the species between ecosystems. This index tells the unique species that are present in each ecosystem. For instance, the diversity of mangroves versus seagrass beds.

Gamma Diversity

Gamma diversity accounts for the diversity of all the organisms in a region as well as other ecosystems. For example, diversity within the coastal region of gowader port in Pakistan.

In an experiment reported by Prasifka et al. (2006), an increase in predator species like Carabidae and a decrease in pests and pathogenic species has been reported, followed by the introduction of mulches in the soil (Prasifka et al., 2006) reported that the highest number of carabid species were reported in woodchip/buckwheat husk used as a combination mulch. Shannon diversity index and richness were significantly high in woodchip/buckwheat husk as compared to individually added cut grass, birchwood, and inorganic black plastic. By the addition of mulch in the soil, a significant increase in the population of order collembola species and nematodes (microfauna) were recorded (Culik et al., 2002; Forge et al., 2003; Burrow, 2018). The abundance and assembly of Carabidae communities are modified when mulch was incorporated in the vegetable garden soil. Likewise, an increase in species diversity was also observed in potato fields (Eyre et al., 2016) and blueberries crops (Renkema et al., 2016). Different kind of mulch materials affects species diversity differently in the soil (Miñarro & Dapena, 2003) their study included 6 types of different mulch materials (pine-bark, plastic, and straw mulches, tillage, herbicide, and natural soil) to check the species diversity of ground beetles (Coleoptera: Carabidae). The results indicated the significant effects of mulching materials on carabid catch and their species diversity. Three species of carabid represented more than 98% population of the catch count, which were *Steropus gallega* Fairmaire (65.8%), *Pseudophonus rufipes* (DeGeer) (18.2%) and *Poecilus cupreus* L. (14.6%). Plastic mulches negatively affected the catch and diversity of *S. gallega* whilst *P. rufipes* was collected in greater numbers in tilled areas and *P. cupreus* in the herbicide treatments. Thus, ground cover in apple orchards may help to build a population of epigenic predators helping to sustain the natural ecosystem. Another study conducted by Leclercq-Dransart et al. (2020) compared the different effect of four types of mulches on the degraded soils addressed the relative attraction of these organic mulches for macrofauna. They concluded that organic mulches highly favoured the pedofauna as compared to bare or plastic-based mulch, which showed almost the least count of the abundance of Coleoptera, Isopoda, and potential preys were also less present there. Mineral mulches (fine gravel) tend to restore the soil biota of mine sites with depleted biota by increasing the microbial diversity of soil (Luna Ramos et al., 2015). Mineral mulches (fine gravel) tend to restore the soil biota of mine sites with depleted biota by increasing the microbial diversity of soil (Luna Ramos et al., 2015).

4.6 Soil Microbial Metabolic Quotient(qCO_2)

The metabolic quotient of soil microbes (respiration: SIR ratio) has been used as an indicator for the assessment of the ecological efficiency of microbial biomass. It elaborates the relationship of the growth phase with the latent phase of MB. Usually, an increase in the ratio of qCO_2 is an indication of the undesirable conditions for the soil microbes, but it is also a very sensitive indicator of soil health and quality (Anderson & Domsch, 1993; Raiesi & Beheshti, 2015).

Certain research experiments involved the determination of qCO_2 as a sensitive index of soil health and quality after the incorporation of mulch in the soil (Wardle & Ghani, 1995) investigated the effects on soil biota by the addition of soil mulch. Soil mulching imposed strong positive effects on substrate-induced reparation rates, as well as CO_2–C release from the soil, treated with chloroform. Similarly, mulching increased the microbial metabolic quotient (qCO_2), bacterial to fungal biomass ratio and temporal variability of the microbial biomass within a span of one year. Furthermore, in the mulched soils breakdown of litter was proportional to microbial biomass, and litter decomposition was highest in mulched soils. This increase in microbial metabolic quotient in mulched plots is likely due to the availability of resources available in the form of saw-dust, indicating less efficiency and more respired CO_2-C. Mulching results in instability of certain indicators for example, bacterial to fungal ratio (Gerson et al., 1981) is elevated and increase in microbial metabolic quotient (Insam & Domsch, 1988), which indicates less efficiency (Rabary et al., 2008) reported in their experiment that mulching induced no significant change was observed in QCO_2 concentration in various treatments (conventionally tilled (CT) and natural fallow (NF) including direct-seeded mulch-based cropping (DMC)). However, certain other experiments also suggest similar kind of results with no significant changes in the metabolic quotient, also suggesting that it is considered as a sensitive indicator of soil quality or management strategies A critique of the microbial metabolic (Alvarez et al., 1998; Wardle & Ghani, 1995). On the other hand, (Culumber et al., 2019) reported the significant effects of different mulches on the metabolic qCO_2 quotient in the orchard soils. qCO_2 varies among the different treatments having a minimum value in trefoil mulch (0.0032) and with living mulch (0.0036). In contrast, the highest values were reported in straw mulch (0.0053) and grass alleyway mulch treatment (0.0048). The treatments which showed the highest qCO_2 were found out to have low biomass microbial biomass (straw mulch (246.3) and alleyway treatments (297.6 mg CO_2–C kg^{-1} soil)). The decrease in qCO_2 ratio along with increased biochemical efficiency, C and N accumulation suggested the growth-promoting role of trefoil alleyways as mulch in orchards. In another study reported by Fu et al. (2020), the effect of straw mulch was accessed on microbial respiration and their communities under high-temperature stress. Straw mulch had a greater amount of soil organic carbon as compared to CK, but respiration rates increased with an increase in the temperature, which led an increase in qCO_2 in both mulched and non-mulched soils, whereas qCO_2 was higher in SM as compared to CK. Whereas at low and optimum temperatures, qCO_2 in straw mulch was higher than CK, indicating the higher efficiency of mulch. Lower value of qCO_2 indicates the more retained microbial biomass C with a larger potential of retaining organic matter and nutrients (Culumber et al., 2019).

5 Conclusions

The above-discussed literature pinpoints the positive impact of soil mulching on soil biota diversity, density, and their activities, more especially soil fauna and microbes. Not only soil biota but activities of enzymes like β-glucosidase, dehydrogenase, cellobiohydrolase, urease, phosphatase, xylanase, phosphomonoesterase, protease, etc. were stimulated under mulches which subsequently exert a positive influence on various processes of cell that are directly related to plant growth and development. Mulches enhance soil biota, which is evident from the changes in the biological indicators in the soil, i.e., microbial biomass, respiration, metabolic quotient, soil nutrient, and dynamics. In particular, organic mulches are better than inorganic mulches. As soil biota is a key factor for accessing soil health and quality, so use of mulches, in particular organic mulches, has been proven to be a suitable strategy for conservation of soil quality, biota, and sustainability in agricultural production.

References

Abawi, G. S., & Widmer, T. L. (2000). Impact of soil health management practices on soilborne pathogens, nematodes and root diseases of vegetable crops. *Impact of Soil Health Management Practices on Soilborne Pathogens, Nematodes and Root Diseases of Vegetable Crops, 15*(1), 37–47.

Abrantes, J. R. C. B., et al. (2018). Effectiveness of the application of rice straw mulching strips in reducing runoff and soil loss: Laboratory soil flume experiments under simulated rainfall. In *Soil and Tillage Research*. Elsevier B.V., 180, pp. 238–249. https://doi.org/10.1016/j.still.2018.03.015.

Acosta-Martinez, V., et al. (1999). The role of tree leaf mulch and nitrogen fertilizer on turfgrass soil quality. *Biology and Fertility of Soils, 29*(1), 55–61 (Springer).

Addison, P., Baauw, A. H., & Groenewald, G. A. (2013). An initial investigation of the effects of mulch layers on soil-dwelling arthropod assemblages in vineyards. *South African Journal of Enology and Viticulture, 34*(2), 266–271. https://doi.org/10.21548/34-2-1104.

Adekalu, K. O., Olorunfemi, I. A., & Osunbitan, J. A. (2007). Grass mulching effect on infiltration, surface runoff and soil loss of three agricultural soils in Nigeria. *Bioresource Technology, 98*(4), 912–917. https://doi.org/10.1016/j.biortech.2006.02.044.

Akhtar, K., et al. (2018). Changes in soil enzymes, soil properties, and maize crop productivity under wheat straw mulching in Guanzhong, China. *Soil and Tillage Research, 182*, 94–102 (Elsevier B.V.). https://doi.org/10.1016/j.still.2018.05.007.

Al-Bayati, H. J. M., & Hamdoon, D. N. (2019). Response of eggplant Solanum melongena L. to soil mulching, organic and inorganic fertilizers on vegetative growth traits and yield grown under unheated plastic house. In *IOP Conference Series: Earth and Environmental Science* (p. 012075). Institute of Physics Publishing. https://doi.org/10.1088/1755-1315/388/1/012075.

Alharbi, A. (2017). *Effect of mulch on soil properties under organic farming conditions in center of Saudi Arabia, Mechanization in agriculture & Conserving of the resources.* Scientific Technical Union of Mechanical Engineering 'Industry 4.0'.

Alvarez, C. R. et al. (1998). Associations between organic matter fractions and the active soil microbial biomass. *Soil Biology and Biochemistry (United Kingdom), 30*(6), 767–773. https://agris.fao.org/agris-search/search.do?recordID=GB1997050225. Accessed 1 Apr. 2021.

Anderson, T.-H., & Domsch, K. H. (1993). The metabolic quotient for CO_2 (qCO_2) as a specific activity parameter to assess the effects of environmental conditions, such as pH, on the microbial biomass of forest soils. *Soil Biology & Biochemistry, 25*(3).

Ashrafuzzaman, M., et al. (2011). Effect of plastic mulch on growth and yield of chilli (Capsicum annuum L.). *Brazilian Archives of Biology and Technology*, Tecpar, *54*(2), 321–330. https://doi.org/10.1590/S1516-89132011000200014.

Bajoriene, K., et al. (2013). Effect of organic mulches on the content of organic carbon in the soil. *Estonian Journal of Ecology, 62*(2), 100–106. https://doi.org/10.3176/eco.2013.2.02.

Bandopadhyay, S., et al. (2018). Biodegradable plastic mulch films: Impacts on soil microbial communities and ecosystem functions. *Frontiers in Microbiology, 9*(APR), 819 (Frontiers Media S.A.). https://doi.org/10.3389/fmicb.2018.00819.

Berglund, R., Svensson, B., & Gertsson, U. (2006). Impact of plastic mulch and poultry manure on plant establishment in organic strawberry production. *Journal of Plant Nutrition, 29*(1), 103–112. https://doi.org/10.1080/01904160500416497.

Bloem, J., Ruiter, P., de & Bouwman, L. (1997). Soil food webs and nutrient cycling in agroecosystems. In *Modern soil microbiology* (pp. 245–278). Marcel Dekker Inc.

Bonilla, N., et al. (2012). Enhancing soil quality and plant health through suppressive organic amendments. *Diversity, 4*(4), 475–491 (Multidisciplinary Digital Publishing Institute). https://doi.org/10.3390/d4040475.

Boyhan, G. E., et al. (2000). Evaluation of virus resistant squash and interaction with reflective and nonreflective mulches. *HortTechnology, 10*(3), 574–580.

Burns, R. G., et al. (2013). Soil enzymes in a changing environment: Current knowledge and future directions. *Soil Biology and Biochemistry*, 216–234. https://doi.org/10.1016/j.soilbio.2012.11.009.

Burrow, C. (2018). Influence of connectivity & topsoil management practices of a constructed technosol on pedofauna colonization: A field study. *Applied Soil Ecology, 123*, 416–419 (Elsevier B.V.). https://doi.org/10.1016/j.apsoil.2017.12.001.

Cabrera, G. (2012). *Edaphic macrofauna as biological indicator of the conservation/disturbance status of soil. Results obtained in Cuba.*

Cabrera, G., Robaina, N., & Ponce De León, D. (2011). *Julio-septiembre, Pastos y Forrajes.*

Chowdhury, M. A. H., et al. (2000). Microbial biomass, S mineralization and S uptake by African millet from soil amended with various composts. *Soil Biology and Biochemistry, 32*(6), 845–852. https://doi.org/10.1016/S0038-0717(99)00214-X.

Culik, M. P., de Souza, J. L., & Ventura, J. A. (2002). Biodiversity of Collembola in tropical agricultural environments of Espírito Santo, Brazil. *Agriculture,Ecosystems & Environment. Applied Soil Ecology, 21*(1).

Culumber, C. M., et al. (2019). Organic orchard floor management impact on soil quality indicators: Nutrient fluxes, microbial biomass and activity. *Nutrient Cycling in Agroecosystems, 115*(1), 101–115 (Springer, Netherlands). https://doi.org/10.1007/s10705-019-10007-2.

Curtin, D., Beare, M. H., & Hernandez-Ramirez, G. (2012). Temperature and moisture effects on microbial biomass and soil organic matter mineralization. *Soil Science Society of America Journal, 76*(6), 2055–2067 (Wiley). https://doi.org/10.2136/sssaj2012.0011.

Dervash, M., et al. (2018). Dynamics and importance of soil mesofauna. *International Journal of Advance Research in Science and Engineering, 7*(4).

Dossou-Yovo, E. R., et al. (2016). Improving soil quality and upland rice yield in northern Benin with no-tillage, rice straw mulch and nitrogen fertilization. *International Journal of Agronomy and Agricultural Research.*

Downer, A. J., Menge, J. A., & Pond, E. (2001). Association of cellulytic enzyme activities in eucalyptus mulches with biological control of Phytophthora cinnamomi. *Phytopathology, 91*(9).

Duda, G. P., et al. (2003). Perennial herbaceous legumes as live soil mulches and their effects on C, N and P of the microbial biomass. *Scientia Agricola, 60*(1), 139–147. https://doi.org/10.1590/s0103-90162003000100021.

Dukare, A., et al. (2017). Root development and nodulation in cowpea as affected by application of organic and different types of inorganic/plastic mulches development of seed priming equipment for selected vegetable seeds view project processing of fruits and vegetables view project Ajinath Dukare Root development and nodulation in cowpea as affected by application of organic and different types of inorganic/plastic mulches. *Article in International Journal of Current Microbiology and Applied Sciences, 6*(11), 1728–1738. https://doi.org/10.20546/ijcmas.2017.611.209.

Espí, E., et al. (2006). Plastic films for agricultural applications. *Journal of Plastic Film and Sheeting, 22*(2), 85–102. https://doi.org/10.1177/8756087906064220.

Eyre, M. D., McMillan, S. D., & Critchley, C. N. R. (2016). Ground beetles (Coleoptera, Carabidae) as indicators of change and pattern in the agro-ecosystem: Longer surveys improve understanding. *Ecological Indicators, 68*, 82–88 (Elsevier B.V.). https://doi.org/10.1016/j.ecolind.2015.11.009.

Farmer, J., et al. (2017). Long-term effect of plastic film mulching and fertilization on bacterial communities in a brown soil revealed by high through-put sequencing. *Archives of Agronomy and Soil Science, 63*(2), 230–241 (Taylor and Francis Ltd.). https://doi.org/10.1080/03650340.2016.1193667.

Follett, R. F., & Schimel, D. S. (1989). Effect of tillage practices on microbial biomass dynamics. *Soil Science Society of America Journal, 53*(4), 1091–1096. https://doi.org/10.2136/sssaj1989.03615995005300040018x.

Foltz, R. B., & Wagenbrenner, N. S. (2010). An evaluation of three wood shred blends for post-fire erosion control using indoor simulated rain events on small plots. *CATENA, 80*(2), 86–94. https://doi.org/10.1016/j.catena.2009.09.003.

Forge, T. A., et al. (2003). Effects of organic mulches on soil microfauna in the root zone of apple: Implications for nutrient fluxes and functional diversity of the soil food web. *Applied Soil Ecology, 22*(1), 39–54 (Elsevier).

Fu, X., et al. (2020). Increasing temperature can modify the effect of straw mulching on soil C fractions, soil respiration, and microbial community composition. *PLOS ONE.*; J. Paz-Ferreiro (Ed.). *Public Library of Science, 15*(8), e0237245. https://doi.org/10.1371/journal.pone.0237245.

Gao, H., et al. (2019). Effects of plastic mulching and plastic residue on agricultural production: A meta-analysis. *Science of the Total Environment*, 484–492 (Elsevier B.V.). https://doi.org/10.1016/j.scitotenv.2018.09.105.

García-Gil, J. C., et al. (2000). Long-term effects of municipal solid waste compost application on soil enzyme activities and microbial biomass. *Soil Biology and Biochemistry, 32*(13), 1907–1913. https://doi.org/10.1016/S0038-0717(00)00165-6.

Ge, T., et al. (2015). Tracking the photosynthesized carbon input into soil organic carbon pools in a rice soil fertilized with nitrogen. *Plant and Soil, 392*(1–2), 17–25 (Kluwer Academic Publishers). https://doi.org/10.1007/s11104-014-2265-8.

Gerson, U., U, G., & I, C. (1981). Are allochthonous and autochthonous soil microorganisms R- and K-selected. *18*(3), 285–289.

Gholami, L., Sadeghi, S. H., & Homaee, M. (2013). Straw mulching effect on splash erosion, runoff, and sediment yield from eroded plots. *Soil Science Society of America Journal, 77*(1), 268–278 (Wiley). https://doi.org/10.2136/sssaj2012.0271.

Goyal, S., et al. (1999). Influence of inorganic fertilizers and organic amendments on soil organic matter and soil microbial properties under tropical conditions. *Biology and Fertility of Soils, 29*(2), 196–200 (Springer-Verlag). https://doi.org/10.1007/s003740050544.

Guan, Q., et al. (2011). Effects of different mulching measures on winter wheat field soil respiration in Loess Plateau dry land region. *Ying Yong Sheng Tai Xue Bao = The Journal of applied Ecology/Zhongguo sheng tai xue xue hui, Zhongguo ke xue yuan Shenyang ying yong sheng tai yan jiu suo zhu ban, 22*(6), 1471–1476. https://europepmc.org/article/med/21941747. Accessed 31 Mar. 2021.

Hai-Ming, T., et al. (2014). Effects of winter cover crops residue returning on soil enzyme activities and soil microbial community in double-cropping rice fields. *PLoS ONE.*; H. Smidt (Ed.) *Public Library of Science, 9*(6), e100443. https://doi.org/10.1371/journal.pone.0100443.

Hättenschwiler, S., & Vitousek, P. M. (2000). The role of polyphenols in terrestrial ecosystem nutrient cycling. *Trends in Ecology and Evolution,* 238–242 (Elsevier Ltd.). https://doi.org/10.1016/S0169-5347(00)01861-9.

Hosseini Bai, S., Blumfield, T. J., & Reverchon, F. (2014). The impact of mulch type on soil organic carbon and nitrogen pools in a sloping site. *Biology and Fertility of Soils, 50*(1), 37–44. https://doi.org/10.1007/s00374-013-0829-z.

Huang, Z., Xu, Z., & Chen, C. (2008). Effect of mulching on labile soil organic matter pools, microbial community functional diversity and nitrogen transformations in two hardwood plantations of subtropical Australia. *Applied Soil Ecology, 40*(2), 229–239. https://doi.org/10.1016/j.apsoil.2008.04.009.

Huera-Lucero, T., et al. (2020). A framework to incorporate biological soil quality indicators into assessing the sustainability of territories in the ecuadorian amazon. *Sustainability (Switzerland), 12*(7). https://doi.org/10.3390/su12073007.

Ingman, M., Santelmann, M. V., & Tilt, B. (2015). Agricultural water conservation in china: Plastic mulch and traditional irrigation. *Ecosystem Health and Sustainability, 1*(4), 1–11 (Taylor and Francis Ltd.). https://doi.org/10.1890/EHS14-0018.1.

Insam, H., & Domsch, K. H. (1988). 'Relationship between soil organic carbon and microbial biomass on chronosequences of reclamation sites. *Microbial Ecology, 15*(2), 177–188 (Springer-Verlag). https://doi.org/10.1007/BF02011711.

Jabran, K. (2019). Mulches for enhancing biological activities in soil (pp. 41–46). Springer, Cham. https://doi.org/10.1007/978-3-030-22301-4_5.

Kahlon, M. S., Lal, R., & Ann-Varughese, M. (2013). Twenty two years of tillage and mulching impacts on soil physical characteristics and carbon sequestration in Central Ohio. *Soil and Tillage Research, 126*(126), 151–158 (Elsevier B.V.). https://doi.org/10.1016/j.still.2012.08.001.

Kashif, M., et al. (2020). Untapped renewable energy potential of crop residues in Pakistan: Challenges and future directions. *Journal of environmental management, 256*, 109924 (NLM (Medline)). https://doi.org/10.1016/j.jenvman.2019.109924.

Kumar, S., & Dey, P. (2011). Effects of different mulches and irrigation methods on root growth, nutrient uptake, water-use efficiency and yield of strawberry. *Scientia Horticulturae, 127*(3), 318–324. https://doi.org/10.1016/j.scienta.2010.10.023.

Lal, R. (1988). Effects of macrofauna on soil properties in tropical ecosystems. *Agriculture, Ecosystems and Environment, 24*(1–3), 101–116. https://doi.org/10.1016/0167-8809(88)90059-X.

Leclercq-Dransart, J., et al. (2020). Comparison of the interest of four types of organic mulches to reclaim degraded areas: A field study based on their relative attractiveness for soil macrofauna. *Ecological Engineering, 158*, 106066 (Elsevier).

Lee, K.-J., Won, H.-Y., & Mun, H.-T. (2012). Contribution of root respiration to soil respiration for Quercus acutissima forest. *Korean Journal of Environment and Ecology. Korean Society of Environment and Ecology, 26*(5), 780–786.

Li, F. M., et al. (2004). Dynamics of soil microbial biomass C and soil fertility in cropland mulched with plastic film in a semiarid agro-ecosystem. *Soil Biology and Biochemistry, 36*(11), 1893–1902. https://doi.org/10.1016/j.soilbio.2004.04.040.

Li, Y., et al. (2010). Organic mulch and fertilization affect soil carbon pools and forms under intensively managed bamboo (Phyllostachys praecox) forests in southeast China. *Journal of Soils and Sediments, 10*(4), 739–747. https://doi.org/10.1007/s11368-010-0188-4.

Li, S. X., et al. (2013). Effect of plastic sheet mulch, wheat straw mulch, and maize growth on water loss by evaporation in dryland areas of China. *Agricultural Water Management, 116*, 39–49. https://doi.org/10.1016/j.agwat.2012.10.004.

Li, Q., et al. (2016). Effects of mulching and planting methods on soil respiration rate and rhizosphere microbes of maize. *Grassland and Turf, 36*(5), 46–51.
Liang, B. C., et al. (1996) Carbon mineralization in soils of different textures as affected by water-soluble organic carbon extracted from composted dairy manure. *Biology and Fertility of Soils, 21*(1–2), 10–16 (Springer Verlag). https://doi.org/10.1007/BF00335987.
Liu, E. K., He, W. Q., & Yan, C. R. (2014). "White revolution" to "white pollution"—Agricultural plastic film mulch in China. *Environmental Research Letters, 9*(9), 091001 (Institute of Physics Publishing). https://doi.org/10.1088/1748-9326/9/9/091001.
Liu, Y., et al. (2016). Differential responses of soil respiration to soil warming and experimental throughfall reduction in a transitional oak forest in central China. *Agricultural and Forest Meteorology, 226*, 186–198 (Elsevier).
López, R., et al. (2014). Long term changes in soil properties and enzyme activities after almond shell mulching in avocado organic production. *Soil and Tillage Research, 143*, 155–163 (Elsevier).
Luna Ramos, L., et al. (2015). Effects of organic amendments and mulches on soil microbial communities in quarry restoration under semiarid climate. In *Geophysical research abstracts*.
Luo, Y., et al. (2001). Acclimatization of soil respiration to warming in a tall grass prairie. *Nature, 413*(6856), 622–625 (Nature Publishing Group). https://doi.org/10.1038/35098065.
Maillard, F. et al. (2019). Soil microbial functions are affected by organic matter removal in temperate deciduous forest. *Soil Biology and Biochemistry, 133*, 28–36 (Elsevier Ltd.). https://doi.org/10.1016/j.soilbio.2019.02.015.
Maughan, T., & Drost, D. T. (2016). *Use of plastic mulch for vegetable production*. https://www.researchgate.net/publication/308886221. Accessed 1 Apr. 2021.
Menta, C. (2012). Soil fauna diversity-function, soil degradation, biological indices, soil restoration. In *Biodiversity conservation and utilization in a diverse world*. https://doi.org/10.5772/51091.
Miñarro, M., & Dapena, E. (2003). Effects of ground-cover management on ground beetles (Coleoptera: Carabidae) in an apple orchard. *Applied Soil Ecology, 23*(2), 111–117 (Elsevier). https://doi.org/10.1016/S0929-1393(03)00025-8.
Monaco, S., et al. (2008). Changes in chemical and biochemical soil properties induced by 11-yr repeated additions of different organic materials in maize-based forage systems. *Soil Biology and Biochemistry, 40*(3), 608–615 (Elsevier). https://doi.org/10.1016/j.soilbio.2007.09.015.
Montenegro, A. A. A., et al. (2013). Impact of mulching on soil and water dynamics under intermittent simulated rainfall. *CATENA, 109*, 139–149. https://doi.org/10.1016/j.catena.2013.03.018.
Ngosong, C., et al. (2016). Comparative advantage of Mucuna and Tithonia residue mulches for improving tropical soil fertility and tomato productivity.
Ngouajio, M., Wang, G., & Goldy, R. (2007). Withholding of drip irrigation between transplanting and flowering increases the yield of field-grown tomato under plastic mulch. *Agricultural Water Management, 87*(3), 285–291 (Elsevier). https://ideas.repec.org/a/eee/agiwat/v87y2007i3p285-291.html. Accessed 1 Apr. 2021.
Nsabimana, D., Haynes, R. J., & Wallis, F. M. (2004). Size, activity and catabolic diversity of the soil microbial biomass as affected by land use. *Applied Soil Ecology, 26*(2), 81–92. https://doi.org/10.1016/j.apsoil.2003.12.005.
Obalum, S., et al. (2011). Short term effects of tillage-mulch practices under sorghum and soybean on organic carbon and eutrophic status of a degraded ultisol in southeastern nigeria. *Tropical and Subtropical Agroecosystems, 14*(2), 393–403. https://www.revista.ccba.uady.mx/ojs/index.php/TSA/article/view/520. Accessed 1 Apr. 2021.
Ocharo, E. N., Korir, N. K., & Gweyi-Onyango, J. (2017). Green pepper growth and yield response to the integration of mulching materials and row plant spacing. *Journal of Agriculture and Crops, 3*(9), 72–77(Academic Research Publishing Group). https://ideas.repec.org/a/arp/jacarp/2017p72-77.html. Accessed 31 Mar. 2021.
Parmar, H. N., Polara, N. D., & Viradiya, R. R. (2013) Effect of mulching material on growth, yield and quality of watermelon (Citrullus Lanatus Thunb) Cv. Kiran. *Universal Journal of Agricultural Research, 1*(2), 30–37. https://doi.org/10.13189/ujar.2013.010203.

Parthasarathi, K., & Ranganathan, L. S. (1999). Longevity of microbial and enzyme activity and their influence on NPK content in pressmud vermicasts. *European Journal of Soil Biology, 35*(3), 107–113 (Elsevier).

Pavan Fernandes, S. A., Bettiol, W., & Cerri, C. C. (2005). Effect of sewage sludge on microbial biomass, basal respiration, metabolic quotient and soil enzymatic activity. *Applied Soil Ecology, 30*(1), 65–77. https://doi.org/10.1016/j.apsoil.2004.03.008.

Pavoine, S., & Ricotta, C. (2019). A simple translation from indices of species diversity to indices of phylogenetic diversity. *Ecological Indicators, 101*, 552–561 (Elsevier B.V.). https://doi.org/10.1016/j.ecolind.2019.01.052.

Paz-Ferreiro, J., & Fu, S. (2016). Biological indices for soil quality evaluation: Perspectives and limitations. *Land Degradation & Development, 27*(1), 14–25 (John Wiley and Sons Ltd.). https://doi.org/10.1002/ldr.2262.

Peacock, A. D., et al. (2001). Soil microbial community responses to dairy manure or ammonium nitrate applications. *Soil Biology and Biochemistry, 33*(7–8), 1011–1019 (Pergamon). https://doi.org/10.1016/S0038-0717(01)00004-9.

Peng, F., et al. (2018). Changes of soil properties regulate the soil organic carbon loss with grassland degradation on the Qinghai-Tibet Plateau. *Ecological Indicators, 93*, 572–580 (Elsevier B.V.). https://doi.org/10.1016/j.ecolind.2018.05.047.

Pervaiz, M. A., Muhammad, I., & Khuram, S. (2009). Effect of mulch on soil physical properties and N, P, K concentration in maize (Zea mays) shoots under two tillage systems. *International Journal of Agriculture and Biology, 11*(2), 119–124 (Friends Science Publishers).

Potthoff, M., et al. (2005). Soil biological and chemical properties in restored perennial grassland in California. *Restoration Ecology, 13*(1), 61–73 (John Wiley & Sons, Ltd.). https://doi.org/10.1111/j.1526-100X.2005.00008.x.

Prasifka, J. R. *et al.* (2006) 'Effects of living mulches on predator abundance and sentinel prey in a corn–soybean–forage rotation', *Environmental Entomology*. Oxford University Press Oxford, UK, 35(5), pp. 1423–1431.

Prats, S. A. et al. (2016). Mid-term and scaling effects of forest residue mulching on post-fire runoff and soil erosion. *Science of the Total Environment, 573*(573), 1242–1254 (Elsevier B.V.). https://doi.org/10.1016/j.scitotenv.2016.04.064.

Qin, S., et al. (2017). Analysis on fungal diversity in rhizosphere soil of continuous cropping potato subjected to different furrow-ridge mulching managements. *Frontiers in Microbiology, 8*(MAY), 845 (Frontiers Media S.A.). https://doi.org/10.3389/fmicb.2017.00845.

Rabary, B., et al. (2008). Effects of living mulches or residue amendments on soil microbial properties in direct seeded cropping systems of Madagascar. *Applied Soil Ecology, 39*(2), 236–243. https://doi.org/10.1016/j.apsoil.2007.12.012.

Raich, J. W., & Tufekcioglu, A. (2000). Vegetation and soil respiration: Correlations and controls. *Biogeochemistry, 48*(1), 71–90 (Springer). https://doi.org/10.1023/A:1006112000616.

Raiesi, F., & Beheshti, A. (2015). Microbiological indicators of soil quality and degradation following conversion of native forests to continuous croplands. *Ecological Indicators, 50*(50), 173–185 (Elsevier B.V.). https://doi.org/10.1016/j.ecolind.2014.11.008.

Ramakrishna, A., et al. (2006). Effect of mulch on soil temperature, moisture, weed infestation and yield of groundnut in northern Vietnam. *Field Crops Research, 95*(2–3), 115–125. https://doi.org/10.1016/j.fcr.2005.01.030.

Ren, C., et al. (2018). Differential soil microbial community responses to the linkage of soil organic carbon fractions with respiration across land-use changes. *Forest Ecology and Management, 409*, 170–178 (Elsevier B.V.). https://doi.org/10.1016/j.foreco.2017.11.011.

Renkema, J. M., et al. (2016). Organic mulches in highbush blueberries alter beetle (Coleoptera) community composition and improve functional group abundance and diversity. *Agricultural and Forest Entomology, 18*(2), 119–127 (Blackwell Publishing Ltd.). https://doi.org/10.1111/afe.12144.

Richter, B. S., et al. (2011). Cellulase activity as a mechanism for suppression of Phytophthora root rot in mulches. *Phytopathology, 101*(2), 223–230. https://doi.org/10.1094/PHYTO-04-10-0125.

Robichaud, P. R., & Ashmun, L. E. (2013). Tools to aid post-wildfire assessment and erosion-mitigation treatment decisions. *International Journal of Wildland Fire, 22*(1), 95 (CSIRO Publishing). https://doi.org/10.1071/WF11162.

Roger-Estrade, J., et al. (2010). Tillage and soil ecology: Partners for sustainable agriculture. *Soil and Tillage Research,* 33–40. https://doi.org/10.1016/j.still.2010.08.010.

Sainju, U. M., et al. (2008). Soil carbon and nitrogen sequestration as affected by long-term tillage, cropping systems, and nitrogen fertilizer sources. *Agriculture, Ecosystems and Environment, 127*(3–4), 234–240. https://doi.org/10.1016/j.agee.2008.04.006.

Scheu, S., Ruess, L., & Bonkowski, M. (2005). Interactions between microorganisms and soil micro- and mesofauna. In *Microorganisms in soils: Roles in genesis and functions* (pp. 253–275). Springer-Verlag. https://doi.org/10.1007/3-540-26609-7_12.

Schirmel, J., et al. (2018). Plasticulture changes soil invertebrate assemblages of strawberry fields and decreases diversity and soil microbial activity. *Applied Soil Ecology, 124*, 379–393 (Elsevier B.V.). https://doi.org/10.1016/j.apsoil.2017.11.025.

Serrano, A., et al. (2009). Evaluation of soil biological activity after a diesel fuel spill. *Science of the Total Environment, 407*(13), 4056–4061. https://doi.org/10.1016/j.scitotenv.2009.03.017.

Shahadat Hossen, M., et al. (2017). Effect of different organic and inorganic mulches on soil properties and performance of Brinjal (Solanum melongena L.) Carbon sequestration view project thermal environment in naturally ventilated classrooms in tropical climates view project effect of different organic and inorganic mulches on soil properties and performance of Brinjal (Solanum melongena L.). *Asian Journal of Advances in Agricultural Research, 3*(2), 1–7. https://doi.org/10.9734/AJAAR/2017/36272.

Siczek, A., & Lipiec, J. (2011). Soybean nodulation and nitrogen fixation in response to soil compaction and surface straw mulching. *Soil and Tillage Research, 114*(1), 50–56. https://doi.org/10.1016/j.still.2011.04.001.

Shaohui, L., & Jingyu, F. (1997). Effect factors of soil respiration and the temperature's effects on soil respiration in the global scale. *Acta Ecologica Sinica, 17*(5), 469–476. https://ci.nii.ac.jp/naid/10020133689. Accessed 31 Mar. 2021.

Skerman, P. J., Cameron, D. G., & Riveros, F. (1988). Tropical forage legumes, 2. rev. edn. In *FAO Plant Production and Protection Series (FAO)* (No. 2). https://agris.fao.org/agris-search/search.do?recordID=XF19900023021. Accessed 1 Apr. 2021.

Smith, J. L., Halvorson, J. J., & Bolton Jr., H. (1994). *Spatial relationships of soil microbial biomass and c and n mineralization in a semi-arid shrub-steppe ecosystem, Biochem.*

Steinmetz, Z., et al. (2016). Plastic mulching in agriculture. Trading short-term agronomic benefits for long-term soil degradation? *Science of the Total Environment,* 690–705 (Elsevier B.V.). https://doi.org/10.1016/j.scitotenv.2016.01.153.

Stenberg, B. (1998). Microbial biomass and activities in soil as affected by frozen and cold storage. *Soil Biology Biochemistry, 30*, 393–402. https://ci.nii.ac.jp/naid/10025850260. Accessed 30 Mar. 2021.

Stirling, G. R., & Eden, L. M. (2008). The impact of organic amendments, mulching and tillage on plant nutrition, Pythium root rot, root-knot nematode and other pests and diseases of capsicum in a subtropical environment, and implications for the development of more sustainable vegetable farming systems. *Australasian Plant Pathology, 37*(2), 123–131. https://doi.org/10.1071/AP07090.

Striegl, R. G., & Wickland, K. P. (2001). Soil respiration and photosynthetic uptake of carbon dioxide by ground-cover plants in four ages of jack pine forest. *Canadian Journal of Forest Research, 31*(9), 1540–1550 (National Research Council of Canada). https://doi.org/10.1139/x01-092.

Sun, L., et al. (2019). Reasonable fertilization improves the conservation tillage benefit for soil water use and yield of rain-fed winter wheat: A case study from the Loess Plateau, China. *Field Crops Research, 242*, 107589 (Elsevier).

Tapia-Báez, R. G. (2015). *Diversidad de Escarabajos Copronecrófagos y Estado de Consevación de la Microcuenca del Río Pindo.* Universisdad Tecnológica Equinoccial.

Teame, G., Tsegay, A., & Abrha, B. (2017). Effect of organic mulching on soil moisture, yield, and yield contributing components of sesame (*Sesamum indicum L.*). *International Journal of Agronomy, 2017*, 1–6 (Hindawi Limited). https://doi.org/10.1155/2017/4767509.

Tian, G., Kang, B. T., & Brussaard, L. (1997). Effect of mulch quality on earthworm activity and nutrient supply in the humid tropics. *Soil Biology and Biochemistry, 29*(3–4), 369–373. https://doi.org/10.1016/S0038-0717(96)00099-5.

Torres Bojórquez, A. I., et al. (2017). Foliar iron and plastic mulch in Capsicum chinense Jacq. Infected with tospoviruses. *Revista Mexicana de Ciencias Agrícolas, 8*(2), 369–380 (Instituto Nacional de Investigaciones Forestales, Agrícolas y Pecuarias (INIFAP)).

Tsanuo, M. K., et al. (2003). Isoflavanones from the allelopathic aqueous root exudate of Desmodium uncinatum. *Phytochemistry, 64*(1), 265–273 (Elsevier Ltd.). https://doi.org/10.1016/S0031-9422(03)00324-8.

Tu, C., Ristaino, J. B., & Hu, S. (2006). Soil microbial biomass and activity in organic tomato farming systems: Effects of organic inputs and straw mulching. *Soil Biology and Biochemistry, 38*(2), 247–255. https://doi.org/10.1016/j.soilbio.2005.05.002.

Tuovinen, T., et al. (2006). Organic mulches versus black plastic in organic strawberry: Does it make a difference for ground beetles (Col., Carabidae)? *Journal of Applied Entomology, 130*(9–10), 495–503 (Wiley). https://doi.org/10.1111/j.1439-0418.2006.01108.x.

Vox, G., et al. (2016). Mapping of Agriculture Plastic waste. *Agriculture and Agricultural Science Procedia, 8*, 583–591 (Elsevier B.V.). https://doi.org/10.1016/j.aaspro.2016.02.080.

Waid, J. S. (1999). Does soil biodiversity depend upon metabiotic activity and influences? *Applied Soil Ecology, 13*(2), 151–158 (Elsevier).

Walters, S. A. (2003). Suppression of watermelon mosaic virus in summer squash with plastic mulches and rowcovers. *HortTechnology, 13*(2), 352–357 (American Society for Horticultural Science).

Wang, Y., et al. (2011). Effects of gravel-sand mulch, plastic mulch and ridge and furrow rainfall harvesting system combinations on water use efficiency, soil temperature and watermelon yield in a semi-arid Loess Plateau of northwestern China. *Agricultural Water Management, 101*(1), 88–92. https://doi.org/10.1016/j.agwat.2011.09.006.

Wang, Y. P., et al. (2016). Multi-site assessment of the effects of plastic-film mulch on dryland maize productivity in semiarid areas in China. *Agricultural and Forest Meteorology, 220*, 160–169 (Elsevier). https://doi.org/10.1016/j.agrformet.2016.01.142.

Wang, J., et al. (2018). Response of soil carbon fractions and dryland maize yield to mulching. *Soil Science Society of America Journal, 82*(2), 371–381 (Wiley). https://doi.org/10.2136/sssaj2017.11.0397.

Wardle, D. A., & Ghani, A. (1995). A critique of the microbial metabolic quotient (qCO_2) as a bioindicator of disturbance and ecosystem development. *Soil Biology and Biochemistry, 27*(12), 1601–1610. https://doi.org/10.1016/0038-0717(95)00093-T.

Whitaker, J., et al. (2014). Microbial community composition explains soil respiration responses to changing carbon inputs along an A ndes-to- A mazon elevation gradient. *Journal of Ecology, 102*(4), 1058–1071 (A. Austin (Ed.), Blackwell Publishing Ltd.). https://doi.org/10.1111/1365-2745.12247.

Wick, B., Kühne, R. F., & Vlek, P. L. G. (1998). Soil microbiological parameters as indicators of soil quality under improved fallow management systems in south-western Nigeria. *Plant and Soil, 202*(1), 97–107 (Springer). https://doi.org/10.1023/A:1004305615397.

Wright, P. J., & Burge, G. K. (2000). Irrigation, saw-dust mulch, and Enhance® biocide affects soft rot incidence, and flower and tuber production of calla. *New Zealand Journal of Crop and Horticultural Science, 28*(3), 225–231 (Taylor & Francis Group). https://doi.org/10.1080/01140671.2000.9514143.

Yang, Y., et al. (2020). Renewable sourced biodegradable mulches and their environment impact. *Scientia Horticulturae, 268*, 109375 (Elsevier).

Yang, Y. J., et al. (2003). Effect of organic mulches on soil bacterial communities one year after application. *Biology and Fertility of Soils, 38*(5), 273–281. https://doi.org/10.1007/s00374-003-0639-9

Yinkun, L., et al. (2013). Dynamics of soil respiration and carbon balance of summer-maize field under different nitrogen addition, Ecol. *Environmental Sciences, 22*, 18–24.

Yu, Y. Y., et al. (2018). Benefits and limitations to straw- and plastic-film mulch on maize yield and water use efficiency: A meta-analysis across hydrothermal gradients. *European Journal of Agronomy, 99*, 138–147 (Elsevier B.V.). https://doi.org/10.1016/j.eja.2018.07.005.

Zaman, M., Di, H. J., & Cameron, K. C. (2006). A field study of gross rates of N mineralization and nitrification and their relationships to microbial biomass and enzyme activities in soils treated with dairy effluent and ammonium fertilizer. *Soil Use and Management, 15*(3), 188–194 (CAB International). https://doi.org/10.1111/j.1475-2743.1999.tb00087.x.

Zhang, Q.-Z., et al. (2005). The effects of crop residue amendment and N rate on soil respiration. *Acta Ecologica Sinica, 25*(11), 2883–2887.

Zhang, F., et al. (2015a). Plastic film mulching increases soil respiration in ridge-furrow maize management. *Arid Land Research and Management, 29*(4), 432–453 (Taylor and Francis Ltd.). https://doi.org/10.1080/15324982.2015a.

Zhang, X., Qian, Y., & Cao, C. (2015b). Effects of straw mulching on maize photosynthetic characteristics and rhizosphere soil micro-ecological environment. *Chilean Journal of Agricultural Research, 75*(4), 481–487 (Instituto de Investigaciones Agropecuarias, INIA). https://doi.org/10.4067/S0718-58392015b.

Zhang, Y. L., et al. (2019). Stoichiometric analyses of soil nutrients and enzymes in a Cambisol soil treated with inorganic fertilizers or manures for 26 years. *Geoderma, 353*, 382–390 (Elsevier B.V.). https://doi.org/10.1016/j.geoderma.2019.06.026.

Zheng, Z.-M., et al. (2009). Temperature sensitivity of soil respiration is affected by prevailing climatic conditions and soil organic carbon content: A trans-China based case study. *Soil Biology and Biochemistry, 41*, 1531–1540. https://doi.org/10.1016/j.soilbio.2009.04.013.

Zheng, Z., et al. (2006). Litter decomposition and nutrient release in a tropical seasonal rain forest of Xishuangbanna, Southwest China[1]. *Biotropica, 38*(3), 342–347 (Wiley). https://doi.org/10.1111/j.1744-7429.2006.00151.x.

Zhou, L. M., et al. (2009). How two ridges and the furrow mulched with plastic film affect soil water, soil temperature and yield of maize on the semiarid Loess Plateau of China. *Field Crops Research, 113*(1), 41–47. https://doi.org/10.1016/j.fcr.2009.04.005.

Mulches Effects on Soil Physical Properties i.e. Porosity, Aggregate Stability, Infiltration Rates, Bulk Density, Compaction

Rafia Mubaraka, Muhammad Baqir Hussain, Hira Tariq, Marina Qayyum, and Iqra Tariq

Abstract Mulch is separating the soil surface from the atmosphere and is consider a layer of dissimilar material. A layer of mulch on the soil surface can affect different soil physical properties, for example enhancing the soil aggregation, reducing water evaporation, increasing infiltration rate and reducing runoff losses. Mulching can be organic and inorganic where organic materials have a positive impact on bulk density, one of the reasons for increased soil bioturbation from earthworm activity. Other positive effects of organic mulching can be improved structural stability and porosity. In this chapter the results of a range of studies concerning the effects of mulching on the physical soil parameters i.e. porosity, pore volume, void ratio, aggregate stability, infiltration, interception and evaporation, bulk density and penetration resistance are discussed.

1 Introduction

Since the nineteenth century, soil and water conservation have been among the key issues (Mekonnen et al., 2015). The need for agricultural land is increasing as the world's population continues to grow, and soil and water losses are becoming increasingly critical. It is big a problem in emerging nations (Mandal & Sharda, 2013; Thomaz & Luiz, 2012), where farming on slope contributes to the loss of water and soil significantly (Li et al., 2019).

R. Mubaraka (✉)
Department of Geography, University of Koblenz-Landau, Universitätsstr. 1, 56070 Koblenz, Germany
e-mail: rafiamubaraka@gmx.de

H. Tariq
Citrus Research Institute, Sargodha, Pakistan

M. B. Hussain · M. Qayyum · I. Tariq
Muhammad Nawaz Shareef University of Agriculture Multan, Multan, Pakistan

K. Akhtar et al. (eds.), *Mulching in Agroecosystems*,
https://doi.org/10.1007/978-981-19-6410-7_3

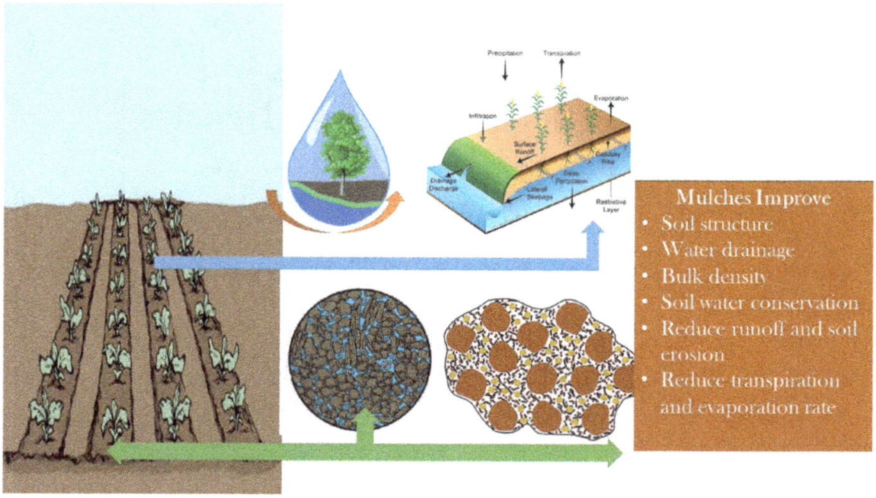

Fig. 1 Mulches effect on soil physical properties

Engineering, biological, and agronomic approaches are examples of traditional soil and water management methods. Although the benefits of biological and engineering approaches for soil and water conservation have been widely acknowledged, due to economic constraints, approaches are still challenging to popularize, particularly in poor nations. Mulching has been practiced as an agricultural procedure to boost plant growth and output since the end of the 17th century. Mulch is gaining popularity as a key agronomic measure due to its low cost and quick effect. The word mulch is most likely derived from a German term "molsch," which means "soft to decay," and reportedly refers to gardeners spreading straw and leaves over the ground surface (Jack et al., 1955). It (mulches) improve plant growth by conserving soil moisture and improving the physical and other qualities of the soil (Fig. 1). Mulch can increase the physical characteristics and organic matter content of soil, hence reducing soil loss and improving land productivity and environmental quality. Mulching can control soil temperature, water content, nutrient availability, and crop growth characteristics. Plant wastes, sawdust, sand, gravel, plastic, and other items were used to cover the soil surface (Table 1). Anikwe et al. (2004) reported that plastic film mulching, among other mulching strategies, raises soil surface temperature by altering the heat balance, therefore raising soil temperature and positively influence crop emergence. Similarly, Anikwe et al. (2007) observed that control had lowest soil temperature (approximately 1–3.8 °C lower) compared to the inorganic or plastic film mulched plots, at various stages from the time of sowing. Mulching boosts nutrient availability by adding organic acids to the soil as organic leftovers decompose under plastic mulch. Tisdale et al. (1985) reported that early crop emergence is induced by plastic mulch, which boosts biomass output during the early phases of crop growth. Li et al. (1999) reported that early seedling emergence and spike differentiation are aided by plastic film mulching, resulting in more spikelets and grains per spike in

Table 1 Positive interactions of mulch types on the soil physical characteristics

Mulch type	Study type	Plant type	Mulching response	References
Different type of mulches were used Inorganic Mulch Cobblestone (CB) Water permeable brick (WPB) Organic Mulch Pine bark (PB) Green waste compost (GWC) Living Turf grass (TG)	The impacts of inorganic and organic living mulches on soil physical parameters at three depths (0–10 cm, 10–20 cm, and 20–40 cm), as well as tree development (*Sophora japonica*) in urban tree pits, were investigated	*Sophora japonica*	At 10–20 cm, soil treated with PB, GWC, and TG improved the total porosity of the soil, while at lower depths, it improved the total porosity of the soil. WPB degraded the soil's bulk density and porosity, raising the pH at lower depths. In comparison to other mulches, organic mulches, particularly GWC, improved soil physical characteristics (0–10 cm depth) and are ideal for covering bare soil in urban tree pits	Bingpeng et al. (2019)
Three type of mulch were used Organic Mulch wood chips (WC) Inorganic Mulch Round gravel (RG) Living Mulch Manila turf grass (MG)	Compare the impact of mulching with inorganic, organic, and living materials on soil qualities at 0–5, 5–10 cm depths, as well as on *Osmanthus fragrans L.* growth and physiological characteristics	(Osmanthus fragrans)	Improve Soil moisture, organic matter, nitrogen availability, root development, soluble sugar, and chlorophyll concentration	Xue et al. (2016)
Cympobogon spp, Panicum spp Cympobogon spp., Panicum spp., Eucalyptus spp., and Cympobogon spp. and mix residue	In coffee growing systems, quantify the impact of various forms of mulch on soil characteristics and erodibility	Coffee plant	Soil organic carbon, wet aggregate stability, and soil erodibility should all be improved	Innocent et al. (2017)

(continued)

Table 1 (continued)

Mulch type	Study type	Plant type	Mulching response	References
Wheat Straw	Calculate the impact of wheat straw mulching on the physical parameters of the soil and runoff		Enhance water infiltration and delay runoff generation by increasing bulk density, porosity aggregate stability, and accessible water capacity	Jordán et al. (2010a, 2010b)
Wheat Straw	Under deep and conventional tillage, assess the impact of mulch on soil physical characteristics and N, P, and K concentrations in maize shoots	Maize	Organic matter, soil moisture content, plant height, and grain yield all increased. The concentrations of N and P in maize shoots were considerably altered by tillage and mulch	Pervaiz et al.(2009)
Rice Straw	Increase sunflower root growth and productivity by removing soil limitations	Sunflower	Improved water content in the soil, root growth, and yield	Paul et al. (2021)
Farmyard manure and Mulch (polyethylene film and Straw mulch)			Manure and mulch boost the water-holding capacity of the soil	Zhang et al. (2014)
	Effects of farmyard manure and mulch on the physical qualities of salt-affected soil in a reclaimed coastal tidal flat		Decreased bulk density, as well as saturated water content and hydraulic conductivity	
Coloured plastic mulches	Colored plastic mulch has an impact on soil characteristics and crop productivity		Increased amount of fruits, roots, tubers, and bulbs; highly significant on water holding capacity, moisture, and temperature	Amare and Desta (2021)

(continued)

Table 1 (continued)

Mulch type	Study type	Plant type	Mulching response	References
Various mulch materials (bark chips, wood chips, wheat straw, cardboard, paper foil, decomposable matting, nonwoven fabric covered by bark chips, and crushed basalt)	The effect of different mulch types on soil parameters that regulate the Haplic Fluvisol's water regime		Improve aggregate stability, saturated and residual water content, saturated hydraulic conductivity, air-entry pressure head reciprocal, and soil quality	Pavlů et al. (2021)
Dried grass and chromolaena shoots, as well as dry teak leaves, are all on a black polythene sheet	The suitability of several mulch materials for soil improvement and sunflower yield was tested in the field (Helianthus annuus, L)	sunflower (Helianthus Annuus, L)	Increased seed yields, improved oil moisture and soil temperature regimes, root and shoot biomass and leaf area growth, leaf area per plant	Agele et a. (2010)
Cellulosic paper and polyethylene	The impacts of biodegradable plastic mulches on soil health were studied	Pumpkin (*Cucurbita pepo*)	Aggregate stability, infiltration, soil pH, electrical conductivity, nitrate-N, and exchangeable potassium should all be improved	Sintim et al. (2019)
Rice straw and polythene sheet	Evaluate the efficiency of different mulching materials on moisture conservation, soil properties and plant growth		Increased 30% infiltration rate, reduce water loss and irrigation requirements. Improve plant height	Chaudhry et al. (2004)
Pueraria phaseoloides, *Mucuna pruriens*, *Pennisetum purpureum* and *Panicum maximum*	To see how different mulch materials affect soil parameters, leaf nutritional composition, okra yield, and growth	Okra (*Abelmoschus esculentus* (L.)	Reduced bulk density, improved porosity, soil moisture content, pH, organic matter, soil and leaf N, P, K, Ca, Mg, pod yield, and okra growth	Adekiya et al. (2017)

(continued)

Table 1 (continued)

Mulch type	Study type	Plant type	Mulching response	References
Plastic film mulching	For a ground cover rice production system, modelling the influence of mulching on soil heat transfer, water flow, and crop growth	Rice	The changes in soil temperature, soil water content, LAI, dry matter, and yield were accurately reproduced using the modified model Reduce irrigation with a field capacity of 80%	Liang et al. (2017)

wheat. Li et al. (2004) reported that mulch prolongs the reproductive time period, resulting in a higher yield.

Mulching can also help to decrease the formation of surface seals, which improves water infiltration and reduce the runoff losses (Ruan et al., 2001), also increased earthworm activity along with soil porosity by the action of roots (Lal et al., 1991). Pervaiz et al. (2009) described that mulch enhanced soil organic matter and soil moisture content up to 1.32 g kg^{-1} and 17%, respectively. Tillage and mulch had a considerable impact on N and P concentrations in maize shoots, but had a minor impact on K concentrations. It improves N and P concentrations up to 1.423 g kg^{-1} and 0.156 g kg^{-1}, respectively, and showed that statistically significant interacting effect of mulch and tillage, but non-significant for K (1.767 g kg^{-1}). Wheat straw mulch with deep tillage improves crop quality and physical health of soil. Similarly, Paul et al. (2021) reported that at a depth of 0–30 cm, the rice straw mulch improved root development, resulting in a 23% increase in sunflower yield. When used as mulching material, cover crops lower the soil bulk density by 4%, enhanced macro pores by approximately 33%, and improve water penetration by as much as 629%. According to reports, these changes have resulted in a 96% reduction in soil loss (Haruna & Nkongolo, 2015). Haruna et al. (2018) revealed that using grain rye as mulch increases saturated hydraulic conductivity by 33% and cumulative water infiltration by 170%. In the same way, Nouri et al. (2019) revealed that use of vetch and wheat as cover crops boosted cumulative infiltration (86 and 116%, respectively) under no-till management. Further discussion on the improvement of soil physical properties is discussed below.

2 Porosity, Pore Volume, Void Ratio

The ratio of soil pore volume that is occupied by non-solid material to total volume of soil is known as soil porosity (Osman, 2013; Blume et al., 2016).

$$\text{Porosity, \%} = \text{Vp/Vt} \times 100$$

Vp Volume of pores
Vt Total volume of soil.

Porosity of soil can be obtained from bulk density and particle density as follow:

$$\text{Porosity} = 1 - (\text{Bulk density / Particle density}) \times 100$$

The relative abundance of each pore size in a representative volume of soil is the **pore-size distribution** (Nimmo, 2013). The **void ratio** is the volume of empty space in relation to the solid particle volume in a given volume of soil (Blume et al., 2016).

$$\text{Void ratio} = \text{Volume of empty space or void / Volume of soil particle}$$

Soil porosity and bulk density are closely related, as an increase in porosity will lead to a decrease in bulk density (Osman, 2013). Mulching effect on the porosity of various soil types and management regimes in various climates was studied. Franzen et al. (1994) reported an increase in void ratio for mulched mechanized and semi-mechanized treatments in comparison to equally managed non-mulched treatments of an Alfisol mulched with 6 tons/ha maize stover mulch. Likewise, Głab and Kulig (2008) found that mulching with fodder radish crop increased the total porosity in combination with reduced tillage. Macroporosity was also increased by mulching in combination with reduced tillage. Macro porosity and total porosity of a mulched Alfisol subjected to different traffic intensities, under straw mulch and litter mulch they found an increase in total porosity and macro porosity, if the soil was subjected to low traffic intensity (Jourgholami et al., 2020). Jordán et al. (2010a, 2010b) also calculated the total porosity for Fluvisols mulched with wheat straw applied at rates of 1, 5, 10, and 15 tons/ha, showed considerable improvements in total porosity (up to 173%) for the 10 and 15 tons/ha treatments. Even though Mulumba and Lal (2008) showed that there was no evidence that mulching had a major influence on soil bulk density, they found that mulching significantly increased (total) porosity by 35–46% with the application of straw mulch at rates of up to 16 tons/ha. Moreover, the pore size distribution was not significantly affected by mulching with 4 tons/ha rice straw on an Ustipsamment (Hulugalle & Palada, 1990).

Most studies found a positive impact of mulching on porosity, void ratio or macro-porosity of different soils. As different parameters were used to describe the porosity of soils, the results of Hulugalle and Palada (1990), who found no significant impact of mulching on the pore size distribution, cannot be compared directly to the findings of studies examining total porosity, as change in pore size distribution is not necessarily an indicator for change in total porosity. The findings of the reviewed studies indicate a positive impact of mulching on total porosity and macro pores distribution. A more extensive literature research as metadata is needed to name a clear trend of the scientific knowledge regarding the effect of mulching on soil porosity.

3 Aggregate Stability

Aggregate stability and **water stable aggregates** both refer to the aggregate‘s resistance to destruction by water (Osman, 2013). Most studies investigating aggregate stability or wet aggregate stability found a positive impact of mulching on soil aggregation and stability of those aggregates. Two methods were used to determine aggregate stability by Jordán et al. (2010a, 2010b) the water-drop test and ultrasonic disruption to determine aggregate stability, where the force needed to disrupt soil aggregates by water-drop impact or ultrasound is measured (Imeson & Vis, 1984). Whereas Kahlon et al. (2013), Mulumba and Lal (2008) and Nzeyimana et al. (2017a, 2017b) used wet sieving method, where air dried soil is sieved in water for, in the case of Mulumba and Lal (2008), 30 min at 30 strokes per minute. In all studies examining the impact of mulching on the aggregate stability of soils, organic mulches were used. An increase in aggregate stability with mulch rate was found by Jordán et al. (2010a, 2010b), who also mentioned that most stable aggregates showed a high proportion of fine roots and organic matter. No crop was planted in this study and the growth of weeds was suppressed by the spraying of glyphosate. Kahlon et al. (2013) observed that addition of organic mulch at rates of 8 and 16 tons/ha resulted in a considerable increase in mean weight diameter and quantity of water stable aggregates. The highest mean weight diameters and number of water stable aggregates were found in the 16 tons/ha wheat straw mulch treatment. Nzeyimana et al. (2017a, 2017b) also found aggregate stability increase in correlation with the application of organic mulches. Mulumba and Lal (2008) described that most stable aggregates were found under the highest mulch rates (16 tons/ha wheat straw mulch) while the least stable aggregates were found when no mulch was used.

Despite the fact that aggregate stability under mulch treatments has been tested in a variety of climates and soils, all reviewed studies found a positive impact on aggregate stability and/or mean weight diameters of organic mulches. The findings of Jordán et al. (2010a, 2010b) suggested that higher contents of organic matter in organically mulched soils could be responsible for the increased aggregate stability. As the number of studies for this parameter is low, further to determine the impact of mulching on aggregate stability, more research is required. Especially the impact of inorganic mulches on aggregate stability is not dealt with sufficiently in the literature.

4 Soil Moisture

Soil moisture describes the volumetric water content of the soil. While it is not the only water-related soil parameter that is relevant to plant growth, it is easily measurable without specialized equipment or complex analysing methods. Most of the studies determined the soil moisture directly via gravimetric method for which sample of soil was oven-dried for 24 h at 105 °C. But other methods such as time domain reflectometry meters (Siczek et al., 2015) or neutron probe moisture meters (Carter et al., 1992) were also used.

As soil temperature greatly affects soil moisture and vice versa (Lakshmi et al., 2003). A positive impact of mulching treatments was expected. Indeed, many of the studies report that mulching increases soil moisture. According to Lal (1974) at both the 0–10 and 10–20 cm layers, mulched plots showed higher soil moisture. Organic mulches in the studies by Agele et al. (2010) increased soil moisture contents significantly. Hulugalle and Palada (1990) reported mulching combined with no-till treatment increased the soil moisture in the 1–15 cm layer but did not affect it significantly in other years. Experiments by Allanov et al. (2019) showed that mulching with 5 tons/ha increased the soil moisture regardless of fertilization level.

Regarding the effectiveness of different mulching materials, the results are less distinct. While Agele et al. (2010) found that plots covered with a mixture of leaves and dry grass achieved better results than mulching with a black polythene sheet. Whereas results by Khatibu et al. (1984) observed that black polythene sheets conserved soil moisture most efficiently. Maurya and Lal (1981) concluded the same results during the second season of the study. They noted that in both the studies mulching with black polythene sheets showed the worst performance in terms of soil temperatures. Cook et al. (2006) found out that soil moisture benefits the most from black horticultural matting over black polythene.

Turf grass as a living mulch was also tested but performed rather poorly. The studies of Qu et al. (2019) and Żelazny and Licznar-Małańczu (2018) both observed similar results because the living plants are the competitors for water and dehydrate the soil due to transpiration. As with other parameters mulching only seems to significantly affect the top layers of the soil. For example, Qu et al. (2019) did not showed positive changes in layer of 20–40 cm. While most crops are shallow-rooted plants, deeper rooting plants such as trees are less likely to benefit.

Alternative tillage treatments as well as its combination with mulch appeared to increase the soil water content (Cook et al., 2006; Hulugalle & Palada, 1990). In Głab and Kulig's tests in 2008 mulching only affected water content of the soil in combination with reduced-tillage treatments having no significant impact when applied to conventionally tilled plots. Pervaiz et al. (2009) compared the interactive effects of mulching with conventional tillage to mulching with deep tillage and came to the conclusion that deep tillage plots combined with 14 tons/ha wheat straw mulch performed the best over the 7 tons/ha while non-mulched, conventionally tilled soil preserved the least amount of water. This also pointed to the fact that the mulch's ability to preserve water raises with increased mulch amount. Similarly, Lal (1978) observed that "the soil moisture reserve was generally proportional to the area of soil surface covered by mulch" and Jordán et al. (2010a, 2010b) reported that low mulching rates did not significantly impact the soil water properties while high mulching rates greatly increased the available water capacity.

Conclusively, mulching can improve soil moisture by reducing direct evaporation from the soil (Cook et al., 2006) and increasing infiltration rates. Due to the more ambiguous nature of the results further and more specific research into the matter is suggested.

5 Infiltration, Interception and Evaporation

Infiltration describes the soil ability to absorb precipitation, interception refers to the water that does not infiltrate the ground but gets intercepted and evaporates before reaching it and evaporation describes the water that directly evaporates from the soil. Infiltration, interception and evaporation are the factors that influence the soil moisture. Change of the soil's physical properties (bulk density, penetration resistance etc.), mulching reduces the impact of falling raindrops and thereby prevents compaction of the surface soil which in turn increases its infiltration capabilities. This is confirmed by many of the studies investigating the parameter (Acharya & Sharma, 1994; Franzen et al., 1994; Jordán et al., 2010a, 2010b; Zhang et al., 2014); however some of the results conflict with this statement. On the other hand, Carter et al. (1992)

recorded no significant increase in infiltration rate on mulched soils compared to control. Research by Cook et al. (2006) showed a net loss of water stored in the soil attributing it to rainfall being intercepted by the mulch and subsequently evaporating even though confirming that mulch reduces direct evaporation of water in the soil. They also came to the conclusion that rainfall interception increased with greater quantities of mulch.

The material used for mulching also impacts infiltration, interception and evaporation. For example, farmyard manure caused 56–80% higher water losses from interception and evaporation than wheat or soybean straw according to Cook et al. (2006). Whereas Zhang et al. (2014) observed that farmyard manure combined with either straw or polythene mulch benefited the hydraulic conductivity—an important indicator of soil infiltration—more than any of the treatments alone. Salau et al. (1992) reported that wood shavings lead to higher infiltration.

It is reported that a quantitative increase in mulching material increased the infiltration rates. In the case of Jordán et al. (2010a, 2010b) the infiltration rates changed to the point where soil absorbed 90% of the precipitation at 5 tons/ha mulch and 100% at 15 tons/ha mulch. Roth et al, (1988) suggested that infiltration rates are mainly affected by surface seal development which in turn is influenced by the degree of surface cover the mulch provides. According to Kahlon et al. (2013) initial infiltration rates as well as steady infiltration rates increased as more mulch was applied.

In regard to different tillage treatments Khatibu et al. (1984) found that different tillage systems influence infiltration. Their experiments showed that no-till plots had the highest infiltration rate (0.01% water run-off compared to 10.2% on untreated soil) while ridged treatments showed the worst infiltration rates as did in terms of soil moisture.

Generally speaking, it can be said that mulching improved the soil's infiltration and evaporation properties, but contradicting results have also been reported. Therefore, specific research is necessary to determine in which situations mulching do not show the expected benefits.

6 Bulk Density

Bulk density (ρB) of a soil sample is used to characterise the packing state of soil (Blume et al., 2016). Bulk density is defined as the mass of a soil sample per volume dried at 105 °C (ρB = mf/Vt with mf = mass of the solid substance and Vt: soil volume) (Hartge & Horn, 2014; Blume et al., 2016). Using bulk density the porosity of a soil sample can be calculated. Some studies did not specify the methodology by which soil bulk density was determined (e.g. Acharya & Sharma, 1994; Carter et al., 1992) while most studies dealing with soil bulk density stated the use of undisturbed soil samples, collected with steel cores between the volumes of 70–137 cm^3, followed by oven drying at 105 °C and the gravimetrical determination of soil bulk density (e.g. Jordán et al., 2010a, 2010b; Kahlon et al., 2013; Siczek et al., 2015). The samples were collected from different soil depths, ranging from 5 cm up to 90 cm.

The literature showed the variable effect of mulching and inconclusive on soil bulk density while several authors, such as Lal (1978), Jordán et al. (2010a, 2010b) and Kahlon et al. (2013) found a decrease of soil bulk density with increase in mulch rate, others, such as Acharya and Sharma (1994), Carter et al., (1992), Hulugalle and Palada (1990), Mulumba and Lal (2008) and Salau et al. (1992) found no significant differences in bulk densities between mulched and non-mulched soils. Lal (1978) found that in the 10–20 cm layer of an Alfisol treated with total mulching and inter-row mulching with rice straw at 4 tons/ha, there was a decrease in soil bulk density. Effect of mulching on Fluvisols was studied by Jordán et al. (2010a, 2010b) by mulching with 1, 5, 10 and 15 tons/ha wheat straw and they found an effect of mulch rates 5, 10 and 15 tons/ha with decreased bulk density almost linear to mulch rate. Jordán et al. (2010a, 2010b) pointed out, that low mulch rates (1 tons/ha wheat straw mulch) had no significant impact on bulk density. Kahlon et al. (2013) found a decrease in bulk density with increased mulch rate of wheat straw and reported the best results, while recorded a decrease in soil bulk density, under mulched no-tillage treatments. The mulching-materials used by Zhang et al. (2014) resulted a decrease in bulk density of soil as a result of farmyard manure, straw- and plastic-mulch applications with the biggest decrease in bulk density in the upper soil layer (0–10 cm) of the farmyard manure + straw mulch treatment.

Other studies reporting a positive effect of mulching on soil bulk density were Allanov et al. (2019), Pervaiz et al. (2009), Głab and Kulig (2008), Qu et al. (2019), Nzeyimana et al. (2017a, 2017b) and Żelazny and Licznar-Małańczuk (2018). They observed positive effects of organic or living mulches on soil bulk density. Qu et al. (2019) used inorganic as well as organic mulches (cobblestones, water permeable bricks, pine bark, woodchips, green waste compost and turf grass) and observed positive effects on soil bulk density in organic mulches (pine bark, green waste compost and turf grass).

In studies that could not find a significant effect of mulching on bulk density, such as Acharya and Sharma (1994), the mulching materials were e.g. pine needles at a rate of 10 tons/ha on an Alfisol, stover at 5 tons/ha on Luvisols and a Cambisol (Carter et al., 1992) or rice straw at 4 tons/ha on an Ustipsamment. Other treatments that did not indicated a significant decrease or increase in bulk density where the application of straw mulch at rates of 2, 4, 8 and 16 tons/ha on a stagnic Luvisol (Mulumba & Lal, 2008). Salau et al. (1992) used a combination of elephant grass at 80 tons/ha fresh weight and plastic sheet mulch on an Ultisol with no significant effect on bulk density. Siczek et al. (2015) conducted experiments on a Haplic Luvisol that was mulched at a rate of 5 tons/ha with wheat straw and could not found significant differences in bulk density due.

The duration of the treatments in studies that found a significant impact on bulk density were 3–22 years (e.g. Jordán et al., 2010a, 2010b; Kahlon et al., 2013) in contrast to a study-period of 2–11 years in studies that did not find a significant impact of mulching on bulk density (Carter et al., 1992; Hulugalle & Palada, 1990; Mulumba & Lal, 2008; Siczek et al., 2015).

Another important difference in the setup of the reviewed studies was the crop-management, as there were different crops planted in different studies, e.g. cowpea (*Vigna unguiculata*), maize-barley (*Hordeum vulgare*) rotations, banana (*Musa sp.*), soybean (*Glycine max*) or no crops. No correlation between crop and soil density was apparent.

Amongst the reviewed studies, a higher number of studies showed a decrease in soil bulk density in organically mulched soils than no change in soil bulk density. None of the studies showed an increase in bulk density as a result of mulching. A more extensive literature review in the form of metadata is required to be able to find a clear trend in the mulching impact on soil bulk density. Mulumba and Lal (2008) also reported inconclusive findings regarding the effect of mulching on soil bulk density in their research experiments.

Acharya et al. (2005) suggested increased earthworm activity under organically mulched soils as a reason for decreased soil bulk density, which correlates with the fact that a significant positive effect of mulching on bulk density was only found with the usage of organic mulches.

The reasons for different findings among the reviewed studies can be varied and are hard to pinpoint as different mulching treatments were used on different types of soils in different climatic conditions. As Zhang et al. (2014) mentioned "the effect of mulch on soil quality may be related to soil type, climate, and land use pattern [etc.]."

7 Penetration Resistance

The force needed to stop the movement of tip of a penetrometer divided by the penetration depth is called the **soil penetration resistance**, which is an indicator for the compaction of the soil in the measured depth (Herrick & Jones, 2002). The penetration resistance can be measured using different kinds of automatic and field penetrometers with different tips. Hulugalle and Palada (1990) used a pocket penetrometer with a 6 mm probe diameter and a blunt tip while Kahlon et al. (2013) used a 12.8 mm diameter cone tipped penetrometer. The penetrologger used by Siczek et al. (2015) also had a cone surface of 1 cm^2 and a top angle of 60°, a similar device was used by Franzen et al. (1994). Lal (1978) used a field penetrometer where the tip-shape was not specified.

A significant decrease in penetration resistance was reported by Franzen et al. (1994) in mulched untilled soils in comparison to unmulched untilled soils in the depth 0–10 cm (mulching rate 6 tons/ha maize stover). In the 10–20 cm soil layer, a similar tendency was observed. Kahlon et al. (2013) also found a decrease in penetration resistance in correlation with straw mulch application of 8 and 16 tons/ha. The straw of rice application at 4 tons/ha correlated with measurements of lower penetration resistance for treatments with mulched inter-row (Lal, 1978). Pervaiz et al. (2009) argued that the application of wheat straw mulch at 7 and 14 tons/ha significantly decreased penetration resistance, with the greatest effect for 14 tons/ha. Although Siczek et al. (2015) studied the impact of mulching and soil deformation on soil properties, they could only found a correlation between increased wheeling and higher penetration resistance, while no impact of mulching on penetration resistance observed (mulching rate: 5 tons/ha wheat straw). Hulugalle and Palada (1990) also found no effect of rice as a mulch on penetration resistance (mulching rate: 4 tons/ha).

Findings of the studies concerning mulching impact on the soil penetration resistance were inconclusive. While four studies found a positive impact of mulching on penetration resistance, hence decreasing it, two studies found no significant impact. The experiments of the previously mentioned studies were carried out with different penetrometers, in different climatic conditions, as well as on different types of soil. Penetration resistance is mainly affected by soil type, soil moisture and bulk density (Costantini, 1996). The bulk of the reviewed studies indicated a positive impact of mulching on penetration resistance, but a more extensive literature research is needed.

8 Soil Erosion and Runoff

Soil erosion is the action that removes or transports soil particles from one location due to the influence of wind or water and deposits them in another. Runoff is water that does not infiltrate into the soil but runs off a location via its surface. Soil erosion is a problem that affects people all around the world and especially affects soils with disturbed ground vegetation as it is the case with fallow agricultural land or fields between crop cycles. While erosion by wind is the biggest problem in arid regions, more humid regions are usually affected by water erosion especially rolling landscapes. Rain that does not infiltrate the soil, runs off and carry soil particles with it. Mulching is expected to diminish these effects by providing a protective layer that prevents the soil from drying out, intercepts runoff water and increases soil transmissivity by alleviating surface soil compaction of rain drop impact.

Evaluation and comparison of the studies led to the conclusion that mulching does indeed reduce the soil erosion. Lal (1974) reported that "unmulched plots eroded severely" while Nzeyimana et al. (2017a, 2017b) reported that mulching decreased soil erodibility significantly. Acharya and Sharma (1994) observed that in mulched soils there was stagnant water found on the testing plots and there was no water standing on the field despite high rainfall. This means that there is less runoff and soil erosion. Research of Khatibu et al (1984) compared the soil loss of mulched and unmulched plots and came to the conclusion that the total soil loss of mulched plots is only 3.6% of that measured in unmulched plots. Experiments of De La Rosa et al. (2019) concluded the results in a similar order of magnitude and determined that mulching decreased the loss of soil by erosion up to 91%. Zhang et al. (2014) observed that farmyard manure incorporated into the soil combined with straw mulch on the surface, shaded the soil from erosion. Results of Jordán et al. (2010a, 2010b) suggested that the amount of mulching also influenced the soil's erodibility.

In summary it can be said that mulching decreased soil erosion and improves the soil's runoff characteristics. None of the reviewed studies observed a negative or negligible impact of mulching on these properties. In practice this knowledge can be applied to give farmers and landscapers an easy and cheap method to prevent disturbed soil from eroding and subsequently degrading in terms of nutrient availability and the ability to preserve water.

9 Conclusion

The soil physical properties such as soil porosity, aggregate stability, soil temperature, soil erosion and runoff are generally positively influenced by mulching. Whereas the effectiveness of mulch treatments is situational and should be assessed on a case-to-case basis when trying to improve these physical parameters. Hence, a more extensive research or metadata analysis is needed to make more reliable statements on the effects of mulching on soil physical properties.

References

Acharya, C. L., & Sharma, P. D. (1994). Tillage and mulch effects on soil physical environment, root growth, nutrient uptake and yield of maize and wheat on an Alfisol in north-west India. *Soil & Tillage Research,* 291–302. https://doi.org/10.1016/0167-1987(94)00425-E.

Acharya, C. L., Hati, K. M. & Bandopadhyay, K. K. (2005). Mulches. In D. Hillel et al. (Eds.), *Encyclopedia of soils in the environment,* (pp. 521–532). Elsevier Publication.

Adekiya, A. O., Agbede, T. M., Aboyeji, C. M., & Dunsin, O. (2017). Response of okra (*Abelmoschus esculentus (L.) Moench*) and soil properties to different mulch materials in different cropping seasons. *Scientia Horticulturae, 217,* 209–216. https://doi.org/10.1016/j.scienta.2017.01.053.

Agele, S. O., Olaore, J. B., & Akinbode, F. A. (2010). Effect of some mulch materials on soil physical properties, growth and yield of sunflower (*Helianthus Annuus,* L). *Environmental Biology, 4,* 368–375. http://www.aensi.org/aeb.html.

Allanov, K., Sheraliev, K., Ulugov, C., Ahmurzayev, S., Sottorov, O., Khaitov, B., & Park, K. W. (2019). Integrated effects of mulching treatment and nitrogen fertilization on cotton performance under dryland agriculture. *Communications in Soil Science and Plant Analysis, 50*(15), 907–1918. https://doi.org/10.1080/00103624.2019.1648496.

Amare, G., & Desta, B. (2021). Coloured plastic mulches: Impact on soil properties and crop productivity. *Chemical and Biological Technologies in Agriculture, 8*(1), 1–9. https://doi.org/10.1186/s40538-020-00201-8.

Anikwe, M. A. N., Mbah, C. N., Ezeaku, P. I., & Onyia, V. N. (2007). Tillage and plastic mulch effects on soil properties and growth and yield of cocoyam (*Colocasia esculenta*) on an ultisol in southeastern Nigeria. *Soil and Tillage Research, 93,* 264–272. https://doi.org/10.1016/j.still.2006.04.007.

Anikwe, N. L., Okereke, O. U., & Anikwe, M. A. N. (2004). Modulating effect of black plastic mulch on the environment, growth and yield of cassava in a derived savannah belt of Nigeria. *Tropicultura, 22,* 185–190.

Bingpeng, Q., Liu, Y., Sun, X., Li, S., Wang, X., Xiong, K., et al. (2019). Effect of various mulches on soil physico—Chemical properties and tree growth (Sophora japonica) in urban tree pits. *PLoS ONE 14*(2), e0210777

Blume, H. P., Brümmer, G. W., Fleige, H., Horn, R., Kandeler, E., Kögel-Knabner, I., Kretzschmar, R., Stahr, K., & Wilke, B. M. (2016). Scheffer/schachtschabel soil science.; Cadavid, L. F., El-Sharkawy, M. A., Acosta, A., Sánchez, T. (1998). Long-term effects of mulch, fertilization and tillage on cassava grown in sandy soils in northern Colombia. *Field Crops Research, 57*(1), 45–56. https://doi.org/10.1016/s0378-4290(97)00114-7.

Carter, D. C., Harris, D., Youngquist, J. B., & Persaud, N. (1992). Soil properties, crop water use and cereal yields in Botswana after additions of mulch and manure. *Field Crops Research, 30*, 97–109. https://doi.org/10.1016/0378-4290(92)90059-I.

Chaudhry, M. R., Malik, A. A., & Sidhu, M. (2004). Mulching impact on moisture conservation, soil properties and plant growth. *Pakistan Journal of Water Resources, 8*(2), 1–8.

Cook, H. F., Valdes, G. S. B., & Lee, H. C. (2006). Mulch effects on rainfall interception, soil physical characteristics and temperature under *Zea mays L. Soil and Tillage Research, 91*(1–2), 227–235. https://doi.org/10.1016/j.still.2005.12.007.

Costantini, A. (1996). Relationships between cone penetration resistance, bulk density, and moisture content in uncultivated, repacked, and cultivated hardsetting and non-hardsetting soils from the coastal lowlands of south-east Queensland. *New Zealand Journal of Forestry Science, 26*(3), 395–412.

De la Rosa, J. M., Jiménez-Morillo, N. T., González-Pérez, J. A., Almendros, G., Vieira, D., Knicker, H. E., & Keizer, J. (2019). Mulching-induced preservation of soil organic matter quality in a burnt eucalypt plantation in central Portugal. *Journal of Environmental Management, 231*(May 2018), 1135–1144. https://doi.org/10.1016/j.jenvman.2018.10.114.

Franzen, H., Lal, R., & Ehlers, W. (1994). Tillage and mulching effects on physical properties of a tropical Alfisol. *Soil and Tillage Research, 28*(3–4), 329–346. https://doi.org/10.1016/0167-1987(94)90139-2.

Głab, T., & Kulig, B. (2008). Effect of mulch and tillage system on soil porosity under wheat (Triticum aestivum). *Soil and Tillage Research, 99*(2), 169–178. https://doi.org/10.1016/j.still.2008.02.004.

Hartge, K. H., & Horn, R. (2014). *Introduction to soil physics* (178 p), 4th revised edn. Schweizerbart.

Haruna, S. I., & Nkongolo, N. V. (2015). Cover crop management effects on soil physical and biological properties. *Procedia Environmental Sciences, 29*, 13–14. https://doi.org/10.1016/j.proenv.2015.07.130.

Haruna, S. I., Anderson, S. H., Nkongolo, N. V., & Zaibon, S. (2018). Soil hydraulic properties: Influence of tillage and cover crops. *Pedosphere, 28*, 430–442. https://doi.org/10.1016/S1002-0160(17)60387-4.

Herrick, J. E., & Jones, T. L. (2002). A dynamic cone penetrometer for measuring soil penetration resistance. *Soil Science Society of America Journal, 66*(4), 1320–1324. https://doi.org/10.2136/sssaj2002.1320.

Hulugalle, N. R., & Palada, M. C. (1990). Effect of seedbed preparation method and mulch on soil physical properties and yield of cowpea in a rice fallow of an inland valley swamp. *Soil & Tillage Research, 17*, 101–113. https://doi.org/10.1016/0167-1987(90)90009-3.

Imeson, A. C., & Vis, M. (1984). Assessing soil aggregate stability by water-drop impact and ultrasonic dispersion. *Geoderma, 34*(3–4), 185–200. https://doi.org/10.1016/0016-7061(84)90038-7.

Innocent, N., Hartemink, A. E., Ritsema, C., Stroosnijder, L., Lwanga, E. H., & Geissen, V. (2017). Mulching as a strategy to improve soil properties and reduce soil erodibility in coffee farming systems of Rwanda. *Catena, 149*, 43–51

Jack C. V., Brind W. D., & Smith, R. (1955). Mulching Tech. Comm. No. 49. *Commnwealth Bulletin of Soil Science.*

Jordán, A., Zavala, L. M., & Gil, J. (2010a). Effects of mulching on soil physical properties and runoff under semi-arid conditions in southern Spain. *CATENA, 81*(1), 77–85. https://doi.org/10.1016/j.catena.2010.01.007.

Jordán, A., Zavala, L. M., & Gil, J. (2010b). Effects of mulching on soil physical properties and runoff under semi-arid conditions in southern Spain. *Catena, 81*(1), 77–85. https://doi.org/10.1016/j.catena.2010b.01.007.

Jourgholami, M., Fathi, K., & Labelle, E. R. (2020). Effects of litter and straw mulch amendments on compacted soil properties and Caucasian alder (*Alnus subcordata*) growth. *New Forests, 51*(2), 349–365. https://doi.org/10.1007/s11056-019-09738-5.

Kahlon, M. S., Lal, R., & Ann-Varughese, M. (2013). Twenty-two years of tillage and mulching impacts on soil physical characteristics and carbon sequestration in Central Ohio. *Soil and Tillage Research, 126*, 151–158. https://doi.org/10.1016/j.still.2012.08.001.

Khatibu, A. I., Lal, R., & Jana, R. K. (1984). Effects of tillage methods and mulching on erosion and physical properties of a sandy clay loam in an equatorial warm humid region. *Field Crops Research, 8*, 239–254. https://doi.org/10.1016/0378-4290(84)90072-8.

Lakshmi, V., Jackson, T. J., & Zehrfuhs, D. (2003). Soil moisture-temperature relationships: Results from two field experiments. *Hydrological Processes, 17*(15), 3041–3057. https://doi.org/10.1002/hyp.1275.

Lal, R. (1974). Soil temperature, soil moisture and maize yield from mulched and unmulched tropical soils. *Plant and Soil, 40*(1), 129–143. https://doi.org/10.1007/BF00011415.

Lal, R. (1978). Influence of within-and between-row mulching on soil temperature, soil moisture, root development and yield of maize (*Zea mays*) in a tropical soil. *Field Crops Research,* 127–139. https://doi.org/10.1016/0378-4290(78)90016-3.

Lal, R., Regnier, E., Eckert, D. J., Edwards, W. M., & Hammond, R. (1991). Expectations of cover crops for sustainable agriculture. In W. L. Hargrove (Ed.), *Cover crops for clean water* (pp. 1–11). Soil and Water Conservation Society of America.

Li, F. M., Wang, J., Xu, J. Z., & Xu, H. L. (2004). Productivity and soil reponse to plastic film mulching duration for spring wheat on entisols in the semiarid loess plateau of China. *Soil and Tillage Research, 78*, 9–20. https://doi.org/10.1016/j.still.2003.12.009.

Li, F. M., GuoA, H., & Wei, H. (1999). Effects of plastic film mulch on yield of spring wheat. *Field Crops Research, 63*, 79–86. https://doi.org/10.1016/S0378-4290(99)00027-1.

Li, R., Wu, Q., Zhang, J., Wen, Y., & Li, Q. (2019). Effects of land use change of sloping farmland on characteristic of soil erosion resistance in typical Karst mountainous areas of South western China. *Polish Journal of Environmental Studies, 28*, 2707–2716. https://doi.org/10.15244/pjoes/94288.

Liang, H., Hu, K., Qin, W., Zuo, Q., & Zhang, Y. (2017). Modelling the effect of mulching on soil heat transfer, water movement and crop growth for ground cover rice production system. *Field Crops Research, 201*, 97–107. https://doi.org/10.1016/j.fcr.2016.11.003.

Mandal, D., & Sharda, V. N. (2013). Appraisal of soil erosion risk in the Eastern Himalayan region of India for soil conservation planning. *Land Degradation and Development, 24*, 430–437. https://doi.org/10.1002/ldr.1139.

Maurya, P. R., & Lal, R. (1981). Effects of different mulch materials on soil properties and on the root growth and yield of maize (Zea mays) and cowpea (*Vigna unguiculata*). *Field Crops Research, 4*, 33–45. https://doi.org/10.1016/0378-4290(81)90052-6.

Mekonnen, M., Keesstra, S. D., Stroosnijder, L., Baartman, J. E., & Maroulis, J. (2015). Soil conservation through sediment trapping: A review. *Land Degradation and Development, 26*, 544–556. https://doi.org/10.1002/ldr.2308.

Mulumba, L. N., & Lal, R. (2008). Mulching effects on selected soil physical properties. *Soil and Tillage Research, 98*(1), 106–111. https://doi.org/10.1016/j.still.2007.10.011.

Nimmo, J. R. (2013). Porosity and pore size distribution. In *Reference module in earth systems and environmental sciences* (pp. 1–10). Elsevier Inc., https://doi.org/10.1016/b978-0-12-409548-9.05265-9.

Nouri, A., Lee, J., Yin, X., Tyler, D. D., & Saxton, A. M. (2019). Thirty-four years of no-tillage and cover crops improve soil quality and increase cotton yield in Alfisols, southeastern USA. *Geoderma, 337*, 998–1008. https://doi.org/10.1016/j.geoderma.2018.10.016.

Nzeyimana, I., Hartemink, A. E., Ritsema, C., Stroosnijder, L., Lwanga, E. H., & Geissen, V. (2017a). Mulching as a strategy to improve soil properties and reduce soil erodibility in coffee farming systems of Rwanda. *CATENA, 149*, 43–51. https://doi.org/10.1016/j.catena.2016.08.034.

Nzeyimana, I., Hartemink, A. E., Ritsema, C., Stroosnijder, L., Lwanga, E. H., & Geissen, V. (2017b). Mulching as a strategy to improve soil properties and reduce soil erodibility in coffee farming systems of Rwanda. *Catena, 149*, 43–51. https://doi.org/10.1016/j.catena.2016.08.034.

Osman, K. T. (2013). Principles, properties and management. In: *Soils*. Springer, Dordrecht. http://doi-org-443.webvpn.fjmu.edu.cn/https://doi.org/10.1007/978-94-007-5663-2_1.

Paul, P. L. C., Bell, R. W., Barrett-Lennard, E. G., & Kabir, E. (2021). Impact of rice straw mulch on soil physical properties, sunflower root distribution and yield in a salt-affected clay-textured soil. *Agriculture, 11*(3), 264. https://doi.org/10.3390/agriculture11030264.

Pavlů, L., Kodešová, R., Fér, M., Nikodem, A., Němec, F., & Prokeš, R. (2021). The impact of various mulch types on soil properties controlling water regime of the Haplic Fluvisol. *Soil and Tillage Research, 205*, 104748. https://doi.org/10.1016/j.still.2020.104748.

Pervaiz, M. A., Iqbal, M., Shahzad, K., & Hassan, A. U. (2009a). Effect of mulch on soil physical properties and N, P, K concentration in maize (Zea mays L.) shoots under two tillage systems. *International Journal of Agriculture and Bio*logy, *11*(2), 119–124.

Qu, B., Liu, Y., Sun, X., Li, S., Wang, X., Xiong, K., & Zhang, H. (2019). Effect of various mulches on soil physic-Chemical properties and tree growth (Sophora japonica) in urban tree pits. *PLoS One, 14* (2), e0210777. https://doi.org/10.1371/journal.pone.0210777.

Roth, C. H., Meyer, B., Frede, H.-G., & Derpsch, R. (1988). Effect of mulch rates and tillage systems on infiltrability and other soil physical properties of an Oxisol in Paran, Brazil*. *Soil & Tillage Research, 11*, 81–91. https://doi.org/10.1016/0167-1987(88)90033-5.

Ruan, H., Ahuja, L. R., Green, T. R., & Benjamin, J. G. (2001). Residue cover and surface-sealing effects on infiltration. *Soil Science Society of America Journal, 65*, 853–861. https://doi.org/10.2136/sssaj2001.653853x

Salau, O. A., Opara-Nadi, O. A., & Swennen, R. (1992). Effects of mulching on soil properties, growth and yield of plantain on a tropical ultisol in southeastern Nigeria*. *Soil & Tillage Research, 23*, 73–93. https://doi.org/10.1016/0167-1987(92)90006-W.

Siczek, A., Horn, R., Lipiec, J., Usowicz, B., & Łukowski, M. (2015). Effects of soil deformation and surface mulching on soil physical properties and soybean response related to weather conditions. *Soil and Tillage Research, 153*, 175–184. https://doi.org/10.1016/j.still.2015.06.006.

Sintim, H. Y., Bandopadhyay, S., English, M. E., Bary, A. I., DeBruyn, J. M., Schaeffer, S. M., & Flury, M. (2019). Impacts of biodegradable plastic mulches on soil health. *Agriculture, Ecosystems & Environment, 273*, 36–49. https://doi.org/10.1016/j.agee.2018.12.002.

Thomaz, E. L., & Luiz, J. C. (2012). Soil loss, soil degradation and rehabilitation in a degraded land area in Guarapuava (Brazil). *Land Degradation and Development, 23*, 72–81. https://doi.org/10.1002/ldr.v23.1.

Tisdale, S. L., Nelson, W. L., & Beaton, J. D. (1985). *Soil fertility and fertilizers* (4th ed.). Macmillan Publishing Company.

Xue, K., M. Yuan, M., J. Shi, Z. et al. (2016). Tundra soil carbon is vulnerable to rapid microbial decomposition under climate warming. *Nature Climate Change 6*, 595–600. https://doi.org/10.1038/nclimate2940

Żelazny, W. R., & Licznar-Małańczuk, M. (2018). Soil quality and tree status in a twelve-year-old apple orchard under three mulch-based floor management systems. *Soil and Tillage Research, 180*(August 2017), 250–258. https://doi.org/10.1016/j.still.2018.03.010.

Zhang, J. B., Yang, J. S., Yao, R. J., Yu, S. P., Li, F. R., & Hou, X. J. (2014). The effects of farmyard manure and mulch on soil physical properties in a reclaimed coastal tidal flat salt-affected soil. *Journal of Integrative Agriculture, 13*(8), 1782–1790. https://doi.org/10.1016/S2095-3119(13)60530-4.

Mulching is an Approach for a Significant Decrease in Soil Erosion

Marwa Tariq and Kashif Akhtar

Abstract Among the applied soil insurance rehearses, mulch has been effectively applied to decrease soil and water misfortune in different conditions, for example, agrarian land, fire-stricken regions, prairies and human sites. In these unique circumstances, a difficult issue is soil erosion from water, particularly in semi-sticky and semi-bone-dry areas of the world. While the valuable impacts of mulch are known, more exploration is expected to evaluate them, particularly in regions where soil erosion from water is a serious threat. There are still a few vulnerabilities in the literature regarding how to boost the adequacy of mulching to lessen the pace of soil and water misfortune. Given the seriousness of soil erosion from water and the vulnerabilities actually connected with the right utilization of mulch, the evaluation of this study aims to (Adekalu et al., 2007) grow a complete and recorded data set on the utilization of mulch with vegetative deposits; (Albaladejo Montoro et al., 2000) evaluate the impacts of mulch on soil and water misfortune dependent on various estimation techniques and consequently unique spatial scales; (Arnáez et al., 2015) survey the impacts of various kinds of cover on soil and water misfortunes dependent on various estimation strategies; and (Badia & Marti, 2000) make ideas for more feasible soil the board. Information have been gathered and distributed in the literature. The outcomes showed the gainful impacts of mulching in the battle against soil disintegration by water in all media considered here, with a decrease in mean residue fixation, soil misfortune and soil volume and overflow that at times added up to over 90%. Be that as it may, the monetary achievability of mulching was not accessible in the literature. Thusly, more exploration should be done to help ranchers and land directors the same by furnishing them with proof based assets to carry out more supportable soil the executives rehearses.

Keywords Soil erosion · Residue fixation · Mulching · Estimation techniques

M. Tariq (✉)
Department of Chemistry, Abdul Wali Khan University Mardan, KPK, Mardan, Pakistan
e-mail: marwatariq2019@gmail.com

K. Akhtar
State Key Laboratory for Conservation and Utilization of Subtropical Agro-Bio-Resources, College of Life Science and Technology, Guangxi University, Nanning 530004, China

K. Akhtar et al. (eds.), *Mulching in Agroecosystems*,
https://doi.org/10.1007/978-981-19-6410-7_4

1 Introduction

More than 50 years after its independence, Pakistan has made enormous strides progress in agriculture and food security. Before the mid-1960s, Pakistan relied upon imports and food help necessities. In any case, two years of extreme dry spell in 1965 and 1966 persuaded Pakistan to change its horticultural arrangement, and that Pakistan couldn't rely upon unfamiliar guide and imported products Food Safety. Pakistan has implemented sweeping policy reforms aimed at with the aim of self-sufficiency of food in grains. This marked the beginning of Pakistan the green revolution. It started with the decision to hire superior wheat varieties resistant to diseases related to them better agricultural knowledge to improve productivity. Developments of irrigated areas, abundant use of fertilizers and pesticides the use of high-yielding varieties made it green possible revolution. But the population of Pakistan continues to increase and we have he realized that this one green revolution alone wouldn't help. We have to use other means to support our agricultural growth and how can we do this? We can do it through conservation agriculture. Organic farming is the answer. We realized that the green revolution has saved us once, but now its dependence on the intensive use of fertilizers and pesticides pollute our environment and affect our environment the terrain. Over-irrigation leads to soil erosion problems and soil salinity. Although many irrigation schedules do developed, we have only 1/3 of our agriculture under irrigation.

So what's the exit? To improve the health of our soils what we are doing? What to do we do everything we can to maintain a good microflora and a nice balance micro-organisms in the soil? What are we doing about it? soil moisture? What do we do to keep the weeds in our fields? The answer is simple: embrace the old practice of mulching in our agricultural fields. So what's mulch? It is a protector blanket, generally organic material like leaves or straw peat, put around plants to forestall dissipation dampness and weed development. The word mulch presumably gets from the German word "Molsch" signifies delicate toecay, obviously alluding to the grounds-keeper utilizes straw and passes on to sprinkle on the ground mulch (Jack et al., 1955). Mulching decreases soil erosion by forestalling it overflow and soils, limits weed invasion and Check the vanishing of the water. For this reason it makes more conceivable holds soil dampness and helps control temperature vacillations, physical, compound and organic improvement properties of the soil, since it adds supplements to the soil lastly further develops crop development and yield. Furthermore, he revealed that cushioning expands execution by 50–60% without cushioning in stormy circumstances (Dilip Kumar et al., 1990). Water erosion is the aftereffect of the association of environment, surface spillover, soil, geography, vegetation cover, soils the executives and protection practices, and appears as a variable over the long haul and space on the ground surface. Loss of soil and supplements and their Subsequent waterway transport is primarily liable for agrarian land debasement, prompting a diminishing in the creation limit of land lastly to the impracticality of rural creation frameworks (Oliveira et al., 2010). Spillover and residue transport are perplexing hydrological peculiarities. Past states of soil dampness, soil cover and precipitation force assume a part a significant job during the time spent precipitation overflow and the subsequent water and soil misfortunes. The transient inconstancy of downpour significantly affects the creation of run-off squander and the related vehicle process particularly in semi-dry regions where tempest transport is common various significant degrees (Römkens et al., 2001; De Lima et al., 2009; De Lima & Grasman, 1999). Numerous analysts have focused on the significance of the blend impact of downpour and wind on the soil vehicle process, which relies upon the properties of the surface, design and grain size of the soil Distribution (Erpul et al., 2004).

2 Mulching

In its least complex structure, mulch is any material that covers the outer layer of the soil. In nature, mulch basically comprises of dead leaves and plant trash. In the nursery, mulch can likewise contain fertilizer, wood chips, and spoiled compost, cardboard or even green growth. As of late have we come to see the value in the biological and practical advantages of mulch. Living microorganisms are feed by mulching in our soil with supplements and the loss from these small organisms makes a solid soil structure for plants decreasing compaction.

2.1 *Advantages and Disadvantages of Mulching*

The influence different types of mulch in crop yield can be positive or negative, related to the weed control effect. A lot of Researchers have shown the positive effects of mulching on crops. The growth and the amounts and qualities of the yield obtained (Ramakrishna et al., 2005). Regardless of the color, non-biodegradable PP and PE film cushions have proven the more effective in preventing seed germination most weeds and their subsequent growth, although they are useful to prevent loss of soil humidity and in the equilibrium of its temperature (Momiroviæ et al., 2010). Its application often brings many other benefits, such as reduced runoff, more rain-water penetration, erosion control, chemical correction soil balance and reduction of damage from pests and diseases. However, they also have certain disadvantages for the environment, regarding the disposal and treatment of your waste (Briassoulis, 2006).

2.2 *Types of Mulching*

The ideal mulch ought to be adequately minimal to impede weed development, however light and open enough for water and air to arrive at the ground. Variables to think about when buying mulch incorporate expense, accessibility, convenience, and how it will examine the nursery. There are numerous materials in various shadings and surfaces to browse.

Both **organic** and **inorganic** mulches can be utilized efficiently in the garden.

2.2.1 Organic Mulch

Organic mulches are normal items produced using leaves, trees, grass and other plant material, frequently from your own nursery. They mirror nature and bit by bit separate over the long haul. The benefit is that they really add organic matter to the soil. The disadvantage is that they should be topped up occasionally.

- **Fertilizer** is promptly accessible and separates rapidly to work on the dirt. Assuming that you don't have yours, urban communities frequently make it accessible for their leaf fertilizing the soil offices. The impediment is that it should be enhanced and may contain pot seeds.
- **Crushed** or **chipped bark**. Softwood bark mulch is appealing, opposes compaction and is delayed to deteriorate. Hardwood bark is alluring, however it decays rapidly and should be appropriately treated the soil to keep away from acidic mulch and inconvenient parasites.
- **Shredded leaves** and leaf shape are promptly accessible, and when cut they deteriorate and furnish the dirt with helpful materials. The drawback is that the leaves can begin to shape when wet, decreasing oxygen and soil dampness. Keep away from tangled layers of wet leaves.
- **Straw** and **marsh** roughage are reasonable and give helpful inclusion; be that as it may, they crumble all the more rapidly, can hold onto rodents and are effectively blown away.

2.2.2 Inorganic Mulch

- **Dark plastic mulch** warms the soil in the spring, decreases water misfortune, and is valuable. It can have a major effect in short developing seasons. Notwithstanding, it isn't porous, making it more hard to water; It additionally breaks down when presented to the sun and the soil underneath the plastic gets extremely blistering in summer if not concealed by leaves or covered with other mulch.
- **Silver plastic mulch** is incredible for warming the dirt in the spring, however won't hold weeds; the soil is additionally warmed by clear plastic in mid-summer and plants can be harmed assuming the plastic isn't in the shade.
- **Broken stone, rock, marble, or chips** make long-lasting mulch around bushes and trees. All things considered, these mulches are costly, hard to move, and can attack your grass. Weed seeds and soil can in any case infiltrate the stones; a base layer of scene texture forestalls this (Sweetser, 2021).

2.3 Impacts of Mulch on Harvest and Weed Development

As well as lessening weed seed germination and development, mulch can work on the development and seriousness of set up crops by holding soil dampness and changing soil temperature (Schonbeck & Evalylo, 1998; Swaider et al., 1992). Warming the dirt under dark plastic or IRT can further develop early season development and aging in heat-adoring harvests while the soil cooling impact of intelligent and natural foil mulch helps the chilly vegetables like potatoes and can assist most yields with flourishing the blistering summer.

Some natural mulch, like roughage, gives slow-discharge supplements or decreases specific vermin by securing their normal hunters. Intelligent or hued manufactured mulch has been found to work on the yields of certain harvests by repulsing vermin or switching the light climate up the yield (Orzolek & Lamont, 2000).

It is essential to take note of that once a weed figures out how to get past the mulch, or arise through an establishing opening in stick film, it partakes in a similar maintenance of soil dampness and soil. Advantages of mulch other than set up culture. Actually, seedlings of harvests arising under mulch will be stifled. In this way, it is normal to apply straw or other natural mulch solely after the harvest is grounded and following development or manual expulsion of existing weeds (Schonbeck, 2015).

2.4 Soil Erosion

Soil erosion is the development and transport of soil through it different assets, particularly water, wind and mass development; in this manner environment is a key variable. Starting around 1930 it has been perceived as a major issue. Furthermore, despite the fact that there has been around 70 years of examination in causes and cycles, increases even more and growing concern. Soil erosion rate in the world have overcome those of the formation of new soils in 10 and 20 times on most of the world's continents. The last decades Increased soil erosion. The date is closely related to the elimination of vegetation, to allow the use of land for herbaceous crops and the use of inappropriate agricultural practices therefor the field in which they are practiced (Holman, 2006). Soil is the basis of agricultural production. Soil erosion happens in the horticultural region has imperiled manageability rural exercises. Sped up soil disintegration has unfavorable results natural and financial impacts (Lal, 1998). The impacts of soil erosion on usefulness are probably going to happen both on location and off-site. The deficiency of on location and off-site efficiency because of soil erosion is credited to three communication impacts: diminished soil quality, long haul usefulness impacts and momentary usefulness impacts (Lal, 2001). With an end goal to control off-site soil erosion and on location impacts analysts have created with regards to agribusiness in Asia Appropriate control practices and techniques to lessen how much land erosion. Controlling soil disintegration comprises of empowering inventive methodologies in land the executives strategies and techniques. Much examination has been led into soil erosion control rehearses as far as the executives consistently in Asia. The investigations are chiefly centered around the impact of various administration rehearses, like culturing, mulching, ground cover and related harvests in overflow creation and erosion process (Li, et al., 2018; Dai et al., 2018; Sharma et al., 2017). Soil disintegration is happening at a disturbing rate and is primarily because of deforestation in the northern areas of Pakistan. The landmass impacted by the breeze and water erosion. Research affirmed that soil erodibility is experiencing inborn properties soil (Liu et al., 2003), wind speed levels (Lopez et al., 2000) and ground cover (Li et al., 2004). Around a billion tons of soil is lost consistently, setting down and shooting important prey in the Arabian Sea. The greatest the recorded disintegration rate is assessed 150–165 tons/ha/year. Wind Erosion is a typical peculiarity with the deserts of Thal, Cholistan and Tharparker and along the Makran coast. Water disintegration is the most well-known risk in the climate, principally caused because of overexposure of uncovered soil as a result of an ineffectively overseen library tasks, self-assertive territory cleaning, cutting of vegetation for fuel and inappropriate use escape. Water disintegration abbreviates the existence segment of the principle supplies, water system framework and lessens proficiency. Water erosion is indispensable on steep slants like the Potohar track and its environmental factors, a local much utilized for developed. Around half of the water is lost as overflow. Assuming that a large portion of this water may be saved, 6 MAF measure of water, which is sufficient to 2/3 of the helpful limit from the Tarbela dam, enough to inundate 4 million hectares of land. Soil erosion includes the misfortune or evacuation of surface soil material through the activity of moving water, wind or ice. The augmentation of the space impacted by water and wind erosion is displayed in Charts 1 and 2 individually. Around 13.05 million hectares of region is experiencing water erosion and around 6.17 million hectares are experiencing water erosion. Soil erosion is happening at a disturbing rate and is principally because of deforestation in the north. Water erosion is unmistakable on steep slants, for example, the Potohar trail and encompassing regions, a region generally utilized for development. The most elevated erosion rate at any point recorded assessed at 150–165 tons/ha/year. The Indus River conveyed the fifth biggest freight of residue in 1990s on the planet is (4.49 tons/ha). As per a few gauges, the Indus adds roughly 500,000 tons silt to the Tarbela supply a day, diminishing the lifetime of the dam by 22% and in this way the limit of it repository by 16%.Wind erosion has generally less effect than water erosion. The mix of the two is, but generally destroying. This decreases the country's efficiency by somewhere in the range of 1.5 and 7.5% each year. This nearly influences one fifth of Punjab.

There are several types of soil erosion, including water erosion, wind erosion, and gravity erosion.

Chart 1 Water erosion affected different Areas of Pakistan (000 ha). *Source* Soil Survey of Pakistan. N.d. Reconnaissance Soil Survey Reports from 1965–88. Lahore

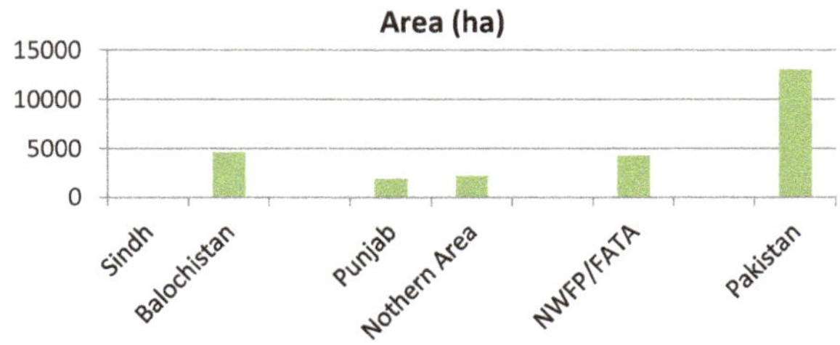

Chart 2 Wind erosion affected different Areas of Pakistan (000 ha). *Source* Soil Survey of Pakistan. N.d. Reconnaissance Soil Survey Reports from 1965–88. Lahore

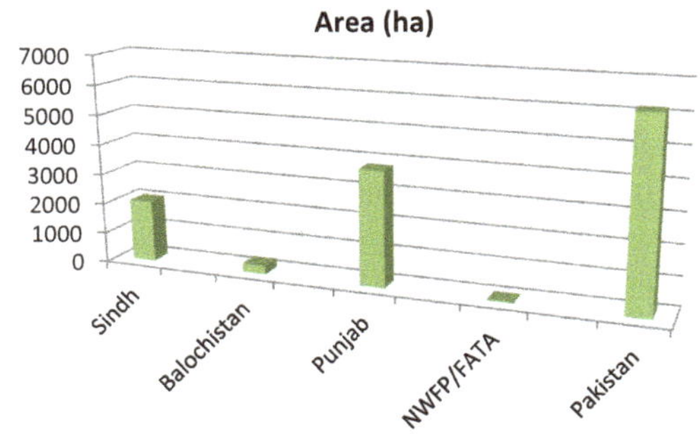

2.4.1 Water Erosion

- Above all else, water erosion is talked about, in which water erosion is isolated into sprinkle erosion, Sheet erosion, Rill erosion, Gully and Tunnel erosion.
- **Sprinkle erosion**; the primary phase of the erosion cycle. At whatever point raindrops tumble to the ground, the soil totals grow so individual soil particles choose the soil surface. Sprinkles can ascend to 60 cm over the ground surface and travel up to 1.5 m from the place of separation. The free particles block the spaces between the bottoms and make a covering which builds waste and lessens invasion.
- **Sheet erosion**; eliminates the slim layer of soil by raindrops and shallow surface flows. Sheet erosion brings about the deficiency of the best soil particles and accessible supplements and natural matter. Soil loss is so gradual that it goes unnoticed, while the accumulated loss leads to great soil loss. The soil is generally helpless against sheet erosion when overgrazed and there is practically no vegetation on arable soils. The primary indications of leaf disintegration are exposed patches, downpour puddles, noticeable foundations of grasses and trees. Vegetable cover shields soil particles from uprooting and expanding soil invasion.
- **Rill erosion**; Shallow seepage pipes around 12 inches deep are called channels. Since surface water gathers in sorrows along glades and dissolves the dirt, these brooks emerge. Rivulet erosion is generally normal on exposed rural land, overgrazing, and free soil. Stream erosion can be decreased when green streams, mulch, and shape takes care of create. It is the moderate stage between the disintegration of the sheet and the erosion of the gorge.
- **Gully erosion**; more than 12 inches further that can't be eliminated by ordinary development are called crevasses, and therefore, the erosion is called ravine erosion.
- **Tunnel erosion**; erosion that happens when surface water moves to and through the spreading soil. Dispersive soils effectively erosion when wet since they are inadequately organized. Tunnel erosion starts when surface water enters the root hollows of trees and bunny hollows and hole in the dirt. Tunnel erosion can be amended by separating existing passages, re-developing and expanding how much natural matter in the dirt.

2.4.2 Wind Erosion

The second most significant type of soil disintegration is wind erosion. Relaxing and development of soil particles in the air at a speed of something like 20 km/h. Wind erosion happens in two ways like suspension and the second is saltation (Nadeem, 2018).

- **Suspension**; The development of fine residue particles under 0.1 mm in breadth while drifting in the air known as suspension. Suspended soil particles settle when the settling power is more prominent than the power that keeps the particles in suspension occurs with decreased breeze speed. The suspension ought to for the most part not surpass 15% of absolute development.

- **Saltation**; Most of the ground conveyed by the triumph is moved in a progression of short bounce back called "saltation". The soil entrained in saline arrangement comprises of fine particles going from 0.1 to 0.5 mm in measurement. Around 50–75% of soil erosion made by wind is expected salt development. Transformation it is brought about by the immediate strain of the breeze on soil particles and their crash with different particles. In the wake of being blown to the surface by the breeze, the particles nearly detonated upward in the beginning phases of saltation (Telkar & Pote, 2014).

Wind erosion happens mostly in regions where precipitation is exceptionally scant and the dampness content is probably going to be with the end result of withering. Wind erosion can lessen by expanding soil vegetation and shelter crops and further developing soil structure. As soil natural matter expands, wind erosion is diminished on the grounds that tall soil particles containing supplements are hard to travel through the air.

2.4.3 Gravity Erosion

Gravitational erosion addresses the development of soil or rock starting with one spot then onto the next because of the power of gravity. At the point when bits of rock tumble to the ground from the side of a mountain, it is on the grounds that gravity has cut them down. At the point when an ice sheet crosses a mountain range, gradually smoothing or forming the Earth's surface around there, it is on the grounds that gravity powers the icy mass down. At the point when avalanches or avalanches happen, streamlining the inclines of mountains or enormous slopes, gravity becomes possibly the most important factor. While geologists perceive water and ice as the primary driver of erosion, it is gravity that drives them both.

2.5 *Mulching to Control Soil Erosion*

Mulch can be inorganic or natural mulch that can be applied to your nursery in Punjab, Pakistan to forestall weeds, soil erosion and hold soil dampness. Covering exposed spaces of your yard or grass with straw or mulch is a viable method for forestalling soil or wind erosion. Straw and mulch are spread over the dirt surface to lessen the effect of downpour on the dirt. From the Wind mulch additionally covers the soil. At the point when the dirt is presented to high breezes, you will require a thick mulch of stones or shakes to hold the breeze back from moving it. Anti-erosion mats (or blankets) or product made of shredded straw, wood fibers, or coconut fibers between layers of jute or UV-degradable plastic net. They designed for a variety of applications reckoning on slope, water speed, and period and desired vegetation. They are offered during a sort of mulch widths, lengths and densities reckoning on the manufacturer. When employed in applications wherever vegetation is obtained from seed, the seeds are planted and Erosion management mats are placed on the planted space. Anti-erosion mats should overlap and anchored with wire staples to manufacturer's specifications. These are effective for erosion management on moderate to steep slopes (NRCS, Michigan September, 2013). Organic or vegetable mulch separates gradually to add natural make a difference to the soil. The microenvironment created by the regular material backings microbial exercises in the soil to build usefulness. Water is fundamental for plant development. Dampness protection is important because of extreme water deficiencies on the planet and Pakistan. Mulch lessens vanishing misfortunes from the dirt surface by rationing water. Water use proficiency is incredibly improved by the utilization of mulch.

Mulches assume a vital part in guideline soil temperature by holding extra hotness in winter to warm the soil surface. They likewise function as a creepy crawly repellent. Soil erosion is a not kidding danger that lessens soil fruitfulness because of dissolving fine soil particles. By giving the soil a defensive covering material, soil debasement can be controlled.

- **Ground cover**

Probably the most effective way to forestall soil erosion is to spread mulch or straw over uncovered soil. Your mulch or straw should make the progress totally to give powerful soil erosion control from wind and downpour. Involving heavier types of mulch in blustery regions or on steep slants will give more successful soil erosion control as heavier mulch is probably not going to move. You should utilize mulch rather than straw as it is heavier than straw and can't be moved effectively in high breezes.

- **Using straw blankets**

Soil Erosion Control Blankets are explicitly intended to forestall soil erosion on uncovered inclines. These covers are made of straw that is connected to mats with a sort of manufactured lattice. The highest point of each mat is put on a trench at the highest point of the slope. You can fill the channel with soil and try to put the covers over the outer layer of the incline. Covers ought to be organized in covering lines and persistent layers. You can likewise utilize modest mulch from Perth, Western Australia for this. Ensure the edges of each sweeping are tucked under the edges of the covers resting close to you. They should be gotten with posts and wooden staples. This plan guarantees that the greater part of the overflow surface and downpour hit the outer layer of the straw mats, dialing it back and forestalling soil erosion.

- **Rolled up straw**

You can utilize fiber rollers to forestall soil erosion on inclines. All you want is a bunch of brakes to introduce along the slant of the slope. The rolls are made of various materials and the straws are tied looking like a square. This is finished with the assistance of a manufactured organization. The loops are put along the incline in various lines dispersed something like four feet separated and not in excess of six feet separated. They are then held set up with wooden stakes. As the water streams down the slant, the trash drifting in the water stops at each line of rollers. This main controls soil erosion from water.

3 Conclusion

Mulches assume a significant part in managing soil temperature by holding more hotness in winter to warm the dirt surface. They additionally function as a bug repellent. Soil erosion is a serious threat that decreases soil ripeness because of dissolving fine soil particles. Assuming the soil is furnished with a defensive covering material, soil degradation can be controlled. The farming community needs to be made aware of the benefits of using mulch in an agrarian production system. The usage of mulch in almost all agrarian production systems is simple, feasible and cheap, and brings great benefits to producers.

References

Adekalu, K. O., Olorunfemi, I. A., & Osunbitan, J. A. (2007). Grass mulching effect on infiltration, surface runoff and soil loss of three agricultural soils in Nigeria. *Bioresources Technology, 98*, 912–917.

Albaladejo Montoro, J., Alvarez Rogel, J., Querejeta, J., Díaz, E., & Castillo, V. (2000). Three hydro-seeding revegetation techniques for soil erosion control on anthropic steep slopes. *Land Degradation and Development, 11*(4), 315–325.

Arnáez, J., Lana-Renault, N., Lasanta, T., Ruiz-Flaño, P., & Castroviejo, J. (2015). Effects of farming terraces on hydrological and geomoprhological processes. A review. *Catena, 128*, 122–134.

Badía, D., Martí, C., (2000). Seeding and mulching treatments as conservation measures of two burned soils in the central Ebro valley, NE Spain. *Arid Soil Research Rehabilitation, 14*, 219–232. https://doi.org/10.1080/089030600406635.

Briassoulis, D. (2006). Mechanical behaviour of biodegradable agricultural films under real field conditions. *Polymer Degradation and Stability, 91*, 1256–1272.

Dai, C., Liu, Y., Wang, T., Li, Z., & Zhou, Y. (2018). Exploring optimal measures to reduce soil erosion and nutrient losses in southern China. *Agricultural Water Management, 210*, 41–48. https://doi.org/10.1016/j.agwat.2018.07.032.

de Lima, M. I. P., & Grasman, J. (1999). Multifractal analysis of 15-min and daily rainfall from a semi-arid region in Portugal. *Journal of Hydrology, 220*(1–2), 1–11.

de Lima, J. L. M. P., Tavares, P., Singh, V. P., & de Lima, M. I. P. (2009). Investigating the importance of rainstorm direction on overland flow and soil loss in a laboratory circular soil-flume. *Geoderma, 152*(1–2), 9–15.

Dilip Kumar, G., Sachin, S. S., & Rajesh, K. (1990). Importance of mulch in crop production. *Indian Journal of Soil Conservation, 18*, 20–26.

Erpul, G., Gabriels, D., & Norton, L. D. (2004). Wind effects on sediment transport by raindropimpacted shallow flow. *Earth Surface Processes and Landforms, 29*(8), 955–967.

Holman, I. P. (2006). Climate change impacts Bullock. Cranfield University-Silsoe, Silsoe, U, Elsevier Ltd.

Jack C. V., Brind W. D., & Smith, R. (1955). Mulching Tech. Comm. No. 49, *Commnwealth Bulletin of Soil Science.*

Lal, R. (1998). Soil erosion impact on agronomic productivity and environment quality. *Critical Reviews in Plant Sciences, 17*(4), 319–464.

Lal, R. (2001). Soil degradation by erosion. *Land Degradation & Development, 12*(6), 519–539.

Li, F. R., Zhao, L. Y., & Zhang, T. H. (2004). Wind erosion and airborne dust deposition in farmland during spring in the Horqin sandy land of eastern Inner Mongolia China. *Soil Tillage Research, 75*, 121–130.

Li, M., Wang, Y. K., Xu, P., Fu, B., Tian, C. S., & Wang, S. (2018). Cropland physical disturbance intensity: Plot-scale measurement and its application for soil erosion reduction in mountainous areas. *Journal of Mountain Science, 15*(1), 198–210. https://doi.org/10.1007/s11629-017-4574-x

Liu, L. Y., Shi, P. J., Zou, X. Y. S., Gao, Y., Erdon, H., Yan, P., Li, X. Y., Dong, Z. B., & Wang, J. H. (2003). Short-term dynamic of wind erosion of three newly cultivated grassland soils in Northern China. *Geoderma, 115*, 55–64.

Lopez, M. V., Gracia, R., & Arrue, J. L. (2000). Effects of reduced tillage on soil surface properties affecting wind erosion in semiarid fallow lands of Central Aragon. *European Journal of Agronomy, 12*, 191–199.

Michigan Technical Note USDA-Natural Resources Conservation Service Agronomy #64 Mulching Materials for Control of Soil Erosion September 2013.

Momiroviæ, N., Oljaèa, V. M., Dolijanović, Ž, & Poštić, D. (2010). Energetska efikasnost proizvodnje paprike u zaštiæenom prostoru u funkciji primene razlièitih tipova polietilenskih folija (PE). *Poljoprivredna Tehnika, 35*(3), 1–13.

Nadeem M. Y. (December 10, 2018). Soil erosion its types and their control.

Oliveira, J. R., Pinto, M. F., Souza, W. J., Guerras, J. G. M., & Carvalho, D. F. (2010). Water erosion in a yellow–red ultisol under different patterns of simulated rain. *Brazilian Journal of Agricultural and Environmental Engineering, 14*(2), 140–147.

Orzolek, M. D., & Lamont, Jr., W. J. (2000). Summary and recommendations for the use of mulch color in vegetable production. http://extension.psu.edu/plasticulture/technologies/plastic-mulches/summary-and-recommendations-for-the-use-of-mulch-color-in-vegetable-production. Accessed 20 Dec. 2011.

Pan, D. L., Zhao, X. N., Gao, X. D., Song, Y. Q., Dyck, M., Wu, P. T., et al. (2018). Application rate influences the soil and water conservation effectiveness of mulching with chipped branches. *Soil Science Society of America Journal, 82*(2), 447–454. https://doi.org/10.2136/sssaj2017.10.0371.

Ramakrishna, A., Tam, H., Wani, S., & Long, T. (2005). Effect of mulch on soil temperature, moisture, weed infestation and yield of ground nut in northern Vietnam. *Field Crops Research, 95*(2–3), 115–125.

Römkens, M. J. M., Helming, K., & Prasad, S. N. (2001). Soil erosion under different rainfall intensities, surface roughness, and soil water regimes. *CATENA, 46*(2–3), 103–123.

Sweetser, R., (Ed.). (January 25, 2021). How to mulch your garden | Types of mulch learn how to mulch your garden in fall. Winter, and Spring.

Schonbeck, M. (2015). Mulching for weed management in organic vegetable production. In *eOrganic Virginia association for biological farming.*

Schonbeck, M. W., & Evalylo, G. E. (1998). Effects of mulches on soil properties and tomato production. I. Soil temperature, soil moisture, and marketable yield. *Journal of Sustainable Agriculture, 13*, 55–81. https://doi.org/10.1300/J064v13n01_06. Accessed 20 Dec. 2011.

Swaider, J. M., Ware, G. W., & McCollum, J. P. (1992). *Producing vegetable crops* (4th ed., p. 626). Interstate Publishers Inc.

Sharma, N. K., Singh, R. J., Mandal, D., Kumar, A., Alam, N. M., & Keesstra, S. (2017). Increasing farmer's income and reducing soil erosion using intercropping in rainfed maize-wheat rotation of Himalaya, India. *Agriculture, Ecosystems & Environment, 247*, 43–53. https://doi.org/10.1016/j.agee.2017.06.026.

Telkar, S. G., & Pote, N. S. (2014). Soil erosion: Types and their mechanism.

Response of Mulching and Tillage Practices on Soil Management

Fasih Ullah Haider, Maqsood Ul Hussan, Kashif Akhtar, and Cai Liqun

Abstract The increase in global population increases the demand in food production, and to satisfy food demand, researcher must increase the cultivated area and or decrease the inorganic fertilizer usage to control over environmental pollution. Effective management practices i.e., reduced or zero-tillage system, the addition of crop residue, crop rotation, and optimum nutrients application enhances the soil properties and promotes agricultural sustainability. In this current chapter, we critically reviewed the effects of mulching and tillage practices on physiological and chemical characteristics i.e., aggregate stability, soil erosion, organic matter content, nutrients availability, and microbial biomass of the soil. It was observed that integrated application of mulches with tillage practices results to improve soil temperature, soil erosion risk, bulk density, and porosity, infiltration rate, soil organic carbon, microbial biomass, nutrients availability, and moisture content of soil which result to improve weed infestation, root morphology, and crop production on sustainable basis. The current chapter can aid in determining the ecological significance of mulches and tillage systems on sustainable basis not only to overcome environmental degradation problems but also improving agricultural soils health on sustainable basis.

Keywords Conservation tillage · Mulches · Sustainability · Soil quality · Crop productivity

F. U. Haider · C. Liqun (✉)
College of Resources and Environmental Sciences, Gansu Agricultural University, Lanzhou 730070, China
e-mail: cailq@gsau.edu.cn

Gansu Provincial Key Laboratory of Arid Land Crop Science, Gansu Agricultural University, Lanzhou 730070, China

M. U. Hussan
Department of Grassland Sciences, Northwest Agriculture and Forestry University Yangling, Xianyang 712100, China

K. Akhtar
State Key Laboratory for Conservation and Utilization of Subtropical Agro-Bio-Resources, College of Life Science and Technology, Guangxi University, Nanning 530004, China

K. Akhtar et al. (eds.), *Mulching in Agroecosystems*,
https://doi.org/10.1007/978-981-19-6410-7_5

1 Introduction

Agriculture is currently facing huge challenges worldwide, including soil erosion, scarcity of water, and low content of organic matter in soil which significantly reduces the soil fertility and crop production (Haider et al., 2019; Quddus et al., 2020; Rafiq, 2016). Wind soil erosion and associated dust emissions can cause major depletion of nutrients and can contribute to soil depletion of agricultural lands (Katra, 2020). Different cultivation approaches have been studied over the few decades to boost the yield efficiency of various crops, and at the same time have the capacity to improve the ecosystem services i.e., increase in soil biodiversity, soil water retention capacity, and soil organic matter in agricultural soil (Chimsah et al., 2020; Ullah, 2015). So, in this aspect soil tillage plays an important role, as it defines both the productivity of the cropping system in terms of production and its environmental effects i.e., carbon sequestration or soil degradation (Lal, 2013; Weber et al., 2017). For centuries, soil tillage has been carried out because it decreases the density of weeds, thereby favorably impacting the supply of nutrients and water (Lal, 2009; Nawaz et al., 2016). Moreover, short exposure to sunlight will cause the germination of deeply buried weed seeds due to soil inversion after conventional tillage practices (Weber et al., 2017).

Comparing with zero tillage or uncultivated land, conventional tillage practices will greatly increase the wind erosion, minimizes the soil organic matter content that ultimately leads to soil depletion. In addition, water scarcity and rise of the price of fuel, and fertilizers would increase the cost of production (Quddus et al., 2020). Growing food demand with unprecedented population has generated big challenges for agricultural sector researchers in recent decades (Salehi et al., 2017). Modern agriculture must seek to optimize efficiencies in cost, labor, and time (Melander et al., 2013). Compared to conventional tillage, reduced tillage systems and in particular no tillage can offer a competitive advantage in agriculture and may leads to an increase in soil quality, e.g., in organic matter content, water holding capacity, and a decrease in topsoil erosion (Nawaz et al., 2016; Zikeli et al., 2013). Conserving soil is the primary reason why farmers follow the conservation tillage practices i.e., no tillage or reduced tillage practices and significantly improves the optimum soil moisture and minimize production costs (Al-Kaisi & Yin, 2005; Faleiros et al., 2018). Moreover, from a sustainable agriculture point of view, preserving crops residues or mulches with an effective tillage scheme is critical because it can enhance the soil structure stabilization and drainage, increase crop yield, and conserve soil water in arid and semi-arid conditions (Wilhelm et al., 2004; Basir, 2014). Contrarily to this, by reducing tillage practices in regions such as temperate areas perennial, annual weed grasses, winter annual weeds species, as well as perennial dicots weeds will become widespread than conventional tillage systems (Faleiros et al., 2018; Weber et al., 2017). Machinery problems of tillage practices, mulching, and stubble maintenance have been intensively studied in the mechanized farming system in recent past (Anderson, 2009; Campbell et al., 2001).

Mulches is a source of highly stable carbon that significantly enhances the organic matter, water holding capacity, nutrient holding capacity, total nutrients availability, and microbial activity and effectively minimizes soil compaction, weeds infestation, soil erosion, and water runoff in soil (Basir, 2014; Hoyle et al., 2006). Mulches are classified as organic mulches (crop straw residues, wood chips, plant leaves, or bark), inorganic mulches (plastic, pebbles, and gravels) (Rafiq, 2016), chemical mulches (resins, hexadecanol, and asphalt) (Basir, 2014), and cultivation of spreading type of crops between the rows of main crop called as living mulches (Haider et al., 2019; Fig. 1). Organic mulches are further classified as ex-situ (crop residues, manure, compost, etc.) and in-situ (annual plants and perennial plants) as mentioned in Fig. 2 (Table 1).

Incorporation of crops stubbles residues as a mulch that are further decomposed by the process called as mineralization, significantly minimized the soil erosion, and enhanced the soil organic matter, soil fertility, and nutrients uptake in plant (Amber

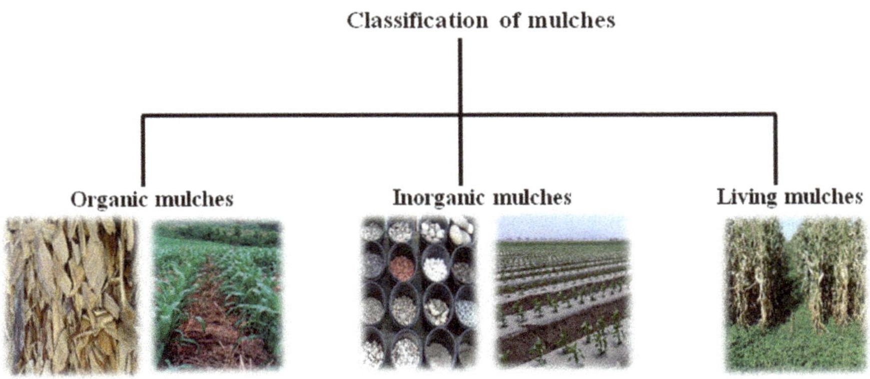

Fig. 1 Classifications of mulches

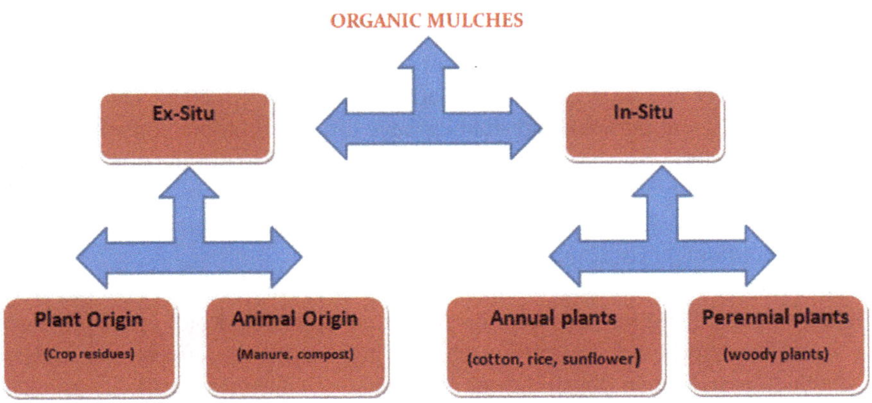

Fig. 2 Classifications of organic mulches

Table 1 Comparison of conventional tillage system and conservation tillage system with crop residues retention

Tillage system	Advantages	Disadvantages
Conventional tillage	Enhanced the residues decomposition, improved the soil texture, water retention, soil organic matter content, carbon stock and its sequestration, decreased the soil bulk density; enhance the scavenging of reactive oxygen species, improved the photosynthetic activity, plant biomass, and delay the senescence of plant at grain filling stage; and beneficial results can be produced in the first year of residue retention	Lead to shallow sowing and poor root penetration; higher incidence of pathogenic bacteria in mulch residue; lower soil moisture and lower soil-seed contact early in crop development, resulting in increased vulnerability to drought and freezing stresses; and a higher carbon/nitrogen ratio, which is determined to microbial growth and slows residue decomposition
Conservation tillage	Minimized soil erosion and runoff of the rainfall; increase the water conservation and infiltration; especially adapted to semiarid and arid regions; enhance the water use efficiency, soil organic carbon content, soil microbial biomass carbon, soil aggregate stability, minimize evapotranspiration and depletion of soil water, and enhance soil fertility and conservation; minimize the bulk density of subsoil; protect the surface layer of by mulching against the action of falling drops of rain; and enhance the microenvironment of crop field	Organic nutrients from the soil linger on the surface, which can reduce crop production in the first few years of residue retention; higher incidence of pathogen bacteria in mulch residues; and in general beneficial effects can be seen 2–3 years later

et al., 2011; Ullah, 2015). In addition, the effect of the level of mulching and stubble management on the mineralization process depends on the quantity of straw mulches, its C:N ratio, the inorganic or organic form of nutrient within the mulch's straw, and the methodology and degree of incorporation (Chalise et al., 2020; Quddus et al., 2020). The different mulches and stubbles have different nutrient content based on soil quality, variety, crop type, fertilizer background, and prevailing conditions and thus can influence both short- and long-term nutrient equilibriums in the soil (Basir, 2014; Rafiq, 2016). Integrated application of various tillage systems with mulches in potato field in Dingxi, China is shown in Fig. 3.

Nevertheless, we still need to emphasize the effect of various tillage practices, mulches, and stubble incorporation on physico-chemical and biological processes in agricultural soils. So, the current chapter highlighted the effects of mulching and tillage practices on physiological and chemical characteristics i.e., aggregate stability, soil erosion, organic matter content, nutrients availability, and microbial biomass of the soil. The information outlined in current chapter would help to clarify

Fig. 3 Integrated application of various tillage systems with mulches in potato field in Dingxi, China. *Photo credit* Cai Liqun

how the mulching and tillage practices regulate the soil biodiversity and soil fertility for sustained the crop production.

2 Significance of Mulching and Tillage Practices Effects on Soil Properties

2.1 Soil Temperature

The temperature of the soil is essentially the indicator of the heat in the soil (Wang et al., 2020). For planting most of the plants, optimal soil temperature ranges between 18 and 24 °C (Gao et al., 2014; Yang et al., 2020). Temperature of soil effects the plant growth directly or indirectly by influencing root growth and nutrient uptake (Anikwe et al., 2007). A reduction in temperature significantly results in a decrease in nutrients uptake and water even at a constant moisture level (Jordán et al., 2011). In addition, at lower temperature, transportation of nutrients and water from roots to shoot and vice versa be decreased (Huang et al., 2009). Generally, the effect of combined mulch and tillage on soil temperature lasted for 75 days after sowing in maize (*Zea mays* L.), (Yang et al., 2020). In Nigeria, combined effects of seedbed preparations and various mulches had significant effects of soil temperature in soil cultivated with cow pea (*Vigna unguiculata* L.), maize (*Zea mays* L.), cassava (*Manihot esculenta* L.), and soybean (*Glycine max* L.), (Lal, 1979). Inorganic mulches i.e., plastic mulch results to increase the soil temperature while, organic mulches i.e., crops stubbles result to reduce the temperature of soil (Jordán et al., 2011). Similarly, in another study combined effect of plastic mulch with tillage practices significantly improves the average temperature of soil by 2 °C in soil cultivated with cocoyam (*Colocasia esculenta* L.), and wheat (*Triticum aestivum* L.) (Anikwe et al., 2007; Li et al., 1999). The increase of soil temperature was observed due to plastic mulch thermal properties such as transmittancy, absorbitivity, or reflectivity (Gao et al., 2014). Soil temperature in soil cultivated with maize by following combined effect of plastic mulch and conventional tillage was higher as compared to combine plastic mulch and no tillage or sole application of mulches or tillage practices (Liu et al., 2009). Similarly, effects were observed in orchards by combine effect of tillage practices and mulching (Huang et al., 2009; Liu et al., 2013). Moreover, it was observed that combine effect of maize straw and no tillage result to minimize the soil temperature i.e., 1.1 °C as compared to conventional tillage practices (Yang et al., 2020). Yet, further studies need to done to evaluate the potential effects of tillage practices and mulches on the temperature of soil.

2.2 Soil Erosion Risk

Soil erosion is a form of soil degradation, generally refers as the displacement of the upper layer of soil (Bationo et al., 2004; Chalise et al., 2020). Erosion also degrades the hydrological conditions in the soil and decreases the water potential available for plants (Akplo et al., 2017; Kong, 2013). Since agriculture started increased soil erosion has become an enduring concern (Atreya et al., 2008; Niziolomski et al., 2020). The monsoon season in sub-continent is more susceptible to soil depletion, when it comes to soil erosion, as 80% of the rainfall occurs during this season (Chalise et al., 2019). Excessive or conventional tillage practices result to loosen the soil, break the soil structure, and makes it vulnerable to soil erosion (Zhang et al., 2007). Mulch and no-tillage are suggested as prospective researchable option to minimize the soil and enhance the production of agricultural crops; however, very few studies have been conducted in this aspect. Chalise et al. (2020) observed that application of combine (no tillage with plastic mulch) in Nepal significantly minimized the soil erosion in soil cultivated with maize by 53.88% as compared to conventional tillage practices having no mulch. In another 24 years of study in Australia, it was observed that direct seeded cultivation of wheat with stubbles retention in soil results to improve the soil porosity and aggregate stability of soil which significantly minimized the soil erosion by 3.7 times as compared to conventional tillage practices (Zhang et al., 2007). Akplo et al. (2017), reported that application of combine (isohypse ridging with plastic mulch) in Benin significantly minimized the soil erosion by 98.85% as compared to isohypse ridging without plastic mulch. Likewise in Ross-on-Wye (United Kingdom), shallow soil disturbance with 5 tones ha^{-1} wheat straw as a mulch significantly minimized the soil erosion by 72% as compared to conventional tillage practices having no mulch (Niziolomski et al., 2020). Still, further research analysis needs to done to explore the potential effects of tillage practices and mulches on the soil erosion.

2.3 Bulk Density and Porosity

Bulk density is referred as the weight of the soil in given volume, while soil porosity refers to the quantity of pores between soil particles or open space (Qamar et al., 2015). Studies reported that bulk density enhances with compaction and soil with higher bulk density i.e., 1.6 g cm^{-3} significantly tend to minimize the root growth and morphology (Kong, 2013; Niziolomski et al., 2020). In recent years, it was reported that combine application of tillage practices with mulches significantly influenced the bulk density and porosity of soil. In Kenya, it was reported that application of crop residues as mulch via conventional tillage practice significantly minimized the bulk density of soil (Gicheru, 1994). Another study in Pakistan, it was observed that application of combine (deep tillage with wheat straw mulch) decreased the bulk density of soil to 1.34 Mg m^{-3} as compared with minimum tillage practices having no mulch i.e., 1.53 Mg m^{-3} (Khurshid et al., 2006). Incorporation of wheat straw as a mulch via conventional tillage practices significantly minimized the bulk density of soil by 1.35 Mg m^{-3} as compared to soil with zero tillage practices having no mulch i.e., 1.53 Mg m^{-3} (Mehmood et al., 2014). Similar findings were observed in Nigeria where application of combine (tillage practices with plastic mulch), significantly reduced the soil bulk density to 1.10 Mg m^{-3} as compared with no tillage soil having no mulch i.e., 1.45 Mg m^{-3} (Anikwe et al., 2007). Similarly, in another 24 years of study in Australia, it was observed that direct seeded cultivation of wheat with stubbles retention in soil results significantly minimized the bulk density of soil to 0.86 g cm^3 as compared to conventional tillage practices having no mulch 1.11 g cm^3 (Zhang et al., 2007). In central Ohio state application of optimum wheat straw i.e., 4 Mg ha^{-1} with zero tillage practices results to improve the bulk density and porosity of soil by 95% as compared with soil having no mulch and tillage practices (Mulumba & Lal, 2008; Fig. 4). Under no-tillage practices in northern part of China, application of 5 years residues mulching trends to improves the subsoil porosity by 13.8% as compared to soil with no mulch (Niu et al., 2007). Incorporation of maize straw via plow tillage has significant potential to minimize the bulk density of soil by 2.9% by as compared to control (no maize mulch and no tillage practices) and trends to improve the soil porosity, soil non-capillary porosity, and capillary of soil (Kong, 2013; Liu et al., 2013). However, contrary to this in some studies it was mentioned that various tillage practices i.e., (deep tillage, conventional tillage, and zero drill tillage) via various mulches i.e., (rice straw, wheat straw, and plastic mulch) had no significant effects on bulk density of soil (Qamar et al., 2015). Further studies and analysis are needed to elucidate detailed mechanisms corresponding to the effects of tillage practices via mulches on bulk density and porosity of soils.

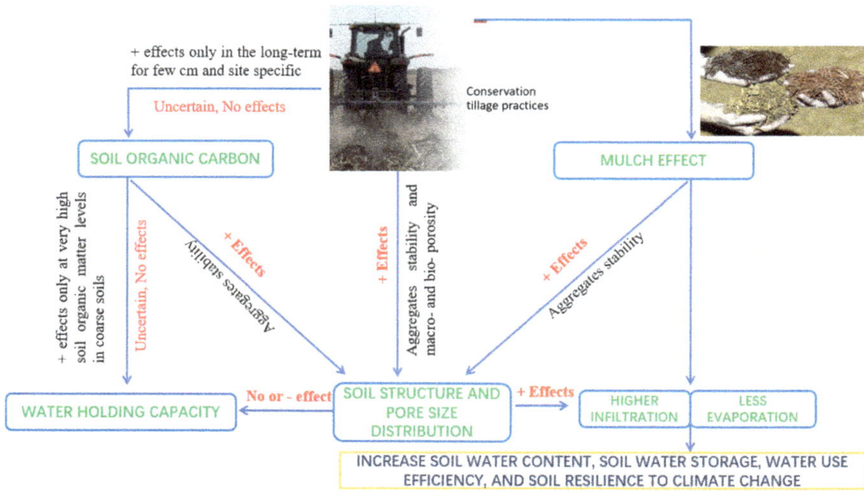

Fig. 4 Flowchart of conservation tillage practices and mulches application on soil organic carbon, water holding capacity, soil, structure, pore size distribution, soil water content, soil water storage and water use efficiency of soil. Modified from Abdallah et al. (2021)

2.4 *Infiltration Rate*

The infiltration rate, usually expressed in inches per hour, is a measure of how quickly water reaches the soil (Nawaz et al., 2016). If the infiltration rate is too slow, it can lead to ponding, erosion, and surface runoff in sloping areas and can leads to floods or insufficient moisture for sustainable crop productivity (Jabran, 2019). For optimal crop production, ample water should penetrate into the soil profile. In Australia, cultivation of wheat by zero tillage and stubbles as a mulch improved the infiltration rate upto 85 mm h^{-1} as compared to conventional tillage practices via no mulch (23 mm h^{-1}) (Zhang et al., 2007). The zero tillage via stubbles mulches soil's higher macro-porosity resulted in more penetration and thus less runoff (Loch et al., 2001; Zhang et al., 2007). In other study, Alliaume et al. (2017) reported that reduce tillage practices via mulching can improved the infiltration rate of soil by 9.5% as compared to conventional practices via no mulch practices. However, further research is needed to determine the possible effects of tillage activities and mulches on soil temperature.

2.5 *Soil Organic Carbon*

Increased soil organic carbon (SOC) encourages soil structure, or tilth, which implies more physical stability (Haider et al., 2019; Nawaz et al., 2016). The increase soil aeration (amount of oxygen in soil), as well as water drainage and preservation, and decreases nutrients leaching and soil erosion (Yang et al., 2020; Fig. 4). Agro-ecosystem productivity and soil fertility are both maintained a satisfactory SOC content. Soil with residue retention can store some of the CO_2 fixed by plants in the atmosphere and help to reduce the greenhouse gas emissions. Yang et al. (2010), reported that when zero tillage with residues mulching system was used for 7 years, SOC and total C content enhanced by 2.04 and 1.56–2.50 g kg^{-1} respectively, when compared to conventional tillage practices without residue retention. The amount of residue retention positively affects the topsoil C-sequestration effect of chemical fertilization, according to a 12-year field experiment, and the topsoil i.e., (0–20 cm) carbon storage increases significantly with the straw retention rate via reduced tillage system (Lou et al., 2011). The increase in carbon storage can be attributed in part to lower soil carbon emissions and a high carbon input due to residue retention (Kong, 2013). Nonetheless, residues preservation in limited, zero-tillage, or deep tillage systems in agricultural soils has the potential to increase carbon sequestration. To boost soil carbon sequestration, researchers must evolve suitable residue management techniques and tillage systems.

2.6 *Microbial Biomass*

The behavior of the soil microflora is one of the most important components of soil quality; hence, the soil ecology regime contributes to agricultural sustainability (Luo et al., 2020; Wang et al., 2012). Soil microbial biomass (i.e., protozoa, fungi, and bacteria), is a determined by the mass of living portion of soil organic matter (Ghimire et al., 2017). The microbial biomass present in soil decomposes the animal and plants residues, and results soil organic matter to release plant available nutrients and carbon dioxide in the soil (Qiang et al., 2004; Zhang et al., 2017). In recent years, much attention has been done to explore the significant effects of tillage practices via plastic or straw mulches on soil organic carbon and microbial biomass. Microorganisms use the steady, uniform availability of carbon from the crop residues or straw used as mulches as energy source (Luo et al., 2020; Wang et al., 2012). When residues or straw as mulches are maintained, the resultant increase in microbial biomass contributes to an increase in soil respiration rate and carbon dioxide release when opposed to residue removal activities (Govaerts et al., 2007). The microbial biomass/activity at the tillering stage of wheat (*Triticum aestivum* L.) was 51.7 and 12.8% greater when straw mulches were combined via tillage and when residues mulching occurred without tillage practices respectively, attributable to improvement in soil temperature and temperature consistency when residues are maintained (Liu et al., 2011). The improvement in microbial biomass was largely attributed to the sustainable enhancement in the bacteria population as compared to tillage activities without straw mulches preservation (Yu et al., 2007). Soil microbial biomass, especially phosphorous-solubilizing bacteria was increased dramatically in reduced tillage practices via residues retention (Fan & Liu, 2005). Although bacterial growth is considerably higher in soil with residues/mulches retention, the improvement in microbial biomass is largely attributed to an increase in bacterial density and population in soil (Liu et al., 2011). Such type of situation may be happening at the early phases of organic mulches decomposition (Luo et al., 2020). In another study it was observed that microbial activity and diversity rise in residues/mulches retention systems, with bacteria dominating; as the residues decomposes, the fungal population increases, and the fungi/bacteria biomass ratio decreases (Cui et al., 2005). The microbial community's composition varies due to changes in the chemical components of the residue and rivalry among various microbes (Bastian et al., 2009; Cui et al., 2005). Bacteria can develop quickly in the early stages of decomposition on readily available compounds those in freshly added plant residues/mulches. In later stages of decomposition, fungi, which grow slowly than bacteria but can decompose more recalcitrant material, are dominant (Ghimire et al., 2017; Poll et al., 2008). The incorporation of microbes has been shown to increase the soil quality and crop growth where residues/mulches are preserved. Inoculation of *Aspergillus niger* L. and *Streptomyces microflavus* L. with residue preservation via reduced tillage, enhances the residues decomposition, and improves carbon, nitrogen, and phosphorous microbial biomass; the release of nitrogen and phosphorous from the straw increases and encourages soil nutrient supply, as well as crop growth and productivity (Zhang et al., 2011). Similarly, the quality of soil organic matter and mineralization of essential nutrients in residues via reduced tillage increases as they are inoculated with fungi (Kong, 2013; Li et al., 2001). Furthermore, this procedure increases the activity of soil phosphatases and urease, as well as the early peaks in activity, resulting in increased microbial biomass in the soil (Zhang et al., 2006). To evaluate the potential effects of tillage practices and mulches on soil microorganisms, further research is required.

2.7 Nutrients Availability

Soil nutrients are very essential for optimum growth of growth of plant (Nawaz et al., 2016). Nutrients availabilities are described by the soil Science Society of America as the quantities of soil nutrients in chemical forms accessible to plant roots or compounds likely to be converted to such forms during the growing seasons (Yadav et al., 2018). Soil enzyme activity is a primary measure of soil health, and it is affected primarily by soil microbes, animals, soil fertility, and plant growth (Ghimire et al., 2017). The addition of straw residues boosts the activity of soil enzymes i.e., catalase, convertases, phosphatases, and urease with deeper tillage system causing more drastic changes than rotary tillage or reduced tillage systems (Yang et al., 2012; Fig. 4). Zero-tillage followed by straw mulching significantly enhanced the total nitrogen and carbon nitrogen in soil by 13.5 and 7.26% respectively higher as compared to conventional tillage practices having no mulches system (Luo et al., 2020). Correspondingly, Guan et al. (2022) reported that residue preservation via tillage systems for two years will increase the total phosphorous content by 10%. Similarly, in another study 11 years of residue retention via reduced tillage, the overall phosphorous content at the 0–10 cm profile reduced by 6.8–6.3 mg kg^{-1} respectively (Wang et al., 2012). Mehmood et al. (2014), observed that zero-tillage via wheat straw mulch significantly improved the soil nitrogen, available phosphorous, and available potassium content in soil by 23.08, 56.84, and 11.32% respectively as compared to conventional tillage practices having no mulch. In another study Quddus et al. (2020) reported that zero-tillage via straw mulches significantly improved the availability of macronutrients and micronutrients in soil as compared to conventional tillage practices via no mulches system.

2.8 *Moisture Content*

The water contained in the soil is known as soil moisture, and is determined by temperature, precipitation, and soil characteristics (Liu et al., 2009). The primary organs that absorb water in plants are plant roots. The moisture conditions of soil impair the of plant root water and leaf transpiration, further influencing the deposition of dry matter and eventually effecting the crop yield (Luo et al., 2020). In recent studies, it was reported that combine application of tillage practices via mulches have significant effects on moisture content of soil. The incorporation of 16 Mg ha^{-1} wheat straw mulch via zero tillage practice significantly improves the water contents in soil by 70% (Mulumba & Lal, 2008). Similarly in another studies, the incorporation of 12 Mg ha^{-1} wheat straw mulch via zero tillage practice significantly improves the water contents in soil by 18.51% (Khurshid et al., 2006). Plow tillage via straw mulching, subsoil tillage via straw mulching, and zero tillage via straw mulching observed a significant enhance in soil moisture and soil saturated water content as compared with conventional tillage via no mulch amendments (Liu et al., 2013). Zero tillage practice via plastic mulch improved the soil moisture content of soil by 57.55% as compared to conventional tillage practice via no mulch (Anikwe et al., 2007; Luo et al., 2020). In Pakistan cultivation of wheat with zero drill tillage with the incorporation of rice straw in wheat-rice cropping system significantly improved the soil moisture content by 63.64% higher as compared to conventional tillage practice having no mulch (Qamar et al., 2015). Incorporation of maize straw mulch via no tillage has significant effect to conserve water by 12.56% as compared with conventional tillage practice with no mulch (Yang et al., 2020). Still, further research analysis needs to be done to explore the potential effects of tillage practices and mulches on moisture content in soil.

3 Economic Analysis

Cost increases, specifically in the content of machinery and fuel, will cause farmers to focus more on abandoning mould-board ploughing if no tillage yields are adequate. Farm consolidation and contracting out tillage operations, which are already popular in some parts of world especially in Europe and Asia (Mikkola et al., 2005), would increases demand for low cost, high-capacity cropping systems based on mulches with zero-tillage practices. Tillage systems have a 1.63 times higher machinery investment than zero-tillage system, while maintenance costs are 4 times higher, fuel costs are 6.5 times higher, working time per unit area is 5 times higher, and combined performance costs are 4.2 times higher as compared with zero-tillage practices. If the crop profit margins fall and labor and fuel costs increase, these disparities are likely to become even more significant in the future. In addition, mineral fertilizers are also a major contributor to total production costs. Increased the levels of SOC in the rhizosphere as a result of implementing zero-tillage via mulches could dramatically reduce the need for mineral nitrogen inputs in cereal production, particularly on agricultural soils with depleted with organic matter. According to Carvalho et al. (2005), an improvement in SOC from 1–2% obtained under long-term zero-tillage conditions via mulches could result in a fertilizer reduction of 62 kg nitrogen per hectare to achieve the optimum amount of economic yield. In arid and semi-arid regions of the world, water and energy cost are becoming increasingly critical for the economic viability of irrigated field crops. Due to increased water penetration, decreased evaporation, and increased its storage; zero-tillage via mulches is commonly discussed as a water-saving strategy. Despite of all the obvious economic benefits of moving away from conventional tillage practices having no mulches, still, Tebrügge (2001), and Yang et al. (2020), considers it difficult to grasp farmers unwillingness to consider zero-tillage. Factors need to consider when comparing tillage systems on a whole-farm basis are such as size of the farm, cropping system patterns, costs of required machinery, use and value of crop residues, labor and fuels cost, and farmer knowledge and skills (Chalise et al., 2020; Qamar et al., 2015). Despite the fact that zero-tillage via mulches is still a small part of the farming landscape, economic conditions are gradually favoring its adoption, which is now strongly supported by some agencies as European Conservation Agriculture Federation (ECAF) and the Spanish Conservation Agriculture Association.

4 Conclusions

Farmers often burn or remove crop residues and practicing of conventional tillage system without residue retention generally decreases the soil quality in terms of soil aggregation and microbial biomass. So, integration of reduced tillage system with residues preservation could be the best option to mitigate the current soil quality challenges and also improving the crop productivity on sustainable basis. Still, retention of crop residues/mulches is not considered as a single practice and should be performed with couple of the strategies i.e., efficient residue cutting, suitable tillage patterns, accurate sowing depth and density, adequate compacting, and proper compacting tools. The proper selection of these components would make a significant contribution to the target area's long-term agricultural sustainability.

Acknowledgements This work was supported by the National Natural Science Foundation of China (Project Nos. 41661049 and 31571594).

References

Abdallah, A. M., Jat, H. S., Choudhary, M., Abdelaty, E. F., Sharma, P. C., & Jat, M. L. (2021). Conservation agriculture effects on soil water holding capacity and water-saving varied with management practices and agroecological conditions: A review. *Agronomy, 11*, 1681. https://doi.org/10.3390/agronomy11091681.

Akplo, T. M., Félix, K. A., Pascal, H., Moncef, B., Naivo, R., Lionel, M., Arcadius, A. A. M., Mathieu, A. F. (2017). Effect of tillage and mulching on soil water erosion in Linsinlin watershed, centre of Benin. *Journal of Experimental Biology and Agricultural Sciences, 5*, 1–11. https://doi.org/10.18006/2017.5(4).515.524.

Al-Kaisi, M. M., & Yin, X. (2005). Tillage and crop residue effects on soil carbon dioxide emission in corn–soybean rotations. *Journal of Environmental Quality, 34*, 437–445.

Alliaume, F., Rossing, W. A. H., Tittonell, P., & Dogliotti, S. (2017). Modelling soil tillage and mulching effects on soil water dynamics in raised-bed vegetable rotations. *European Journal of Agronomy, 82*, 268–281. https://doi.org/10.1016/j.eja.2016.08.011.

Amber, L. H., Stetson, S. J., Osborne, S. L., Schumacher, S. E., & Pikul, J. L., Jr. (2011). Corn residue removal impact on soil aggregates in a no-till corn/soybean rotation. *Soil Science Society of America Journal, 76*, 1390–1398.

Anderson, G. (2009). *The impact of tillage practices and crop residue (stubble) retention in the cropping system of Western Australia* (pp. 1–79), Department of Agriculture and Food, Govt. of W. Australia. Bulletin Number 4786. ISSN:1833-7236.

Anikwe, M. A. N., Mbah, C. N., Ezeaku, P. I., & Onyia, V. N. (2007). Tillage and plastic mulch effects on soil properties and growth and yield of cocoyam (*Colocasia esculenta*) on an ultisol in southeastern Nigeria. *Soil till Res, 93*, 264–272. https://doi.org/10.1016/j.still.2006.04.007.

Atreya, K., Sharma, S., Bajracharya, R. M., & Rajbhandari, N. P. (2008). Developing a sustainable agro-system for central Nepal using reduced tillage and straw mulching. *Journal of Environmental Management, 88*, 547–555.

Basir, A. (2014). Tillage and stubble management in cereal based cropping system. Ph.D. thesis. Department of Agronomy, Faculty of Crop Production Sciences, The University of Agriculture.

Bastian, F., Bouziri, L., Nicolardot, B., & Ranjard, L. (2009). Impact of wheat straw decomposition on successional patterns of soil microbial community structure. *Soil Biology & Biochemistry, 41*, 262–275. https://doi.org/10.1016/j.soilbio.2008.10.024.

Bationo, A., Kimetu, J., Ikerra, S., Kimani, S., Mugendi, D., Odendo, M., Silver, M., Swift, M. J., Sanginga, N. (2004). The African network for biology and fertility: New challenges and opportunities. In *Managing nutrient cycles to sustain soil fertility in Sub-Sahara Africa* (pp. 1–23). Academy Science Publisher.

Campbell, C. A., Selles, F., LaFond, G. P., & Zentner, R. P. (2001). Adopting zero tillage management: Impact on soil C and N under long-term crop rotations in a thin Black Chernozem. *Canadian Journal of Soil Science, 81*, 139–148.

Carvalho M, Basch G, Alpendre P, Brandão M, Santos F, Figo M (2005) A adubação azotada do trigo de sequeiro: o problema da sua eficiência. *Melhoramento, 40*, 5–37 (in Portuguese).

Chalise, D., Kumar, L., Sharma, R., & Kristiansen, P. (2020). Assessing the impacts of tillage and mulch on soil erosion and corn yield. *Agronomy, 10*, 63. https://doi.org/10.3390/agronomy10010063.

Chalise, D., Kumar, L., Spalevic, V., & Skataric, G. (2019). Estimation of sediment yield and maximum outflow using the IntErO model in the Sarada river basin of Nepal. *Water, 11*, 952.

Chimsah, F. A., Cai, L., Wu, J., & Zhang, R. (2020). Outcomes of long-term conservation tillage research in Northern China. *Sustainability, 12*, 1062. https://doi.org/10.3390/su12031062.
Cui, J. T., Dou, S., Zhang, W., & Liu, Y. D. (2005). Effects of maize stalk on microbiological characteristics of soil. *Journal of Jilin Agriculture University, 27*, 424–442.
Faleiros, G. D., Santos, D. F. L., Corá, J. E. (2018). Analysis of profitability of conservation tillage for a soybean monoculture associated with corn as an off-season crop. *Cogent Food & Agriculture*. https://doi.org/10.1080/23311932.2018.1429699.
Fan, B. Q., & Liu, Q. L. (2005). Effect of conservation tillage and straw application on the soil microorganism and P dissolving characteristics. *Chinese Journal of Eco-Agriculture, 13*, 130–133. (in Chinese with English abstract).
Gao, Y., Xie, Y., Jiang, H., Wu, B., & Niu, J. (2014). Soil water status and root distribution across the rooting zone in maize with plastic film mulching. *Field Crops Research, 156*, 40–47.
Ghimire, R., Lamichhane, S., Acharya, B. S., Bista, P., & Sainju, U. M. (2017). Tillage, crop residue, and nutrient management effects on soil organic carbon in rice-based cropping systems: A review. *Journal of Integrative Agriculture, 16*, 1–15. https://doi.org/10.1016/s2095-3119(16)61337-0.
Gicheru, P. T. (1994). Effects of residue mulch and tillage on soil moisture conservation. *Soil Technology, 7*, 209–220.
Govaerts, B., Mezzalama, M., Unno, Y., Sayre, K. D., Luna-Guido, M., Vanherck, K., Dendooven, L., & Deckers, J. (2007). Influence of tillage, residue management, and crop rotation on soil microbial biomass and catabolic diversity. *Applied Soil Ecology, 37*, 18–30. https://doi.org/10.1016/j.apsoil.2007.03.006.
Guan, Y., Xu, B., Zhang, X., Yang, W. (2022). Tillage practices and residue management manipulate soil bacterial and fungal communities and networks in maize agro ecosystems. *Microorganisms, 10*, 1056. https://doi.org/10.3390/microorganisms10051056
Haider, F. U., Cheema, S. A., & Farooq, M. (2019). Impact of cover crops in improving agro-ecosytems including sustainable weed suppression—A review. *Pakistan Journal of Weed Science Research, 25*, 47–62.
Hoyle, F. C., Murphy, D. V., & Fillery, I. R. P. (2006). Temperature and stubble management influence microbial CO_2–C evolution and gross N transformation rates. *Soil Biology and Biochemistry, 38*, 71–80.
Huang, J. H., Liao, Y. C., Gao, M. S., & Yin, R. J. (2009). Effects of tillage and mulching on orchard soil moisture content and temperature in Loess Plateau. *Chinese Journal of Applied Ecology, 20*, 2652–2658. (in Chinese with English abstract).
Jabran, K. (2019). Role of mulching in pest management and agricultural sustainability. *Springer Briefs in Plant Science*. https://doi.org/10.1007/978-3-030-22301-4
Jordán, A., Zavala, L. M., Muñoz-Rojas, M. (2011). Mulching, effects on soil physical properties. *The Encyclopedia of Earth Sciences Series*, 492–496. https://doi.org/10.1007/978-90-481-3585-1_275.
Katra, I. (2020). Soil erosion by wind and dust emission in semi-arid soils due to agricultural activities. *Agronomy, 10*, 1–10.
Khurshid, K., Iqbal, M., Arif, M. S., & Nawaz, A. (2006). Effect of tillage and mulch on soil physical properties and growth of maize. *International Journal of Agriculture and Biology, 8*, 1–5.
Kong, L. (2013). Maize residues, soil quality, and wheat growth in China a review. *Agronomy for Sustainable Development, 34*, 405–416. https://doi.org/10.1007/s13593-013-0182-5.
Lal, R. (2009). The plow and agricultural sustainability. *Journal of Sustainable Agriculture, 33*, 66–84.
Lal, R. (2013). Enhancing ecosystem services with no-till. *Renewable Agriculture and Food Systems, 28*, 102–114.
Lal, R. (1979). Soil and micro-climate considerations for developing tillage systems in the tropics. In E. Lal (Ed.), *Soil tillage and crop production. Proceeding series* (No. 2). International Institute for Tropical Agriculture.
Li, F.-M., Guo, A.-H., & Wei, H. (1999). Effect of plastic-film mulch on yield of spring wheat. *Field Crops Research, 63*, 79–86.

Li, L. J., Liu, K. Y., Hu, Y. N., Yang, Z. F., Xu, Z. B., Miao, C. L., & Xiao, F. T. (2001). Effect of the straw returned to field on the long-term improvement of Shajiang black soil. *Journal of Anhui Agricultural Sciences, 29*, 765–766.

Liu, C. A., Jin, S. L., Zhou, L. M., Jia, Y., Li, F. M., Xiong, Y. C., & Li, X. G. (2009). Effects of plastic film mulch and tillage on maize productivity and soil parameters. *European Journal of Agronomy, 31*, 241–249. https://doi.org/10.1016/j.eja.2009.08.004.

Liu, D. H., Shu, L., Chen, Q., Chen, S. H., Chen, H. L., & Zhu, Z. L. (2011). Effects of straw mulching and little- or zero-tillage on microbial diversity and biomass C and N of alluvial soil in Chengdu Plain, China. *The Chinese Journal of Applied and Environmental Biology, 17*, 158–161. (in Chinese with English abstract).

Liu, Y., Gao, M., Wu, W., Tanveer, S. K., Wen, X., & Liao, Y. (2013). The effects of conservation tillage practices on the soil water-holding capacity of a non-irrigated apple orchard in the Loess Plateau, China. *Soil and Tillage Research, 130*, 7–12. https://doi.org/10.1016/j.still.2013.01.012.

Loch, R. J., Robotham, B. G., Zeller, L., Masterman, N., Orange, D. N., Bridge, B. J., Foley, J. L., Sheridan, G., & Bourke, J. J. (2001). A multipurpose rainfall simulator for field infiltration and erosion studies. *Australian Journal of Soil Research, 39*, 599–610.

Lou, Y., Xu, M., Wang, W., Sun, X., & Zhao, K. (2011). Return rate of straw residue affects soil organic C sequestration by chemical fertilization. *Soil and Tillage Research, 113*, 70–73. https://doi.org/10.1016/j.still.2011.01.007.

Luo, Y., Iqbal, A., He, L., Zhao, Q., Wei, S., Ali, I., Ullah, S., Yan, B., & Jiang, L. (2020). Long-term no-tillage and straw retention management enhances soil bacterial community diversity and soil properties in Southern China. *Agronomy, 10*, 1233. https://doi.org/10.3390/agronomy10091233.

Mehmood, S., Zamir, S. I., Rasool, T., & Akbar, W. (2014). Effect of tillage and mulching on soil fertility and grain yield of sorghum. *Science Agriculture, 4*, 31–36. https://doi.org/10.15192/PSCP.SA.2014.4.1.3136.

Melander, B., Munier-Jolain, N., Charles, R., Wirth, J., Schwarz, J., Vander-Weide, R., Bonin, L., Jensen, P. K., & Kudsk, P. (2013). European perspectives on the adoption of nonchemical weed management in reduced-tillage systems for arable crops. *Weed Technology, 27*, 231–240.

Mikkola, H. J., Alakukku, L., Känkänen, H., Jalli, H., Lindroos, M., Huusela-Veistola, E., Nuutinen, V., Lätti, M., Puustinen, M., Turtola, E., Myllys, M., Regina, K. (2005, May). Direct drilling in Finland, a review. *Proceedings 4th International Scientific and Practical Conference, Ecology and Agricultural Machinery, 2,* 25–26 St Petersburg, 141–151.

Mulumba, L. N., & Lal, R. (2008). Mulching effects on selected soil physical properties. *Soil and Tillage Research, 98*, 106–111. https://doi.org/10.1016/j.still.2007.10.011.

Nawaz, A., Farooq, M., Lal, R., Rehman, A., Hussain, T., & Nadeem, A. (2016). Influence of sesbania brown manuring and rice residue mulch on soil health, weeds and system productivity of conservation rice wheat systems. *Land Degradation & Development,* 1–13.

Niu, X. S., Ma, Y. L., Niu, L. A., Hao, J. M., & Zhang, S. K. (2007). Effects of no-tillage planting for winter wheat with maize straw mulching on soil physicochemical properties. *Acta Agriculturae Boreali Sinica, 22*(suppl), 158–163.

Niziolomski, J. C., Simmons, R. W., Rickson, J. R., & Hann, M. J. (2020). Efficacy of mulch and tillage options to reduce runoff and soil loss from asparagus interrows. *CATENA, 191*, 104557. https://doi.org/10.1016/j.catena.2020.104557.

Poll, C., Marhan, S., Ingwersen, J., & Kandeler, E. (2008). Dynamics of litter carbon turnover and microbial abundance in a rye detritusphere. *Soil Biology & Biochemistry, 40*, 1306–1321. https://doi.org/10.1016/j.soilbio.2007.04.002.

Tebrügge, F. (2001). No-tillage visions – protection of soil, water and climate and influence on management and farm income. In L. Garcia-Torres., J. Benites., & A. Martĩnez-Vilela (Eds.), *Conservation Agriculture – A Worldwide Challenge.* World Congress on Conservation Agriculture, (Vol 1, pp. 303–316).

Qamar, R., Ehsanullah, S. M., Muhammad, H., Javeed, R., Rehman, A., & Atiqueur-Rehman, A. A. (2015). Influence of tillage and mulch on soil physical properties and wheat yield in rice-wheat system. *West African Journal of Applied Ecology, 23*, 21–38.

Qiang, X. C., Yuan, H. L., & Gao, W. S. (2004). Effect of crop residue incorporation on soil CO_2 emission and soil microbial biomass. *Chinese Journal of Applied Ecology, 15*, 469–472. (in Chinese with English abstract).
Quddus, M. A., Naser, H. M., Siddiky, M. A., Ali, M. R., Mondol, A. T. M. A. I., & Islam, M. A. (2020). Impact of zero tillage and tillage practice in chickpea production. *Journal of Agricultural Science, 12*, 106–118.
Rafiq, M. H. (2016). Development of techniques for crop residues and nitrogen management in no-till wheat sown under rice based system. Ph.D. thesis. Department of Agronomy, Faculty of Agriculture University of Agriculture.
Salehi, S., Rokhzadi, A., Abdolahi, A., Mohammadi, K., & Nourmohammadi, G. (2017). Effect of soil tillage systems on chickpea yield and moisture of soil. *Bioscience Biotechnology Research Communications, 10*, 404–409. https://doi.org/10.21786/bbrc/10.3/11.
Ullah, Z. (2015). Conservation vs conventional tillage system studies for cereal-legume forage productivity. Ph.D. thesis. Department of Agronomy, Faculty of Crop and Food Sciences, Pir Mehr Ali Shah Arid Agriculture University Rawalpindi.
Wang, H., Guo, Q., Li, X., Li, X., Yu, Z., Li, X., Yang, T., Su, Z., Zhang, H., & Zhang, C. (2020). Effects of long-term no-tillage with different straw mulching frequencies on soil microbial community and the abundances of two soil-borne pathogens. *Applied Soil Ecology, 148*, 103488. https://doi.org/10.1016/j.apsoil.2019.103488.
Wang, X., Wu, H., Dai, K., Zhang, D., Feng, Z., Zhao, Q., Wu, X., Jin, K., Cai, D., Oenema, O., & Hoogmoed, W. B. (2012). Tillage and crop residue effects on rainfed wheat and maize production in northern China. *Field Crops Research, 132*, 106–116. https://doi.org/10.1016/j.fcr.2011.09.012.
Weber, J., Kunz, C., Peteinatos, G., Zikeli, S., & Gerhards, R. (2017). Weed control using conventional tillage, reduced tillage, no-tillage, and cover crops in organic soybean. *Agriculture, 7*, 43. https://doi.org/10.3390/agriculture7050043.
Wilhelm, W. W., Johnson, J. M. F., Hatfield, J. L., Voorhees, W. B., & Linden, D. R. (2004). Crop and soil productivity response to corn residue removal: a literature review. *Agronomy Journal, 96*, 1–17.
Yadav, G. S., Das, A., Lal, R., Babu, S., Meena, R. S., Patil, S. B., Saha, P., & Datta, M. (2018). Conservation tillage and mulching effects on the adaptive capacity of direct-seeded upland rice (*Oryza sativa* L.) to alleviate weed and moisture stresses in the North Eastern Himalayan Region of India. *Archives of Agronomy and Soil Science, 64*, 1254–1267. https://doi.org/10.1080/03650340.2018.1423555.
Yang, H., Wu, G., Mo, P., Chen, S., Wang, S., Xiao, Y., Ma, H. A., Wen, T., Guo, X., & Fan, G. (2020). The combined effects of maize straw mulch and no-tillage on grain yield and water and nitrogen use efficiency of dry-land winter wheat (*Triticum aestivum* L.). *Soil and Tillage Research, 197*, 104485. https://doi.org/10.1016/j.still.2019.104485.
Yang, J., Shen, Y., Nan, Z., Gao, C., Niu, Y., Wang, X., Luo, C., & Li, G. (2010). Effects of conservation tillage on crop yield and carbon pool management index on top soil within a maize-wheat-soybean rotation system in the Loess Plateau. *Acta Prataculturae Sinica, 19*, 75–82.
Yu, X. L., Wu, P. T., Wang, Y. K., Zhang, L. Q., Yun, X. F., & Zhang, J. X. (2007). Effects of different quantity of straw mulching on physiological character of winter wheat and soil moisture and temperature. *Journal of Irrigation and Drainage, 26*, 41–44.
Yang, K., Zhu, J. J., Yan, Q. L., Zhang, J. X. (2012). Soil enzyme activities as potential indicators of soluble organic nitrogen pools in forest ecosystems of Northeast China. *Annals Forest Science, 69*, 795–803. https://doi.org/10.1007/s13595-012-0198-z
Zhang, D. X., Han, Z. Q., Liu, W., Gao, S. G., Chang, L. S., Hou, D. J., & Li, G. F. (2006). The effect of maize stock returning on the dynamic changes of soil enzyme activities under different decay conditions. *Journal of Soil Science, 37*, 475–478. (in Chinese with English abstract).
Zhang, F., Li, B., Zhang, F. Y., Li, X. H., Sun, M. Z., & Gao, G. Q. (2011). Effects of maize stalk returning on yield and quality of different types of wheat. *Shandong Agriculture Science, 3*, 30–32.

Zhang, G., Chan, K., Oates, A., Heenan, D., & Huang, G. (2007). Relationship between soil structure and runoff/soil loss after 24 years of conservation tillage. *Soil and Tillage Research, 92*, 122–128. https://doi.org/10.1016/j.still.2006.01.006.

Zhang, Q., Wang, Z., Miao, F., & Wang, G. (2017). Dryland maize yield and water-use efficiency responses to mulching and tillage practices. *Agronomy Journal, 109*, 1196. https://doi.org/10.2134/agronj2016.10.0593.

Zikeli, S., Gruber, S., Teufel, C. F., Hartung, K., & Claupein, W. (2013). Effects of reduced tillage on crop yield, plant available nutrients and soil organic matter in a 12-year long-term trial under organic management. *Sustainability, 5*, 3876–3894.

Response of Mulching on Soil Physical and Biochemical Properties and Functions

Adeel Ahmad, Muhammad Yaseen, Imtiaz Ahmed, Bushra Niamat, Aqarab Husnain Gondal, Amir Aziz, Muhammad Irfan, Muhammad Faizan Ilyas, and Muhammad Jafir

Abstract The covering of soil surface, for creating suitable conditions for the growth of plants, either by organic or inorganic material is termed mulching. Mulching is found effective in soil and water conservation, weeds and salinity control, temperature moderation, reducing the cost of fertilizer and creating a suitable soil microenvironment for better crop growth and yield. This chapter highlighted different aspects of mulching including their types and nature. Secondly, the effect of mulching on the physical, biological and physicochemical properties of soil are considered parameters for determining the health of the soil. Thirdly, the impact of mulching on environmental conditions is also highlighted. This discussion made a skeleton for measuring the quality and health improvement of soil by mulching.

1 Introduction

The semi-arid and arid areas of the world are mostly facing a shortage of water due to several reasons like increment in population pressure, degradation of natural resources, change in rainfall pattern, global warming, evaporation of soil water

A. Ahmad (✉) · M. Yaseen · B. Niamat · A. H. Gondal · M. F. Ilyas
Institute of Soil and Environmental Sciences, University of Agriculture Faisalabad, Faisalabad, Pakistan
e-mail: Adeelgujjar046@yahoo.com

I. Ahmed
Institute of Agricultural Resources and Regional Planning, Graduate School of Chinese Academy of Agricultural Sciences, Beijing, China

A. Aziz
Department of Soil and Environmental Sciences, College of Agriculture, University of Sargodha, Sargodha, Pakistan

M. Irfan
School of Environmental Science and Engineering, Tianjin University, Tianjin 300354, PR China

M. Jafir (✉)
Department of Entomology, University of Agriculture, Faisalabad, Pakistan
e-mail: m.jafir.uaf@gmail.com

K. Akhtar et al. (eds.), *Mulching in Agroecosystems*,
https://doi.org/10.1007/978-981-19-6410-7_6

content and climate change (Colak et al., 2015; Deng et al., 2006; Li et al., 2000; World Water Assessment programme, 2012). Such factors deteriorate the soil quality and health also, which are considered the key factors of sustainable agriculture (El Chami et al., 2020). To overcome this situation, several farming practices are being initiated in an area where soil moisture conservation is necessary. Various practices help in conserving water for the sustainable farming of agricultural land. For the conservation of soil on agricultural lands in every climatic zone, the shielding effects of plant covers have been used (Gyssels et al., 2005).

The word mulch is a derivative of the German word molsch which means initiative of decay process (Jacks et al., 1955). In contrast to the materials incorporated in the soil profile, the term mulch is meant for those materials that are added to the soil surface (Jacks et al., 1955; Chalker-scott, 2007). Mulching is a technique for reducing runoff and water infiltration to save water and a mulch is a material spreaded on the surface of the soil (Adekalu et al., 2007; Ghawi & Battikhi, 1986). Mulching is the most suitable technique for providing suitable crop environment to obtain quality production and higher yield as mulching stabilize soil moisture, control temperature and decrease the process of evaporation (Chakraborty & Sadhu, 1994). Mulches can provide resistance against weeds to germinate, can control soil erosion, nutrient loss, moisture loss and can provide a shield against disease and insect injury while promising crop and plant establishment of potentially higher quality. The study also stated that plastic or artificial mulches are fully resistant to water threats (flood and drought) and minimize the losses of soil and water through runoff. Organic mulches help to improve soil structure and decrease production cost (Mugalla et al., 1996).

Moreover, mulching involves spreading a cover of material on the surface soil around the desired crop to modify the micro-environment to improve the productivity of the crop. Typically, mulching blocks the light or create such environmental conditions which can inhibit germination or suppress the growth of weeds just after germination. Though, further benefits including moisture conservation, reduce nutrient leaching, temperature regulation, improved soil organic matter and some cases reduce soil compaction. So, typically mulches applications results in to obtain higher yield and quality of crops increasing profitability for the growers (Malik et al., 2018). Crop and soil interactions mainly influenced due to such factors as clay contents, soil moisture content, O_2 availability in the rhizosphere and temperature (Powlson et al., 2011). Mulched soil by affecting the chemical and physical properties of the soil improved soil hydrologic characteristics. Soil water environment is directly influenced by temperature and soil moisture that has significant impacts on soil microbiology and soil physics (Smets et al., 2008). By reducing soil loss and preventing runoff mulches reduce soil deterioration, which modifies aeration, structure, physical properties of soil and organic matter content (Jordán et al., 2010).

2 Types of Mulches and Materials

Mulches are categorized broadly into three main groups (Organic, inorganic and living mulch) and each type of mulch has specific characteristics according to the nature of its material. The application of mulch practice depends on the soil and environmental conditions also according to the nutritional requirements of plants (Wang et al., 2015). Organic mulches are produced by the means of organic substances like green wastes (wood chips and leaves), wood industries waste (Barks and sawdust) and agricultural waste (rice husks and straw) (Kader et al., 2017). Inorganic mulches can be comprised of polyethylene film bricks, gravel and cobblestones (Qian et al., 2015). The very first attempt to mulch the soil in Poland was taken with polyvinylchloride (PVC) film in cucumber cultivation (Libik, 1976). Accordingly, the use of synthetic/plastic mulch may be a potential practice and prospective to conserve water in modern agriculture. Living mulches consist of Manila grass, dwarf lilyturf, clover, ryegrass and other types of grasses (Qian et al., 2015). Living mulches like cover crops undersown or interplanted with the main crop usually enhances the natural predators of pests for that specific crop (Hartwig & Ammon, 2002).

Currently, an intensive explorative approach going on the production technology, properties and implementation of films. Biodegradable mulches are synthetic polymers made from such substances responsible for biodegradation (Lopez-Tolentino et al., 2017), polysaccharides (Moreno et al., 2017), thermosetting polymers obtained from vegetable oils (Adekunle, 2015) as well as nonwoven fabrics made from polylactide (Zawiska & Siwek, 2014). The addition of iron, zinc, magnesium, manganese and cobalt facilitates oxidation, as influenced by light, air and heat to degrade the long-chain polymers (Zenner de Polania & Peña Baracaldo, 2013). Micro-organisms under aerobic conditions can decompose these polymers into water, methane or carbon dioxide, biomass and other organic substances and have been observed no adverse effects occurring in the soil during nitrification processes (Ardisson-Araújo et al., 2014).

3 Effects on Soil Physical Properties

Mulching induces very important effects on the physical properties of the soil. Some important physical properties are discussed below. Figure 1 also supports the comparison of mulched and non-mulched soil for some important physical properties of the soil.

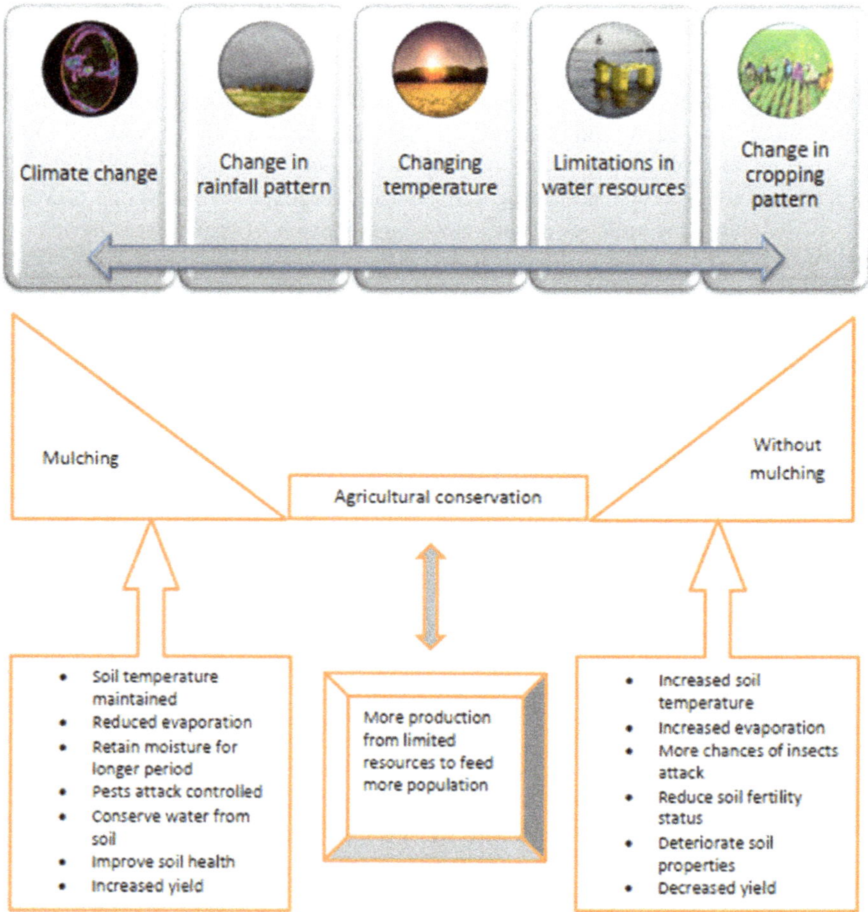

Fig. 1 A comparative analysis of mulched and non-mulched soil for some important properties

3.1 Soil Structure

The structure of the soil is enormously significant for maintaining the quality and health of the soil. Density and elongation of roots, air and water flow and erosion are affected by aggregate stability (Amézketa, 1999) which is determined via unified forces among particles. Thus, it can be utilized as a catalogue of the structure as well as corporal constancy of soil. Crucial factors controlling aggregation are the texture of the soil, mineralogy of clay, positive ions and amount plus the quality of soil organic material. Roots of plant, microbes and carbon-based materials are too tangled in the development and stability of aggregates. Aggregate stability may fluctuate seasonally or during ploughing (Hillel, 1998). After mulching, augmented soil organic matter content adds to improve aggregation under a variety of climatic areas (Jordán et al., 2010; Mulumba & Lal, 2008) even in small period (Hermawan & Bomke, 1997). Inert mulches like plastic film exhibit no or partial influence on the structure of the soil. Geotextiles may enhance soil organic matter contents, refining the structure of top-soil as well as stability of aggregates (Bhattacharyya et al., 2010).

3.2 Soil Porosity

There is an extreme variability in the bulk density of the soil under the influence of mulching. Even though greater bulk density has been detected underneath mulching comparative to tillage system (Bottenberg et al., 1999), reduced bulk densities have also been described by Ghuman and Sur (2001) and Oliveira and Merwin (2001). No correlation between the rate of mulching and bulk density exists in other cases which may be attributable to alterations in managing practices, type of mulch material and soil utilized in the experimentations. Nonetheless, Mulumba and Lal (2008) reported that when 0 and 5 mg ha^{-1} wheat straw was applied as mulch material, bulk densities were enlarged but lessened in case of greater rates. Pores of dissimilar shape, size and permanency are formed by biotic and abiotic features (Kay & Bygaart, 2002). Mulumba and Lal (2008) observed that with increment in the rate of mulching, total porosity enhanced considerably with an eleven-year treatment in the USA. Afterwards, in shorter periods, mulching has been attributable to enhanced porosity (Jordán et al., 2010; Oliveira & Merwin, 2001).

3.3 Soil Temperature

The rate of chemical and biotic processes of soil is affected by temperature. The amount of energy that enters into the soil be determined by soil colour, aspect as well as the vegetative cover. Mulching can distress or entirely alter the temperature regime of soil via decreasing the amount of energy entering the soil as a result of shielding temperature variants, shielding the soil surface and radiation capture. The soil temperature range is typically less significant in mulched soils compared to unmulched soils. Wheat straw decreases the input of solar energy as it has a lower thermal conductivity and a higher albedo compared to unembellished soil (Horton et al., 1996). In contrast, during cooler periods, the shielding of the soil from cooler atmosphere is done by wheat-straw-mulch on the surface of soil (Zhang et al., 2009). Afterwards depending on mulch-material a discrepancy is shown by the application of inorganic mulches. Afterwards field experimentations in the United Kingdom, it was validated by Cook et al. (2006) that higher rates of mulching declined soil temperature. In contrast, the usage of inert mulches can upturn the temperature of the soil. Nachtergaele et al. (1998) described that gravel-mulch caused an increment in temperature of soil and decline in evaporation in vineyard soils in Switzerland.

Carbon-based geotextiles can also decrease loss of water through evaporation mitigate and thril temperature variations,. In horticulture, mineral materials for example plastic-mulches are frequently utilized to upturn the temperature of soil which leads to greater vintage. Apart from other ecological issues, some restrictions were observed with the rigorous use of plastic-mulches for enhancing soil temperature. Higher rates of mineralization can cause a collapse of soil organic matter, affecting durable soil physical as well as chemical productiveness (Li et al., 2004).

3.4 Soil Moisture

Mulching adjusts the evaporation and transpiration rates to cope with the heat stress, drought stress and salinity stress, which (JimEnez et al., 2017; Wang et al., 2011). In diverse climatic situations, the conservation of soil moisture varies following the type of mulch. Conversely, in comparison to the unprotected soil conservation of moisture in soil having mulch is considerably greater (Zhao et al., 2014). Mulching also has a prodigious influence on surface water and soil water. Mulching improves infiltration rate, increases storage capability of water and recovers its retention and thus declines runoff. Moreover, abridged rates of evaporation help to prolong the duration for which soil remains wet. Mulching recovers substantial water characteristics of soil, while diverse results have been testified. Carbon-based mulches on the soil surface make ideal environments of soil for crop growth, augmenting water-retention as well as the availability of soil and increasing macro-porosity (Martens & Frankenberger, 1992).

Bhardwaj (2013) said that the application of carbon-based mulches impedes the growth of weeds, increases intrusion of rainwater and causes reductions in evaporation. Though the use of carbon-based mulch declines moisture contents of the soil parallel to control by striving with the plants for moisture (Watson, 1988), nonetheless, for sesame (*Sesamum indicum* L.) and additional crops it was documented that the paper-mulch and carbon-based mulch induce greater moisture as paralleled to no-mulch (Teame et al., 2017). Earlier studies suggested that increase in soil moisture contents was also noticed with the use of gravel (Nachtergaele et al., 1998) and wood chips (Sinkeviciene et al., 2009; Watson, 1988) mulches. Use of straw-mulch considerably enhanced moisture of soil up to forty-centimetre depth (Zhao et al., 2014). Research has revealed that mulch application can enhance infiltration and reduce evaporation, which results in more water stored and abridged rates of runoff (Smika & Unger, 1986).

An excellent approach to improve the maintenance of water in the soil and decreasing evaporation of soil is the wheat-straw mulch. Underneath high rates of mulching as well as reduced-till practices, high aptitudes of accessible water have been testified. Jordán et al. (2010) and Mulumba and Lal (2008) described that even fewer degrees of mulching has a durable influence on accessible water content. Most visible, ultraviolet radiation (UV) and infrared solar-radiation are immersed by black-polyethylene-mulch which reradiates infrared-radiation. Effects on the microclimate of a plant and energy scorching behaviour could be estimated by colour-mulch

(Lamont, 2005). Soil moisture content is enhanced using black polyethylene mulch (Liu et al., 2014). Likewise, Gao et al. (2014) stated that under the full-film mulching systems moisture contents were considerably higher up to the deeper depth of the soil.

4 Mulching Effect on Soil Chemical Properties

4.1 Total Organic Carbon

Afterwards decomposition, carbon-based mulches return carbon-based material and plant nutrients to the soil and improve the biological, chemical and physical properties of the soil, which consecutively upturns crop produce (Ray & Biswasi, 2016). The soil underneath the mulch remains friable, loose and leading to an appropriate environment for penetration of root. Carbon-based mulches not only retain the soil moisture but also enhance nutrients of soil via adding carbon-based material (Kumar et al., 1990). Khurshid et al. (2006) and Saroa and Lal (2003) determined that carbon-based material was considerably higher when mulch was applied up to a greater extent. Least organic-C contents (0.48%) were recorded in the non-mulched plot while a higher soil organic-C content was recorded with paddy straw (0.66%), silkworm bed waste (0.68%) and sun hemp mulch (0.71%) mulched plots (Shashidhar et al., 2009).

4.2 Soil Nutrition Status

As mulching ingredients, carbon-based resources are not only the sources of all essential macro and micronutrients, but also accomplish several other functions such as sustaining soil water, improving the accessible nutrient status of soil, amending organic matter pool as well as the quality of soil, and decreasing the accretion of leached nitrate (Abarchi et al., 2009). In addition to enhancing the nutrient retaining capacity of the soil, the application of organic mulch provides a supportable source of macro as well as micronutrients to crops (Kumar et al., 2017). Thus, it helps in reducing the fertilizer leaching for efficient utilization of nutrient by keeping the nutrient in the plant root zone.

Underneath plastic mulch, the breakdown of carbon-based residues supplements organic-acids to the soil which results in low pH of soil increasing the bio-accessibility of micronutrients like iron, copper, zinc and manganese. Tisdale et al. (1985) also testified the enhanced iron and zinc contents in the soil underneath plastic mulch. Due to the mineralization of carbon-based-N with time, the availability of soil-N increases as mineral-N contents such as nitrate and ammonium ion in the soil is high. The decay of carbon-based material release essential nutrients like NO_3,

NH_4^+, Ca^{2+}, Mg^{2+}, K^+ and fulvic acid to the soil and increases nutrient accessibility of soil underneath plastic mulch (Sharma & Bhardwaj, 2017).

5 Mulching and Soil Biodiversity

5.1 Soil Microfauna

Mulching boosts the microbial population of soil which is responsible for the discharges plant-nutrients for growth of crop, enough temperature and moisture of soil resulting in speedy microbial-decay and production of the well aerobic environs (Wang et al., 2018). The variations in the microbial-population underneath diverse carbon-based mulches produced inconsistency in the chemical configuration as well as decay rates (Wu et al., 1993). Enhanced moisture contents underneath mulching affects the obtainability of nutrients producing an affirmative influence on the soil biota and the biological nitrogen fixation (Surya et al., 2000). Mulches amplified the soil-biota by providing nitrogen (N) for protein formation, energy from C-compounds and other nutrients. In nutrient cycling processes, mulching plays an imperative role, thus, provide the crop with plenty of nutrients (Holland, 2004). By increasing microbial biomass and breakdown of carbon-based materials, mulching ensures extreme nutrient accessibility to crops (Tu et al., 2006; Wang et al., 2016).

5.2 Soil Macrofauna

The use of mulches accelerates rate of infiltration, recovers the soil aggregate constancy and supports the propagation of earthworms. By providing proper environmental situations and food for soil macro-invertebrates, carbon-based mulch boost their vegetative-growth (Sugiyarto, 2009).

6 Conclusion

It is clear from the above discussion that mulching has a strong influence on the physical, chemical and biological properties of the soil. Mulching not only improves soil structure, nutrient status, soil temperature, soil microbial activities and density of soil but also improves soil moisture retention which is of great concern to any crop. Thus, it stimulates crop production attributes. Among mulching types, organic mulches have more influence on soil characteristics than inorganic material. As organic mulches also add up organic matter to the soil, while living mulches are also of great concern because this kind of mulching material enhances the nutrient status of the soil. All type of mulches improves soil quality in relation to crop growth and yield. As soil quality and health are very important for producing an optimum yield of all crops, mulching is proved beneficial in enhancing not only the health of soil but also the yield of many crops. In the future attention could be focused to accelerate agricultural production by further improving soil characteristics by using organic and inorganic mulches in combination, as organic mulches will enhance the organic matter of soil in addition to the improvements made by inorganic mulches.

References

Abarchi, I., Zhang, Z. Y., Vanlauwe, B., Guo, X. P., Wang, W. M., Ongor, B. T. I., & Timbely, D. (2009). Effects of plant age and rock phosphate on quality and nutrient release of legume residue. *Pedosphere, 19*, 78–85.

Adekalu, K. O., Olorunfemi, I. A., & Osunbitan, J. A. (2007). Grass mulching effect on infiltration, surface runoff and soil loss of three agricultural soils in Nigeria. *Bioresource Technology, 98*, 912–917.

Adekunle, K. F. (2015). Surface treatments of natural fibres—A review: Part 1. *Open Journal of Polymer Chemistry, 5*(03), 41.

Amézketa, E. (1999). Soil aggregate stability: A review. *Journal of Sustainable Agriculture., 14*, 83–151.

Ardisson-Araújo, D. M. P., Melo, F. L., de Souza, A. M., Brancalhão, R. M. C., Báo, S. N., & Ribeiro, B. M. (2014). Complete genome sequence of the first non-Asian isolate of Bombyx mori nucleopolyhedrovirus. *Virus Genes, 49*(3), 477–484.

Bhardwaj, R. L. (2013). Effect of mulching on crop production under rainfed condition-a review. *Agricultural Review, 34*(3), 188–197.

Bhattacharyya, R., Smets, T., Fullen, M. A., Poesen, J., & Booth, C. A. (2010). Effectiveness of geotextiles in reducing runoff and soil loss: A synthesis. *CATENA, 81*, 184–195.

Bottenberg, H., Masiunas, J., & Eastman, C. (1999). Strip tillage reduces yield loss of snapbean planted in rye mulch. *HortTechnology, 9*, 235–240.

El Chami, D., Daccache, A., & El Moujabber, M. (2020). How can sustainable agriculture increase climate resilience? A systematic review. *Sustainability, 12*, 3119.

Chakraborty, R. C., & Sadhu, M. K. (1994). Effect of mulch type and color on growth and yield of tomato (*Lycopersicon esculentum*). *Indian Journal of Agricultural Sciences, 64*, 608–612.

Chalker-Scott, L. (2007). Impact of mulches on landscape plants and the environment-a review. *Journal of Environmental Horticulture, 25*, 239–249.

Colak, Y. B., Yazar, A., Çolak, I., & Duraktekin, G. (2015). Evaluation of crop water stress index (CWSI) for eggplant under varying irrigation regimes using surface and subsurface drip systems. *Agriculture Science Procedia, 4*, 372–382.

Cook, H. F., Valdes Gerardo, S. B., & Lee, H. C. (2006). Mulch effects on rainfall interception, soil physical characteristics and temperature under *Zea mays* L. *Soil and Tillage Research, 91*, 227–235.

Deng, X. P., Shan, L., Zhang, H., & Turner, N. C. (2006). Improving agricultural water use efficiency in arid and semiarid areas of China. *Agricultural Water Management, 80*, 23–40.

Gao, Y., Xie, Y., Jiang, H., Wu, B., & Niu, J. (2014). Soil water status and root distribution across the rooting zone in maize with plastic film mulching. *Field Crops Research, 156*, 40–47.

Ghawi, I., & Battikhi, A. (1986). Water melon production under mulch and trickle irrigation in the Jordan valley. *Journal of Agronomy and Crop Science, 157*, 145–155.

Ghuman, B. S., & Sur, H. S. (2001). Tillage and residue management effects on soil properties and yields of rainfed maize and wheat in a subhumid subtropical climate. *Soil and Tillage Research, 58*, 1–10.

Gyssels, G., Poesen, J., Bochet, E., & Li, Y. (2005). Impact of plant roots on the resistance of soils to erosion by water: A review. *Progress in Physical Geography, 29*, 189–217.

Hartwig, N. L., & Ammon, H. (2002). Cover crops and living mulches. *Weed Science, 50*, 688–699.

Hermawan, B., & Bomke, A. A. (1997). Effects of winter cover crops and successive spring tillage on soil aggregation. *Soil and Tillage Research, 44*, 109–120.

Hillel, D. (1998). *Environmental soil physics*. Academic.

Holland, J. M. (2004). The environmental consequences of adopting conservation tillage in Europe: Reviewing the evidence. *Agriculture, Ecosystems and Environment, 103*(1), 1–25.

Horton, R., Bristow, K. L., Kluitenberg, G. J., & Sauer, T. J. (1996). Crop residue effects on surface radiation and energy balance—Review. *Theoretical and Applied Climatology, 54*, 27–37.

Jacks, C. V., Brind, W. D., & Smith, R. (1955). Mulching technology comm., no. 49, common wealth. *Bulletin of Soil Science*, 118.

JimEnez, M. N., Pinto, J. R., Ripoll, M. A., SAnchez-Miranda, A., & Navarro, F. B. (2017). Impact of straw and rock-fragment mulches on soil moisture and early growth of holm oaks in a semiarid area. *Catena, 152*, 198–206.

Jordán, A., Zavala, L. M., & Gil, J. (2010). Effects of mulching on soil physical properties and runoff under semi-arid conditions in southern Spain. *CATENA, 81*, 77–85.

Kader, M. A., Senge, M., Mojid, M. A., & Ito, K. (2017). Recent advances in mulching materials and methods for modifying soil environment. *Soil and Tillage Research, 168*(5), 155–166.

Kay, B. D., & Bygaart, A. J. V. (2002). Conservation tillage and depth stratification of porosity and soil organic matter. *Soil and Tillage Research, 66*, 107–118.

Khurshid, K., Iqbal, M., Arif, M. S., & Nawaz, A. (2006). Effect of tillage and mulch on soil physical properties and growth of maize. *International Journal of Agriculture & Biology, 8*, 593–596.

Kumar, G. D., Sachin, S. S., & Kumar, R. (1990). Importance of mulch in crop production. *Indian Journal of Soil Conservation., 18*, 20–26.

Kumar, V., Sharma, J. C., Kumar, M., & Singh, S. K. (2017). Soil and Plant Nutrient Status as Modified by Different Types of Mulching in Cauliflower. *Current Journal of Applied Science and Technology, 24*(1), 1–8.

Lamont, W. J. (2005). Plastics: Modifying the microclimate for the production of vegetable crops. *HortTechnology, 15*(3), 477–481.

Li, X. Y., Gong, J. D., & Wei, X. H. (2000). In-situ rainwater harvesting and gravel mulch combination for corn production in the dry semi-arid region of China. *Journal of Arid Environments, 46*, 371–382.

Li, F. M., Wang, J., Xu, J. Z., & Xu, H. L. (2004). Productivity and soil response to plastic film mulching durations for spring wheat on entisols in the semiarid Loess Plateau of China. *Soil and Tillage Research, 78*, 9–20.

Libik, A. (1976). Wpływ ściółkowania gleby folią i papierem silosowym na wzrost i plonowanie ogórka gruntowego. *Acta Agraria Silvestria, Series Agraria, XVI*(2), 69–84.

Liu, X. E., Li, X. G., Hai, L., Wang, Y. P., & Li, F. M. (2014). How efficient is film fully-mulched ridge—Furrow cropping to conserve rainfall in soil at a rainfed site? *Field Crops Research, 169*, 107–115.

López-Tolentino, G., Ibarra-Jiménez, L., Méndez-Prieto, A., Lozano-del Río, A. J., Lira-Saldivar, R. H., Valenzuela-Soto, J. H., Lozano-Cavazos, C. J., & Torres-Olivar, V. (2017). Photosynthesis, growth, and fruit yield of cucumber in response to oxo-degradable plastic mulches. *Acta Agriculturae Scandinavica, Section B—Soil & Plant Science, 67*(1), 77–84.

Malik, N. S., Subhani, A., Bibi, R., & Naseem, W. (2018). Do organic and inorganic mulches affects soil moisture conservation and crop yield?. *Journal of Applied Agriculture and Biotechnology, 3*(1), 43–52.

Martens, D. A., & Frankenberger, W. T. (1992). Modification of infiltration rates in an organic-amended irrigated soil. *Agronomy Journal, 84*, 707–717.

Moreno, F. J., Corzo, N., Montilla, A., Villamiel, M., & Olano, A. (2017). Current state and latest advances in the concept, production and functionality of prebiotic oligosaccharides. *Current Opinion in Food Science, 13*, 50–55.
Mugalla, C. I., Jolly, C. M., & Martin, N. R. (1996). Profitability of black plastic mulch for limited resource farmers. *Journal of Production Agriculture, 9*, 175–176.
Mulumba, L. N., & Lal, R. (2008). Mulching effects on selected soil physical properties. *Soil and Tillage Research, 98*, 106–111.
Nachtergaele, J., Poesen, J., & van Wesemael, B. (1998). Gravel mulching in vineyards of southern Switzerland. *Soil and Tillage Research, 46*, 51–59.
Oliveira, M. T., & Merwin, I. A. (2001). Soil physical conditions in a New York orchard after eight years under different groundcover management systems. *Plant and Soil, 234*, 233–237.
Powlson, D. S., Gregory, P. J., Whalley, W. R., Quinton, J. N., Hopkins, D. W., Whitmore, A. P., Hirsch, P. R., & Goulding, K. W. T. (2011). Soil management in relation to sustainable agriculture and ecosystem services. *Food Policy, 36*, 72–87.
Qian, X., Gu, J., Pan, H. J., Sun, W., Wang, X. J., & Gao, H. (2015). Effects of living mulches on the soil nutrient contents, enzyme activities, and bacterial community diversities of apple orchard soils. *European Journal of Soil Biology, 70*, 23–30.
Ray, M., Biswasi, S. (2016). Impact of mulching on crop production: A review. *Trends in Biosciences, 9*(14).
Saroa, G. S., & Lal, R. (2003). Soil restorative effects of mulching on aggregation and carbon sequestration in a Miamian soil in Central Ohio. *Land Degradation Development, 14*, 481–493.
Sharma, R., & Bhardwaj, S. (2017). Effect of mulching on soil and water conservation—A review. *Agricultural Reviews, 38*(4), 311–315.
Shashidhar, K. R., Bhaskar, R. N., Priyadharshini, P., & Chandrakumar, H. L. (2009). Effect of different organic mulches on pH, organic carbon content and microbial status of soil and its influence on leaf yield of M-5 mulberry (*Morus indica* L.) under rainfed condition. *Current Biotica, 2*, 405–412.
Sinkeviciene, A., Jodaugiene, D., Pupaliene, R., & Urboniene, M. (2009). The influence of organic mulches on soil properties and crop yield. *Agronomy Research, 7*(1), 485–491.
Smets, T., Poesen, J., & Knapen, A. (2008). Spatial scale effects on the effectiveness of organic mulches in reducing soil erosion by water. *Earth-Science Reviews, 89*, 1–12.
Smika, D. E., & Unger, P. W. (1986). Effect of surface residues on soil water storage. *Advanced Soil Science, 5*, 111–138.
Sugiyarto, S. (2009). The effect of mulching technology to enhance the diversity of soil macroinvertebrates in Sengon-based agroforestry systems. *Biodiversitas Journal of Biological Diversity, 10*(3), 129–133.
Surya, J. N., Puranik, R. B., Zadode, S. D., & Deshmukh, S. D. (2000). Effect of wheat straw incorporation on yield of green gram and wheat, soil fertility and microbiota. *Journal of Maharashtra Agriculture Universities, 25*(2), 158–160.
Tisdale, S. L., Nelson, W. L., & Beaton, J. D. (1985). *Soil fertility and fertilizers* (4th ed.). Macmillan Publishing Company.
Teame, G., Tsegay, A., & Abrha, B. (2017). Effect of organic mulching on soil moisture, yield, and yield contributing components of sesame (*Sesamum indicum* L.). *International Journal of Agronomy*. https://doi.org/10.1155/2017/4767509.
Tu, C., Ristaino, J. B., & Hu, S. (2006). Soil microbial biomass and activity in organic tomato farming systems: Effects of organic inputs and straw mulching. *Soil Biology and Biochemistry, 38*(2), 247–255.
Wang, T. C., Wei, L., Wang, H. Z., Ma, S. C., & Ma, B. L. (2011). Responses of rainwater conservation, precipitation-use efficiency and grain yield of summer maize to a furrow planting and straw-mulching system in northern China. *Field Crops Research, 124*(2), 223–230.
Wang, H., Wang, C., Zhao, X., & Wang, F. (2015). Mulching increases water-use efficiency of peach production on the rainfed semiarid Loess Plateau of China. *Agriculture Water Management, 154*, 20–28.

Wang, J., Lv, S., Zhang, M., Chen, G., Zhu, T., Zhang, S., & Luo, Y. (2016). Effects of plastic film residues on occurrence of phthalates and microbial activity in soils. *Chemosphere, 151*, 171–177.

Wang, X., Fan, J., Xing, Y., Xu, G., Wang, H., Deng, J., Wang, Y., Zhang, F., Li, P., & Li, Z. (2018). The effects of mulch and nitrogen fertilizer on the soil environment of crop plants. *Advances in Agronomy, 153*, 121–173.

Watson, G. W. (1988). Organic mulch and grass competition influence tree root development. *Journal of Arboriculture, 14*(8), 200–203.

World Water Assessment Programme. (2012). The United Nations World Water Development Report 4: Manageing Water Under Uncertainty and Risk. UNESCO.

Wu, J., O'Donnell, A. G., & Syers, J. K. (1993). Microbial growth and sulphur immobilization following the incorporation of plant residues into soil. *Soil Biology and Biochemistry, 25*(11), 1567–1573.

Zawiska, I., & Siwek, P. (2014). The effects of PLA biodegradable and polypropylene nonwoven crop mulches on selected components of tomato grown in the field. *Folia Horticulturae., 26*(2), 163–167.

Zenner de Polanía, I., Peña Baracaldo, F. (2013). Plastic products in agriculture: Benefice and ambient cost: A review. *Revista UDCA Actualidad & Divulgación Científica, 16*(1), 139–150.

Zhang, S., Lövdahl, L., Grip, H., Tong, Y., Yang, X., & Wang, Q. (2009). Effects of mulching and catch cropping on soil temperature, soil moisture and wheat yield on the Loess Plateau of China. *Soil and Tillage Research, 102*, 78–86.

Zhao, Y., Pang, H., Wang, J., Huo, L., & Li, Y. (2014). Effects of straw mulch and buried straw on soil moisture and salinity in relation to sunflower growth and yield. *Field Crops Research, 161*, 16–25.

Plant Section

Mechanistic Insights into Mulching and Plant Physiological Attributes Under Abiotic Stresses

Naheeda Begum, Rana Roy, Hafeez Ur Rahim, Fangguo Chang, and Tuanjie Zhao

Abstract Drought, salinity, temperature extremes, and heavy metals are the major environmental factors that limit sustainable crop production worldwide and consequently restrict crop yield. There is a dire need for environment-friendly agricultural practices to achieve long-term food production for the growing population. Mulching has become a common method in modern agricultural practices because of its numerous benefits, such as moisture conservation, augmentation temperature of the soil, reduction of insect pests, weed management, escalation of crop yield, and the effective use of nutrients present in the soil, as well as decreased soil salinity. Mulching also enhances plants' resistance to pests and diseases and various stress factors like heat, salinity, drought, metals, and high or low temperatures. In addition to this, mulches could also provide economic, aesthetic, and other ecological benefits to agriculture. Mulching markedly increased the growth, yield, nutrient use, and water use efficiency in crop plants under stressed and non-stressed conditions. As a result, future research could also focus on economic, environment-friendly, and, more importantly, biodegradable materials on plant growth, balanced nutrition, yield, and quality under various abiotic stress conditions. This chapter focuses on the many essential aspects of mulches on the productivity and establishment of multiple crops under stressful environments.

Keywords Abiotic factors · Mulching · Plant growth regulation · Crop yield · Stress tolerance

---

N. Begum · F. Chang · T. Zhao (✉)
National Center for Soybean Improvement, Key Laboratory of Biology and Genetics and Breeding for Soybean, Ministry of Agriculture, State Key Laboratory for Crop Genetics and Germplasm Enhancement, Nanjing Agricultural University, Nanjing 210095, China
e-mail: tjzhao@njau.edu.cn

R. Roy
Department of Agroforestry and Environmental Science, Sylhet Agricultural University, Sylhet 3100, Bangladesh

H. U. Rahim
Key Laboratory of Industrial Ecology and Environmental Engineering (Ministry of Education), School of Environmental Science and Technology, Dalian University of Technology, Dalian 116024, China

K. Akhtar et al. (eds.), *Mulching in Agroecosystems*,
https://doi.org/10.1007/978-981-19-6410-7_7

1 Introduction

Mulches can be applied to the soil's surface to aid in water conservation and other related factors to nourish the soil for sustainable crop production. Mulch is most likely being derived from the German word "mulch," which means "soft to rot" and seemingly refers to gardeners spreading straw and leaves over the ground as mulch. (Jacks et al., 1995). Recently Kader et al. (2019) presented that mulching is mentioned as the dispersal of numerous surface cover materials on the soil to reduce losses of moisture and decrease soil temperature and soil salinity and augment crop production. According to the previous study, the mulches would be applied instantly after crop germination to benefit from it. Mulches have the potential to minimize water runoff, increase soil infiltration capability, shade weed populations, and act as an evapotranspiration barrier (Rathore et al., 1998). Various reasons to use mulches include; modifying the temperature of the soil, maintenance of the soil and adding nutrients already present in the soil, improvement of soil structure and texture, and plant tolerance to different environmental factors. As a result, it aids in preserving moisture in the soil and thus helps in temperature regulation and improves the physical, biological, and chemical properties of the soil (Dilipkumar et al., 1990; Shirzadi et al., 2020). Other benefits of the mulches include; environmental impacts by regulating salinity and nutrient losses, also reducing soil erosion and compactness, and improving plant roots morphology and growth (Lamont, 2005; Shan et al., 2020). In addition to this, mulches are also a good source of protecting plant roots from environmental stresses like drought, salinity, and even heat in the arid regions (Kazemi & Safari, 2018). Mulching mostly releases organic matter, trace elements, and other substances during straw decomposition, thus increases physico-chemical properties, organic matter, soil fertility, and texture, which enhances the crop yield significantly (van Asten et al., 2005; Abdelraouf et al., 2019). Mulches have been shown to enhance crop yield by increasing the water use efficiency (WUE) and thus conserving soil moisture by changing the microclimate within the field (Kannan et al., 2020). Different mulches have been studied on a variety of plants, i.e., eggplant (Sabatino et al., 2018), wheat (Qi et al., 2020), rice (Gangaiah et al., 2019), and tomato (Todd-Searle et al., 2020). According to Abdrabbo et al. (2017), plant response to various types of mulches is reliant on the cultivars, materials as well as climate variation.

Mulches can be natural (organic) or synthetic (inorganic). Organic mulches are made up of animal and plant wastes. Straws, husks, compost, live mulches of cover crops, sawdust, as well as manure are the most widely used organic mulches (Rathore et al., 1998). In contrast, polyethylene plastic mulch is the most commonly used inorganic mulch worldwide (Rathore et al., 1998). Crop residue mulching (Erenstein, 2002) is a method in which organic residue is used to cover the soil surface mainly taken from the harvested crop, such as maize stalks, palm fronds or leafy organic material (Bot & Benites, 2005). These organic mulches include; bark or sawdust and compost that are made up of woody material, which are becoming more readily accessible in communities with green waste recycling systems. (Merfield, 2002). The research studies reported that different types of mulches have different effects under different conditions. Organic mulch products, such as straw, papers, dry clips, leaves, compost, and so on (Kader et al., 2019; Shojaei et al., 2019), have significant advantages over plastic mulch. They are inexpensive and eco-friendly and adding nutrients to the soil (Kader et al., 2019). Furthermore, these mulches enhance germination by almost upto 70% while decreasing soil erosion upto 30% (Massa et al., 2019; Shojaei et al., 2019).

The current chapter focuses on the effect of mulching on crop production with a major focus on crop yield under stressful abiotic conditions by focussing the importance of mulches and their implication on agriculture at large.

2 Role of Mulching in Plant Growth Regulation

Implication of mulches on plants is mediated by their direct effects on soil temperature, moisture, soil structure, and erosion. Mulch contributes to plant growth and, as a result, high crop production by reducing evaporation. Mulching creates an ideal growing climate for plants. Mulching materials are commonly used to help many herbs and tree species to grow them. Mulches have been shown to improve seed germination, freshly grown plant survival, seedling transplantation, and crop plant's overall efficiency compared to non-mulched plants in numerous studies (Table 1). Mulching is advantageous for optimum yield with minimal input resources in this way (Ahmed et al., 2013; Chalker-Scott, 2007; Kader et al., 2019; Kwambe et al., 2015). Plants that are more robust, stronger, and resistant to pest injury are the result of a combination of the above and possibly other factors. Mulched plants, on the other hand, tend to grow and mature more uniformly than un-mulched plants. Plant root growth is stimulated by increased soil temperature and moisture content, resulting in increased plant growth and yield (Zhang et al., 2020). Furthermore, using black plastic mulch, can improve growth, flowering, and tuberous root formation while also reducing weeding in different vegetables (Kader et al., 2019). Furthermore, according to Singh et al. (2019) and Zhang et al. (2020), the application of mulch increases the vegetative growth of pepper and maize crops (Fig. 1).

3 Role of Mulching in Plant Quality and Yield

Mulch can increase not only the vegetative growth of crops but also the quality and yield. Moreover, mulch keeps fruits safe by preventing them from touching the ground and, in many cases, prevents soil rot, fruit cracking, and blossom end rot. During heavy rains, plastic mulches help to prevent soil from splashing onto the crop plants, especially tomatoes, cucumbers, sesame eggplant, and orange (Abdelraouf et al., 2019; Behzadnejad et al., 2020; Král et al., 2019; Shehata et al., 2019; Wang et al., 2019). Straw mulch dramatically improves the quality of early growing cabbage, potatoes, and other off season vegetables. Recent research has shown that

Table 1 Effect of Mulching on plant growth enhancement and stress tolerance

Plant species	Stress type		Influence on plant	Response	References
Common names	Scientific names				
Onion	(*Allium cepa* cv.)	Drought	Increased the leaf number, fresh and dry weight of the plant, plant height, bulb length and diameter, early yield, and final bulb yield	+	Shirzadi et al. (2020)
Sesame	(*Sesamum indicum* L.)	Drought	Improved the photosynthetic pigment and relative water content, Enhanced the catalase, superoxide dismutase activity, proline content, and water use efficiency were increased with increases in water deficiency	+	Behzadnejad et al. (2020)
Sweet pepper	(*Capsicum annuum*)	Cold	Improved leaves membrane stability index and relative water content and helped reduce leaf water loss and electrolyte leakage, indicating better plant physiological responses. Significant improvement in leaf chlorophyll content	+	Singh et al. (2019)
Zinnia	(*Zinnia elegans*)	Drought	Mulching increased the fresh and dry weight of plant and water use efficiency	+	Kazemi and Safari (2018)

(continued)

Table 1 (continued)

Plant species	Stress type		Influence on plant	Response	References
Eggplant	(*Solanum melongena*)	Drought	Mulching increased the carotenoids contents, fruit weight, protein content	+	Rodan et al. (2020)
Potatoes	(*Cyphomandra betacea*)	Drought	Mulch application increased total plant yield	+	Král et al. (2019)
Tamarillo	(*Cyphomandra betacea*)	Heavy metals	Mulching significantly increased the contents of nutrients in both shoot and roots	+	Liu et al. (2018)
Squash	*Cucurbita*	Salinity	Mulching enhanced relative water content, canopy temperature, fruit quality yield, and water use efficiency, Photosynthesis efficiency, total soluble sugars, leaf area index, harvest index, and total yield	+	Abd El-Mageed et al. (2016)
Cotton	(*Gossypium* (*hirsutum* L)	Salinity	Cotton yield and water productivity were increased due to mulching	+	Bezborodov et al. (2010)
Gallant soldier	(*Galinsoga parviflora*)	Heavy metals	Application of mulching improves CD stress tolerance	+	Lin et al. (2014)

(continued)

Table 1 (continued)

Plant species	Stress type		Influence on plant	Response	References
Winter wheat	(*Triticum aestivum*)	Salinity	Plastic film mulch significantly inhibited soil evaporation and salt accumulation, promoted aboveground biomass, and increased grain yield and water use efficiency	+	Zhang et al. (2018)
Sunflower	(*Helianthus*)	Salinity	With mulching, the plant has the highest seed and biomass yield and improved salt tolerance	+	Zhao et al. (2016)
Winter wheat-summer maize	*Triticum aestivum*) and *Zea mays L)*	Salinity	Mulch enhances crop yields and reduces the risk of soil salinization	+	Pang et al. (2010)
Sunflower	(*Helianthus*)	Salinity	It increased plant growth and biomass and effective saline soil management practice	+	Zhao et al. (2014)
Cotton	(*Gossypium hirsutum* L)	Salinity	Mulching substantially reduced Na + accumulation in root and leaf tissues, inhibited peroxidation of lipids, and improved leaf photosynthesis and dry matter production	+	Dong et al. (2008)
Rice	(*Oryza sativa)*	Alkaline Soil	Mulching significantly improved nitrogen availability	+	van Asten et al. (2005)

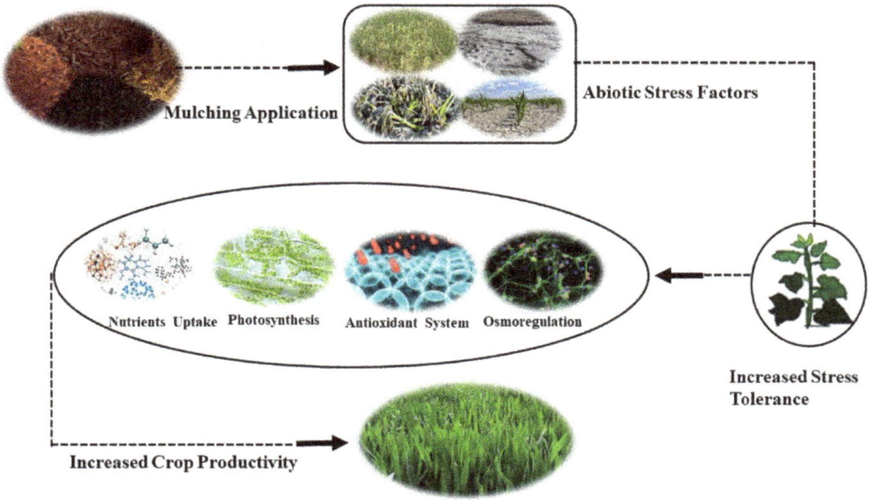

Fig. 1 Schematic representation illustrating the application of mulches to regulate plants morphological and physiological traits under abiotic stress conditions

the combination of high precision planting and straw mulching is an effective way to compensate the losses in wheat yield while maintaining its quality (Ahmad et al., 2015, 2020; Iqbal et al., 2019; Tan et al., 2019). The use of sorghum stover mulching and planting basin tillage is a good technology specially developed for limited crop areas with limited access to animal draft control. This can considerably enhance dry-land sorghum crop yield (Masaka et al., 2019).

In addition to this, the application of straw mulch considerably increases the yield of tomato and okra, respectively (Gupta & Gupta, 1987; Silva & Godawatte, 2016). Recently, Behzadnejad et al. (2020) reported that the application of mulch of about 7.5 tons ha^{-1} improves the morphological features, crop yield, and oil content of sesame. It is further proved that potatoes yield and starch contents increased by 27.9 and 18.18% by applying paddy straw mulches (Dixit & Majumdar, 1995). Zhao et al. (2016) obtained increasing sunflower yield and the average seed yield of about 3198 kg ha^{-1} due to plastic mulch. Additionally, polythene mat, and black mulching also increased the flower yield of the Anna and Sari rose varieties by 29 and 56%, respectively (Rodrigues et al., 1999). Mahato et al. (2020) reported the use of mulching on the growth and development of wood apple; they have observed that under different irrigation levels, the plant with mulching application gave the highest fruit yield fruit per plant by 78.25, yield per plant of 19.01 kg and yield per hectare of 7.60 ton respectively. Gao et al. (2001) found that paper mulching increased the fruit quality and yield of tomatoes. Nagalakshmi et al. (2002) found that comparatively organic mulch, black LLDPE mulch produced the highest number of fruits plant^{-1} such as 97.67, fruit circumference by 3.57 cm, fresh fruit length by 6.93 cm, and yield of chili by 8.60 tons ha^{-1}, respectively. Furthermore, it is reported that the straw mulching gives a good number of branches per plant (8.7), fruit biomass (28.08 g), and total yield (49.63 tons ha^{-1}) as compared to no mulch that was recorded with 8.1, 27.86 g, and 47.85 tons ha^{-1}, respectively in tomato (Gandhi & Bains, 2006). Dubey et al. (2019) found the highest total yield of 28.71 and 29.47, average fruit weight of 43.65 and 44.81 g, marketable fruit yield by 295.69 and 303.54 q ha^{-1}, compared to control with a bio-mulch application, unmarketable fruit yield by 32.85 and 33.72 q ha^{-1} and total yield by 328.54 and 337.26 q ha^{-1}. In comparison to control treatments, Shashidhar et al. (2009) found that paddy straw mulched plots of 15.20 tons ha^{-1} had the highest total leaf yield of mulberry, such as 15.20 and 11.78 tons ha^{-1}.

4 Mulching and Abiotic Stresses

Abiotic factors primarily limit crop productivity (Begum et al., 2019; Zafar et al., 2018). Drought, salinity, flooding, temperature and heavy metal stresses, all have adverse effects on crops (Vaughan et al., 2018; Zafar et al., 2018). Almost 90% of cultivated land is vulnerable to the above mentioned stresses that result in huge yield losses, up to 70% in major crops (Mantri et al., 2012). Tigchelaar et al. (2018) estimated that based on the amalgamation of climate change and its consequences on crop yield models, major crops such as rice, wheat, and maize would lose productivity even further, posing a significant threat to global food security. Increased salinity (37%) was recorded in irrigated lands during the last two decades (199–2013) (Ghassemi et al., 1995; Qadir et al., 2014). Drought stress has become more common and severe due to changes in precipitation patterns and a rise in evapotranspiration caused by global warming (Dai, 2011). According to a recent meta-analysis report, the global average temperature will rise by 2.0–4.9 °C by 2100 (Raftery et al., 2017). Increased heavy metal pollution of arable lands limits crop production while also posing significant health risks to humans (Rehman et al., 2018). Therefore, plants need to be supplied with certain additives to protect the plant from these stress factors and increase plant productivity.

Improving yield in the unfavorable environmental conditions in the prevailing changing climate is a great challenge for growers. Several agronomic methods have been practiced to minimize abiotic stress constraints, among which mulching has been shown to reduce abiotic stress in a variety of crops, including wheat, barley, corn, soybean, strawberry, onion, and squash (Table 1). Mulching plays an important role in conserving soil moisture during dry periods by reducing water loss through evaporation, providing better nutrient availability, avoiding soil erosion, and improving water and fertilizer use efficiency. Mulching also suppresses weed growth and gives nutrients to the plant, and maintains soil temperature. There are many reports regarding the beneficial effect of mulching on plant growth and quality under abiotic stress conditions.

4.1 Mitigation of Drought Stress Through Mulches

Abiotic stresses, such as drought, are the critical challenges for growth and production, leading to huge monetary losses amounting to billions of dollars annually (Pereira, 2016). Drought stress is a major environmental threat that limits crop productivity, and global climate change has a major role in persuading this menace. Drought stress down-regulates the functioning of critical metabolic pathways, including photosynthesis (Chastain et al., 2014), nitrogen assimilation, and osmolyte accumulation by affecting crucial enzymes' functioning (Ahanger & Agarwal, 2017; Khan et al., 2015). Drought stress causes plants to adapt in various ways, including changes in growth patterns, the morphology of plants, and defensive mechanisms (Zandalinas et al., 2018). Mulching has been shown to significantly boost plant protection systems, which aids in the improvement of plant photosynthesis process in crops and, as a result, persuades salinity and drought tolerance in cereals (Baumhardt & Jones, 2002; Zhang et al., 2011). Plastic mulch is increasingly being used to boost plant growth and precocity (Caruso et al., 2019; Sekara et al., 2019) and improving water quality. Shirzadi et al. (2020) recently observed that the water deficit stress declined all vegetative and reproductive growth and yield mechanisms in onion plants; however, mulch and PGRs application increased the onion growth and declined the damaging effect of water stress. Furthermore, the study reported that under severe drought stress, regulators affected increasing water use efficiency and RWC, and compared to the results, mulch and PGRs further increased water use efficiency by 12–14%. Besides this, recently, Behzadnejad et al. (2020) have demonstrated that mulching mediated the enhancement of physiological characteristics, grain yield, and oil content in Sesame (*Sesamum indicum* L.) plant species through up-regulation of the antioxidant system. According to a study comparing the effects of three forms of plastic mulch on cabbage pepper, the transparent mulch had the highest weed population while the dark mulch had the lowest (Ashraf et al., 2010). Mulching has been studied extensively on tomato, wheat, corn, and other crops (Table 1). It has also been stated that using plastic mulch to preserve soil moisture effectively increases growth and yield (Bajguz & Hayat, 2009) and increases water-use quality (Steinmetz et al., 2016).

4.2 Role of Mulches in Mitigation of Salt Stress

Healthy soils will help to ensure food security by providing long-term crop production. However, as a result of anthropogenic perturbations and a variety of other causes, the issue of soil salinity is spreading along a third part of the land surface every day. Soil erosion reduces crop production and lowers soil water holding ability, decreases the soil biodiversity, desertification, and carbon depletion resulting from the misuse and mismanagement of these soils. Irrigating plants with untreated urban waste water containing high salt levels can negatively affect crop growth and production. Furthermore, the widespread use of pesticides, chemical fertilizers, and detergents can significantly increase salt levels in soils (Chalker-Scott, 2007). Mulching has the potential to effectively solve salinity issues by increasing soil moisture retention and decreasing evapotranspiration. Mulching has been shown in many experiments to reduce the toxicity of salts (Yobterik & Timmer, 1994; Zhao et al., 2016). Organic mulches have been found to be more beneficial for soil reclamation and soil desalinization (Pang et al., 2010). Microbes can decompose organic mulches, result in several harmful residues degradation and reducing salt adulteration (Gan et al., 2003).

On the other hand, plastic mulches are not widely used to reduce salt stress (Sun et al., 1994). Mulching can improve crop yield by increasing soil temperature, retaining soil moisture, and decreasing soil evaporation, improving water use efficiency, reducing salt damage to crops, improve physical characteristics of soil, enhancing seed germination and development (Li et al., 2013). Plastic mulch on wheat has been shown in studies to increase yield, minimize water usage, and enhance water-use performance in salinity conditions (Zhang & Yang, 2001). Furthermore, research has shown that straw mulching is a promising management choice for farmers in terms of controlling soil salinity, reducing soil water evaporation, and regulating soil water and salt movement (Pang et al., 2010; Qiao et al., 2006; Table 1).

4.3 Remediation of Heavy Metals by Using Muclhes

Another stress is heavy metal stress that has emerged as a major problem in different terrestrial habitats around the world. By accumulating heavy metals, widespread industrialization has a negative impact on crops and soil (Sharma et al., 2007). Losses to soil texture, e.g., soil pH, soil nutrients, and the build-up of heavy metals, reduce the growth of the affected plants directly/indirectly by affecting them with different physiological activities (Hassan et al., 2017; Panuccio et al., 2009).

It is evident that heavy metals are detrimental to the well-being of both animals and humans. Mulches are an excellent source to remove these metals from soils (Chalker-Scott, 2007). Poplar and eucalyptus leaves are often used to remove heavy metals from soil (Salim & El-Halawa, 2002). Consequently, compost and woodchips can form a complex with copper metal in forest areas and transform it into a non-toxic form for crop plant growth (Kiikkila et al., 2002).

Heavy metals have a great impact on the rhizosphere environment under heavy metal contamination and thus help the plant to flourish (Tatár et al., 1998; Yang et al., 2000). Organic acids and other organic matter derived from straw via decay and decomposition will alter rhizosphere pH, redox potential, and nutrient availability, affecting heavy metal bioavailability in the rhizosphere soil (Blanco-Canqui & Lal, 2007; Ge et al., 2014). As compared to the monitor, the soil pH was increased by *R. sieboldii* and *C. confine* straws but decreased by *P. asiatica* and *M. japonicus* straws in the current pot experiment. According to Rao et al. (2016), with increased straw mulching, total heavy metal concentrations in the LM and HM treatments decreased by 79.90–82.84 and 81.90–90.07%, respectively, and combined total heavy metals decreased by 86.5–87.0 and 90.3–94.6%. As a result, straw mulching on soil may reduce sediment yields as well as the loss of particulate-bound heavy metals, especially Cd and Ni, as well as cumulative total heavy metals in the runoff. As a result, it can be used to monitor heavy-metal-contaminated soil that poses a contamination risk to the ecosystem due to surface runoff.

4.4 Mitigation of Temperature Stress Using Different Mulches

Mulch helps to regulate the temperature of the soil by shading it in summer, making it cooler, and helps to isolate it in the winter from the cold. This systematic effect of temperature promotes plant root growth, preventing soil erosion. Wheat straw mulch elevated the temperature of soil by 2–3 °C in the peak winter season (Sarolia & Bhardwaj, 2012). Further, the studies showed that the temperature of the soil can be increased up to 7 °C under clear mulch than the bare soil in cold weather (Lamont, 2005). Park et al. (1996) observed an increase of 2.4 °C in average soil temperature at 15 cm depth under the transparent film and a rise of 0.8 °C under the black film. At night, strengthening the mulch's underside absorbs the long wave radiation emanated by the soil, thus slowing the soil's cooling (Lamont, 2005).

Mulching influences numerous characteristics of the soil environment and crop requirements, especially in arid and desert regions. Mulches improve the properties and conditions of soil, either directly or indirectly.

Mulching provides many advantages, which include: moisture conservation, soil temperature regulation, reduction in weed-crop interference, and subsequently improved crop yield and quality (Murungu et al., 2011). Retention of soil moisture and increased soil temperatures due to mulch application can also affect soil fertility by influencing the biological activity in soils (Marinari et al., 2015) by increasing the level of enzyme and mineralization rates, making it available to plants. Since straw mulch increases residue accumulation and decreases soil disturbance on the soil surface, it can also preserve soil water and reduce temperature (Zhang et al., 2011).

Key benefits of straw mulch treatment can reduce soil temperature, quick application, day temperature fluctuation mitigation, and soil moisture (Adamchuk et al., 2016; Dudás et al., 2016; Elbl et al., 2014).

The effects on soil temperature, on the other hand, are extremely diverse. Clear plastic mulches increase soil temperature, while white mulches decrease it. According to Chen and Katan (1980), plastic mulching increased soil temperature by 0.9–4.3 °C at the seedling level, 1.6–2.3 °C at the bud initiation stage, and 0.8–1.9 °C at the flowering stage. Duhr and Dubas (1990) discovered that clear, photodegradable polythene film mulching increased soil temperatures by 2.9–3.3 °C. Wheat straw mulch increased soil temperature by 2–3 degrees Celsius (Dayal et al., 1991). Under transparent mulch, the soil temperature can be up to 7 °C higher than bare soil (Ham et al., 1993). Park et al. (1996) found a 2.4 °C increase in average soil temperature at 15 cm depth when using transparent film and a 0.8 °C increase when using black film. Thermometers mounted at the soil surface reported a rise in soil temperatures of 2.8–9.4 and 0.9–7.3 °C at 5 cm depth, according to Choi and Chung (1997). In addition, plastic mulching has numerous agricultural benefits, such as maintaining soil temperature and humidity, preventing soil-borne diseases, fighting soil pests, and accelerating growth. Condensation on the underside of the mulch absorbs the longwave radiation released by the soil at night, delaying its cooling. The soil solarization process is based on clear mulches' ability to generate soil temperatures high enough to suppress weeds, plant pathogens, and nematodes.

4.5 Nutrients Nourishment and Mulches

Mulching has also been related to soil nutrient supply chains, lower soil temperature due to shading, and increased infiltration (Buerkert et al., 2000). After decay, nutrients and organic mulches return organic matter and at the same time improve biological, physical, and chemical properties of the soil and thus increases the yield of the crops. Patil and Singh (1983) found that applying sunflower stover mulch at a 20 tons/ha rate significantly increased N, P, and K uptake compared to not mulching under the hot and dry season. Organic mulches help to preserve soil moisture and improve soil nutrients by adding organic matter (Dilip Kumar et al., 1990). Mulching increased the amount of moisture and organic matter in the soil, creating an ideal atmosphere for root penetration. Crop residue mulches increase the amount of organic carbon in the soil (Duiker & Lal, 1999). By increasing mulches quantity can increase the organic matter (Khurshid et al., 2006; Lal et al., 1980). Increased organic carbon was recorded through crop residues (Ghuman & Sur, 2001).

Liu et al. (2018) showed that the contents of phosphorus and potassium in shoots of *C. betacea* seedlings were higher than those of control after mulching four tolerant plant straw, indicating that mulching tolerant plant straw promoted the translocation of phosphorus and potassium to the shoots of *C. betacea* seedlings. Further, Shehata et al. (2019) reported that polyethylene mulch significantly enhanced the contents of nitrogen, phosphorus, and potassium.

5 Conclusion and Future Prospects

This book chapter concluded that different abiotic factors declined the plant's growth and yield; however, the application of mulches' mitigated the adverse effects of the referred environmental stresses and thus helped in crop production. In the current review, we have presented; the efficacy of mulching on soil status, plant growth, yield, conservation of soil moisture content, and its beneficial effect on nutrient addition and stress tolerance modulation. From different studies, it is concluded that mulching is not only valuable entities for the enhancement of soil temperature, weed pressure reduction, moisture conservation, insect pests reduction, the addition of nutrients in the soil, higher crop yields, but it can also efficiently protect plants from numerous abiotic factors, like salinity, drought, nutrient stress, and extreme temperatures, heavy metals and thus helping to increase per hectare yield of different crops and vegetable world-wide. Encouragement of the use of mulching is critical for the long-term viability of modern global agricultural systems. The future research primary focus in this field should be on identifying genes that regulate mulching-mediated growth and productivity regulation under stressful conditions and the main cellular and metabolic pathways under various environmental stresses. Understanding the modulations in tolerance mechanisms caused by mulching and the crosstalk activated to control plant output can help increase crop productivity. Mulches must be investigated at all levels to learn more about their function in nature as an eco-friendly agricultural practice for long-term agricultural production.

References

Abd El-Mageed, T. A., Semida, W. M., & Abd El-Wahed, M. H. (2016). Effect of mulching on plant water status, soil salinity and yield of squash under summer-fall deficit irrigation in salt affected soil. *Agricultural Water Management, 173*, 1–12. https://doi.org/10.1016/j.agwat.2016.04.025.

Abdelraouf, R. E., Azab, A., Tarabye, H. H. H., & Refaie, K. M. (2019). Effect of pulse drip irrigation and organic mulching by rice straw on yield, water productivity and quality of orange under sandy soils conditions. *Plant Archives, 19*, 2613–2621.

Abdrabbo, M. A. A., Saleh, S. M., & Hashem, F. A. (2017). Eggplant production under deficit irrigation and polyethylene mulch. Egypt. *Journal of Applied Sciences, 32.*

Adamchuk, V., Prysyazhnyi, V., Ivanovs, S., & Bulgakov, V. (2016). Investigations in technological method of growing potatoes under mulch of straw and its effect on the yield. In *Proceedings of the 15th International Scientific Conference Engineering for Rural Development* (pp. 1098–1103).

Ahanger, M. A., & Agarwal, R. M. (2017). Salinity stress induced alterations in antioxidant metabolism and nitrogen assimilation in wheat (*Triticum aestivum* L) as influenced by potassium supplementation. *Plant Physiology Biochemistry, 115*, 449–460. https://doi.org/10.1016/j.plaphy.2017.04.017.

Ahmad, S., Raza, M. A. S., Saleem, M. F., Zahra, S. S., Khan, I. H., Ali, M., Shahid, A. M., Iqbal, R., & Zaheer, M. S. (2015). Mulching strategies for weeds control and water conservation in cotton. *ARPN Journal of Agricultural and Biological Science, 8*, 299–306.

Ahmad, S., Raza, M. A. S., Saleem, M. F., Zaheer, M. S., Iqbal, R., Haider, I., et al. (2020). Significance of partial root zone drying and mulches for water saving and weed suppression in wheat. *Journal of Animal and Plant Sciences, 30*, 154–162.

Ahmed, M., Baiyeri, K. P., & Echezona, B. C. (2013). Effect of coloured polyethylene mulch and harvesting stage on growth and yield of industrial sugarcane in Nigeria. *African Journal of Biotechnology, 12*, 10–18.

Ashraf, M., Akram, N., Arteca, R., & Foolad, M. (2010). The physiological, biochemical and molecular roles of brassinosteroids and salicylic acid in plant processes and salt tolerance. *Critical Reviews in Plant Sciences, 29*(3), 162–190. https://doi.org/10.1080/07352689.2010.483580.

Bajguz, A., & Hayat, S. (2009). Effects of brassinosteroids on the plant responses to environmental stresses. *Plant Physiology and Biochemistry, 47*(1), 1–8. https://doi.org/10.1016/j.plaphy.2008.10.002.

Baumhardt, R. L., & Jones, O. R. (2002). Residue management and tillage effects on soil-water storage and grain yield of dry land wheat and sorghum for a clay loam in Texas. *Soil and Tillage Research, 68*, 71–82.

Begum, N., Qin, C., Ahanger, M. A., et al. (2019). Role of arbuscular mycorrhizal fungi in plant growth regulation: Implications in abiotic stress tolerance. *Frontiers in Plant Science, 19*(10), 1068. https://doi.org/10.3389/fpls.2019.01068.

Behzadnejad, J., Tahmasebi-Sarvestani, Z., Aein, A., & Mokhtassi-Bidgoli, A. (2020). Wheat straw mulching helps improve yield in sesame (*Sesamum indicum* L.) under drought stress. *International Journal of Plant Production, 14*, 389–400. https://doi.org/10.1007/s42106-020-00091-8.

Bezborodov, G. A., Shadmanov, D. K., Mirhashimov, R. T., et al. (2010). Mulching and water quality effects on soil salinity and sodicity dynamics and cotton productivity in Central Asia. *Agriculture, Ecosystems & Environment, 138*, 95–102. https://doi.org/10.1016/j.agee.2010.04.005.

Blanco-Canqui, H., & Lal, R. (2007). Soil structure and organic carbon relationships following 10 years of wheat straw management in no-till. *Soil and Tillage Research, 95*(1–2), 240–254. 32.

Bot, A., & Benites, J. (2005). The importance of soil organic matter. Key to drought-resistant soil and sustained food roduction. *FAO Soils Bulletin, 80*.

Buerkert, A., Bationo, A., & Dossa, K. (2000). Mechanisms of residue mulch-induced cereal growth increases in West Africa. *Soil Science Society of America Journal, 64*, 346–358.

Caruso, G., Stoleru, V., De Pascale, S., Cozzolino, E., Pannico, A., Giordano, M., Teliban, G., Cuciniello, A., & Rouphael, Y. (2019). Production, leaf quality and antioxidants of perennial wall rocket as affected by crop cycle and mulching type. *Agronomy, 9*(4), 194. https://doi.org/10.3390/agronomy9040194.

Chalker-Scott, L. (2007). Impact of mulches on landscape plants and the environment—A review. *Journal of Environmental Horticulture, 25*, 239–249.

Chastain, D. R., Snider, J. L., Collins, G. D., Perry, C. D., Whitaker, J., & Byrd, S. A. (2014). Water deficit in field-grown *Gossypium hirsutum* primarily limits net photosynthesis by decreasing stomatal conductance, increasing photorespiration, and increasing the ratio of dark respiration to gross photosynthesis. *Journal of Plant Physiology, 171*, 1576–1585.

Chen, Y., & Katan, J. (1980). Effect of solar heating of soils by transparent polyethylene mulching on their chemical properties. *Soil Science, 130*, 271–277.

Choi, B. H., & Chung, K. Y. (1997). Effect of polythene-mulching on flowering and yield of groundnut in Korea. *International Arachis Newsletter, 17*, 49–51.

Dai, A. (2011). Drought under global warming: A review. *Wiley Interdisciplinary Review of Climate Change, 2*, 45–65.

Dayal, D., Naik, P. R., & Dongre, B. N. (1991). Effect of mulching on soil temperature and groundnut yield during rabi-summer season. *Groundnut News, 3*, 4–5.

Dilipkumar, G., Sachin, S. S., & Rajesh, K. (1990). Importance of mulch in crop production. *Indian Journal of Soil Conservation, 18*, 20–26.

Dixit, C. K., & Majumdar, A. M. (1995). Effect of mulching on water use efficiency, yield and quality of potato (*Solanum tuberosum* L.). *Haryana Journal of Horticultural Science., 24*, 286–291.

Dong, H., Li, W., Tang, W., & Zhang, D. (2008). Furrow seeding with plastic mulching increases stand establishment and lint yield of cotton in a saline field. *Agronomy Journal, 100*, 1640–1646. https://doi.org/10.2134/agronj2008.0074.

Dubey, A. K., Tomar, S., & Tripathi, V. K. (2019). Effects of transplanting schedule and types of mulching on yield and quality of tomato (*Lycopersicon esculentum* Mill.). *Progressive Horticulture, 51*, 177. https://doi.org/10.5958/2249-5258.2019.00028.9.

Dudás, P., Menyhárt, L., Gedeon, C., Ambrus, G., & Tóth, F. (2016). The effect of hay mulching on soil temperature and the abundance and diversity of soil-dwelling arthropods in potato fields. *European Journal of Entomology, 113*, 456–461.

Duhr, E., & Dubas, A. (1990). Effect of covering the soil with plastic film on the dynamics of plant development and yield of maize sown on different dates. *Prace Komisji Nauk Rolniczych Ikomisji Nauk Lesnych, 69*, 9–18.
Duiker, S. W., & Lal, R. (1999). Crop residue and tillage effects on carbon sequestration in a Luvisol in central Ohio. *Soil and Tillage Research, 52*, 73–81.
Elbl, J., Plošek, L., Kintl, A., Hynšt, J., Záhora, J., Javoreková, S., Charou-sová, I., Kalhotka, L., & Urbánková, O. (2014). Effects of drought on microbial activity in rhizosphere, soil hydrophobicity and leaching of mineral nitrogen from arable soil depending on method of fertilization. *International Journal of Agricultural and Biosystems Engineering, 8*, 844–850.
Erenstein, O. (2002). Crop residue mulching in tropical and semi-tropical countries: An evaluation of residue availability and other technological implications. *Soil and Tillage Research, 67*, 115–133.
Gan, J., Zhu, Y., Wilen, C., Pittenger, D., & Crowley, D. (2003). Effect of planting covers on herbicide persistence in landscape soils. *Environmental Science & Technology, 37*, 2775–2779.
Gandhi, N., & Bains, G. S. (2006). Effect of mulching and date of transplanting on yield contributing characters of tomato. *PAU Agricultural Research Journal, 43*, 6–9.
Gangaiah, B., Latha, P. C., Singh, T. V., & Rao, P. R. (2019). Rice cultivation using plastic mulch under saturated moisture regime and its implications on weed management, water saving, productivity and profitability. *Indian Journal of Weed Science, 51*, 198–202.
Gao, G. X., Jin, L. Z., Guo, F. C., Gu, Z. H., Yu, Y., & Chen, B. (2001). Effect of nutrient paper mulching on tomato (*Lycopersicon esculantum*) cultivation. *China Vegetables, 6*, 6–7.
Ge, X., Wei, S. Y., Guo, H. N., Chu, Y. C., Ding, J., et al. (2014). Variation of humus in content during composting and its influence on heavy metal distribution. *Journal of Ecology and Rural Environment, 30*(3), 369–373.
Ghassemi, F., Jakeman, A. J., & Nix, H. A. (1995). *Salinisation of land and water resources: human causes, extent, management and case studies*. Sydney, Australia, and CAB International, UNSW Press.
Ghuman, B. S., & Sur, H. S. (2001). Tillage and residue management effects on soil properties and yields of rainfed maize and wheat in a sub humid subtropical climate. *Soil and Tillage Research, 58*, 1–10.
Gupta, J. P., & Gupta, G. N. (1987). Response of tomato and okra crops to irrigation and mulch in arid region of India. *Agrochemica, 31*, 183–202.
Ham, J. M., Kluitenberg, G. J., & Lamont, W. J. (1993). Optical properties of plastic mulches affects the field temperature regime. *Journal of the American Society for Horticultural Science, 118*, 188–193.
Hassan, T. U., Bano, A., & Naz, I. (2017). Alleviation of heavy metals toxicity by the application of plant growth promoting rhizobacteria and effects on wheat grown in saline sodic field. *International Journal of Phytoremediation, 19*, 522–529. https://doi.org/10.1080/15226514.2016.1267696.
Iqbal, R., Muhammad, A. S. R., Muhammad, F. S., Imran, H. K., Salman, A., Muhammad, S. Z., Muhammad, U., & Imran, H. (2019). Physiological and biochemical appraisal for mulching and partial rhizosphere drying of cotton. *Journal of Arid Land, 11*, 785–794.
Jacks, C. V., Brind, W. D., & Smith, R. (1995). Mulching technology comm., no. 49, common wealth. *Bulletin of Soil Science*, 118.
Kader, M. A., Singha, A., Begum, M. A., Jewel, A., Khan, F. H., & Khan, N. I. (2019). Mulching as water-saving technique in dry land agriculture. *Bulletin of the National Research Centre, 43*, 1–6.
Kannan, R., Solaimalai, A., Anandan, P., & Suthin, R. T. (2020). Uses of mulching in agriculture: A review (pp. 01–32). https://doi.org/10.22271/ed.book.733.
Kazemi, F., & Safari, N. (2018). Effect of mulches on some characteristics of a drought tolerant flowering plant for urban landscaping. *Desert (biaban), 23*, 75–84. https://doi.org/10.22059/jdesert.2018.66353.
Khan, M. I. R., Nazir, F., Asgher, M., Per, T. S., & Khan, N. A. (2015). Selenium and sulfur influence ethylene formation and alleviate cadmium-induced oxidative stress by improving proline and

glutathione production in wheat. *Journal of Plant Physiology, 173*, 9–18. https://doi.org/10.1016/j.jplph.2014.09.011.
Khurshid, K., Iqbal, M., Arif, M. S., & Nawaz, A. (2006). Effect of tillage and mulch on soil physical properties and growth of maize. *International Journal of Agriculture & Biology, 8*, 593–596.
Kiikkila, O., Derome, J., Brugger, T., Uhlig, C., & Fritze, H. (2002). Copper mobility and toxicity of soil percolation water to bacteria in metal polluted forest soil. *Plant and Soil, 238*, 273–280.
Král, M., Dvořák, P., & Capouchová, I. (2019). The straw as mulch and compost as a tool for mitigation of drought impacts in the potatoes cultivation. *Plant, Soil & Environoment, 65*, 530–535. https://doi.org/10.17221/493/2019-PSE.
Kwambe, X. M., Masarirambi, M. T., Wahome, P. K., & Oseni, T. O. (2015). The effects of organic and inorganic mulches on growth and yield of green bean (*Phaseolus vulgaris* L.) in a semi-arid environment. *Agriculture and Biology Journal of North America, 6*, 81–89.
Lal, R., Vleeschauwer, D. D., & Nganje, R. M. (1980). Changes in properties of a newly cleared tropical Alfisols as affected by mulching. *Soil Science Society of American Journal, 44*, 827–833.
Lamont, W. J. (2005). Plastics: Modifying the microclimate for the production of vegetable crops. *Hort Technology, 15*, 477–481.
Li, S. X., Wang, Z. H., Li, S. Q., Gao, Y. J., & Tian, X. H. (2013). Effect of plastic sheet mulch, wheat straw mulch, and maize growth on water loss by evaporation in dryland areas of China. *Agriculture Water Management, 116*, 39–49.
Lin, L., Liao, M., Ren, Y., et al. (2014). Effects of mulching tolerant plant straw on soil surface on growth and cadmium accumulation of galinsoga parviflora. *PLoS ONE, 9*, 1–13. https://doi.org/10.1371/journal.pone.0114957.
Liu, P., Lian, H., Li, H., & Lin, L. (2018). Effects of mulching tolerance plant straw on phosphorus and potassium uptakes of cyphomandra betacea seedlings under cadmium. In *2018 7th International Conference on Energy and Environmental Protection (ICEEP 2018).*
Mahato, S., Ghosh, S., & Ghosh, S. N. (2020). Effect of Mulching and Supplemented Irrigation on Growth, Yield and Fruit Quality of Wood Apple (*Feronia limonia Swingle*). *International Journal of Current Microbiology and Applied Sciences, 9*, 2479–2484. https://doi.org/10.20546/ijcmas.2020.910.297.
Mantri, N., Patade, V., Penna, S., Ford, R., & Pang, E. (2012). Abiotic stress responses in plants: present and future. In *Abiotic stress responses in plants* (pp. 1–19). Springer.
Marinari, S., Mancinelli, R., Brunetti, P., & Campiglia, E. (2015). Soil quality, microbial functions and tomato yield under cover crop mulching in the Mediterranean environment. *Soil and Tillage Research, 145*, 20–28.
Masaka, J., Dera, J., & Muringaniza, K. (2019). Dry land grain Sorghum (Sorghum bicolor) yield and yield component responses to tillage and mulch practices under subtropical African conditions. *Agricultural Research*, 1–9.
Massa, D., Benvenuti, S., Cacini, S., Lazzereschi, S., & Burchi, G. (2019). Effect of hydro-compacting organic mulch on weed control and crop performance in the cultivation of three container-grown ornamental shrubs: Old solutions meet new insights. *Scientia Horticulturae, 252*, 260–267.
Merfield, C. (2002). Organic weed management: A practical guide (pp. 16–30). Lincoln University. http://researcharchive.lincoln.ac.nz/bitstream/10182/4902/1/Merfield_organic_weed_2002.pdf.
Murungu, F. S., Chiduza, C., Muchaonyerwa, P., & Mnkeni, P. N. S. (2011). Mulch effects on soil moisture and nitrogen, weed growth and irrigated maize productivity in a warm-temperate climate of South Africa. *Soil and Tillage Research, 112*(1), 58–65.
Nagalakshmi, S., Palanisamy, D., Eswaran, S., & Sreenarayanan, V. V. (2002). Influence of plastic mulching on chilli yield and economics. *South Indian Horticulture, 50*, 262–265.
Pang, H., Li, Y., Yang, J., & Liang, Y. (2010). Effect of brackish water irrigation and straw mulching on soil salinity and crop yields under monsoonal climatic conditions. *Agricultural Water Management, 97*, 1971–1977. https://doi.org/10.1016/j.agwat.2009.08.020.

Panuccio, M. R., Sorgona, A., Rizzo, M., & Cacco, G. (2009). Cadmium adsorption on vermiculite, zeolite and pumice: Batch experiment studies. *Journal of Environmental Management, 90*, 364–374. https://doi.org/10.1016/j.jenvman.2007.10.005.

Park, K. Y., Kim, S. D., Lee, S. H., Kim, H. S., & Hong, E. H. (1996). Differences in dry matter accumulation and leaf area in summer soybeans as affected by polythene film mulching. *RDA Journal of Agricultural Sciences, 38*, 173–179.

Patil, J. C., & Singh, M. C. (1983). Yield and nutrient uptake of sunflower (*H. annuus* L.) as affected by irrigation, mulch and cycocel. *Indian Journal of Agronomy, 28*, 205–210.

Pereira, A. (2016). Plant abiotic stress challenges from the changing environment. *Frontier in Plant Science, 7*, https://doi.org/10.3389/fpls.2016.01123.

Qadir, M., Quillérou, E., Nangia, V., Murtaza, G., Singh, M., Thomas, R. J., et al. (2014). Economics of salt-induced land degradation and restoration. *Natural Resources Forum*, 282–295 (Wiley Online Library). https://doi.org/10.1111/1477-8947.12054.

Qi, Y., Ossowicki, A., Yang, X., Lwanga, E. H., Dini-Andreote, F., Geissen, V., & Garbeva, P. (2020). Effects of plastic mulch film residues on wheat rhizosphere and soil properties. *Journal of Hazardous Materials, 387*, 121711.

Qiao, H., Liu, X., Li, W., Huang, W., Li, C., & Li, Z. (2006). Effect of deep straw mulching on soil water and salt movement and wheat growth. *Chinese Journal of Soil Science, 37*, 885–889.

Raftery, A. E., Zimmer, A., Frierson, D. M. W., Startz, R., & Liu, P. (2017). Less than 2 °C warming by 2100 unlikely. *Nature Climate Change, 7*, 637. https://doi.org/10.1038/nclimate3352.

Rao, Z. X., Huang, D. Y., Zhu, H. H., Zhu, Q. H., Wang, J. Y., Luo, Z. C., et al. (2016). Effect of rice straw mulching on migration and transportation of cd, cu, zn, and ni in surface runoff under simulated rainfall. *Journal of Soils and Sediments, 16*(8), 2021–2029.

Rathore, A. L., Pal, A. R., & Sahu, K. K. (1998). Tillage and mulching effects on water use, root growth, and yield of rain-fed mustard and chickpea grown after lowland rice. *Journal of the Science of Food and Agriculture, 78*, 149–161.

Rehman, K., Fatima, F., Waheed, I., & Akash, M. S. H. (2018). Prevalence of exposure of heavy metals and their impact on health consequences. *Journal of Cellular Biochemistry, 119*(1), 157–184.

Rodan, M. A., Hassandokht, M. R., Sadeghzadeh-Ahari, D., & Mousavi, A. (2020). Mitigation of drought stress in eggplant by date straw and plastic mulches. *Journal of the Saudi Society of Agricultural Sciences, 19*, 492–498. https://doi.org/10.1016/j.jssas.2020.09.006.

Rodrigues, E. J. R., Minami, K., & Farina, E. (1999). Mulching in soilless systems of the rose crop: Productivity, water consumption, temperature and solarisation. *Scientia Agricola, 56*, 785–795.

Sabatino, L., Iapichino, G., Vetrano, F., Moncada, A., Miceli, A., De Pasquale, C., D'Anna, F., & Giurgiulescu, L. (2018). Effects of polyethylene and biodegradable starch-based mulching films on eggplant production in a Mediterranean area Carpathian. *Journal of Food Science and Technology, 10*(3), 81–89.

Salim, R., & El-Halawa, R. A. (2002). Efficiency of dry plant leaves (mulch) for removal of lead, cadmium and copper from aqueous solutions. *Process Safety and Environmental Protection, 80*, 270–276.

Sarolia, D. K., & Bhardwaj, R. L. (2012). Effect of mulching on crop production under rainfed condition: A Review. *International Journal of Research in Chemistry and Environment, 2*, 8–20.

Sekara, A., Pokluda, R., Cozzolino, E., Del Piano, L., Cuciniello, A., & Caruso, G. (2019). Plant growth, yield, and fruit quality of tomato affected by biodegradable and non-degradable mulches. *Horticultural Science, 46*(3), 138–145. https://doi.org/10.17221/218/2017-HORTSCI.

Shan, S., Tang, W., Peng, X, et al. (2020). Effects of DA-6 on phosphorus and potassium uptakes of tomato seedlings under cadmium stress. In *E3S Web of Conferences* (Vol. 165, pp. 1694–1697). https://doi.org/10.1051/e3sconf/202016502003.

Sharma, R. K., Agrawal, M., & Marshall, F. (2007). Heavy metal contamination of soil and vegetables in suburban areas of Varanasi, India. *Ecotoxicology and Environmental Safety, 66*, 258–266.

Shashidhar, K. R., Bhaskar, R. N., Priyadharshini, P., & Chandrakumar, H. L. (2009). Effect of different organic mulches on pH, organic carbon content and microbial status of soil and its influence on leaf yield of M-5 mulberry (*Morus indica* L.) under rainfed condition. *Current Biotica, 2*, 405–412.

Shehata, M., Abd El-Hady, M., & El-Magawry, N. (2019). Effect of irrigation, some plant nutrients with mulching on growth and productivity of cucumber. *Journal of Plant Production, 10*, 231–239. https://doi.org/10.21608/jpp.2019.36253.

Shirzadi, M. H., Arvin, M. J., Abootalebi, A., & Hasandokht, M. R. (2020). Effect of nylon mulch and some plant growth regulators on water use efficiency and some quantitative traits in onion (*Allium cepa* cv.) under water deficit stress. *Cogent Food Agriculture, 6*, 1779–2562. https://doi.org/10.1080/23311932.2020.1779562.

Shojaei, S., Ardakani, M. A. H., Sodaiezadeh, H., Jafari, M., & Afzali, S. F. (2019). Optimization of parameters affecting organic mulch test to control erosion. *Journal of Environmental Management, 249*, 109414.

Silva, C. S. D., & Godawatte, V. N. A. (2016). Impact of different mulches on growth and yield of red Okra (*abelmoschus esculentus*) indigenous variety exposed to temperature stress. *Open University of Sri Lanka Journal, 10*, 41. https://doi.org/10.4038/ouslj.v10i0.7334.

Singh, R. K., Acharya, S., & Chaurasia, O. P. (2019). Effects of mulching and zinc on physiological responses and yield of sweet pepper (*Capsicum annuum*) under high altitude cold desert condition. *Indian Journal of Agricultural Sciences, 89*, 300–306.

Steinmetz, Z., Wollmann, C., Schaefer, M., Buchmann, C., David, J., Troger, J., Munoz, K., Fror, O., & Schaumann, G. L. (2016). Plastic mulching in agriculture. Trading short-term agronomic benefits for long-term soil degradation. *Science of the Total Environment, 550*, 690–705. https://doi.org/10.1016/j.scitotenv.2016.01.153.

Sun, D., Dickinson, G., & Bragg, A. (1994). The establishment of Eucalyptus camaldulensis on a tropical saline site in north Queensland. Australia. *Agriculture, Ecosytem and Environment, 48*, 1–8.

Tan, D., Fan, Y., Liu, J., Zhao, J., Ma, Y., & Li, Q. (2019). Winter wheat grain yield and quality response to straw mulching and planting pattern. *Agricultural Research*, 1–5.

Tatár, E., Mihucz, V. G., Varga, A., et al. (1998). Determination of organic acids in xylem sap of cucumber: Effect of lead contamination. *Microchemical Journal, 58*, 306–314.

Tigchelaar, M., Battisti, D. S., Naylor, R. L., & Ray, D. K. (2018). Future warming increases probability of globally synchronized maize production shocks. *Proceedings of the National Academy of Sciences of the United States of America, 115*, 6644–6649. https://doi.org/10.1073/pnas.1718031115.

Todd-Searle, J., Friedrich, L. M., Oni, R. A., Shenge, K., LeJeune, J. T., Micallef, S. A., Danyluk, M. D., & Schaffner, D. W. (2020). Quantification of Salmonella enterica transfer between tomatoes, soil, and plastic mulch. *International Journal of Food Microbiology, 316*, 108–480.

van Asten, PJA., van, Bodegom. PM., Mulder, LM., Kropff, MJ. (2005). Effect of straw application on rice yields and nutrient availability on an alkaline and a pH-neutral soil in a Sahelian irrigation scheme. *Nutrient Cycling in Agroecosystems*, 72(3), 255–266.

Vaughan, M. M., Block, A., Christensen, S. A., Allen, L. H., & Schmelz, E. A. (2018). The effects of climate change associated abiotic stresses on maize phytochemical defenses. *Phytochemistry Reviews, 17*, 37–49. https://doi.org/10.1007/s11101-017-9508-2.

Wang, L., Coulter, J. A., Palta, J. A., et al. (2019). Mulching-induced changes in tuber yield and nitrogen use efficiency in potato in China: A meta-analysis. *Agronomy, 9*(12), 7–93. https://doi.org/10.3390/agronomy9120793.

Yang, Y. Y., Jung, J. Y., Song, W. Y., Suh, H. S., & Lee, Y. (2000). Identification of rice varieties with high tolerance or sensitivity to lead, and characterization of the mechanism of tolerance. *Plant Physiology, 124*(3), 1019–1026.

Yobterik, A. C., Timmer, V. R. (1994). Nitrogen mineralization of agroforestry tree mulches under saline soil conditions. In R. B. Bryan (Ed.) *Advances in Geo-ecology, 27*, 181–194.

Zafar, S. A., Noor, M. A., Waqas, M. A., Wang, X., Shaheen, T., & Raza, M. (2018). Temperature extremes in cotton production and mitigation strategies. In *Past, present and future trends in cotton breeding*. IntechOpen.
Zandalinas, S. I., Balfagón, D., Arbona, V., & Gómez-Cadenas, A. (2018). Modulation of antioxidant defense system is associated with combined drought and heat stress tolerance in citrus. *Frontier in Plant Science, 8*, 9–53.
Zhang, B. J., & Yang, W. P. (2001). Studies on the dynamic change of soil water of dibbling wheat in film-mulched field. *Journal of Northwest Science Technology University of Agriculture and Forestry, 29*, 70–73.
Zhang, S. L., Lövdahl, H., Grip, Y. A., Shuqin, W., Wei, H., Shiping, L., & Shuhui, L. (2011). Salt distribution and the growth of cotton under different drip irrigation regimes in a saline area. *Agriculture Water Management, 100*, 58–69.
Zhang, M., Dong, B., Qiao, Y., et al. (2018). Effects of sub-soil plastic film mulch on soil water and salt content and water utilization by winter wheat under different soil salinities. *Field Crop Research, 225*, 130–140. https://doi.org/10.1016/j.fcr.2018.06.010.
Zhang, L., Meng, Y., Li, S., & Yue, S. (2020). Film mulching optimizes the early root and shoot development of rain-fed spring maize. *Agronomy Journal, 112*, 309–326. https://doi.org/10.1002/agj2.20039.
Zhao, Y., Pang, H., Wang, J., et al. (2014). Effects of straw mulch and buried straw on soil moisture and salinity in relation to sunflower growth and yield. *Field Crop Research, 161*, 16–25. https://doi.org/10.1016/j.fcr.2014.02.006.
Zhao, Y., Li, Y., Wang, J., et al. (2016). Buried straw layer plus plastic mulching reduces soil salinity and increases sunflower yield in saline soils. *Soil and Tillage Research, 155*, 363–370. https://doi.org/10.1016/j.still.2015.08.019.

Living Mulches for Sustainable Pest Management

Hina Gul, Arzlan Abbas, Farman Ullah, Nicolas Desneux, Kaleem Tariq, Asad Ali, and Xiaoxia Liu

Abstract The need for non-chemical insect pest management is the demand of present-day lifestyle emphasized by the demand for organic food production and environmental degradation caused by synthetic insect control measures. Mulches are a well-known environmental diversification technique that has been used in agriculture for decades. They have become commonplace in cultivating a wide range of crops worldwide, gaining significance in organic systems as a sustainable management strategy for controlling weeds and pests and providing other ecological benefits. Mulches can also improve the soil's structure, porosity, and fertility, making it more suppressive and difficult for insect pests to survive. Mulches come in various shapes and compositions, including organic, non-organic, alive, synthetic, dead, biodegradable, and non-biodegradable mulches. In this chapter, we discussed the potential for using live mulches in biological pest management. Moreover, we discussed the

H. Gul · X. Liu (✉)
MOA Key Laboratory of Pest Monitoring and Green Management, Department of Entomology, College of Plant Protection, China Agricultural University, Beijing, China
e-mail: liuxiaoxia611@cau.edu.cn

H. Gul
e-mail: gulhina@cau.edu.cn

A. Abbas
Department of Entomology, College of Plant Protection, Jilin Agricultural University, Changchun 130118, PR China

F. Ullah
Department of Plant Biosecurity, College of Plant Protection, China Agricultural University, Beijing 100193, China
e-mail: farmanullah@cau.edu.cn

N. Desneux
Université Côte d'Azur, INRAE, CNRS, UMR ISA, 06000 Nice, France
e-mail: nicolas.desneux@inrae.fr

K. Tariq · A. Ali (✉)
Department of Entomology, Abdul Wali Khan University Mardan, Khyber Pakhtunkhwa, Pakistan
e-mail: drasadali@awkum.edu.pk

K. Tariq
e-mail: drkaleem@awkum.edu.pk

K. Akhtar et al. (eds.), *Mulching in Agroecosystems*,
https://doi.org/10.1007/978-981-19-6410-7_8

problems caused by living mulches, such as insect pest infestation, weed infestation, disease transmission, and bird infestations.

Keywords Integrated pest management · Insect pest control · Mulches · Agriculture

1 Introduction

The important component of the sustainable ecosystems is biodiversity (Pimentel, 1961), The management of insects depends on crop modification. The non-host plant supported the susceptibility of plants, and therefore host plant density is decline, while enhance the occurrence of natural enemies (Andow, 1991). Sans and Altieri (2005), and Jones and Sieving (2006) reported that weed control, plant pest and diseases could significantly impacted by living mulch and intercropping. Further, in the north part of USA a case study reported that significant lower cucumber beetle were noted on the leaves if Zucchini plants in the interplanted plots compared with fallow plots. The report of Schmidt et al. (2007) revealed the higher rate of natural enemies with a decrease in Aphis glycine's in a soybean field under alfalfa living mulch conditions. While Frank and Liburd (2005) reported that compared with mulch treated plots, living mulch plots have less adult whiteflies and aphids.

Mulches are known for their numerous benefits, for instance, improving grain yield and consistency, enhancing soil quality, soil water retention, and weed suppression (Jabran & Chauhan, 2018; Nawaz et al., 2017). Mulches can also play a vital role in managing insect pests and disease pathogens (Brown & Tworkoski, 2004; Farooq et al., 2011). In traditional and organic farming, common mulches are often observed to combat insect pests and disease pathogens (Quintanilla-Tornel et al., 2016). There are sufficient evidence and proof that mulches effectively manage target insects without chemical applications (Brown & Tworkoski, 2004). Because the misapplication and/or degradation (after initial applications) of chemical insecticides causes long term sublethal effects (Desneux et al., 2007; Ullah et al., 2019b, 2019c), hormesis effects (Cutler et al., 2022; Ullah et al., 2019a, b; 2020c) and facilitate resistance development in insects (Ullah et al., 2020a, 2020b). Organic agricultural operations utilize pest-control tactics that eliminate or minimize the use of high-risk synthetic pesticides and benefit from IPM programs (Nottingham et al., 2016). The core object of these approaches is to diminish pests' impact, interrupt the progress of insecticide resistance, reduce the threats to human health, and eliminate air emissions (Pedigo & Rice, 2009).

Mulches are a well-known environmental diversification technique that alters the soil habitats of arthropods and other organisms, affecting crop yields and other variables. Mulches come in a variety of shapes and compositions, including organic, non-organic, alive, synthetic, dead, biodegradable, and non-biodegradable mulches. Insect species and crop loss are partially decreased owing to increased natural enemy behavior (Schmidt et al., 2004; Thomson & Hoffmann, 2007). Pests and their associated damage are increased due to the pest's increased micro-environment, interfering with natural enemy operation. Mulching is the technique of covering the soil surface with mulch to reduce moisture loss, weed development, and insect pests and disease pathogens (Mulvaney et al., 2011). It is also considered one of the most efficient means of providing shelter for predatory insects (Brown & Tworkoski, 2004; Gill & McSorley, 2010; Johnson et al., 2004). Furthermore, pine bark mulch (*Panax quinquefolius* L.) has been observed due to an increase in the control of weeds and diseases in ginseng (Reeleder et al., 2004). Mulches can significantly promote plants ability to tolerate the attack by insect pests and help them preserve the temperature and soil moisture, which is vital for plant strength (Johnson et al., 2004). Against certain plant species, mulch residues may have allelopathic efficiency. Weston et al. (1989) observed that sorghum (*Sorghum bicolor* (L.) Moench) has the efficiency to control weeds and also has allelopathic potential. Adler and Chase (2007) reported that Allelopathic residues of cowpea (*Vigna unguiculata* (L.) Walp) impaired the weed species, including *Amaranthus lividus* L. (livid amaranth) and *Amaranthus hydridus* L (smooth amaranth). It is necessary to produce chemical mulches from cotton, seed residues, hay, pine needles, shredded bark, or other readily available plant materials (Campiglia et al., 2010). Cultural management strategies, which include mulches and cover crops, are non-chemical approaches to controlling various pest species. Wheat (*Triticum aestivum* L) is a winter cover crop that reduces the number of insects, including Aphid (Aphididae), thrips (Thysanoptera), and plant bugs (Miridae), and leafhoppers (Cicadellidae) (Tremelling et al., 2002). In contrast to broccoli (*Brassica oleracea* L. var. botrytis) monoculture, living mulches increased the abundance of spiders and decreased the number of lepidopteran eggs and larvae (Hooks & Johnson, 2004).

To combat invasive soybean aphid outbreaks, *Aphis glycines* Matsumura (Hemiptera: Aphididae), alfalfa living mulch (*Medicago sativa* L.) has tremendously enlarged the aphidophagous insect predator community (Schmidt et al., 2007). Alfalfa mulches and kura clover (*Trifolium ambiguum* M. Bieb) are ingested by European maize borer (*Ostrinia nubilalis* Hubner) and predator organisms (Prasifka et al., 2006). Sweet potato (*Ipomoea batatas* (L.) Lam.) gained an increased population of fire ants, rove beetles, and carabid beetles using pitfall traps in parcels loaded with killed crop cover (Jackson & Harrison, 2008). Numerous factors, including sampling dates and soil cover sort, including plastic mulch and straw mulch (Minarro & Dapena, 2003), relied on the dominance of several carabid species in an apple (*Malus domestica* Borkh.) orchard. Natural enemy predation on beet armyworm pupae, *Spodoptera exigua* (Hubner), was 33% greater in killed cover grain mulch than conventional production plots (Pullaro et al., 2006). Sunshine hemp mulch (*Crotalaria juncea* L.) protects the seed with a reduced incidence of lower cornstalk borer (*Elasmopalpus lignosellus* Z.) in bush beans (*Phaseolus vulgaris* L.) (Gill et al., 2010). Mulches can be used to control airborne insect pests by affecting their capacity to find food or hosts (Brown & Tworkoski, 2004; Gill et al., 2010; Hooks & Johnson, 2004; Prasifka et al., 2006; Pullaro et al., 2006; Reeleder et al., 2004; Schmidt et al., 2007; Tremelling et al., 2002).

2 Uses of Living Mulches in Biological Pest Management

The occurrence and dominance of predators in mulching environments proved to be effective against insect pest control (Flint, 2018; Vincent et al., 2003). The predator population was enhanced under the mulching residues (Quintanilla-Tornel et al., 2016). Schmidt et al. (2007) discovered that 45% of natural enemies were increased in soybean field under alfalfa living mulch condition, while took longer for *Aphis glycines* to establish resulting in lower peak populations (Schmidt et al., 2007). Natural enemies will slow the spread of *A. glycines*, and an increase in the aphidophagous predator, plenty vulnerable the spread of *A. glycines*, avoiding economic populations from developing. Living mulch showed an unintended effect on *A. glycines* population development, reducing the inherent growth rate of aphids and offering a bottom-up suppression of the aphid (Fig. 1).

Compared with no mulch plot, the increase in predator abundance and intake of corn borer were in noted in corn and soybean plots under alfalfa and kura clover mulches (Prasifka et al., 2006) (Fig. 2).

An indirect adverse impact is caused by wheat straw mulch on the Colorado potato beetle (*Leptinotarsa decemlineata*) (Brust, 1994). After the mulch application, a substantial concentration of rodents may occur within 15–20 days fed on the Colorado potato beetle eggs, 1st, and 2nd instars. Several major predators were *Coleomegilla maculate*, *Chrysoperla carnea* and *Perillus bioculatus* (Brust, 1994). In the bush bean (*Phaseolus vulgaris* L.), sunshine hemp straw (*Crotalaria juncea* L.) mulch was used to control the disease of lesser cornstalk borer (*Elasmopalpus lignosellus*) (Gill et al., 2010). By manuring mulch in the apple orchard, the number of predators increased, and the population of spotted tentiform leafminer (*Phyllonorycter blancardella*) and woolly apple aphid (*Eriosoma lanigerum*) was controlled (Brown & Tworkoski, 2004). In agricultural production, plastic mulch gained its demand specifically to overcome the farming of high-value crops. Insect pests, including aphids, thrips, and whiteflies, can be suppressed using plastic mulch due to their smell, color, or surface (Diaz & Fereres, 2007; Vincent et al., 2003). Colored plastic mulch has its light reflectance pattern to suppress the insect pest population (Vincent et al., 2003).

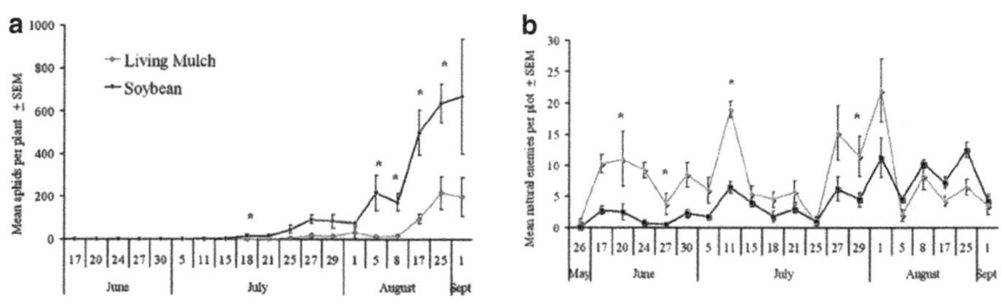

Fig. 1 Mean natural *A. glycines* infestation in soybean-growing alone or alfalfa living mulch (A) and mean natural enemies in soybean with living mulch and alone (B). *Two treatments significant changes (df = 1,3, $P \leq 0.05$). Reproduced from Schmidt et al. (2007)

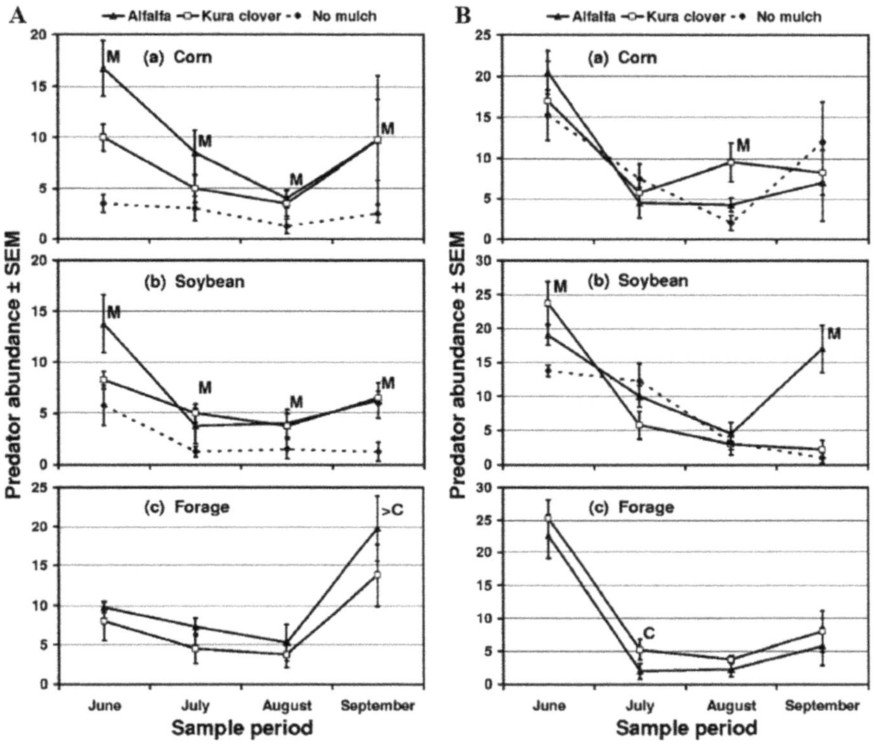

Fig. 2 During June to September 2004 (**a**) and 2005 (**b**) the predator abundance in different crops

Silver mist and black plastic mulch usage in squash will keep the aphid population away from it and boost the marketable squash yield (Summers et al., 1995).

3 Living Mulching Problems

3.1 Insect Pest Infestation

There is inconsistency in the usage of sustainable mulching practices. Certain researchers have noticed that organic kinds of mulches also attract rodents, but in contrast, several researchers also observed that such organic mulches are being used as pest repellent (Anderson et al., 2002). For several pests, such as clothing moths, cockroaches, and repelling types of ants, carpet beetles, and termites with certain varieties of Thuja are utilized as a natural repellent (Chalker-Scott, 2007). Sometimes it is considered that woody types of mulching are attractions for the termites. An experiment that was performed using organic (grass) and inorganic (gravel) mulches (Long et al., 2001) disclosed termite activity at a higher rate under gravel mulch as compared to woody mulch. The organic mulches used for controlling termites cause a higher mortality rate than other mulches. Duryea et al. (1999) showed that mulches with more significant nitrogen and phosphorus content cause higher mortality in termites. Therefore, in places where termites are the primary pests, organic mulches with less nutritional content could be used (Martin & Poultney, 1992). Landscape fabrics and black plastic mulch can minimize different pests (Duncan et al., 1992). By nature, numerous mulch products can repel organisms by their distinct odors, spines, or unique texture characteristics (Chaudhary et al., 2003; Szwedo & Maszczyk, 2000). Thick mulches may also attract pests such as rats trying to find protection (Siipilehto, 2001).

3.2 *Weed Infestation*

Composts or residues that are practically used as mulches are assumed to bear different weed species' seeds if such mulches are not sufficiently decomposed (Chalker-Scott, 2007). Research has shown that weed destruction is closely connected with mulch depth (McDonald et al., 1996). Compared to those mulches applied at shallow depths, organic mulches used at a higher depth have increased potential to minimize weed infestation (Horowitz & Thomas, 1994).

3.3 *Disease Transmission*

Pathogens are present in the mulch materials of diseased plants from where they can spread to healthy plants. As a result, mulch products should be thoroughly composted before being used (Hoitink & Krause, 1999). Temperature treatment is very important to kill all useful and detrimental organisms present in mulch products. To stop pathogens, we should focus on commercially available mulches that are sterile (Chalker-Scott, 2007). Honey locust canker, commonly used as mulch, contains toxic pathogens (Koski & Jacobi, 2004). Different studies have shown conflicting findings, claiming that using diseased mulch as contaminated maple trees does not result in disease transmission (Dochinger, 1956). According to a six-year investigation, no pathogen transmission was identified when canker infected mulches were put to robust plants. Infected Austrian pine foliage was used as mulch for the same species and was the only proof of tip blight transmission (Jacobs, 2005). The Austrian pine tip blight pathogen did not affect other crop plants and may be susceptible to the pathogen (Chalker-Scott, 2007). Epidemiology, rather than the pathogenic source, may cause the disease (Chalker-Scott, 2007). Pathogens can also be transmitted by soil. In healthy soils, pathogens or microorganisms are still present. These pathogens become active when soil conditions are low or anaerobic, causing significant losses to vigorous crop plants (Foreman et al., 2002). The use of uncomposted mulch materials will result in pathogen transmission (Koski & Jacobi, 2004).

3.4 *Bird Infestations*

Pesticide and herbicide use has been shown to have both beneficial and harmful effects on agricultural bird species diversity, reproduction, and food resources such as insects and weeds (Geiger et al., 2010). According to theory (Tscharntke et al., 2012), the spatial heterogeneity will reduce the destruction of bird habitats caused by the use of foils on crops. Depending on the species studied, most of the factors studied (crop diversity, forest and human settlement cover, cabbage crop cover) had different (positive or negative) effects, so they are of limited practical use in bird conservation landscape planning. Grassland patches scattered around an intensively cultivated landscape with controlled crops are critical for farmland biodiversity (Batary et al., 2007). Grassland cover in the landscape presents an ideal breeding and foraging environment for birds (Concepcion et al., 2012). Grassland, which is intensively mowed or grazed by cows, is ideal for ground-feeding birds because, in these conditions, the prey becomes easier (Morris & Thompson, 1998). In habitats with small fields, there were more bird species. This was also valid for experts in farmland. More species live in agricultural habitats with smaller fields (Herzon & O'Hara, 2007). Fields in Poland are separated by narrow grass strips, increasing the density of potentially useful microhabitats for birds (Concepcion et al., 2012).

4 Conclusion

The need for non-chemical insect pest management is the demand of present-day lifestyle emphasized by the demand for organic food production and environmental degradation caused by synthetic insect control measures. With this idea, the potential for using various types of live mulches to manage insect pests has been assessed in this chapter. Resultantly, increased predator and parasitoids activity or populations will aid in the consumption of more insect pests. Similarly, inter-cropping, dense vegetation, and border copings patterns alter the spectrum of the incident light, which negatively affects insect pest activity via changes in temperatures and humidity of their habitats. This provides a means of repelling or deflecting many insect pests, particularly in high-value crops. Organic mulches are likely to improve biocontrol agent activity or increase the concentration and activity of certain enzymes that can suppress the pest population (by hardening the cell wall consumed by insects). Mulches can also have a beneficial effect on soil structure, porosity, and fertility, making it more suppressive and difficult to survive for insect pests.

References

Adler, M. J., & Chase, C. A. (2007). Comparison of the allelopathic potential of leguminous summer cover crops: Cowpea, sunn hemp, and velvet bean. *Horticultural Science, 42*, 289–293.
Anderson, J. T., Thorvilson, H. G., & Russell, S. A. (2002). Landscape materials as repellents of red imported fire ants. *Southwest. Entomologist, 27*, 155–163.
Andow, D. A. (1991). Vegetational diversity and arthropod population response. *Annual Review of Entomology, 36*(1), 561–586.
Batary, P., Baldi, A., & Erdos, S. (2007). Grassland versus non-grassland bird abundance and diversity in managed grasslands: Local, landscape and regional scale effects. *Biodiversity and Conservation, 16*, 871–881.
Brown, M. W., & Tworkoski, T. (2004). Pest management benefits of compost mulch in apple orchards. *Agriculture, Ecosystems and Environment, 103*(3), 465–472.
Brust, G. E. (1994). Natural enemies in straw-mulch reduce Colorado potato beetle populations and damage in potato. *Biological Control, 4*(2), 163–169.
Campiglia, E., Caporali, F., Radicetti, E., & Mancinelli, R. (2010). Hairy vetch (*Vicia villosa* Roth.) cover crop residue management for improving weed control and yield in notillage tomato (*Lycopersicon esculentum* Mill.) production. *European Journal of Agronomy, 33*, 94–102.
Chalker-Scott, L. (2007). Impact of mulches on landscape plants and the environment—A review. *Journal of Environmental Horticulture, 25*, 239–249.
Chaudhary, R. S., Patnaik, U. S., & Dass, A. (2003). Efficacy of mulches in conserving monsoonal moisture for the Rabi crops. *Journal of Indian Society of Soil Science, 51*, 495–498.
Concepcion, E. D., Diaz, M., Kleijn, D., Baldi, A., Batary, P., & Clough, Y. (2012). Interactive effects of landscape context constrain the effectiveness of local agri-environmental management. *Journal of Applied Ecology, 49*, 695–705.
Cutler, G. C., Amichot, M., Benelli, G., Guedes, R. N. C., Qu, Y., & Rix, R. R., et al. (2022). Hormesis and insects: Effects and interactions in agroecosystems. *Science of the Total Environment, 825*, 153899.
Desneux, N., Decourtye, A., & Delpuech, J. M. (2007). The sublethal effects of pesticides on beneficial arthropods. *Annual Review of Entomology, 52*, 81–106.
Diaz, B. M., & Fereres, A. (2007). Ultraviolet-blocking materials as a physical barrier to control insect pests and plant pathogens in protected crops. *Pest Technology, 1*(2), 85–95.
Dochinger, L. S. (1956). New concepts of Verticillium wilt disease of maple. *Phytopathology, 46*, 467 (abstract).
Duncan, R. A., Stapleton, J. J., & McKenry, M. V. (1992). Establishment of orchards with black polyethylene film mulching: Effect on nematode and fungal pathogens, water conservation, and tree growth. *Journal of Nematology, 24*, 681–687.
Duryea, M. L., Huffman, J. B., English, R. J., & Osbrink, W. (1999). Will subterranean termites consume landscape mulches? *Arboricultural Journal, 25*, 143–149.
Farooq, M., Jabran, K., Cheema, Z. A., Wahid, A., & Siddique, K. H. (2011). The role of allelopathy in agricultural pest management. *Pest Management Science, 67*(5), 493–506.
Flint, M. L. (2018). Pests of the garden and small farm: a grower's guide to using less pesticide (Vol. 3332). UCANR Publications.
Foreman, G. L., Rouse, D. I., & Hudelson, B. D. (2002). Wood chip mulch as a source of Verticilliumdahliae. *Phytopathology, 92*, S26 (abstract).
Frank, D. L., & Liburd, O. E. (2005). Effects of living and synthetic mulch on the population dynamics of whiteflies and aphids, their associated natural enemies, and insect-transmitted plant diseases in zucchini. *Environmental Entomology, 34*(4), 857–865.
Geiger, F., Bengtsson, J., Berendse, F., Weisser, W. W., Emmerson, M., & Morales, M. B. (2010). Persistent negative effects of pesticides on biodiversity and biological control potential on European farmland. *Basic and Applied Ecology, 11*, 97–105.

Gill, H. K., & McSorley, R. (2010). Integrated impact of soil solarization and organic mulching on weeds, insects, nematodes, and plant performance. *Proceedings of the Florida State Horticultural Society, 123*, 308–311.
Gill, H. K., McSorley, R., Goyal, G., & Webb, S. E. (2010). Mulch as a potential management strategy for lesser cornstalk borer, *Elasmopalpus lignosellus* (Insecta: Lepidoptera: Pyralidae), in bush bean (*Phaseolus vulgaris*). *Florida Entomologist, 93*(2), 183–191.
Herzon, I., & O'Hara, R. B. (2007). Effects of landscape complexity on farmland birds in the Baltic States. *Agriculture, Ecosystems and Environment, 118*, 297–306.
Hoitink, H. A. J., & Krause, M. S. (1999). Control of nuisance and detrimental molds (Fungi) in mulches and composts. *Special Circular Ohio Agricultural Research and Development Center, 165*, 66–69.
Hooks, C. R., & Johnson, M. W. (2004). Using undersown clovers as living mulches: Effects on yields, lepidopterous pest infestation, and spider densities in a Hawaiian broccoli agroecosystem. *International Journal of Pest Management, 50*, 115–120.
Horowitz, M., & Thomas, J. M. (1994). Couverture du sol pour la gestion des mauvaises herbes (Soil cover for weed management). In *Maitrise des adventices par voie non chimique. Communications de la quatrieme conference internationale international federation of organic agriculture movements* (pp. 149–154).
Jabran, K., & Chauhan, B. S. (2018). *Non-chemical weed control*. Elsevier.
Jackson, D. M., & Harrison, H. F. (2008). Effects of killed-cover crop mulching system on sweet potato production, soil pests, and insect predators in South Carolina. *Journal of Economic Entomology, 101*, 1871–1880.
Jacobs, K. A. (2005). The potential of mulch to transmit three tree pathogens. *Arboricultural Journal, 31*, 235–241.
Jones, G. A., & Sieving, K. E. (2006). Intercropping sunflower in organic vegetables to augment bird predators of arthropods. *Agriculture, Ecosystems & Environment, 117*(2–3), 171–177.
Johnson, J. M., Hough-Goldstein, J. A., & Vangessel, M. J. (2004). Effects of straw mulch on pest insects, predators, and weeds in watermelons and potatoes. *Environmental Entomology, 33*, 1632–1643.
Koski, R., & Jacobi, W. R. (2004). Tree pathogen survival in chipped wood mulch. *Arboricultural Journal, 30*, 165–171.
Long, C. E., Thorne, B. L., Breisch, N. L., & Douglass, L. W. (2001). Effect of organic and inorganic landscape mulches on subterranean termite (Isoptera: Rhinotermitidae) foraging activity. *Journal of Environmental Entomology, 30*, 832–836.
Martin, P. J., & Poultney, R. (1992). Survival and growth of clove seedlings in Zanzibar. Effects of mulching and shade crops. *Journal of Tropical Agriculture, 69*, 365–373.
McDonald, H. G., Smith, J. M., & Britt, C. P. (1996). The effectiveness of organic mulches on weed control in farm woodlands. *Aspects of Applied Biology, 44*, 63–68.
Minarro, M., & Dapena, E. (2003). Effects of groundcover management on ground beetles (Coleoptera: Carabidae) in an apple orchard. *Applied Soil Ecology, 23*, 111–117.
Morris, L. P., & Thompson, F. R. (1998). Effects of habitat and invertebrate density on abundance and foraging behavior of Brown-headed Cowbirds. *The Auk, 115*, 376–385.
Mulvaney, M. J., Price, A. J., & Wood, C. W. (2011). Cover crop residue and organic mulches provide weed control during limited-input no-till collard production. *Journal of Sustainable Agriculture, 35*, 312–328.
Nawaz, A., Farooq, M., Lal, R., Rehman, A., Hussain, T., & Nadeem, A. (2017). Influence of sesbania brown manuring and rice residue mulch on soil health, weeds and system productivity of conservation rice–wheat systems. *Land Degradation and Development, 28*(3), 1078–1090.
Nottingham, L. B., Dively, G. P., Schultz, P. B., Herbert, D. A., & Kuhar, T. P. (2016). Natural history, ecology, and management of the mexican bean beetle (Coleoptera: Coccinellidae) in the United States. *Journal of Integrated Pest Management, 7*, 2.

Prasifka, J. R., Schmidt, N. P., Kohler, K. A., O'Neal, M. E., Hellmich, R. L., & Singer, J. W. (2006a). Effects of living mulches on predator abundance and sentinel prey in a corn–soybean–forage rotation. *Environmental Entomology, 35*(5), 1423–1431.
Pimentel, D. (1961). Species diversity and insect population outbreaks. *Annals of the Entomological Society of America, 54*(1), 76–86.
Pedigo, L. P., & Rice, M. E. (2009). *Entomology and pest management* (6th ed., pp. 255–284). Prentice Hall.
Pullaro, T. C., Marino, P. C., Jackson, D. M., Harrison, H. F., & Keinath, A. P. (2006). Effects of killed cover crop mulch on weeds, weed seed, and herbivores. *Agriculture, Ecosystems and Environment, 115*, 97–104.
Quintanilla-Tornel, M. A., Wang, K. H., Tavares, J., & Hooks, C. R. (2016). Effects of mulching on above and below ground pests and beneficials in a green onion agroecosystem. *Agriculture, Ecosystems and Environment, 224*, 75–85.
Reeleder, R. D., Capell, B. B., Roy, R. C., Grohs, R., & Zilkey, B. (2004). Suppressive effect of bark mulch on weeds and fungal diseases in ginseng (*Panax quinquefolius* L.). *Allelopathy Journal, 13*, 211–232.
Schmidt, M. H., Thewes, U., Thies, C., & Tschamthe, T. (2004). Aphid suppression by natural enemies in mulched cereals. *Entomologia Experimentalis et Applicata, 113*, 87–93.
Schmidt, N. P., O'Neal, M. E., & Singer, J. W. (2007). Alfalfa living mulch advances biological control of soybean aphid. *Environmental Entomology, 36*(2), 416–424.
Siipilehto, J. (2001). Effect of weed control with fiber mulches and herbicides on the initial development of spruce, birch and aspen seedlings on abandoned farmland. *Silva Fennica, 35*, 403–414.
Summers, C. G., Stapleton, J. J., Newton, A. S., Duncan, R. A., & Hart, D. (1995). Comparison of sprayable and film mulches in delaying the onset of aphid-transmitted virus diseases in zucchini squash. *Plant Disease, 79*(11), 1126–1131.
Sans, F. X., & Altieri, M. A. (2005). *Effects of intercropping and fertilization on weed abundance, diversity and resistance to invasion* (pp. 31–34). International Society of Organic Agricultural Research (ISOFAR).
Szwedo, J., & Maszczyk, M. (2000). Effects of straw-mulching of tree rows on some soil characteristics, mineral nutrient uptake and cropping of sour cherry trees. *Journal of Fruit and Ornamental Plant Research, 8*, 147–153.
Thomson, L. J., & Hoffmann, A. A. (2007). Effects of ground cover (straw and compost) on the abundance of natural enemies and soil macro invertebrates in vineyards. *Agricultural and Forest Entomology, 9*, 173–179.
Tremelling, M. J., McSorley, R., & Gallaher, R. N. (2002). Effect of winter cover crops on soil surface invertebrate community. *Proceedings of Soil and Crop Science Society of Florida, 62*, 77–82.
Tscharntke, T., Tylianakis, J. M., Rand, T. A., Didham, R. K., Fahrig, L., Bataty, P., et al. (2012). Landscape moderation of biodiversity patterns and processes—Eight hypotheses. *Biological Reviews, 87*, 661–685.
Ullah, F., Gul, H., Desneux, N., Qu, Y., Xiao, X., Khattak, A. M., Gao, X., & Song, D. (2019a). Acetamiprid-induced hormetic effects and vitellogenin gene (*Vg*) expression in the melon aphid *Aphis Gossypii*. *Entomologia Generalis, 39*(3–4), 259–270. https://doi.org/10.1127/entomologia/2019/0887.
Ullah, F., Gul, H., Desneux, N., Tariq, K., Ali, A., Gao, X., & Song, D. (2019b). Clothianidin-induced sublethal effects and expression changes of vitellogenin and ecdysone receptors genes in the melon aphid *Aphis Gossypii*. *Entomologia Generalis, 39*(2), 137–149. https://doi.org/10.1127/entomologia/2019/0865.
Ullah, F., Gul, H., Yousaf, H. K., Xiu, W., Qian, D., Gao, X., Tariq, K., Han, P., Desneux, N., & Song, D. (2019c). Impact of low lethal concentrations of buprofezin on biological traits and expression profile of chitin synthase 1 gene (*CHS1*) in melon aphid *Aphis Gossypii*. *Scientific Reports, 9*(1), 12291. https://doi.org/10.1038/s41598-019-48199-w.

Ullah, F., Gul, H., Tariq, K., Desneux, N., Gao, X., & Song, D. (2020a). Fitness costs in clothianidin-resistant population of the melon aphid Aphis Gossypii. *Plos One, 15*(9), e0238707.
Ullah, F., Gul, H., Tariq, K., Desneux, N., Gao, X., & Song, D. (2020b). Functional analysis of cytochrome P450 genes linked with acetamiprid resistance in melon aphid Aphis Gossypii. *Pesticide Biochemistry and Physiology, 175*, 104687.
Ullah, F., Gul, H., Tariq, K., Desneux, N., Gao, X., & Song, D. (2020c). Thiamethoxam induces transgenerational hormesis effects and alteration of genes expression in *Aphis gossypii. Pesticide Biochemistry and Physiology, 165*, 104557.
Vincent, C., Hallman, G., Panneton, B., & Fleurat-Lessard, F. (2003). Management of agricultural insects with physical control methods. *Annual Review of Entomology, 48*(1), 261–281.
Weston, L. A., Harmon, R., & Mueller. (1989). Allelopathic potential of sorghum-sudan grass hybrid (sudex). *Journal of Chemical Ecology, 15*, 1855–1865.

Effects of Mulching Practices on the Management of Weeds

Zahid Hussain and Luqman

Abstract Application of mulches is effective in reducing soil erosion through air and water, in retaining the soil moisture content, and also in suppressing the weed growth, which ultimately altogether improve the soil health. In the tropical production systems, the crop yields are drastically declined by the weed competition which has severely affected the small land holding farmers who cannot afford to purchase herbicides and other yield improving chemicals. Thus, the mulching practice of leaving the crop residues in the field after crop harvest can be a solid tool to suppress weeds, especially in the conservation agriculture systems. As a result, the weed emergence and biomass are reduced by the increased amount of the post-harvest crop residue. More than 10,000 tons ha^{-1} residue is required to effectively manage reduce the weed emergence and biomass together, as compared with the treatment of the bare soil having no crop residues. There are several physical methods used for suppression of weeds in cultivation. However, here the main focus is on the application of numerous mulches for the purpose of weed management and soil fertility. Mulches are a strong too for weed management in organic farming. Either the biodegradable mulching materials are used or various mulch films are utilized in the process of mulching. Ideally the organic mulches can be conveniently collected from the surrounding nature, which becomes a cheaper source of crop production. Also, the use of biodegradable mulches has a positive effect on environment unlike the chemical weed control measures. Generally, the physical weed control methods may result both in the positive or negative effect on the growth and yield of herbs and vegetables; however the target of weed suppression can be easily achieved which indirectly ends with a desirable production i.e. when weeds are suppressed the available soil resources will be utilized by the crop plants. Yes, if these mulches are used as 'living mulches' in crop production, then the living mulches do compete with the main crop for essential resources available in the soil. Apart from the weeds, the living cover crop (used as mulch) also affects the main crop at the same time. Consequently, a significant heed must be given during the selection of the most suitable 'living mulch' with the aim to achieve effective weed suppression in a crop production

Z. Hussain (✉) · Luqman
Department of Weed Science and Botany, Faculty of Crop Protection Sciences, The University of Agriculture Peshawar, Peshawar 25130, Khyber Pakhtunkhwa, Pakistan
e-mail: zhussainws@aup.edu.pk

K. Akhtar et al. (eds.), *Mulching in Agroecosystems*,
https://doi.org/10.1007/978-981-19-6410-7_9

strategy. Various scientists have well described their experiences with a variety of biodegradable mulches including chopped newspapers, straw, photodegradable and biodegradable films, compost and gravel etc., but this special attention is given only to the to the use of mulches in medicinal plants cultivation. Therefore, the use of physical methods (mulches in our case) for weed control in the agronomic crops should also be made effective as in the medicinal plants cultivation. Nowadays, there is a steady increase in the human and livestock populations, while reciprocally the per capita arable land is declining with the passage of time. The farmers since long have been utilizing numerous methods to meet the requirements of the demand for enhancement in the production of the food crops. Worldwide, the infesting weeds are the key challenge in getting the desirable production of a crop. Among the major weeds in the crop production, about 10 weeds are there which grow in almost every crop and severely influence the target yields of these crops. The list of these problematic weeds comprises of *Cannabis sativa, Chenopodium* spp., *Medicago lupulina, Taraxacum officinale, Cirsium arvense, Stellaria meidia, Cynodon dactylon,* thistles, grasses, etc. The mulch material physically suppresses the growth of emerging weeds because the soil surface is covered and the growing weeds are shadowed due to which they fail to get suitable conducive environment to compete with the target crop. Mulching is eco-friendly, safe, feasible, less expensive and has shown fruitful results, when compared with other weed management techniques. Mulching also helps in early maturity of the main crop plants which ultimately enhances the crop yields. However, it is a pre-requisite for using the mulching process to have the prior knowledge of the site of cultivation of the crop.

Keywords Crop residues · Herbicides · Mulching · Crop production · Farmers income

1 Introduction

Mulching is actually the process of covering the soil surface with the help of organic materials like straw, husks, stocks, or any biological biomass which could be the biomass of a weed that is uprooted and spread on the soil surface. Or mulch could be any inorganic material like black or white plastic etc. which should bio-degradable.

Mulches generally limit the weeds growth using the resource limitation principle, thus the available light is hampered and the growing weeds are smothered. Rice straw used as mulch controlled the weeds growth effectively. Also the living mulches substantially decreased the biomass of the weeds growing with the main crop. Also, the biomass of weeds was effectively decreased in vetch plots that were treated with oat plants as mulch and grazing was also conducted, in comparison with the bare plots with no mulch applied. Further, the different mulching materials have a variety of effects on the growth of weeds. For instance, the straw mulch and use of polythene significantly suppressed the weeds growth, however the other mulch types resulted in no effect.

In sericulture, mulching gives a greatly favorable environment to the mulberry trees to effectively utilize the organic manures as well as the bio fertilizers so that the quantity and quality of the mulberry leaves are improved. The bio and organic amendments in the mulched treatments in addition to the reduced doses of inorganic fertilizers resulted in the improvement of the chemical, physical and biological features of the concerned soils.

Many of the cover crops growing in winter such as peas, hairy vetch, crimson clover, and rye can effectively manage the plant diseases like the fruit rots in tomato. Further, the hairy vetch and rye as cover crops substantially improved the fruit yield of tomato, as compared with the fallow plots (control). The polyethylene as mulch and the spent mushroom compost both effectively enhanced the fruit yield of tomato and significantly suppressed the growing weeds in organic production of tomato crop.

2 Pros and Cons of Mulches for the Management Weeds

Pros

The benefits of mulching are numerous. Many of the possible benefits are listed below in detail.

- **Soil moisture conservation**
  Mulches prevent rapid evaporation and make water and nutrients available to the crop roots and thereby create conditions favorable for the new root growth.
- **Improved rain infiltration**
  The rain water is sufficiently absorbed into the soil by capturing and slowing down the movement of water through mulches.
- **Inhibition of weed seed germination**
  Mulches reduce seed germination of weeds by preventing light from hitting the soil surface.
- **Reduction in weed biomass**
  This can also be termed as smothering of weeds or weeds suppression. Field trials in commercial vineyards showed an annual loss of 50–65% of the covered weed biomass. Some follow-up with hand crews or spot herbicide applications is often necessary to achieve complete weed control with mulches. To be effective, mulches must be applied in a uniform layer to soil before weed germination and replenished on a regular basis.

Mulches are capable of suppressing the weeds growth in several ways. These include, (1) there is physical influence of the mulch and its depth. In this way, there are several factors involved which make a combined effect. These factors include decrease in soil temperature, moisture, and also lack of light availability. (2) the mulch inhibits the seed germination due to the phyto-toxic constituents produced by the micro-flora living within the composted material. This is probably because the immature composts are greatly effective in suppression of weeds.

The phyto-toxic components mentioned above include a variety of the volatile organic constituents, which include phenolic acids, ammonia, and organic acids (i.e. acetic and propionic acids, isobutyric and isovaleric acids together). The compost maturity and its C:N ratio determine the levels of these chemicals. A very higher C-N ratio always results in the excessive accumulation of the organic acids. On the other hand, a very lower C-N ratio ends in the amassing of ammonia compound. These chemical compounds effectively inhibit the seeds germination; can cause injury to these plants, especially when the composts are very immature. Thus, a rationale is lying there for using these composts in the distances between the plants, but should be put very adjacent to them.

- **Crop seed protection**
  Mulches are spread over the surface of the soil to protect seed from the erosive effects of water and fluctuations in soil temperature and moisture. Mulches can sometimes be applied together with seed or fertilizers. However, too much mulch mixed with seed can result in poor stands. Netting, disking, and asphalt spray are among the methods used to anchor loose mulch in order to prevent loss via erosion.
- **Improving soil chemical environment**
  Organic mulches add organic matter and plant nutrients to soil upon decomposition. Thus, they improve carbon sequestration.
- **Improvement in soil structure**
  Mulches improve the soil structure through addition of organic matter to the soil.
- **Reduction in nitrogen loss**
  The volatilization and leaching loss of nitrogen is reduced under mulched condition. During decomposition of organic mulches soil mineral nitrogen is immobilized by microbes and thus its loss is minimized.
- **Increase in soil fertility**
  Cation exchange capacity is substantially influenced by organic matter content in soils containing predominantly low activity clays. Improvement in the cation exchange capacity of soil improves the fertility status of these soils. Mulches also influence the availability of nutrients through their effect on physical conditions, hydrothermal regime and biological activity of the soil.
  The weed growth is indirectly promoted by the N fertilizer application, and as a result the weed competition with rice crop has decline the rice yield to much extent. However, the effect of the placement of nitrogen fertilizer on the weeds population has not been studied yet.
- **Reduction in soil pH**
  Furthermore, decomposition of organic mulches add organic acids to the soil resulting in low soil pH which influences the bioavailability of many plant nutrients viz., Fe, Mn, Zn, Cu, etc.

- **Soil protection from rain splash**
  Mulches protect the soil from rain splash which reduces the soil erosion and splashing of pathogen inoculum on to the plants.

Cons

- Cultivation or irrigation practices that disturb the mulch barrier limit the effective life of the mulch.
- Mulches made up of organic materials will decompose in response to normal microflora and microfauna activity.
- Mulches have little or no impact on re-growth of perennial weeds such as field bindweed.
- Vertebrate pest activity may increase or be hidden by mulches.

3 Types of Mulches Used for the Management Weeds and Diseases

Generally, there are two types of mulches including organic and inorganic mulches.

3.1 Organic Mulch

There are various positive attributes of the organic mulch, which effectively conserves the soil moisture through reduction in the water loss via evaporation, diminishes the soil erosion through wind and water, keeps the soil temperatures quite normal for the crop plants, hampers the growth and number of weeds, helps in development of beneficial soil microbes, and last but not the least decreases the spread of soil borne pathogens due to protection of the soil from splashing upon the crop plants during watering or rainstorms. Further, the mulching process can help in avoiding the mowing practices around trees and shrubs, which is also helpful in preventing mechanical injury to the tree trunks. Mulch also stops the process of heaving (i.e. upward protrusion of the plant roots out of soil) when it is used as a winter protection tool, during freeze and/or thaw periods. The eroded areas can also be stabilized, which helps prevent the soil erosion through wind and water both. Organic mulches get decomposed with the passage of time, which improve the soil quality and structure, and the nutrients are returned to the soil. The inorganic mulches are mostly utilized for creation of barriers to the growing weeds; and are sometimes used for the purpose of decoration. Inorganic mulches include rocks or gravels which do not decompose readily. The rocks absorb and then reflect heat that can be fatal to the plants in hot and dry weather conditions. Due to extremely slow decomposition, the inorganic mulches are not capable of improving the quality of the soil.

4 Different Types of the Organic Mulches

1. **Tree barks (softwood)**
   The softwood bark is also one of the by-products of the lumber and paper industries. The pine bark is one of the common examples of softwood and is mostly utilized under the larger trees and also the shrubs. The pine bark is somewhat acidic in nature i.e. having lower pH. It takes sufficient time to get decomposed, and does not required replacement quite often, unlike the other types of organic mulches do.
2. **Tree barks (hardwood mulch)**
   The shredded and torn hardwood is considered a by-product of the lumber and paper industries. The size of this hardwood ranges from shredded smaller chips to somewhat larger nuggets. This kind of mulching is mostly applied around the trees and/or shrubs and in the perennial beds too. The natural or dyed varieties contain this shredded bark. The dyed varieties are mostly a mixture of the hardwood or the recycled wood waste that contain the artificial dyes.
3. **Cocoa bean hulls/cocoa bean mulch**
   The cocoa hulls are one of the by-products of the chocolate industry. The weight of the hulls is light due to which they are easy to handle and can be used for any planting area. These hulls smell pleasantly as well. These should be applied up to one inch deep and must be watered lightly to hold them firmly. These hulls get decomposed rapidly and thus should be applied annually. These chocolate by-products are detrimental to the grazing animals if it is consumed; therefore it is better to choose another mulch type if there are pets frequent moving in the area. We can develop the leaf mulch at home with composting of the shredded and aged leaves that can be used in all types of garden beds. The leaves infected with scab, anthracnose, or leaf spot should be eliminated and must not be used in the compost for mulch purpose. The grass clippings should be spread in layers across the perennial beds and should be turned over when growing season is getting close to end. Dry up each layer before adding any additional layer. In addition, the application should not be made in thick layers because clippings get mat. Also, the clippings from lawns should not be used if the lawn is already treated with a herbicide or an insecticide. Further, the grass clippings that started seeding should not be applied in order to stop the growth of undesirable turf grass in the garden beds.
4. **Municipal tree waste**
   The utility companies and the city or village arborists mostly prepare mulches and make them available free of cost for the home owners. This type of mulch is normally composed of somewhat larger and not-aged wooden chunks. Thus, the fresh material use higher amounts of the soil nitrogen as it gets decomposed, and this mulch type is particularly useful in making pathways.
5. **Composted animal manure**
   Well-composted animal manure can also be utilized as mulch or as a soil amendment. This animal manure is a best option for any new planting beds because

it adds nutrients to the soil and boosts the soil quality. Fresh animal manures must not be applied in garden beds as the plant roots might be burnt with it. Extreme care must be taken when the animal manure is being used in the vegetable gardens. The manure must be well-composted before using it as mulch in temperatures between the ranges of 130–140°F for a period not more than a week, and be composted for about 4–6 months in order to eradicate the potential disease causing organisms. The layers of white and black newspapers may be applied for weeds suppression. Two to three layers can be applied at a time and must be covered with an organic material i.e. leaf mulch or grass clippings for the purpose of holding it in place. The newspapers will finally get decomposed which may then be incorporated into the soil.

5 Types of Inorganic Mulches

1. **Landscape fabric**
   The landscape fabric is the best choice for a long-term effect on the weeds suppression because the landscape fabric allows the water and air to pass through it. It is also used along with other organic mulches and gets decomposed more rapidly than any other inorganic mulch. Examples of stones include crushed gravel, marble chips, and volcanic rock. Stones cannot sustain moisture and thus cause heat stress on the nearby plants. The light reflection and ground heating causes the roots to burn.
2. **Plastic film (polyethylene)**
   The plastic film is an impermeable film i.e. in which the water and nutrients cannot penetrate. To warm the soil in spring vegetables the plastic film is used along the rows; however for long-term use, it is not the best choice. Plastic film gets deteriorated upon exposure to the sunlight and can be used for one season only.
3. **Rubber as mulch**
   The rubber mulch comprises of recycled or ground tires of the vehicles. This type of product is still under research; however, the initial studies showed some possibility of toxicity levels and also there is a risk of flammability as well. Also, the rubber mulch can be functional in the soil for an indefinite time. It is therefore not recommended for application in the home landscape practices.

6 Instances Where Mulches Worsened the Practice of Weeds Management

The organic mulch can enhance the soil acidity (i.e. lowering the soil pH) which negatively affects the crop productivity (Tables 1, 2, 3 and 4).

Table 1 Different types of mulches and their effect on weeds and crops (Chopra & Koul, 2020)

S. No	Mulch type	Crops	Weeds targeted	Impact	References
Organic mulch					
1	Alfalfa	Corn cropping system	*D. aegyptium*, *D. arvensis*	Rich in nitrogen; durable for long time	Lal (1974)
2	Ash	Garlic	*A. vineale*, *A. candanse*	Controls weed growth	Tiedjens (1940)
3	Bark	Used in vegetation and landscaping	*T. officinale*, *E. prostrate*	Moisture content remains high for long duration	
4	Dry leaves	Natural forest area	*D. bipinnata*,	Enrich soil with nutrients	David (2007)
5	Grass clipping	Gardens, groundnut, legumes	*L. aphaca*	Enhances the %age of the soil nitrogen because of its quick decomposition	Ashrif and Thornton (1963)
6	Newspaper	Vegetable garden Potting cups	*M. indica*	Controls weed; time saving; biodegradable	Ashrif and Thornton (1963) and Lal (1974)
7	Saw dust	Conifers, blueberries, strawberries	*C. album*, *S. media*	C-N ratio is higher, thus decomposes slowly; somewhat less nutritive, having longer moisture retention period	Kumar and Dey (2011)
8	Seaweed	Gardens	*C. arvensis*, *C. rotundus*	Provide minerals; reduce water requirement	Robinson (1988)
9	Stubble	Stevia, vegetables, wheat	*E. helioscopia*, *C. album*	Decrease water demand; reduces weed	Bilalis et al. (2002)
Synthetic mulch					
10	Plastic mulch	Cauliflower, stevia	*E. colonum* *A. viridis*	Controls weed; increases warmth	Isenberg and Odland (1950)
A	Black plastic mulch	Muskmelon	*A. retroflexus* *C. esculentus*	Large scale use; warmth is provided to the crop in winters; crop production is improved	Tiffany and Drost (2016)

(continued)

Table 1 (continued)

S. No	Mulch type	Crops	Weeds targeted	Impact	References
B	Clear plastic mulch	Muskmelon	*E. crusgalli*	Less solar radiations is absorbed; the soil borne diseases are reduced	Chakraborty and Sadhu (1994)
C	Red plastic mulch	Tomato, zucchini honeydews	*P. oleracea*	Reduce effect of early blight in many crops	Angima (2009)
D	Other colors plastic mulch (yellow, blue, grey mulch)	Black pepper	*C. esculentus, A. spinosus*	Used in winter crops; insect repellent; resulted in 2nd green revolution in India	
11	Gravels, pebbles, crushed stones	Gardens	*C. arvensus C. rotundud*	An inch layer can effectively control the emerging weeds	Sanders (2001)
12	Rubber	Home gardens	*P. lanceolata D. sanguinalis*	Can be applied easily in gardens, and easily available as well	Gupta and Yadav (1979)

Table 2 Mulching treatment for weed management in different crops (Chopra & Koul, 2020)

Crop	Types of mulches	Region of Experimentation	Outcomes	References
Blueberry	Black polythene mulch	Pennsylvania	Black plastic mulch effectively increases the yield and controlled the weeds	Gupta and Yadav (1979)
	Corn cob, sawdust, and wood chips	Pennsylvania	The wood chips and saw dust from the beech and red maple can effectively control the weeds growth	
Brinjal	Straw mulch, 30μ silver bicoloured, black plastic as mulch	The tropical and subtropical regions of Iindia (Bhagalpur, Bihar)	Highest yield-480.24 quintal ha^{-1} was achieved from a 30 μ bi-colored silver mulch Weeds were effectively controlled	Kumar et al. (2019)

(continued)

Table 2 (continued)

Crop	Types of mulches	Region of Experimentation	Outcomes	References
Carrot	leaf mulch, black polythene mulch, sugarcane straw mulch, paddy straw mulch, grass mulch, blue polythene mulch, white polythene mulch	Gwalior, M.P., India	54.69 tons ha^{-1} of yield was achieved with the help of black polythene mulch, as it is the maximum yield obtained from the various mulch plots Weeds were managed	Jaysawal et al. (2018)
Cauliflower	RD of N and K Polythene mulch with different conc. of OPE i.e. open pan evaporation	Assam, India	About 282.53 quintal ha^{-1} yield was achieved by using a bi-layer polythene mulch with 24.96 lit of OPE supplemented through drip irrigation plant^{-1}; 125% fertigation of N & K was done. Weeds were controlled too	Bhoutekar et al. (2017)
Garlic	Grass mulch and black polyethylene mulch	Addis Ababa, Africa	Reduced weed growth, increased soil moisture content and crop yield	Mahdiesh et al. (2012)
Lemon	Maize straw, Bajra straw, brankad, grasses, black polythene and FYM	At rainfed research sub-station, India	Maximum yield (1848 kg ha^{-1}) and soil content was obtained under black polythene mulch, followed by FYM mulch (1780 kg) and brankad mulch yield (1744 kg ha^{-1})	Kumar et al. (2015)
Maize	Plastic and straw mulch	Arid and semi-arid regions	Weeds were managed and increase in yield was 60%; plastic-mulch better than organic mulch in efficacy	Qin et al. (2015)
	Legume mulch: Leucaena twig mulch, Sunhemp, Sunhemp + Leucaena	Selakui, Dehradun India	Control soil erosion; reduce number of weeds; 59.3 kg ha^{-1} increase in N uptake recorded for Leucaena + sunhemp mulch, and 2.36 tons ha^{-1} yield increase	Sharma et al. (2009)

(continued)

Table 2 (continued)

Crop	Types of mulches	Region of Experimentation	Outcomes	References
Onion	Water hyacinth, Rice straw mulch	Dhaka, Bangladesh	Yield was increased with water hyacinth mulch and weeds infestation was decreased 10.46 tons ha^{-1} yield was obtained	Larentzaki et al. (2008)
Musk melon	Grass mulch, black polythene mulch, polythene (transparent) mulch	MPKV, Rahuri, India	Black polythene mulch and grass mulch were more effective than transparent mulch causing 80.02% weed control; increase in yield ha^{-1} was achieved	Johnson et al. (2004)
Potato	Straw as mulch	Northern Hessen, Germany	Aphids and weeds infestation was much reduced	Saucke and Doring (2004)
	Plastic mulch	North China Plain, China	Decreased weeds and increased yield	Wang et al. (2008)
Stevia	Poplar leaf, pine needles, tree leaf mulch, silver oak	Western Himalaya, India	poplar mulch and silver oak mulch both enhance the nutrition matter of the soil, also decrease the weed growth	Kumar and Dey (2011)
Strawberry	Blackpolythene,transparent polythene, plastic mulch	Punjab, India	Black polythene proved to be most useful in attaining 41% higher fruit yield and effective weed control	Rajbir et al. (2006)
	Wheat straw, paddy straw, saw dust, cut grass, black polythene, and transparent polythene	Chatha, Jammu, India	The black plastic as mulch highly effective causing 7% increase in total sugar content, 11.83 g increase in fruit weight, 3.93% increase in fruit length, and 143.38 g increase in total yield/plant	Bakshi et al. (2014)

(continued)

Table 2 (continued)

Crop	Types of mulches	Region of Experimentation	Outcomes	References
Sugarcane	Straw, no straw, straw burnt on soil	Bandeirantes Paraná, Brazil	Highest suppression of weeds was recorded under 75 and 100% straw mulch treatments	Hoshino et al. (2017)
Tomato	Black polythene mulch, Straw mulch Strawmulch + different combinations of drip system	Bangladesh	Black mulch increased the yield, also increased C-content upto 27.07%. Weeds were decreased	Cong et al. (2005)
Wheat	Plastic and straw mulch	Arid and semi-arid regions	Weed control and increase in yield Plastic mulch was more effective than straw mulch	Qin et al. (2015)
	Legume mulching: Leucaena twig as mulch, Sunhemp as mulch, Sunhemp + Leucaena as mulch	Selakui, Dehradun, India	Weeds were controlled 69.5 kg ha^{-1} increase in N uptake through sunhemp + Leucaena mulch; 2.38 t ha^{-1} increase in yield	Sharma et al. (2009)
Winter Pigeon pea	Paddy straw mulching, sugarcane trash@ 8 t ha^{-1}	B.C.K.V, Jaguli, Naida, West Bengal, India	the yield was increased upto 2.07 t ha^{-1} with sugarcane trash mulch. Weed management was also effective	Basu et al. (2009)

Table 3 Type of mulch and soil used in accordance with the existing conditions

S. no	Type of mulch used	Soil type/area/crop stage	References
1	Thinner film	Early germination	Ngouajio (2011)
2	Clear plastic mulch	Field prone to Soil born diseases	Chakraborty and Sadhu (1994)
3	Silver color film	Insect repellent	Penn State Extension (2015)
4	Stubble	Nutrient deficient	Orzolek and Lamont (2015)
5	Thicker mulch	Orchard mulch	Orzolek and Lamont (2015)

(continued)

Table 3 (continued)

S. no	Type of mulch used	Soil type/area/crop stage	References
6	Black film plastic mulch	Saline water area	Sanders (2001)
7	Black film	Sandy soil	Orzolek and Lamont (2015)
8	Thin and transparent film	Soil solarization	Ngouajio (2011)
9	White film	Summer cropped land	Orzolek and Lamont (2015)
10	Sea weeds	Water deficient area	Robinson (1988)
11	Black film plastic mulch	Weed control in cropped land	Penn State Extension (2015)
12	Transparent film	Weed control through solarization	Sanders (2001)

Table 4 Enhancement in soil moisture content under different mulches

Crops	Types of mulches used	Mulched plots	Non-mulched plots	Percent increase	References
Allium cepa	Grass mulched	18.20	17.1	0.30	Larentzaki et al. (2008)
Sugar beet	Peat	19.70	17.5	2.6–7.3	Lal (1974)
Gardens	Dry leaves	12.42	10.13	6–8	Ashrif and Thornton (1963)
Mustard	Sawdust	22.70	17.2	3.8–6.1	Kumar and Dey (2011)
Maize	Sunhemp + Leucaena	14.62	12.54	19.90	Sharma et al. (2009)
Tomato	Straw	100	86.1	16–27	Cong et al. (2005)
Potted shrubs	Bark chips	12.17	9.34	20–23	Stelli et al. (2000)

7 Conclusion and Recommendations

Mulching provides a favourable environment for effective utilization of plant biomasses, food wastes, organic manures and other biological materials for the purpose of improving the growth and yield of crops. Bio and organic amendments in mulched plots along with reduced dose of inorganic fertilizers play a vital role in improving the physical, chemical and biological characteristics of soil, thereby optimizing the yield potential of crops. The use of organic mulches under water stress condition along with the systematic inputs of organic fertilizers including bio fertilizers may be recommended for sustainable farming (Fig. 1).

Fig. 1 Various fields where different mulches are used: **a** Pebbles as mulch, **b** Dry leaves as mulch, **c** Ash used as mulch, **d** Plastic used as mulch, **e** Stubbles as mulch

The role of mulch is not ignorable in agricultural practices. Like a coin having two sides, mulches also have some positive as well as negative impacts on the associated crops, weeds and soils. In addition, mulching helps in soil moisture retention, in controlling weeds growth, and in enhancing the crop yields. It also affects the soil pH, making the soil acidic that indirectly diminishes the fertility of soil and then decline the crop yield ultimately. Therefore, the selection of proper mulches by farmers is an important aspect with the purpose to increase the crop productivity. Thus, mulch type must be chosen by considering the soil type, the crop that is to be cultivated, the climatic conditions of the area, and the target weeds to be managed.

The physical methods also play major role in the organic and integrated plant protection strategies. Like the vegetable crop, mulches must also be regularly applied in the medicinal plants under row cultivation. There are numerous similarities in growing the vegetables and the medicinal plants including the fact that both are grown in smaller areas, pesticides are limitedly used in both and the weeds are the major trouble in their cultivation. Therefore, out of the various physical weed control methods, mulches seem to be the best option.

References

Angima. (2009). Season extension using mulches. Oregon State University Extension: Small Farms *4*, http://smallfarms.oregonstate.edu/sfn/f09SeasonMulches.

Ashrif, M. I., & Thornton. (1963). Effects of grass mulch on groundnuts in the Gambia. *Journal of Experimental Agriculture, 1*, 145–152.

Bakshi, P., Bhat, D., Wali, V. K., Sharma, A., & Iqbal, M. (2014). Growth, yield and quality pf strawberry (*Fragaria x ananassa* Duch) cv. Chandler as influenced by various mulching materials. *African Journal of Agricultural Research, 9*, 701–706.

Basu, T. K., & Bandyopadhyay, S. R. (2009). Productivity of rabi pigeon pea (*Cajanus cajan l. Milsp.)* as influenced by scheduling of irrigation. *Journal of Crop and Weed, 5*, 90–91.

Bhoutekar, S., Luchon, S., Bonti, G., & Sonbeer, C. (2017). Fertigation level and mulching in Cauliflower (*Brassica oleraceae* L. var. botrytis) cv. Snowball White. *International Journal of Agriculture Sciences, 9*, 4226–4228.

Bilalis, D., Sidiras, N., Economo, G., & Vakali, C. (2002). Effect of different levels of wheat straw soil surface coverage on weed flora in *Vicia faba* crops. *Journal of Agronomy and Crop Sciences, 189*, 233–241.

Chakraborty, R. C., & Sadhu, M. K. (1994). Effect of mulch type and color on growth and yield of Tomato. *Indian Journal of Agricultural Sciences, 64*, 608–620.

Chopra, M., & Koul, B. (2020). Comparative assessment of different types of mulching in various crops: A review. *Plant Archives, 20*, 1620–1626.

Cong, T., Jean, B. R., & Hu, S. (2005). Soil microbial biomass and activity in organic tomato farming system: Effect of organic inputs and straw mulching. *Journal of Soil Biology and Biochemistry, 20*, 1–9.

David, A.B. (2007). *A guide for desert and dryland restoration: New hope for arid lands* (p. 239). Island Press. ISBN 978-1-61091-082-8.

Gupta, J. P., & Yadav, R. C. (1979). A note on the efficiency of rubber mulch in conserving soil moisture. *Journal of Indian Forestry, 105*, 816–817.

Hoshino, A. T., Hata, F. T., Aquino, G. S. D., Junior, A. D. O. M.,Ventura, M. U., & Medina, C. D. C. (2017). Mulching with sugarcane straw reduces weed density in sugarcane field. *International Journal of Agriculture & Biology, 19*, 121–124.

Isenberg, F. M., & Odland, M. L. (1950). Comparative effects of various organic mulches and clean cultivation on yields of certain vegetable crops. Pennsylvania Agriculture Experimental Station Progress Rpt. No. 35.

Jaysawal, N., Singh, G., Kanojia, A., & Debbarma, B. (2018). Effect of different mulches on growth and yield of carrot (*Dacus carota* L.). *International Journal of Chemical Studies, 6*, 381–384.

Johnson, J. M., Goldstein, J. A., & Vangessel, M. J. (2004). Effect of straw mulch on pest insects, predators and weeds in watermelons and potatoes. *Journal of Environmental Entomology, 33*, 1632–1643.

Kumar, P., Kumar, S., Kumari, M., & Kumar, V. (2019). Effect of mulching on brinjle cultivation. *International Journal of Science, Environment and Technology, 8*(3), 624–629.

Kumar, S., & Dey, P. (2011). Effects of different mulches and irrigation methods on root growth, uptake, water- use efficiency and yield of strawberry. *Journal of Horticulture Science, 127*, 318–324.

Kumar, V., Bhatt, A. K., Sharma, V., Gupta, N., Sohan, P., & Singh, V. P. (2015). Effect of different mulches on soil moisture, growth and yield of Eureka Lemon under rainfed conditions. *Indian Journal of Dryland Agriculture Research and Development, 30*, 83–88.

Lal, R. (1974). Soil temperature, soil moisture and maize yield from mulched and unmulched tropical soils. *Journal of Plant and Soils Sciences, 40*, 129–145.

Larentzaki, E., Plate, J., Nault, B. A., & Shelton, A. M. (2008). Impact of straw mulch on population of onion thrips in onion. *Journal of Economic Entomology, 101*, 1317–1324.

Mahdiesh, N. M. B., Peyvast, G., Asil, M. H., Olfati, J. A., & Rabiee, M. (2012). Mulching effects on the yield and quality of garlic as second crop in rice fields. *International Journal of Plant Production, 6*, 1735–8043.

Ngouajio, M. (2011). Managing plastic mulches profitably. Michigan State University Extension. http://msue.anr.msu.edu/news/managingplasticulchesprofitably.

Orzolek, M. D., & Lamont, W. J. (2015). Summary and recommendation for the use of mulch color in vegetable production. Penn State Ext.

Penn State Extension. (2015). Plastic mulches, Penn State Extension, College of Agriculture Sciences.

Qin, W., Hu, C., & Oenema, O. (2015). Soil mulching significantly enhances yields and water and nitrogen use efficiencies of maize and wheat: A meta analysis. *Scientific Reports, 5*, 16210.

Rajbir, S., Ram, A., & Kumar, S. (2006). Effect of plastic tunnel and mulching on growth and yield of strawberry. *Indian Journal of Horticulture, 16*, 18–20.

Robinson, D. W. (1988). Mulches and herbicides in ornamental plantings. *Journal of Horticultural Sciences, 23*, 547–552.

Sanders, D. (2001). Using plastic mulches and drip irrigation for home vegetable gardens. Horticulture information leaflet. North Carolina Extension Resources. http://content.ces.ncsu.edu/using-plastic-mulches-anddrip-irrigation-for-vegetable-gardens/.2001.

Saucke, H., & Doring, T. F. (2004). Potato virus Y reduction by straw mulch inorganic potatoes. *Journal of an Appied. Biology, 144*, 347–355.

Sharma, A. R., Singh, R., Dhyani, S. K., & Dube, R. K. (2009). Moisture conservation and nitrogen recycling through legume mulching in rainfed maize-wheat cropping system. *Journal of Nutrient Cycle Agroecosystem, 87*, 187–197.

Stelli, S., Hoy, L., Hendrick, R., & Taylor, M. (2000). Effects of different mulches on soil moisture content in potted shrubs. *Journal of Applied Behavioural Ecology and Ecosystem Research Unit* (UNISA, 0003, South Africa).

Tiedjens, V. A. (1940). Mulching vegetable crops. *Journal of New Jersey Agricultural Experiment Station Mimeo.*

Tiffany, M., & Drost, D. (2016). Use of plastic mulch for vegetable production. *Journal of Horticulture.*

Wang, F. X., Feng, S. Y., Hou, X.-Y., Kang, S., & Han, J. (2008). Potato growth with and without plastic mulch in two typical regions of Northern China. *Journal of Field Crop Research, 110*, 123–129.

Agronomic and Economic Valuation of Mulches

Rashid Mahmood, Sajid Farooq, Aqib Hameed, and Muhammad Riaz

Abstract A good soil health ensures more, healthy, environment friendly and economical crop production. Weeds cause serious loss of crop yields and economic return by competing with the plants for space, nutrients, and water. Crops cultivated with more line to line and plant to plant spaces can get more benefit from mulching and hence mulching can add more to the economic return. It is because of the reason that large bare soil surface in between crop plants evaporates more water and gives more space to weed germination and growth. Moreover, mulches boost up economic return by reducing the cost of multiple herbicide and insecticide sprays, and by eliminating competition of weeds with crop plants. Further, high value crops like potato, tomato and pea etc. give more economic return on mulch application because per unit increase in economic yield is more valuable. On the other hand low value crops like coriander, toria etc. cannot afford mulching cost and ultimately result in low benefit to cost ratio in comparison to un-mulched crop.

Keywords Weeds · Nutrients · Water · Mulch · Crop production · Economic return

1 Contribution of Mulching in Economic Return

Mulching can contribute to economic return through various ways that are discussed under the following headings and are graphically presented in Fig. 1.

R. Mahmood (✉)
Department of Soil Science, University of the Punjab, Lahore, Pakistan
e-mail: rashid.ss@pu.edu.pk

S. Farooq · A. Hameed
Department of Agronomy, University of the Punjab, Lahore, Pakistan

M. Riaz
Department of Environmental Sciences & Engineering, Government College University, Faisalabad, Pakistan

K. Akhtar et al. (eds.), *Mulching in Agroecosystems*,
https://doi.org/10.1007/978-981-19-6410-7_10

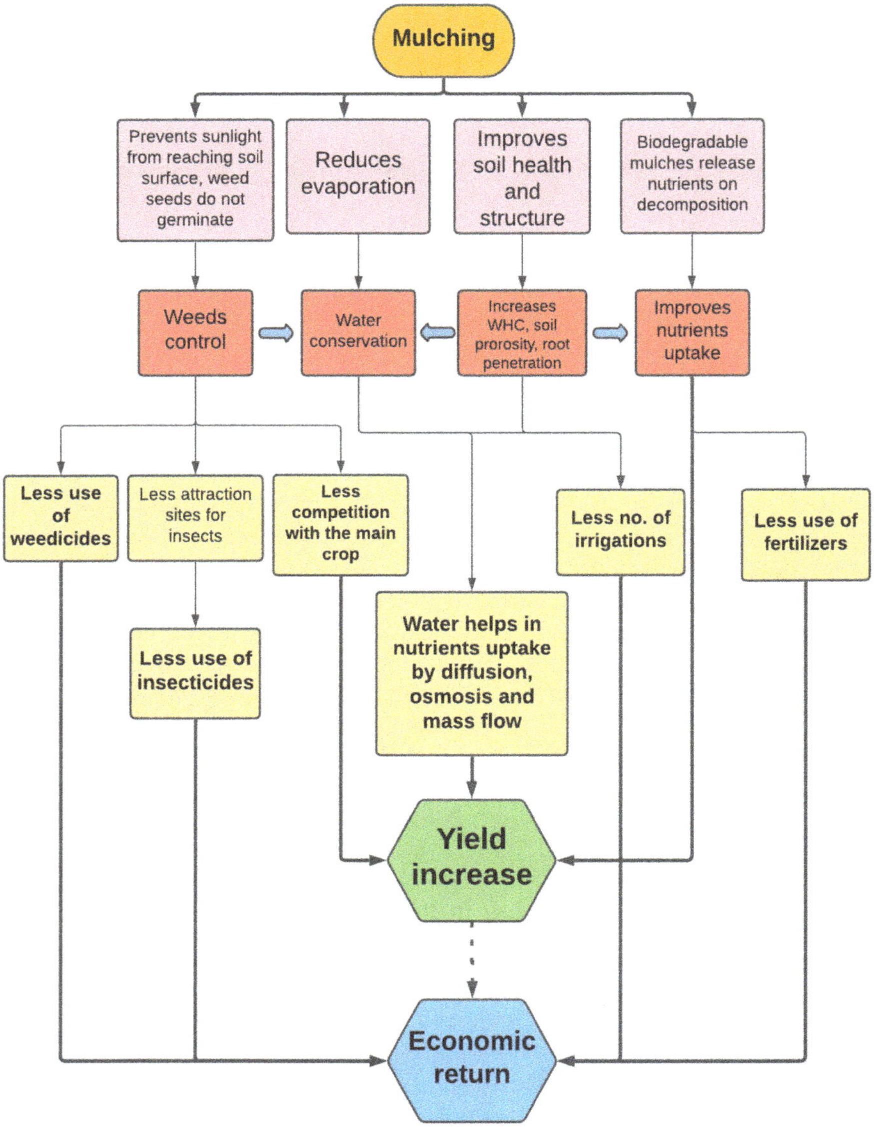

Fig. 1 Schematic contribution of mulching to crop economic return

1.1 *Economic Benefit of Mulching Through Weed Control*

Weeds cause serious loss of crop yields and economic return by competing with the plants for space, nutrients, and water. They also act as alternate host for insect pest and diseases and a hurdle in field operations (Patterson et al., 1999). According to an estimation of Weed Science Society of America (WSSA), weeds can cause a yield loss up to 47% in wheat, 52% in corn and 49.5% in soybean (Anonymous, 2021). In case of major crops like wheat, rice, cotton and maize, farmers spend a reasonable input cost on pre and post emergence weedicide sprays. Application of an effective mulch can control weed infestation by acting as physical barrier and preventing sunlight from reaching the soil surface. In comparison to weedicides, cost on mulches, particularly that of plastic films, farmers bear once during crop growth period and provide an effective weed control. This is a benefit of mulches over the repeated use of pre and post emergence, narrow and broad leaf herbicides. Mulches boost up economic return by reducing the cost of multiple herbicide and insecticide sprays, and by eliminating competition of weeds with crop plants.

1.2 Economic Benefit of Mulching Through Water Conservation

Depending upon nature of the crop, soil type, weather conditions, water availability, irrigation method and other agronomic practices adopted, there is a great variation in share of input cost spend on irrigation. Whatsoever the case is, evaporation of moisture from the soil surface is of major concern in reducing irrigation use efficiency of crop plants and ultimately result in decrease in economic return. These losses exceed to a maximum level when plant to plant distance is too high to cover the soil surface with crop canopy particularly at early crop growth stages (Al-Kaisi et al., 2009). Mulches act as a physical barrier over the soil surface and reduce evaporation and increase irrigation efficiency by decreasing irrigation frequency. This ultimately boost up economic return by saving input cost of frequent irrigation (Gao et al., 2019).

1.3 Economic Benefit of Mulching Through Soil Health Improvement

Soil is increasingly recognized as an important non-renewable natural asset that should be properly managed to ensure sustainable development. Mulching improves soil health in many ways. It increases water absorption by reducing evaporation and increasing moisture retention in the soil. It regulates soil temperature in favor of soil biology and soil organic matter retention, which in turn improves soil structure and other physical properties (Leroy et al., 2008). By reducing evaporation, mulching discourages upward movement of salts in the soil profile and their accumulation in the root zone (Dong et al., 2009). Mulching with crop residues not only becomes a part of soil organic matter but also improves soil fertility on decomposition (Jodaugienė et al., 2010). A good soil health ensures more, healthy, environment friendly and economical crop production (Gamliel & Bruggen, 2016).

2 Factors Affecting Mulch Based Economic Return of a Crop

The factors which can affect mulch based economic return are grouped into following four categories.

2.1 *Amount of Irrigation*

Besides other factors, change in economic benefit of a crop due to mulching also depends on the amount of irrigation applied to a crop. This becomes even more important in scenarios where irrigation cost is high, and the main purpose of mulching is water conservation. According to the data presented in Table 1 minimum benefit to cost ratio was noted where irrigation was applied at zero, or at or near 100% of crop irrigation requirement. At minimum irrigation, someone can save maximum input cost but obviously it would be at the expense of crop yield and economic output. On the other hand, irrigation at about 100% crop irrigation requirement increases cost and due to the availability of sufficient amount of water it reduces the utilization of water conserved by mulching. It is concluded from the data of the two trials presented in Table 1 that water application at 50–60% crop irrigation requirement is the best to gain maximum economic return by mulching (Biswas et al., 2015; Zhang et al., 2010) (Fig. 2).

Table 1 Impact of amount of irrigation on mulch based economic return of crops

Crop	Mulch	Irrigation (% of requirement)	Change in BCR due to mulching (%)	Country	References
Tomato	Plastic film	100	0.37	Bangladesh	Biswas et al. (2015)
		75	1.16		
		50	3.02		
	Paddy straw	100	4.18		
		75	6.14		
		50	7.03		
Wheat	Plastic film	0	−0.12	China	Xie et al. (2005)
		40	0.36		
		50	0.46		
		60	0.55		
		70	0.47		
		85	0.41		

BCR: benefit to cost ratio

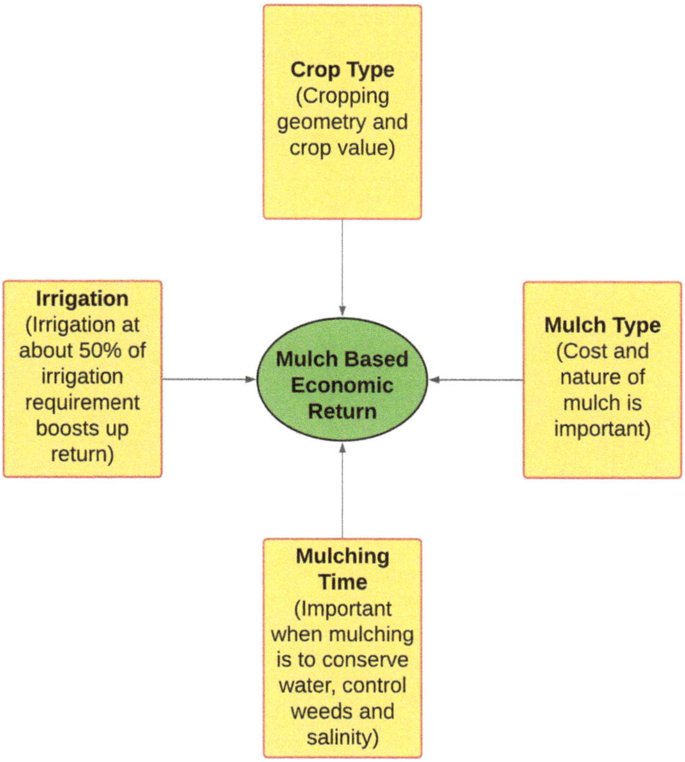

Fig. 2 Factors affecting mulch based economic return

2.2 *Time of Mulching*

Time of mulch application is critical in some cases particularly when in a rotation a crop is cultivated on residual moisture of a previous crop. For example, in wheat maize rotation application of mulch at 60 days of maize cultivation was found best in improving economic return of wheat cultivated thereafter. Compared to that mulching at 30 days of maize cultivation was found less effective probably due to early disintegration or decomposition of biological mulches. Similarly, a mulch applied after harvesting of maize was too late to conserve enough moisture for successful cultivation of wheat crop (Table 2; Sharma et al., 2011).

A second case presented in Table 2 is control of salinity through mulching. A pre-sowing mulch application significantly reduced evaporation and hence salt accumulation in the root zone. However, post sowing application of plastic mulch was less effective (Table 2; Dong et al., 2009).

Table 2 Impact of time of mulching on economic return of crops

Crop	Mulch	Time of mulching	Change in economic return/BCR due to mulching (%)	Country	References
Wheat in rotation with maize	*Peuraria hirsute* applied at	30 days of maize	2	India	Sharma et al. (2011)
		60 days of maize	13		
		After maize harvest	−5		
		After wheat sowing	−1		
	Lantana camara applied at	30 days of maize	2		
		60 days of maize	10		
		After maize harvest	−3		
		After wheat sowing	2		
	Leucaena leucocephala applied at	30 days of maize	5		
		60 days of maize	12		
		After maize harvest	9		
		After wheat sowing	12		
Cotton at saline soil	Plastic film	After sowing	2	China	Dong et al. (2009)
		Pre-sowing	5		

BCR: benefit to cost ratio

2.3 Crop Type

Crops cultivated with more line to line and plant to plant spaces can get more benefit from mulching and hence mulching can add more to the economic return. It is because of the reason that large bare soil surface in between crop plants evaporates more water and gives more space to weed germination and growth.

High value crops like potato, tomato and pea etc. give more economic return on mulch application because per unit increase in economic yield is more valuable. On the other hand low value crops like coriander, toria etc. can not afford mulching cost and ultimately result in low benefit to cost ratio in comparison to un-mulched crop (Table 3).

Table 3 Effect of crop and mulch types on mulch based economic return

Crop	Mulch	Scenario	Change in economic return/BCR due to mulching (%)	Country	References
Potato	Paddy straw	–	46	India	Rautaray (2016)
Tomato	Paddy straw	–	90		
Pea	Paddy straw	–	15		
Toria	Paddy straw	–	−13		
Lentle	Paddy straw	–	12		
Gram	Paddy straw	–	26		
Coriander	Paddy straw	–	−73		
Potato	Water hyacinth	–	56		
Maize (no tillage)	Wheat straw	Cultivated after no tillage wheat	−25	India	Ram et al. (2012)
Pepper	Plastic sheet	Non-degradable	80	Spain	Marí et al. (2019)
		Degradable	83		
	Paper	–	19		

2.4 *Mulch Type*

In most of the cases plastic film mulches better increased economic return obviously due to effective coverage of the soil surface to reduce evaporation and weed seed germination. However, disintegrated fragments of the sheet are difficult to remove from the field at the end of crop season. This may cause soil pollution. On the other hand, biodegradable straw mulches even if have effective coverage of the soil surface can lower yield parameters due to wider C:N ratio. Live mulches of leguminous plants are the best option for being having l ow C:N ratio and no contamination of the soil as in case of plastic film mulch (Sharma et al., 2011) (Table 3).

3 Mulching in Economics of Direct Seeded Rice (DSR) Cultivation; A Special Case

Rice is an important cash crop cultivated in most areas of the world by puddling method. The puddling is very suitable with crop growth and yield perspective. However, it is time consuming, labor intensive, requires large amount of water and expensive. To avoid puddling, various resource conservation technologies for rice are being developed and used in different areas of the world. Direct seeded rice (DSR) under un-puddled conditions is one of such techniques which is more water efficient, and labor and cost-effective. The major problem in DSR is weed infestation, which is otherwise controlled due to tillage and submergence in case of puddling. According to an estimation, transplantation of rice after puddling can reduce weed infestation up to 86%, obviously due to submergence maintained due to continuous flooding of the field, which is not the case in DSR.

In DSR, mulching is found effective in controlling weed infestation and the plastic film mulch is noted to be the most effective in covering inter row space and making the temperature and light conditions unfavorable for weeds growth. As studied by Ehsanullah et al. (2014), control of weed infestation with plastic mulching in DSR can increase plant height (9%), number of fertile tillers (15%), panicle length (11%), number of grains per panicle (11%), 1000 grains weight (12%) and paddy yield (29%). They also tried biodegradable mulches in the form of crop residues of maize, berseem, wheat and sunflower applied at 5 tons ha^{-1}. All biodegradable mulches were less effective than plastic film mulch and they did not effectively control weed infestation. However, owing to high cost of plastic film, economic return as benefit to cost ratio of plastic mulch DSR decreased in comparison to no mulch DSR (Table 4).

Alternatively, in areas where rice-wheat rotation is in practice, large amounts of surplus wheat and rice residues are available in the fields for mulching at no or nominal cost. Straw mulching can increase benefit to cost ratio of DSR particularly where DSR already gives better economic return than puddled rice probably due to less weed seed inoculum in the soil (Table 4; Yadav et al., 2016). In less weed seed load, wheat straw mulch can increase paddy yield of DSR up to 10%, not only due to weed control and water saving but also due to better nitrogen, phosphorus, potassium, and iron uptake. However, residual beneficial effects of wheat straw mulch are rarely shifted to next wheat crop yield (Yadav et al., 2016).

In some areas where straw is a very valuable feed for livestock, use of live mulches like *Sesbania aculeata* is a very good option. These mulches are in most of the cases are good green manuring legume crops which can be cultivated between the lines of DSR and after 30 days or so these are chopped and spread on the soil surface. Being a green manure crop, Sesbania can increase DSR yield and economic return even more than that of wheat straw mulch (Table 4).

Table 4 Impact of mulching on yield and benefit to cost ratio of direct seeded rice (DSR)

Mulch type	Increase in yield due to mulching	Increase/decrease in benefit to cost ratio in comparison to puddled Rice	References	Country
No mulch	–	−0.18	Ehsanullah et al. (2014)	Pakistan
Maize straw	25%	−0.07		
Plastic sheet	31.25%	−0.24		
Wheat straw	21.87%	−0.15		
Sunflower straw	18.75%	0.01		
Berseem clippings	25%	−0.08		
No mulch	–	−0.47	Jabran et al. (2016)	Pakistan
Plastic mulch	19.85%	−0.41		
Straw mulch	6.50%	−0.58		
No mulch	–	0.32	Yadav et al. (2016)	India
Wheat straw	10.65%	0.33		
Sasbania aculeata mulch	13.15%	0.50		

4 Conclusion

Mulches boost up economic return by reducing the cost of multiple herbicide and insecticide sprays, and by eliminating competition of weeds with crop plants. Further, high value crops like potato, tomato and pea etc. give more economic return on mulch application because per unit increase in economic yield is more valuable. On the other hand low value crops like coriander, toria etc. cannot afford mulching cost and ultimately result in low benefit to cost ratio in comparison to un-mulched crop. Mulch also helps plant to sustain in drought condition and therefore improve crop production.

References

Al-Kaisi, M. M., Broner, I., & Andales, A. A. (2009). *Crop water use and growth stages.* Colorado State University.

Anonymous. (2021). 'Crop loss due to weeds' Weed Science Society of America. https://wssa.net/wssa/weed/croploss-2/.

Biswas, S. K., et al. (2015). Effect of drip irrigation and mulching on yield, water-use efficiency and economics of tomato. *Plant, Soil and Environment, 61*(3), 97–102.

Dong, H., et al. (2009). Early plastic mulching increases stand establishment and lint yield of cotton in saline fields. *Field Crops Research, 111*, 269–275. https://doi.org/10.1016/j.fcr.2009.01.001.

Ehsanullah, R. Q., et al. (2014). Growth and economic assessment of mulches in aerobic rice (Oryza sativa L.). *Journal of Agricultural Research (03681157), 52*(3).

Gamliel, A., & Van Bruggen, A. H. C. (2016). Scientia horticulturae maintaining soil health for crop production in organic greenhouses. *Scientia Horticulturae, 208*, 120–130 (Elsevier B.V.). https://doi.org/10.1016/j.scienta.2015.12.030.

Gao, H., et al. (2019). Exploring optimal soil mulching to enhance yield and water use e ffi ciency in maize cropping in China : A meta-analysis. *Agricultural Water Management, 225*(July), 105741 (Elsevier). https://doi.org/10.1016/j.agwat.2019.105741.

Jabran, K., et al. (2016). Economic assessment of different mulches in conventional and water-saving rice production systems. (February). https://doi.org/10.1007/s11356-016-6162-y.

Jodaugienė, D., et al. (2010). The influence of organic mulches on soil biological properties. *Zemdirbyste-Agriculture, 97*(2), 33–40.

Leroy, B. L. M., et al. (2008). The quality of exogenous organic matter: short-term effects on soil physical properties and soil organic matter fractions. (June), 139–147. https://doi.org/10.1111/j.1475-2743.2008.00142.x.

Marí, A. I., et al. (2019). Economic evaluation of biodegradable plastic films and paper mulches used in open-air grown pepper (Capsicum annum L.) crop. *Agronomy, 9*(1), 36 (Multidisciplinary Digital Publishing Institute).

Patterson, D. T., et al. (1999). Weeds, insects, and diseases. *Climatic Change, 43*(4), 711–727 (Springer).

Ram, H., et al. (2012). Agronomic and economic evaluation of permanent raised beds, no tillage and straw mulching for an irrigated maize-wheat system in northwest India. *Experimental Agriculture, 48*(1), 21 (Cambridge University Press).

Rautaray, S. K. (2016). Benefits of mulching with dried water. (August).

Sharma, A. R., et al. (2011). Agronomic and economic evaluation of mulching in rainfed maize—Wheat cropping system in the Western Himalayan Region of India, *7528*. https://doi.org/10.1080/15427528.2011.574498.

Xie, Z., Wang, Y., & Li, F. (2005). Effect of plastic mulching on soil water use and spring wheat yield in arid region of northwest China, *75*, 71–83. https://doi.org/10.1016/j.agwat.2004.12.014.

Yadav, G. S., et al. (2016). Agronomic evaluation of mulching and iron nutrition on productivity, nutrient uptake, iron use efficiency and economics of aerobic rice-wheat cropping system. *Journal of Plant Nutrition, 39*(1), 116–135 (Taylor & Francis). https://doi.org/10.1080/01904167.2015.1084323.

Zhang, F., et al. (2010). Potassium nutrition of crops under varied regimes of nitrogen supply. *Plant and Soil, 335*(1–2), 21–34 (Springer).

Mulching and Nutrients Use Efficiencies in Plant

Mukkram Ali Tahir, Noor-us-Saba, Amir Aziz, and Adeel Ahmad

Abstract Rapid urbanization and industrialization have resulted in an increased global temperature over the year. Consequently, the agro-ecological system disturbing worldwide. Therefore, new agricultural practices that are eco-friendly are needed. Mulching could potentially serve the purpose by conserving moisture, reducing weed growth, reducing soil evaporation, improving microbial activities and controlling soil temperature. Additionally, mulches could provide environmental and economical advantages to agriculture and landscape and enhance the nutrient status in soil. This review chapter focuses on multiple significant impacts of mulches on nutrient use efficiencies in the plant. Secondly, discuss problems regarding nutrients use efficiencies and loss of nutrient from soil system and also discussed strategies to improving nutrient use efficiencies. This discussion leads to improve the nutrients use efficiencies in the plant by mulching.

1 Background

The Agricultural and food industry made huge progress in the last five decades throughout the world (Alexandratos, 1999). In the coming fifty years it is estimated that the population will increase continuously and consequently the demand for soil, water and nutrients will also increase to fulfil the food requirements of people (Godfray et al., 2010; Tilman et al., 2001). Therefore, it is a need of the hour to boost the productivity of agricultural goods. Mulching is an important practice used in agriculture for increased productivity and nitrogen-containing fertilizers have also been used (Qin et al., 2015; Tilman et al., 2002; Wang et al., 2016). It is also necessary to minimize the environmental risks caused by modern agriculture. That's why it is

M. A. Tahir · Noor-us-Saba · A. Aziz (✉)
Department of Soil and Environmental Sciences, College of Agriculture, University of Sargodha, Sargodha, Pakistan
e-mail: amieazizuos@gmail.com

A. Ahmad
Institute of Soil and Environmental Science, University of Agriculture Faisalabad, Faisalabad, Pakistan

K. Akhtar et al. (eds.), *Mulching in Agroecosystems*,
https://doi.org/10.1007/978-981-19-6410-7_11

essential to discuss the impacts of nitrogen fertilizers as well as mulching on the soil along with the requirement of the plants. Presently mulching is used for the cultivation of many crops namely maize, rice, wheat, potatoes, barley, sunflowers, groundnuts, hairy vetch, coffee, okra, turmeric, green grams, rosemary, mint, fruit and vegetable trees (Alliaume et al., 2017; Li et al., 2001; Liuet al., 2014a, 2014b; Nzeyimana et al., 2017; Qin et al., 2014; Singh, 2013).

In general, nitrogen fertilizer and mulch application show the best performance in terms of agricultural production (Fan et al., 2005; Liu et al., 2014a, 2014b; Mo et al., 2017; Rahman et al., 2005; Wang et al., 2015). Application of N fertilizers plays a vital role in improved crop yield but excessive use is imparted negative effects on soil health (Han et al., 2015; Xu et al., 2012). Therefore, it is necessary to consider the impacts of N fertilizers on the crop, soil and the environment. The main sources of nitrogen are both organic and inorganic namely crop stubbles, farmyard manure, compost, biological fixation, urea, ammonium bicarbonate, ammonium nitrate, ammonium hydroxide and ammonium sulfate respectively (Agehara & Warncke, 2005; Crews & Peoples, 2004; Das & Adhya, 2014). The available forms of N are NO_{3^-} and NH_{4^+} in dry soil and flooded soils respectively (Krapp, 2015; Xu et al., 2012).

2 Importance of Nutrition for Plants

The nutrient requirement of plant is similar as for animals. The nutrients are important for seed germination, plant growth, and insect pest resistance and reproduction. For plant, health nutrients are required in variable concentrations (Sainju et al., 2008). In plants nutrients play different roles like these are the building blocks of cells, it modifies the osmotic as well as turgor pressure, metabolic reactions and enzymatic activities. These all functions must be performed smoothly to improve the plant yield (Zhao et al., 2016). The nutrients that are essential for plants are seventeen in number. Among these nutrients that are required in higher quantities are called macronutrients (N, P, K, S, Mg, and Ca. However, those required in lower quantity are called micronutrients like Fe, Mn, B, Mo, Cu, Zn, Cl, Co. It is not possible to recognize the knowledge of plant nutrition. This is because plants have many species and all species are variable from each other. If the nutrients are not available according to the need of the plants this may lead to nutrient deficiency. On the other hand, if these nutrients are present in excess of plant need this may lead to nutrient toxicity. Moreover, there is the possibility that excess of one nutrient may suppress the other nutrient like excess of ammonium ion may suppress the uptake of potassium ion (Norman & Hunter, 2008). The concentration of nitrogen in the atmosphere is 78% and this atmospheric nitrogen is now changing to available form through nitrogen fixation. But plants mostly meet their nitrogen requirement from the soil. The nitrogen present in the soil is in most plant-available form. Though the nitrogen in the atmosphere is in higher concentration its utilization demands a lot of energy to change it into a plant-available form. The plants that fix atmospheric nitrogen are mostly beans, gram, alfalfa etc. However, rice, wheat, cotton and other commercial crops uptake nitrogen from soil (Wang et al., 2017).

3 Problems Regarding Plant Nutrition

In developing countries, the loss of soil fertility is now becoming serious environmental degradation and affect the crop yield to a higher extent. The main causes of fertility losses are nutrient diminution and nutrient depletion. This may affect food production and consequently the lives of many people. Due to fertility loss water holding capacity of oil is affected and results in drought condition (Bodner et al., 2007). Soil fertility is an important factor for farmers and the whole ecosystem. The main purpose of a farmer is to sustain the fertility of his soil. This purpose can only be achieved by improving soil structure, proper airing, appropriate soil moisture, suitable pH and favourable nutrient concentrations. To manage such a vast system is very difficult. The soil fertility can be enhanced or decreased depending upon the cropping pattern, number of animals rearing on the farm and management practices. It is concluded that if we want to sustain soil fertility, a balance should be maintained between nutrient removed and replaced in each crop rotation (Souri & Hatamian, 2019).

4 Losses of Macronutrients

The deficiency of nitrogen causes slow growth, chlorosis and stunted growth. Plants having nitrogen-deficient accumulate anthocyanin pigments and result in the appearance of the purple stem, underside of leaves and petioles (Norman & Hunter, 2008). Deficiency of phosphorus shows the same symptoms as nitrogen deficiency characterized by more reddish or green colouration due to lack of chlorophyll in leaves. The plant leaves become denatured and show sign of death if the plant faces a high deficiency of phosphorus and leaves of the plant appear purple due to anthocyanin accumulation. According to Russel, deficiency of phosphorus fifer from deficiency of nitrogen and it is very difficult to diagnose the phosphorus deficiency. The deficiency of potassium causes necrosis or interveinal chlorosis. This deficiency leads to wilting, pathogens, brown spotting, and chlorosis and plant damage from heat and frost. Potassium deficiency affects older tissues and then progress toward growing points. Acute deficiency of potassium affects growing points, reduce growth in diameter and height, and reduced the needle length (Heiberg & White, 1950). Calcium deficiency affects the newly developed cell in the root system. Biological and root functions disrupt even short term disruption in calcium supply. Leaf curling is common symptoms of calcium deficiency that moves toward the centre of the leaf. Sometimes leaves have a blackened appearance. Leaves tips mat appear cracking and burned by the deficiency of calcium if they face sudden humidity increase. Calcium deficiency mostly arises in tissues causing blossom end rot (watermelons), tomatoes and peppers, bitter pits in apple and empty peanut (White & Broadley, 2003).

5 Losses of Micronutrients

Molybdenum (Mo) deficiency usually occurs in older growth. Iron (Fe), Copper (Cu) and Manganese (Mn) affect new growth, causing yellow or green veins. Zinc (Zn) can affect new or old leaves and Boron (B) seems on terminal buds. Due to reduced internodal expansion, plant leaves on top of each other in zinc deficiency. For industrial crop cultivation, the most widely deficient plant nutrient is zinc, followed by boron. The deficiency of boron affecting pollen fertility and seed yields are common in laterite soils. For cell wall strengthening and proper forming, boron is essential. Due to boron deficiency, short thick cells produced stunted roots and fruiting bodies. Boron deficiency results in stunted growth and death of the terminal growing points. The deficiency of boron can be fined by analysis of plant material. Boron deficiency in strawberries will produce lumpy fruit and apricots will drop their fruit or not fruit depending on boron deficient level. Foliar application of boron is immediate but must be repeated and broadcast supplements of boron are very effective and long term (Heiberg & White, 1950).

6 Strategies for Improving Nutrient Use Efficiencies

6.1 Crop Variety/Species

With the help of plant breeding and genesis, we can increase the nutrient use efficiency by selecting those species/genotype of plant that is more efficient to the uptake of plant nutrients from the soil system. Generally, genotype closely linked with extensive and efficient root system or effective associations with mycorrhizal fungi in order to access the volume of soil (Ramaekers et al., 2010).

6.2 Rate and Time of Fertilizer Application

With the help of the right time and right rate application of fertilizer, we can enhance nutrient use efficiency. The best time of application of fertilizer (P) is at sowing time. In sandy soil, fertilizer can be applied in the split application. The rate of application of fertilizer is very important to nutrient use efficiency. Adding fertilizer to a soil system that already has a sufficient amount of plant is wasteful and lead to nutrient losses to water bodies. Soil testing is the only way to find the correct rate of fertilizer requirement along with other agronomic considerations (Roberts, 2007).

7 Mulching and Its Importance in Plant Nutrition

Mulching is important in plant nutrition, it is because:

- Mulch prevents the compaction of the soil.
- It also reduces lawnmower damage.
- It also keeps weeds out to help to prevent root compaction.
- It also retains water that helps to moist the roots.
- It also provides a buffer from cold and heat temperature.

Soil moisture conditions may improve by the use of mulch on the top surface of the soil. Mulch also increases the yield of crop by improving soil physical conditions. As compared to um-mulched soil, different types of mulch material increase the soil moisture and ultimately decreased the evaporation rate from the surface of the soil (Maged, 2006). Mineral mulch is more effective to impervious water vapour and expected to conserve the moisture of soil as compared to organic mulch (Lei et al., 2004). Tillage and mulching used in combination also increased soil water conservation. Mulched soil, almost 0–60 cm soil layer contains more moisture content as compare to un-mulch soil (Ramakrishna et al., 2006). By the use of mulch, the greatest reduction in soil moisture content showed in soil; 92% soil moisture content at 10 cm soil, 83% soil moisture content at 5 cm soil and 52% soil moisture content at 2 cm soil (Diaz et al., 2005). Some researchers also experimented in the laboratory to check the effect of gravel mulch on evaporation (Mellouli et al., 2000). By this research, soil surface covering with gravels and coarse sand can reduce 10–20% evaporation as compared to un-mulched soil. Soil surface area available for evaporation decreases by the use of gravel mulch material. In many crops, mulching increased productivity by conserving the soil moisture (Huang et al., 2005; Rahman et al., 2005; Zhang et al., 2005). While mulching material also controlling the growth of weeds (Erenstein, 2002). In wheat crop, uptake of water is increased by using mulch combined with an irrigation system (Li et al., 2004). Mulch decreases capillary diffusion during the first stage of evaporation and water moves mostly vapour phase from the soil surface to mulch surface (Li, 2003). On the other hand, mulch also reduces the evaporation of water from the soil by shading the surface of the soil from the sun (Shading is most effective when soil is wet during the first stage of evaporation). Different type of mulch also affects soil temperature. Furthermore, researchers found that the mulch influenced the temperature of soil (Kar & Singh, 2004).

7.1 Decline of Weeds

In nursery and field conditions, mulching is an important tool for controlling the population of weeds. However, the weeds reduction phenomenon is not fully understood till now. But weeds the population reduces 92% as compared to non-mulched soil. Mulch act as a barrier and cannot light pass, resulting in reduced the small seed of weed germination after the spread of mulch on the surface of the soil. Researchers also found that almost fifteen different types of mulches were used and results that there were no significant differences between all types of mulch but a huge difference exists for weeds reduction when compare with non-mulched soil (Kader et al., 2019). Mulches act as a physical barrier to the emergence of weeds. However, weeds seed quickly come out of the surface of the soil when decomposed the organic mulches (Ahmad et al., 2015, 2020). Some mulches create an environment that is very beneficial for microbes (Chalker-Scott, 2007).

In different mulch materials, by using organic and inorganic mulch materials at adequate soil depth are used widely for the reduction of weed control and these materials help to prevent the soil from compaction. On the other hand, inorganic mulch materials (gravel and stone mulch) avoid weed species colonization when used at 4 cm depth of soil. Organic mulches control the colonization of weed in different ways. Compost (organic mulch) does not control the weeds because compost is full of nutrients and fertile the soil (Maclean et al., 2003). A thick layer of mulch material reduces the seed germination otherwise, a thin layer of mulch material enhance the germination of weed seed (Rokich et al., 2002). Sawdust thick layer will be helpful for the exchange of gas and water (Stenn, 2005). Mulching also reduces the penetration of light that helps to stop the photosynthesis process in weeds and ultimately weeds cannot use nutrients from the soil. So mulching is the best strategy to save the pant nutrients to the uptake of weeds (Ahmad et al., 2015, 2020).

7.2 Soil Moisture Conservation

Some factors (abiotic) are responsible for the loss of plant nutrients and soil moisture and convert it into barren land. These factors including harsh climate conditions, high winds, temperature level elevation and competing plantation. It has seemed that up to 25% water loss due to the presence of weeds by the process of evapotranspiration. On the other hand, straw mulch decreases the evaporation rate by 35% (Harris, 1992). A few advantages of mulching have been shown in Fig. 1.

Organic and inorganic mulches have conserved the water of soil as compared to synthetic and barren soil (Lakatos et al., 2000). Generally, plant residues, livestock wastes and different types of stone gravels are used to retain soil moisture (Siipilehto, 2001). The irrigation requirement of pants can be reduced and sometimes the need for irrigation can be finished by the use of mulches (Kader et al., 2019; Iqbal et al., 2019). Surface runoff of water also reduces by used straw mulch up to 43%. Supplemental

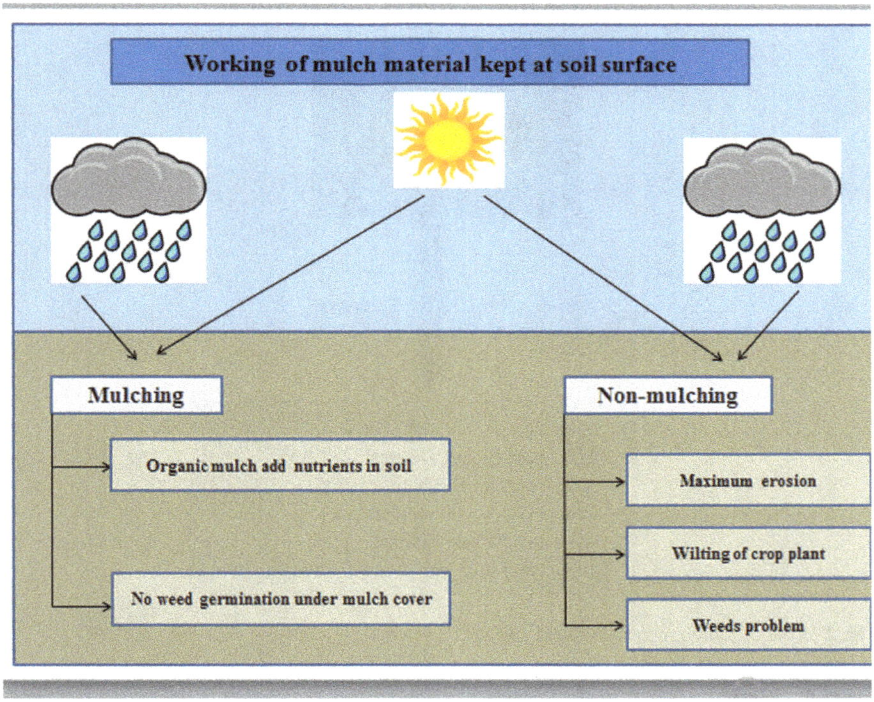

Fig. 1 Working of mulch material kept at the soil surface

irrigation is also reduced by the use of mulches. It is because water runoff decreases and water retention ability increases (Smith, 2000).

8 Role of Mulching to Reduce Nutrient Losses

8.1 Minimizing Soil Compaction and Erosion

Soil can be protected from water and wind erosion with the help of mulching material and also reduce the soil compaction which can negatively affect the plant roots and consequently decrease plant growth. Some legumes and grasses are used as living mulch. Grass growing (Living mulch) on slopes reduces the erosion of soil by aggregating the soil particles and makes a complex unit (Tanavud et al., 2001). Mulch material increases the rate of infiltration and also maintain the slope stabilization n hilly areas (Chalker-Scott, 2007). We can solve the compaction problem by the addition of organic mulch (Fig. 2). The researcher suggested that before the development of soil compaction, mulching should be performed. There will be no improvement in the soil after the compaction of soil by mulching (Oliveira & Merwin, 2001).

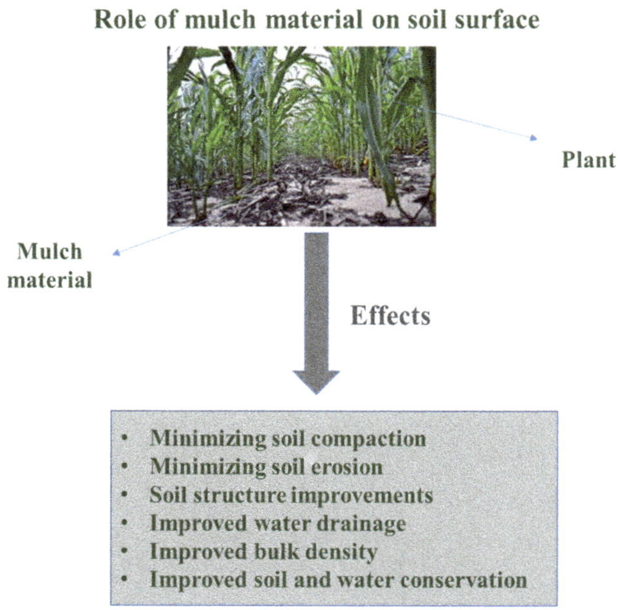

Fig. 2 Role of mulch material on soil surface

8.2 Soil Temperature Regulation

Soil surface covers by mulch and it's helpful to maintain the temperature of soil that is beneficial for the growth of the crop. Researchers found that soil remains cool during a very hot climate by the application of mulch (Kader et al., 2019; Long et al. 2001) while on chilling days soil remain warm/normal (Kader et al., 2019). High temperature adversely affects the nutrients and water uptake and also affects the newly growing roots. Newly plant roots are not able to uptake plant nutrients and adequate amount of water at high-temperature condition (Chalker-Scott, 2007). Therefore, soil temperature regulation is a very important factor for the growth of the plant. However, mulch decreases the 10 °C temperature in a dry and hot climate as compared to barren soil. To control the soil temperature, course mulch is more favourable as compared to fine ones. Different type of mulches is used to control the soil temperature. Some mulch (living mulches) increase the soil temperature due to solar radiation absorption as compared to barren soil (Montague & Kjelgren, 2004). Moreover, Researchers observed that organic and living mulch materials perform better in maintaining soil temperature as compared to other types of mulches. Living mulch decreases the soil temperature through increase evapotranspiration due to its cooling effect by evaporation effect (Montague & Kjelgren, 2004).

9 Influence of Mulching on Nutrient Use Efficiencies

For mulch selection, it is very important to know that how soil will be explored by the application of mulching. By organic mulches applications, we achieved more root density and development as compared to synthetic mulch, living mulch and un-mulch soil (Fausett & Rom, 2001). Some synthetic mulch material (film and sheet) perform as barriers to air and water which enhance root growth. Organic mulch performs best as compared to other mulches, it is because organic mulch provides water and nutrients to newly grown plant roots. If the root of plants successfully grows, then increase the survival of transplanted seedling under nursery and field conditions (Ansari et al., 2001). In early study, some researchers perform experiments and the outcomes of the study are the mulched crop performed better as compared to the control treatment. Turf mulch used as a competitive cover crop and also reduced plant growth rate as compared to un-mulched soil (Cahill et al., 2005).

10 Conclusion and Further Directions

It is clear from the above discussion that mulching has a strong influence on nutrients use efficiencies in the plant. The application of mulch not only conserves soil moisture, reduces soil evaporation, influences soil microorganisms, control soil structure and temperature but also improves soil moisture retention which is of great concern to any crop. It is necessary to understand the effect of various mulching material on crop yield, soil environment and nutrient use efficiency. Plastic mulching materials are more efficient than organic mulches. However, organic mulching materials are environmentally friendly, inexpensive and beneficial to soil microorganisms. Furthermore, mulching mitigates disease, insects and weeds and can further improve nutrients use efficiencies. In the future attention could be focused to accelerate crop production by further improving nutrients use efficiencies by using organic and inorganic mulches in combination, as organic mulches will enhance the organic matter of soil in addition to the improvements made by inorganic mulches.

References

Agehara, S., & Warncke, D. D. (2005). Soil moisture and temperature effects on nitrogen release from organic nitrogen sources. *Soil Science Society of America Journal, 69*, 1844–1855.

Ahmad, S., Raza, M. A. S., Saleem, M. F., Zahra, S. S., Khan, I. H., Ali, M., & Zaheer, M. S. (2015). Mulching strategies for weeds control and water conservation in cotton. *Journal of Agriculture and Biological Sciences, 8*, 299–306.

Ahmad, S., Raza, M. A. S., Saleem, M. F., Zaheer, M. S., Iqbal, R., Haider, I., Aslam, M. U., Ali, M., & Khan, I. H. (2020). Significance of partial root zone drying and mulches for water saving and weed suppression in wheat. *Journal of Animal and Plant Sciences, 30*, 154–162.

Alexandratos, N. (1999). World food and agriculture: Outlook for the medium and longer term. *Proceedings of the National Academy of Sciences of the United States of America, 96*, 5908–5914.

Alliaume, F., Rossing, W. A. H., Tittonell, P., & Dogliotti, S. (2017). Modelling soil tillage and mulching effects on soil water dynamics in raised-bed vegetable rotations. *European Journal of Agronomy, 82*, 268–281.

Ansari, R., Marcar, N. E., Khanzada, A. N., Shirazi, M. U., Crawford, D. F. (2001). Mulch application improves survival but not growth of Acacia ampliceps Maslin, Acacia niloticaL. and Conocarpuslancifolius L. on a saline site in southern Pakistan. *International Journal of Review, 3*, 158–163.

Bodner, G., Loiskandl, W., & Kaul, H. P. (2007). Cover crop evapotranspiration under semi-arid conditions using FAO dual crop coefficient method with water stress compensation. *Agricultural Water Management, 93*(3), 85–98.

Cahill, A., Chalker-Scott, L., & Ewing, K. (2005). Wood-chip mulch improves plant survival and establishment at no-maintenance restoration site (Washington). *Ecological Restoration, 23*, 212–213.

Chalker-Scott, L. (2007). Impact of mulches on landscape plants and the environment—A review. *Journal of Environmental Horticulture, 25*(4), 239–249.

Crews, T. E., & Peoples, M. B. (2004). Legume versus fertilizer sources of nitrogen: Ecological tradeoffs and human needs. *Agriculture, Ecosystems & Environment, 102*(3), 279–297.

Das, S., & Adhya, T. K. (2014). Effect of combine application of organic manure and inorganic fertilizer on methane and nitrous oxide emissions from a tropical flooded soil planted to rice. *Geoderma, 213*, 185–192.

Diaz, F., Jimenez, C. C., & Tejedor, M. (2005). Influence of the thickness and grain size of tephra mulch on soil water evaporation. *Agricultural Water Management, 74*(1), 47–55.

Erenstein, O. (2002). Crop residue mulching in tropical and semi-tropical countries: An evaluation of residue availability and other technological implications. *Soil and Tillage Research, 67*(2), 115–133.

Fan, M., Jiang, R., Liu, X., Zhang, F., Lu, S., Zeng, X., & Christie, P. (2005). Interactions between non-flooded mulching cultivation and varying nitrogen inputs in rice–wheat rotations. *Field Crops Research, 91*, 307–318.

Fausett, J. B., & Rom, C. R. (2001). The effects of transitioning a mature high-density orchard from standard herbicide ground-cover management system to organic ground-cover management systems. *Arkansas Agricultural Experiment Station Research Series, 483*, 33–36.

Fraedrich, S. W., & Ham, D. L. (1982). Wood chip mulching around maples: Effect on tree growth and soil characteristics. *Journal of Arboriculture, 8*, 85–89.

Godfray, H. C. J., Beddington, J. R., Crute, I. R., Haddad, L., Lawrence, D., Muir, J. F., Pretty, J., Robinson, S., Thomas, S. M., & Toulmin, C. (2010). Food security: The challenge of feeding 9 billion people. *Science, 327*, 812–818.

Han, M., Okamoto, M., Beatty, P. H., Rothstein, S. J., & Good, A. G. (2015). The genetics of nitrogen use efficiency in crop plants. *Annual Review of Genetics, 49*, 269–289.

Harris, R. W. (1992). *Arboriculture: Integrated management of landscape trees, shrubs, and vines* (No. 2). Prentice-Hall International.

Heiberg, S. O., White, D. P. (1950). Potassium deficiency of reforested Pine and spruce stands in northern New York. In *Potassium deficiency of reforested Pine and spruce stands in northern New York.*

Huang, Y., Chen, L., Fu, B., Huang, Z., & Gong, J. (2005). The wheat yields and water-use efficiency in the Loess Plateau: Straw mulch and irrigation effects. *Agricultural Water Management, 72*(3), 209–222.

Iqbal, R., Muhammad, A. S. R., Muhammad, F. S., Imran, H. K., Salman, A., Muhammad, S. Z., Muhammad, U., & Imran, H. (2019). Physiological and biochemical appraisal for mulching and partial rhizosphere drying of cotton. *Journal of Arid Land, 11*, 785–794.

Kader, M. A., Singha, A., Begum, M. A., Jewel, A., Khan, F. H., & Khan, N. I. (2019). Mulching as water-saving technique in dryland agriculture. *Bulletin of the National Research Centre, 43*(1), 1–6.

Kar, G., & Singh, R. (2004). Soil water retention—Transmission studies and enhancing water use efficiency of winter crops through soil surface modification. *Indian Journal of Soil Conservation, 8*, 18–23.

Krapp, A. (2015). Plant nitrogen assimilation and its regulation: A complex puzzle with missing pieces. *Current Opinion in Plant Biology, 25*, 115–122.

Lakatos, T., Buban, T., Muller, W., Polesny, F., Verheyden, C., & Webster, A. D. (2000). Effectiveness of different groundcover materials to preserve soil water content in a young apple orchard. *Acta Horticulture, 525*, 425–426.

Lei, Y., Hidenori, T., & Weiqiang, L. (2004). Effects of concrete mulch on soil thermal and moisture regimes. *Journal of Agricultural Meteorology, 60*(1), 17–23.

Li, X., Gong, J., Gao, Q., & Li, F. (2001). Incorporation of ridge and furrow method of rainfall harvesting with mulching for crop production under semiarid conditions. *Agricultural Water Management, 50*, 173–183.

Li, F. M., Wang, P., Wang, J., & Xu, J. Z. (2004). Effects of irrigation before sowing and plastic film mulching on yield and water uptake of spring wheat in semiarid Loess Plateau of China. *Agricultural Water Management, 67*(2), 77–88.

Li, X. Y. (2003). Gravel-sand mulch for soil and water conservation in the semiarid loess region of northwest China. *52*(2), 105–127.

Liu, C., Zhou, L., Jia, J., Wang, L., Si, J., Li, X., Pan, C., Siddique, K. H. M., & Li, F. (2014a). Maize yield and water balance is affected by nitrogen application in a film-mulching ridge–furrow system in a semiarid region of China. *European Journal of Agronomy, 52*, 103–111.

Liu, J., Zhu, L., Luo, S., Bu, L., Chen, X., Yue, S., & Li, S. (2014b). Response of nitrous oxide emission to soil mulching and nitrogen fertilization in semi-arid farmland. *Agriculture, Ecosystems & Environment, 188*, 20–28.

Long, C. E., Thorne, B. L., Breisch, N. L., & Douglass, L. W. (2001). Effect of organic and inorganic landscape mulches on subterranean termite (Isoptera: Rhinotermitidae) foraging activity. *Journal of Environmental Entomology, 30*, 832–836.

MacLean, R. H., Litsinger, J. A., Moody, K., Watson, A. K., & Libetario, E. M. (2003). Impact of Gliricidia sepium and Cassia spectabilis hedgerows on weeds and insect pests of upland rice. *Agriculture, Ecosystems & Environment, 94*(3), 275–288.

Maged, A. F. (2006). Radon concentrations in elementary schools in Kuwait. *Health Physics, 90*(3), 258–262.

Mellouli, H. J., Van Wesemael, B., Poesen, J., & Hartmann, R. (2000). Evaporation losses from bare soils as influenced by cultivation techniques in semi-arid regions. *Agricultural Water Management, 42*(3), 355–369.

Mo, F., Wang J. Y., Zhou, H., Luo, C. L., Zhang, X. F, Li, X. Y., Li, F. M., Xiong, L. B., Kavagi, L., Nguluu, S. N., & Xiong, Y. C. (2017). Ridge-furrow plastic-mulching with balanced fertilization in rainfed maize (Zea mays L.): An adaptive management in east African Plateau. *Agricultural and Forest Meteorology, 236*, 100–112.

Montague, T., & Kjelgren, R. (2004). Energy balance of six common landscape surfaces and the influence of surface properties on gas exchange of four containerized tree species. *Scientia Horticulturae, 100*, 229–249.

Norman, P. A., Huner, W. H. (2008). *Introduction to plant physiology*, 4th edn. Wiley. ISBN: 978-0-470-24766-2.

Nzeyimana, I., Hartemink, A. E., Ritsema, C., Stroosnijder, L., Lwanga, E. H., & Geissen, V. (2017). Mulching as a strategy to improve soil properties and reduce soil erodibility in coffee farming systems of Rwanda. *CATENA, 149*, 43–51.

Oliveira, M. T., & Merwin, I. A. (2001). Soil physical conditions in a New York orchard after eight years under different groundcover management systems. *Journal of Plant and Soil, 234*, 233–237.

Qin, S., Zhang, J., Dai, H., Wang, D., & Li, D. (2014). Effect of ridge–furrow and plastic-mulching planting patterns on yield formation and water movement of potato in a semi-arid area. *Agricultural Water Management, 131*, 87–94.

Qin, W., Hu, C., & Oenema, O. (2015). Soil mulching significantly enhances yields and water and nitrogen use efficiencies of maize and wheat: A meta-analysis. *Scientific Reports., 5*(1), 1–13.

Rahman, M. A., Chikushi, J., Saifizzaman, M., & Lauren, J. G. (2005). Rice straw mulching and nitrogen response of no-till wheat following rice in Bangladesh. *Field Crops Research, 91*(1), 71–81.

Ramaekers, L., Remans, R., Rao, I. M., Blair, M. W., & Vanderleyden, J. (2010). Strategies for improving phosphorus acquisition efficiency of crop plants. *Field Crops Research., 117*(2–3), 169–176.

Ramakrishna, A., Tam, H. M., Wani, S. P., & Long, T. D. (2006). Effect of mulch on soil temperature, moisture, weed infestation and yield of groundnut in northern Vietnam. *Field Crops Research, 95*(2–3), 115–125.

Roberts, T. L. (2007). Right product, right rate, right time and right place, the foundation of best management practices for fertilizer. *Fertilizer Best Management Practices, 29*, 1–8.

Rokich, D. P., Dixon, K. W., Sivasithamparam, K., & Meney, K. A. (2002). Smoke, mulch, and seed broadcasting effects on woodland restoration in Western Australia. *Journal of Restoration Ecological, 10*, 185–194.

Sainju, U. M., Senwo, Z. N., Nyakatawa, E. Z., Tazisong, I. A., & Reddy, K. C. (2008). Soil carbon and nitrogen sequestration as affected by long-term tillage, cropping systems, and nitrogen fertilizer sources. *Agriculture, Ecosystems & Environment, 127*, 234–240.

Siipilehto, J. (2001). Effect of weed control with fiber mulches and herbicides on the initial development of spruce, birch and aspen seedlings on abandoned farmland. *Silva Fennica, 35*, 403–414.

Singh, M. (2013). Influence of organic mulching and nitrogen application on essential oil yield and nitrogen use efficiency of rosemary (Rosmarinus officinalis L.). *Archives of Agronomy and Soil Science, 59*, 273–279.

Smith, M. W. (2000). Cultivar and mulch affect cold injury of young pecan trees. *Journal of American Pomological Society, 54*, 29–33.

Souri, M. K., & Hatamian, M. (2019). Aminochelates in plant nutrition: A review. *Journal of Plant Nutrition, 42*(1), 67–78.

Stenn, H. (2005). *Woody mulch research review, professional users and product availability surveys*. Seattle Public Utilities.

Tanavud, C., Kheowvongsri, P., Yongchalermchai, C., Leowarin, W., Densrisereekul, O., Bennui, A., Muraseand, J., & Kimura, M. (2001). Effects of land use patterns on soil and water quality in Khlong U-Taphao Basin. *Thai Journal of Agriculture Science, 34*, 15–31.

Tilman, D., Fargione, J., Wolff, B., Dantonio, C., Dobson, A., Howarth, R., Schindler, D., Schlesinger, W. H., Simberloff, D., & Swackhamer, D. (2001). Forecasting agriculturally driven global environmental change. *Science, 292*, 281–284.

Tilman, D., Cassman, K. G., Matson, P. A., Naylor, R., & Polasky, S. (2002). Agricultural sustainability and intensive production practices. *Nature, 418*, 67–677.

Wang, X., Li, Z., & Xing, Y. (2015). Effects of mulching and nitrogen on soil temperature, water content, nitrate-N content and maize yield in the Loess Plateau of China. *Agricultural Water Management, 161*, 53–64.

Wang, J., Lv, S., Zhang, M., Chen, G., Zhu, T., Zhang, S., Teng, Y., Christie, P., & Luo, Y. (2016). Effects of plastic film residues on occurrence of phthalates and microbial activity in soils. *Chemosphere, 151*, 171–177.

Wang, Z., Chen, J., Mao, S., Han, Y., Chen, F., Zhang, L., Li, Y., & Li, C. (2017). Comparison of greenhouse gas emissions of chemical fertilizer types in China's crop production. *Journal of Cleaner Production, 141*, 1267–1274.

White, P. J., & Broadley, M. R. (2003). Calcium in plants. *Annals of Botany, 92*(4), 487–511.

Xu, G., Fan, X., & Miller, A. J. (2012). Plant nitrogen assimilation and use efficiency. *Annual Review of Plant Biology, 63*, 153–182.

Zhang, X. Y., Chen, S. Y., Pei, D., Liu, M. Y., & Sun, H. Y. (2005). Evapotranspiration, yield and crop coefficient of irrigated maize under straw mulch. *Pedosphere, 15*(5), 576–584.

Zhu, Y., Lv, G., Chen, Y., Gong, X., Peng, Y., Wang, Z., Ren, A., & Xiong, Y. (2017). Inoculation of arbuscular mycorrhizal fungi with plastic mulching in rainfed wheat: A promising farming strategy. *Field Crops Research, 204*, 229–241.

Zhao, J., Ni, T., Li, J., Lu, Q., Fang, Z., Huang, Q. & Shen, Q. (2016). Effects of organic–inorganic compound fertilizer with reduced chemical fertilizer application on crop yields, soil biological activity and bacterial community structure in a rice-wheat cropping system. *Applied soil ecology, 99*, 1–12.

Mulching and Micronutrient Synergisms for Sustainable Crop Production

Sajid Ali, Ammara Fatima, Adnan Zahid, Sheraz Shoukat, Bisma, Robina Khaliq, Nimra Khan, Usman Ali, and Ayesha Akram

Abstract Soil is progressively being recognized as a natural resource which is nonrenewable that should be closely handled to guarantee long-term growth. Population growth is projected to be natural over the next 50 years, with intense competition for water and supplements to meet the increasing demand for food While soil quality is critical for long-term growth, the structure must also conserve property, sustain the climate, and be socio-culturally, economically, and environmentally beneficial. Mulch is said to have come from the German term molsch. This simply translated as "fragile to decay," and it led to landscapers' use of paint and leaves as mulch when strewn around the ground Mulching was a popular horticultural practice, and many nitrogen manures were used to grow crops Organic mulches are sometimes used in efforts to increase soil health. Nitrogen fertilization will significantly boost soil productivity and crop yields. On the other side, excessive nitrogen fertilizer usage will jeopardize the efficiency of nitrogen use (NUE) and the climate. Improving NUE in field management was a critical priority for more efficient usage of useful N fertilizers. Organic mulching products are more environmentally friendly and cost less than inorganic mulches. Which change soil content in order to improve NUE and crop yield more effectively. Precision nitrogen fertilizer control on plantations, as well as the cultivation of plants with strong NUE and output, would be quite feasible in the future. Supplement accessibility is improved by mulching with seed accumulation. Suffocating weed development, improving production quality, and harvesting, controlling soil temperature, enable large-scale experiments in microbial soils, worms, natural soil matter, and carbon stock, suffocating weed growth, improving output nature, and harvesting.

S. Ali (✉) · A. Zahid · S. Shoukat · Bisma · R. Khaliq · N. Khan · U. Ali · A. Akram
Department of Agronomy, University of the Punjab, Lahore, Pakistan
e-mail: sajid.agronomy@pu.edu.pk

A. Fatima
Department of Environmental Science, Lahore College for Women University, Lahore, Pakistan

Bisma · R. Khaliq · A. Akram
Department of Agronomy, University of Agriculture, Faisalabad, Pakistan

K. Akhtar et al. (eds.), *Mulching in Agroecosystems*,
https://doi.org/10.1007/978-981-19-6410-7_12

Keywords Mulch · Nitrogen use efficiency · Soil functionality · Sustainable crop production

1 Introduction

The importance of soil as a nonrenewable resource for long-term growth is becoming increasingly apparent. Soil health relates to a soil's ability to function as a critical tool for crop development even in the face of adversity (Pompili et al., 2006). Safe soils are defined by many primary interactions between physical, chemical and biological characteristics and their capacity to manage pests and diseases to sustain fertility and productivity. Healthy soils foster natural ecological processes by increasing water penetration and productivity, reducing compaction and depletion, recycling nutrients, and increasing water infiltration (Lakaria et al., 2019).

Soil health is a subjective evaluation focused on agricultural soil fitness parameters, and it is a primary predictor of the long-term sustainability of agricultural operations. Fertilizers, mulching, compost, and chemicals are examples of farming practices that may affect soil quality and sustainability (Xu et al., 2020).

Mineral fertilizers and chemicals, in general, have a host of negative environmental consequences and are prohibitively expensive for many low-income resource producers. This necessitates the use of alternative natural soil management techniques that are both cost-effective and adaptable to farmers' needs while presenting no environmental or human risks (Oelofse et al., 2015).

The integrated soil fertility management (ISFM) is a soil fertility management scheme that incorporates natural soil fertilizers, locally available nutrients (such as fertilizer, seed residues, livestock, and green manure), and soil productivity mineral engravings to preserve the environment's fragile natural resource base. Physical, chemical, biological, societal, fiscal, and policy considerations are all taken into account by the ISFM when determining soil fertility (Vanlauwe & Giller, 2006). A gradual change from conventional agriculture to organic farming is needed to enhance sustainable development processes (Azadi et al., 2011). ISFM also argues that agricultural productivity is important for long-term development, since long-term growth can only be achieved through a resource management scheme that is socially desirable, commercially viable, and environmentally secure.

Sustainable agricultural practices such as nitrogen fertilizer usage, breeding advances, organic and disease-resistant crops, irrigation, and land management development would be needed to satisfy, if not surpass, food demand if the world's population continues to grow at its current pace (Alexandratos, 1999).

More water and nitrogen fertilizers have been used in agriculture to close the demand-yield gap, but with an uncertain soil temperature impact and potential risks, food production in the future could be jeopardised (Liu et al., 2016). While nitrogen fertilizers are essential for crop development, excessive quantities are damaging to the environment (Han et al., 2015). It is therefore essential to study the impact of nitrogen fertilizers on crop surface structures. Nitrogen fertilizers are commonly produced from crop residues, fertilizer, green manure, livestock waste, or manure, which can include bacterial aggregation. They may be used as commercial organic fertilizers or as nitrogen fertilizers in agricultural settings (Wang et al., 2017).

Nitrate (aerobic) and ammonia (in flooded lakes or acidic) are the two primary nitrogen sources (Krapp, 2015). The effect of nitrogen fertilizer usage on the soil system has been extensively studied. A medium volume of nitrogen fertilizer improves crop yields, soil organic matter content, and soil biological operation (Haynes, 1998). Furthermore, nitrogen fertilizer has influenced relative lipid-fungal concentrations, phenol quality, fungal biomass, and soil behaviour (Frey et al., 2014). Fertilizers application of nitrogen to soils damaged the chemical composition of solar solution, and the flocculation and rearrangement phenomenon harmed the integrity of soil aggregates (Bronick & Lal, 2005).

Instead of being linear, the association between increased nitrogen fertilizer usage and increased oxide pollution is exponential (Shcherbak et al., 2014).

Even, before we get down to it, let's have a peek at the different types of mulch.

2 Types of Mulches

2.1 Organic Mulches

By soil maturity, humidity, and temperature, natural mulch improves soil quality, minimising surface erosion and supplementary loss (Montenegro et al., 2013). The birthplaces of natural plants or animals that have been discovered to increase soil health are normally natural plants or animals (Teame et al., 2017). Natural mulch increases land fertility and accelerates soil regeneration, resulting in higher yields (Kader et al., 2017). Cowpea, bracharia, and leguminous C are live mulches, while plants (including cereal straws and palm) and livestock (such as chickens, pigs, donkey, pony, and cow compost) are non-living natural mulches (Akhtar et al., 2018; Henschke & Politycka, 2016). Crop residues have been found to be useful as mulch for enhancing soil quality and facilitating a realistic turning point in events (Adekiya, 2018; Berglund et al., 2006; Liang et al., 2002; Payam et al., 2013).

2.1.1 Inorganic Mulches

Plant protection to soil preservation (e.g., against scorching temperatures and decay), as well as mitigating water draws from arable land, non-living materials (such as plastics, tiles, gravel, and coverings) are used in cultivation for a variety of purposes (Ingman et al., 2015). As a result of water shortages in arable frameworks, the whole agricultural area has recently spread around the globe under plastic mulch, which is a huge stumbling block to a financial aspect turn of events. It's crucial to persuade people to use scarce water resources wisely, particularly in dry areas with little soil moisture.

2.1.2 Biodegradable Mulches

Plant protection to soil preservation (e.g., against scorching temperatures and decay), as well as mitigating water draws from arable land, non-living materials (such as plastics, tiles, gravel, and coverings) are used in cultivation for a variety of purposes (Ingman et al., 2015). As a result of water shortages in arable frameworks, the whole agricultural area has recently spread around the globe under plastic mulch, which is a huge stumbling block to a financial aspect turn of events. It's crucial to persuade people to use scarce water resources wisely, particularly in dry areas with little soil moisture. Plastic mulch is commonly used as a water-saving technique in China's arable farms.

2.2 Non-biodegradable Film Mulches

Using distinctive polypropylene (PP) dark films, weed management output was tested, and the findings confirmed substantial plant growth (Fontana et al., 2006). Dark and other film shades are widely used for the production of strawberry and watermelon since they need higher soil temperatures to achieve desired satisfaction. Polara and Viradiya (2013) have launched unrivalled yields and production highlights for watermelon in silver black PE films, regardless of the fact that conventional PE films are a significant ecological concern. Costa et al. (2014) contrasted PE film to five biodegradable films and found no major variations in strawberry quality or design.

By examining various forms of mulch, we recommend a number of methods for raising nitrogen productivity by mulching and nitrogen fertilizer treatment.

2.3 Sustainable Soil Health Management

Soil health, as defined by field management practices, is the foundation for agriculture's long-term sustainability. Around half of the solids in soil are fused together to create aggregates, with air and water making up the remainder. Water does not readily spread into good aggregates in well-structured soils, and there are plenty of pores for easy root penetration and aeration. A lack of soil water and nutrients is a major problem in agriculture, particularly in high-temperature tropical regions.

The problem is exacerbated by the absence of farming practices in areas vulnerable to wind and/or flood erosion. Mulching is often used to categorise optimum soil management practices depending on climatic conditions and crop type (Kader et al., 2017).

Ideal soils sustain ecological existence by encouraging natural soil development and a sufficient supply of critical components for increased crop production while avoiding contamination of plants and the environment. The pH of good soils is therefore optimal, as the cation exchange potential is increased, meaning that plants get enough nutrients. In addition, stable soils host a number of species that contribute to the physical, chemical, and biological wellbeing of the soil (such as fungi, microbes, nematodes, collembola, and earthworms). Minerals in bioform that are not suitable for plant uptake are transformed by soil biota into mineral types that are not suitable for plant uptake.

Soil biota can aid in the stabilisation of soil particle deposition (structure), improving soil water retention and reducing erosion. Soil biota may also increase crop health by preventing and reducing the effects of soil-borne pests and diseases. Symbiotic interactions between plants and microbes improve surface fertility, protection, and crop quality (e.g., Mycorrhiza and Rhinobia) (e.g. Beauveria and Trichoderma).

Meanwhile, by growing surface erosion and soil degradation, the painting approach was successful in promoting long-term soil quality (Lin et al., 2018; Rahma et al., 2017). Straw mulch increases soil moisture, soil salinity, and evaporation by promoting water accumulation and restoration (Jimenez et al., 2017; Zhao et al., 2014) (Fig. 1).

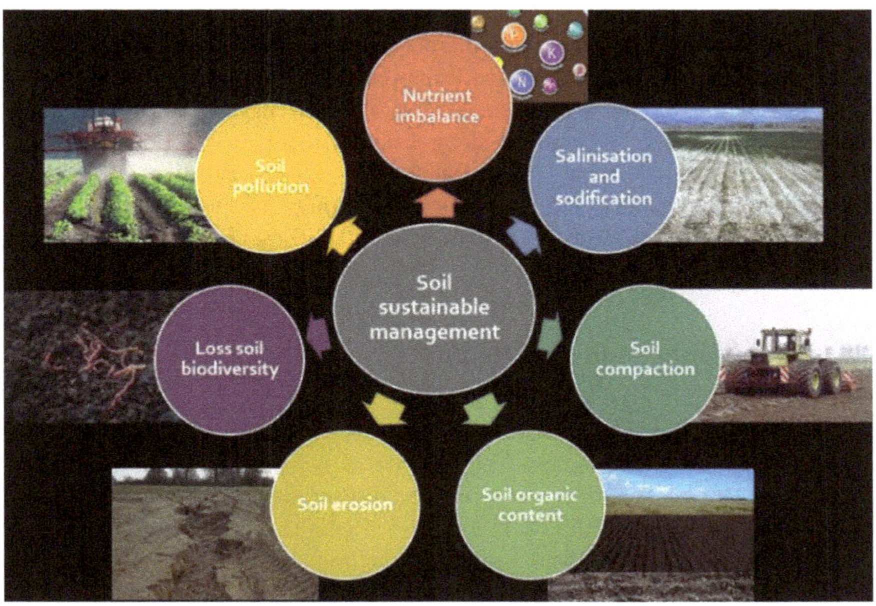

Fig. 1 Sustainable soil management and factors of soil quality (Augier et al., 2020)

2.4 Soil Fertility Management

Soil health, as defined by field management practices, is the foundation for agriculture's long-term sustainability. Around half of the solids in soil are fused together to create aggregates, with air and water making up the remainder. Water does not readily spread into good aggregates in well-structured soils, and there are plenty of pores for easy root penetration and aeration. A lack of soil water and nutrients is a major problem in agriculture, particularly in high-temperature tropical regions. The problem is exacerbated by the absence of farming practices in areas vulnerable to wind and/or flood erosion. Mulching is often used to categorize optimum soil management practices depending on crop types and climatic conditions (Kader et al., 2017).

Ideal soils sustain ecological existence by encouraging natural soil development and a sufficient supply of critical components for increased crop production while avoiding contamination of plants and the environment. The pH of good soils is therefore optimal, as the cation exchange potential is increased, meaning that plants get enough nutrients. In addition, stable soils host a number of species that contribute to the physical, chemical, and biological wellbeing of the soil (such as fungi, microbes, nematodes, collembola, and earthworms). A stable soil biota aids nitrogen cycling by decomposing plant and animal waste, which contributes biomass to the soil and produces humus. Minerals in bioform that are not suitable for plant uptake are transformed by soil biota into mineral types that are not suitable for plant uptake.

Soil biota can aid in the stabilization of soil particle deposition (structure), improving soil water retention and reducing erosion. Soil biota may also increase crop health by preventing and reducing the effects of soil-borne pests and diseases. Symbiotic interactions between plants and microbes improve surface fertility, protection, and crop quality (e.g., Mycorrhiza and Rhinobia).

Meanwhile, by growing surface erosion and soil degradation, the painting approach was successful in promoting long-term soil quality (Lin et al., 2018; Linnell et al., 2000; Prosdocimi et al., 2016; Rahma et al., 2017). Straw mulch increases soil moisture, soil salinity, and evaporation by promoting water accumulation and restoration (Zhao et al., 2014, 2016; Zribi et al., 2015; Jimenez et al., 2017).

2.5 Mulch Impact on Soil Health

Mulching is a soil-quality-improvement technique that promotes moisture preservation, temperature management, vegetation control, erosion mitigation, fertility enhancement and plant nutrition, and pest and disease prevention (Groen & Woods, 2008; Groen & Woods, 2008; He et al., 2016; Robichaud et al., 2013). Mulching increases the roughness of land surfaces, reducing surface water movement and distribution potential while further collecting water and soil (Prats et al., 2016; Shi et al., 2013). Mulching is classified as organic or inorganic depending on the products used. On the other side, the form of mulch used is determined by its supply, price, degradation rate, durability, and soil properties and functions. Manure, plant and animal residues, and field covering are examples of commonly used organic mulch products that have been shown to be effective in arable systems. Organic mulch improves soil productivity by raising soil nitrogen and humidity while also lowering evaporation and nutrient depletion in the region (Montenegro et al., 2013).

These plants coexist with the main crop, providing massive amounts of nitrogen through bacterial fixation and affecting microbial biomass ponds for food, nitrogen, and phosphorus (Duda et al., 2003). Straw mulch increases nodulation, nitrogenase quality, and yield parameters such as grain legume seed and protein content by reducing soil compaction (Siczek & Lipiec, 2011). Straw mulch reduces salt concentration at the optimum depth of soil, depending on the type of straw mulch used (Abd El-Mageed et al., 2016). In certain soils, soil organic matter is thought to be the most important determinant of soil productivity, and organic mulch can help enhance the physical, chemical, and biological properties of the soil (Thy & Buntha, 2005).

2.6 Use of Organic Mulch

Organic mulches have a track record of helping to improve seed and poultry soil quality (Teame et al., 2017). Organic mulch improves plant nutrients while still protecting degraded soils, resulting in increased crop yields (Kader et al., 2017). Non-living agricultural mulches include farm residues (such as rice or wheat straw and palm tree), as well as livestock (such as pigs, donkeys, horses, and dung), whereas rising agricultural mulches include cowpea, leguminous C., and baracharia grass (Abrantes et al., 2018). When used as a mulch, plant materials have been shown to increase soil quality and encourage long-term development (Adekiya, 2018).

Crop extracts from Tithonia diversifolia and Mucuna spp. may be used to improve soil fertility and agricultural crops production (Ngosong et al., 2017). Tithonia has a lot of biomass and nutrients that include 3.5% nitrogen, 0.37% phosphorus, and 4.1% potassium (Agbede & afolabi, 2014; Olabode et al., 2007) 0.5 Tithonia also produces lignin and polyphenol recalcitrant compounds in concentrations of 6.6 and 1.5%, respectively, in the form of lignin and polyphenol recalcitrant compounds (Jama et al., 2000).

Species of Mucuna are living mulches that have a strong propensity for nitrogen fixation and a large amount of biomass, making them ideal for improving soil quality and revival. Mucuna biomass has a nitrogen content of 3%, a phosphorus content of 0.2%, and a potassium content of 1.4%, rendering it suitable for soil rejuvenation (Chiu & Bisad, 2006). Organic mulch also improves soil bioavailability by regulating soil temperature, increasing nutrient supply and root absorption, and increasing soil bioactivity (Payam et al., 2013). Due to the large amount of bulk that is often required and because sourcing adequate mulch content externally is not feasible or cost efficient, organic mulch is best manufactured at the facility (Fig. 2).

Fig. 2 Effect of mulch on moisture retention and yield (Fix.com accessed on 04.04.2022)

3 Effect of Mulching on Crop Diseases

3.1 Organic Mulches

Matkovic et al. (2015) define living mulches as plants that are cultivated alongside the existing crop and provide numerous ecological benefits to the main crop, such as disease, pathogen, and soil species mitigation. Cold butterfly (Pieris brassicae) larvae parasitize living mulch at a higher pace (88%) than non-living mulch, according to Burgia et al. (2014) (63%). In contrast, pepper pest colonies were successfully suppressed using cowpea live mulch (Mochiach et al., 2012). By reducing dispersal, causing unfavourable soil conditions, and producing alllelopathic compounds, decking plants may inhibit the growth of weeds and soil pests (Brown & Tworkoski, 2004).

3.1.1 Non-living Organic Mulches

Most non-living mulches consist of non-lived compostable plant residues (e.g. straws, and saw dust etc.). Mulches such as this are also used in agricultural systems, since they are abundant, cheap and simple to get and add to crops. Non-living mulch, unlike live mulch which competes for main crops (for example for energy, moisture and nutrients), consumes a broad range of beneficial rhizobic biota, and provides nutrients that enhance soil productivity and crop productivity (Kolota & Sowinska, 2013).

Straw mulches, according to Mochiach et al., provided better shelter and enhanced plant yield for natural pepper pest enemies (2012). According to Gill et al., cowpea mulch breaks down quickly, allowing weeds to thrive and plant-feeding insect species to thrive (2011). Plant pests have attracted several predators and natural enemies as a result of their rapid development. In the cowpea mulch, which had a higher weather density, a higher population of insects and spiders, and a higher population of insects and spiders, more plant pests (grasshoppers, sap feeders, and crickets) were noticed (Gill & Goyal, 2014).

Since this gall-forming nematode is thermophile and has an increase in soil temperature caused by a mulch, maple bladder, along with physical or biological modifications, reduced the root tomato of Meloidogyne incognita. Overall, the biochemical elements of organic mulch content determine its effectiveness (Oroka & Omovbude, 2016).

3.2 *Mulching and Micronutrient Synergisms*

Mulching and nitrogen fertilisation have unexpected or noticeable effects on soil fertility. Mulch and nitrogen fertilizer, when used together, have been shown to increase fertility of soil. Mulching within 30–60 days after planting increases soil moisture, foliage, microbial cycling, and nutrient supply, leading to a better soil microenvironment (Li et al., 2004a, 2004b). As supplies become more available, mulch and nitrogen fertilizers are used as compost to help the soil support itself (Pinamonti, 1998). In a short period of time, mulch increased total soil organic C and N, as well as microbial C and N material, demonstrating mulch's role in increasing fertility of soil (Duda et al., 2003). Mulch from crop residues improves soil productivity greatly at a depth of 0–5 cm, with an inorganic nitrogen content of 84 mg N kg^{-1} relative to 64 mg N kg^{-1} in bare soil treatments (Murungu et al., 2011).

Wheat fiber blue mulch also improves surface consistency and chemistry physical properties at high mulching intensities (Jordan et al., 2010). Numerous research, on the other hand, have found no connection between soil fertility, mulch, and N fertilizers usage (Fan et al., 2005a). Mostly in rice rhizosphere, organic nitrogen fertilizers increase soil nutrient supplies (Steiner et al., 2007). Mulch increased soil fertility in understory forests by raising nitrate–N levels, while wood chip mulch improved soil water supply by lowering nitrogen levels (Hoagland et al., 2008).

The numbers of fluorescent pseudomonads and culturable heterotrophic bacteria in plots treated with composted yard waste as a nitrogen source was significantly higher than in sterile land plots treated with chemical fertilizers and mulches (Tiquia et al., 2002). Mulch alone lifts soil C and N in a shaded agroecosystem's top 20 cm even more than no mulch (Youkhana & Idol, 2009). Organic mulches improve soil quality in general, and nitrogen in mulch and non-mulch treatments improves soil productivity (Chakraborty et al., 2010). Despite an increase in the total fraction of mineralized nitrogen, increased nitrogen fertilizer application volumes were found to reduce biological nitrogen fixation (Chu et al., 2004). Rasmussen and colleagues (1998).

In comparison to the 23-year usage of inorganic fertilizers alone, the combined long-term use of inorganic fertilizers and farm waste significantly improved total levels of soil nitrogen, available nitrogen, phosphorus, and potassium (Su et al., 2006). In sandy loam, mulch crops fertilized with 200 kg nitrogen fertilizer outperformed unmulched plots in terms of soil fertility (Ram et al., 2006). As compared to non-mulched rice—what crop systems flooding in the soil–mulching has been shown to increase land exchangeable K and Olsen P, as well as higher field gross nitrogen immobilization rates and gross nitrogen mineralization (Liu et al., 2003; Huang and colleagues, 2008).

4 Effects of Mulching and Nitrogen Fertilizer on WUE, NUE and Grain Yield

4.1 Water Use Efficiency

Water demand is generally measured by grain yield or total biomass generated per unit of water consumed by crops. The average volume of water utilized by plants and soil surfaces, as well as that preserved inside plant systems, is used to measure the quantity of water used by crops. Despite this, the volume of water required by plant systems is calculated to be less than 1% of total water consumed during a typical growing season. As a consequence, plant transpiration and surface evaporation play a significant role in water absorption. The continuing rise of WUE is a major concern for food protection and survival as droughts intensify and the world's population increases (Bu et al., 2013).

Mulching has been shown to improve grain sweating and increase yield and WUE by reducing soil evaporation (Jia et al., 2006; Zhou et al., 2009). Direct surface water evaporation by mulching retains a comparatively steady soil water content in the top soil by reducing the movement of water from deeper soil layers to the top soil by steam transfer and capillary motion (Wang et al., 2009). Mulching increases yield and WUE by lowering water use. Mulching channels precipitation into capillaries in low-lying areas, providing enough soil moisture near plants (Arora et al., 2011). In addition, nitrogen fertilizers are commonly used to improve WUE (Arora et al., 2011). Increased WUE improves crop yield response to nitrogen fertilizer application rates in general. The volume of water in the soil has a big influence on how much nitrogen plants consume (Walsh et al., 2012).

The optimal nitrogen fertilizer application rates, on the other hand, are uncertain. The interactive impact of water and nitrogen on grain yield and WUE must be taken into consideration in several studies with optimum rates. Mulching and no-mulching treatments produced 1406 WUE in maize, wheat, and rice, respectively. The mulching and non-mulching treatments were determined separately, and each WUE value was calculated as an average of one specific article of the same nitrogen fertilizer quantity. Mulching increased the WUE by 6.73%, 24.31%, and 8.27%, respectively, and nitrogen fertilizers for wheat, maize, and weeds increased the WUE significantly. Mulching, on the other hand, did not often improve WUE, and in maize and wheat, non-mulching therapy generated higher WUE than low-nitrogen fertilization. Increased WUE has been linked to high soil water degradation in the 140–200 cm soil layer, indicating that it does not occur over long periods of time, particularly during dry seasons (Liu et al., 2009a; Zhang et al., 2011).

For all three cultures, the WUE in mulching response, particularly wheat, was higher than in non-mulching treatments. A mismatch between soil water supplies and drip irrigation requirements, compounded by a mismatch between soil water excess and evapotranspiration (Huang et al., 2005; Zhang et al., 2009), results in poor crop yields and nitrogen fertilizer WEU applications of rice, maize, and wheat above 100 kg ha^{-1} (Huang et al., 2005; Zhang et al., 2009). Figure 3 shows an example of a another hypothesis contends that mulch improves dryland wheat yields by retaining rainwater in the winter, improving soil moisture, and, of course, storing rainwater (Passioura, 2006; Vohland & Barry, 2009). Sharma et al. (2010) discovered that no-mulching therapies improved ground water quality (2–2.3%), seed yield (15.1%), and WUE dramatically. The soil moisture in the soil profile improved as more mulch was added to improve nitrogen fertilizer use (Gao et al., 2009; Hai et al., 2015). In this analysis, WUE increases in wheat were 1,67 times greater after mulching relative to non-mulching procedures (Fig. 4).

Fig. 3 A comparison of "health" and "quality" emphasizes the dynamic, living nature of soil (ensia.com accessed on 04.04.2022)

Fig. 4 Use of opaque plastic mulch in small crop production (Agritech.tnau.ac.in accessed on 04.04.2022)

4.2 Nitrogen Use Efficiency

The grain output or biomass produced is usually used to calculate water demand per plant consumption unit of water. Applying the overall water volume utilized by plants and soil surfaces, as well as that held inside plant systems, is measured water used for planting. Nevertheless, less than 1% of the water utilized in a typical growing season is expected to be available for the systems. As a consequence, water absorption plays a significant role in plant transpiration and surface evaporation. The continuing rise of WUE is a big problem for food safety and biodiversity as droughts get worse and the world's population increases (Bu et al., 2013).

It was interesting for several years that we learned how to develop WUEs in rainfed and irrigated farming and how to adjust agricultural systems. Mulching shows that output and WUE improve by reducing soil evaporation and grain sweating (Zhou et al., 2009; Jia et al., 2006). The direct evaporation of the surface waters by a mattressing system by decreasing the movement of water from the deeper soil to the top soil ensures a comparatively steady soil water level in the top soil (Tian et al., 2003; Wang et al., 2009, 2011). By minimizing water intakes, mulching increases yield and WUE. Mulching in low fields causes capillary precipitation, which ensures that soil moisture around plants is adequate (Arora et al., 2011; Kar & Kumar, 2007; Wang et al., 2005). In addition, nitrogen fertilizer is also used to raise the WUE (Arora et al., 2011; Kar & Kumar, 2007; Wang et al., 2005). Increased WUE generally improves the response from crops to the application of nitrogen fertilizers. The water level in the ground has a significant influence in plant usage of nitrogen (Martin et al., 1982; Walsh et al., 2012), and because of the high intake of nitrogen and decent water resources, solid grain yields and WUE are (Fan et al., 2005b; Zhang et al., 2004). In high water quality, the optimal strength of the application of nitrogen improves WUE in nitrogen-deficient soils. WUE increases if the volume of nitrogen used in crops is greater than predicted, whereas the probability of nitrate–N flushing and of root zone flushes would rise (Al-Kaisi & Yin, 2003; Zotarelli et al., 2008).

The optimum application rates for nitrogen fertilizers, however, are not certain. 1406 WUE were raised in maize, wheat, and rice, respectively, with mulching and no-mulching care. A different determination was made of the Mulching and non-Mulching treatments and each WUE amount was averaged by one individual article for the same quantity of nitrogen fertilizer. The WUE increased by 6.73%, 24.31%, and 8.27% respectively, and the WUE increased markedly with nitrogen fertilizers for wheat, maize and weeds. On the other side, mulching did not increase WUE often, but non-mulching therapy generated higher WUE than low-nitrogen fertilization for maize and wheat. Increased WUE was associated with a high deterioration of soil water in the 140–200 cm layer soil suggesting a lack of water for long periods of time (Liu et al., 2009a; Zhang et al., 2011).

For all crops, the WUE was higher in mulching reactions than in non-mulching treatments, especially wheat. A mismatch between soil water sources and drip irrigation needs exacerbated by a mismatch between soil water excess and evapotranspiration (Huang et al., 2005; Zhang et al., 2009), resulting in low crop yields and nitrogen fertilizer WEU applications of rice, maize, and wheat above 100 kg ha^{-1} (Huang et al., 2005; Zhang et al., 2005, 2009) (Take a look at Fig. 3). Another theory claims that mulch will help dryland wheat yields by storing rainwater in the winter, improving soil moisture, and, of course, storing rainwater (Passioura, 2006; Vohland & Barry, 2009). Sharma et al. (2010) discovered that no-mulching therapies significantly increased ground water production (2–2.3%), seed yield (15.1%), and WUE. When more mulch was applied to increase nitrogen fertilizer use, the soil moisture in the soil profile increased (Gao et al., 2009; Hai et al., 2015). WUE raises in wheat were 1,67 times greater after mulching compared to non-mulching procedures in this study.

4.3 Grain Yield

Mulching is a common farm technique to increase crop yields by alteration of soil properties in arid and semiarid areas. For instance, mulching can reduce the evaporation of water and boost the availability of crop water. In cold environments, mulching may often assist in plant growth by increasing the temperature of the top soil. In one sample, the most solar energy was allowed to pass through the pelic and heat up the air and soil below the film in the case of plastic mulch (Liu et al., 2009a), while the other study showed that the top soil warmed up more rapidly all day long as the soil had "greenhouse effects" (Wang et al., 2005). In the course of the study, water under the film decreased the radiation from the long waves, causing slower cooling at night (Li et al., 2013; Liu et al., 2009a). In addition, the use of plastic film decreases the latent heat flow and the exchange of sensitive heat between the soil and the air (Bu et al., 2013; Ham & Kluitenberg, 1994). Different mulching materials have different effects on soils. In the case of stubble residue conservation, stubble residue conservation boost water supplies and crop yields, for example, soil quality is increased by lower water evaporation and increased organic soil content.

The usage of straw mulch in semi-arid areas in the Loess Plateau of northwest China is restricted however, as the soil surface temperature is likely to decrease and grain yields would decrease.

Plastic film mulching technologies have proved to be effective in raising the yield of grain in these areas to solve this problem (Li et al., 2004b). Because of mulch's exceptional effectiveness in maintaining soil moisture and improving nutrient transfers and abundance, grain yield is higher under mulching than in bare fields; however, mulching is expensive and time-consuming for farmers (Qin et al., 2006). As a part of this, the economic benefits of crops and the environmental costs of mulching materials may be measured. Increase in the early growing season, however, higher soil temperatures caused by a mulch will ease crop senescence and decrease dry matter accumulation, decreased crop yields later in the growing season (Liu et al., 2014; Yi et al., 2010).

Due to the fact that different crops need different soil temperatures, mulching time may be crucial for the advancement of new mulching technologies. For example, farmers have used techniques of microbial degradation to track decomposition progress. Drylands occupy almost 45% of the world's land region and are vital for the processing of global food (Schimel, 2010). Increased field management to improve dryland agriculture's production and yield is vital to global food safety (Cassman, 1999; Lele, 2010). Present grain yields increase, the determination of optimum mulching and nitrogen speeds are both highly efficient ways of achieving the target. We have been able to examine the effects of mulching and nitrogen use on yield by collected 1516 grain yield values for maize, wheat and rice. Results suggest a marginally higher mean grain yield than in non-fertilizer treatments where fertilizer was used.

Overall, the application in nitrogen fertilizer enhances the grain returns of all three species. The importance of 1516 grain yield values in maize, wheat and rice was derived from mulching and non-mulching treatments. The mean yield of rice, maize and weeat in mulching treatments was 2.6%, 28.5% and 9.6% higher respectively than in non-mulching treatments (Fig. 5). Mulched rice was 28.5% higher than non-mulched maize, but mulched rice was 2.6% higher than non-mulched rice.

Two potential reasons for the high yield of maize and low rice yield are given. First, the high return of maize in the early growing season was largely due to increased soil moisture and mulching temperatures, which led to significant under-ground growth and root system productivity in cultivation. Secondly, organic decomposition and the development of mulched rice roots through the soils and microbial breathing and excision, both dictated by soil, nutrients and temperature.

5 Suggestions for Future Research

It's more difficult to increase yields by applying more nitrogen as part of advanced nutrient resource control than it is to increase NUE. In both crops, increased nitrogen

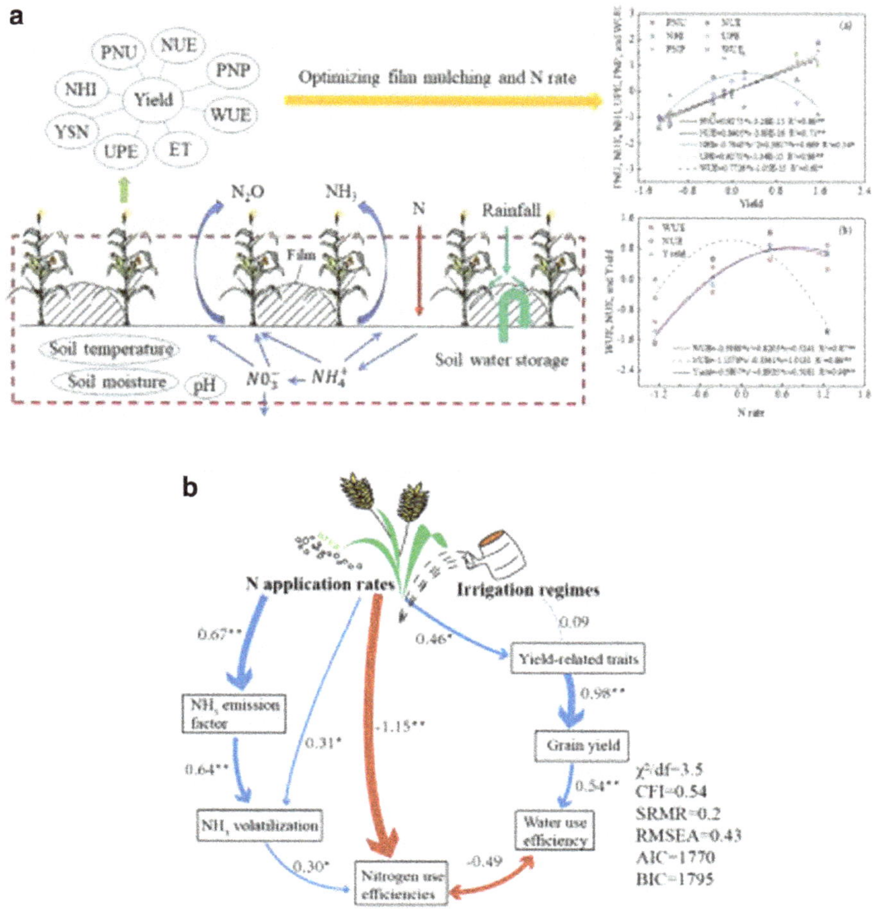

Fig. 5 **a** Diagrammatic presentation of mulching and nitrogen dynamics (Fang et al., 2022). **b** Impact of irrigation and nitrogen management for reducing NH_3 emission and increasing NUE and WUE (Wan et al., 2021)

fertilizer application lowers NUE. On the one hand, differences in nitrogen absorption and nitrogen intake efficiency among crops have a major impact on the effects of NUE nitrogen fertilizer application at different levels. The concentration of nitrate–N in the root zone increases as nitrogen is added more often, as does the degree to which it spills through deeper soil layers. Mulching in agriculture has many benefits, including reducing surface-water evaporation, raising top-level soil temperature, shifting microbial biomass, preserving soil organic carbon balance, maximising mineral nutrient cycling, encouraging the activity of soil enzymes, eliminating weed infestation, and improving soil-based stabilisation.

Seed residues on the ground's surface are used to monitor soil degradation and drainage in a number of ways. Mulching crop residues increases water infiltration and protects the surface soil from rain-induced crusting and settling. Mulching semi-arid slopes with planted residues was also discovered to be very efficient. As a consequence, we assume that combining mulching and nitrogen application is the most effective way to increase NUE and grain yields thus reducing N_2O pollution and other N loss pathways in agriculture.

It's tough to choose mulching materials and nitrogen application rates, and it raises a lot of scientific concerns. From one crop to the next, the best nitrogen fertilizer, fertilizer operation, fertilizer depth, mattress, and mulching material differ. It would be surprisingly easy in the future to manage precise nitrogen fertilizer applications on farms, as well as planting plants with a strong NUE and yield.

5.1 *Mulching Technologies in Combination with Fertilization*

The primary objectives of good agriculture, particularly in rainfed areas, are increased field productivity and soil fertility. Small-scale farmers make up the bulk of the population in certain parts of the world, so creating innovative projects that combine simplicity of management, low prices, and conservation is a realistic necessity for them. Banded basal fertilisation and dressing, in which the basal fertilizer injects the necessary volume of nitrogen fertilizer into the surface soil during planting in rows adjacent to or below the sowing plants, and the surface and side dressing are inserted in the subsurface soil during post-appearance, is a good type of nitrogen fertilizer application. This strategy will be effective when used in conjunction with mulching, but the best nitrogen fertilizer application speeds, mulch products, and mulch application spacing in different regions must be investigated further.

5.2 *Legume Mulching and Precise Fertilization*

If the amount of atmospheric CO_2 continues to increase, significant environmental changes have occurred, and have changed the usage of water and energy in manufacturing, either explicitly or indirectly. Legumes are a great way to enhance productivity and crop yields by increasing soil humidity and nutrient retention in maize-white cultivation fields in arid and semi-arid zones' plains, valleys, and hills. Lentil mulching in combination with specific fertilisation, as well as the identification of critical and optimum values for spectral diagnostic indexes in different plant growth stages, are examples of new developments in the efficient use of nitrogen methods in precision agriculture.

5.3 The Combination of Fertilizer Application and Ridge-Furrow Mulching

The primary objectives of good agriculture, particularly in rainfed areas, are increased field fertility and soil fertility. Small-scale farmers make up the bulk of the population in certain parts of the world, so creating innovative projects that combine simplicity of management, low prices, and sustainability is a realistic necessity for them. Banded basal fertilisation and dressing, in which the basal fertilizer injects the necessary volume of nitrogen fertilizer into the surface soil during planting in rows adjacent to or below the sowing plants, and the surface and side dressing are inserted in the subsurface soil during post-appearance, is a good type of nitrogen fertilizer application. This strategy will be effective when used in conjunction with mulching, but the best nitrogen fertilizer application speeds, mulch products, and mulch application spacing in different regions must be investigated further.

5.4 Mechanisms of Nitrogen Use Efficiency Enhancement

Understanding the impact of mulching and nitrogen fertilizer application on crop yield is important not only for understanding nitrogen absorption, utilisation, translocation, deposition, and leaching, but also for optimising global high yield and high NUE strategies. More research into the effects of mulching and nitrogen fertilizer application on physiological nitrogen needs is needed, and determining the best solution for mulching and nitrogen fertilizer applications for farmers is more scientific, precise, and practical.

5.5 Sustainable Soil Health Management

The capacity of soil to act as an underlying element that promotes crop efficiency under environmental constraints is referred to as soil well-being (Doran & Zeiss, 2000; Pompili et al., 2006). Eventually, low and diminishing soil maturity is a critical condition for decreasing productivity and potentially turning things around as a consequence of continued expansion (Tening et al., 2013).

On the farm, the board brings the case for a viable agricultural development, including soil well-being, into practice. Around half of the rigid material of soil is bound to full forms, while the rest is made up of water and air. Soils with a lot of structure have predictable water amounts, don't spread it well, and have enough openings to enable rapid root infiltration while increasing air circulation and water penetration. In agriculture, groundwater and excessive failure are major concerns, especially in tropical areas with high temperatures. Mulching is often used to test various varieties and climates to find the most productive land practices (Kader et al.,

2017). In optimal soils, organic life thrives, allowing for everyday soil regeneration and adequate supplies of essential nutrients to boost crop yields without poisoning plants or the atmosphere. Strong soils also improve the cation exchange limits and raise the optimum pH, allowing plants more available.

The strong phase, fluid amount, and vaporous stage are the three main components of soil. The inventory of plant root supplements is specifically affected by one of the three stages. The strong stage is the primary supplemental repository. Cationic supplements such as K, Na, Ca, Mg, Fe, Mn, Zn, and Cu are now included in the inorganic particles, while the regular particles have N as their theory and P and S as their primary targets.

6 Conclusion

Soil has been recognised as a valuable non-renewable natural capital that must be managed carefully for long production, paradigms, and concepts on sustainable land management. They are designed to boost soil biodiversity and sustainability, while safeguarding the planet's limited natural capital. Mulching is therefore a necessary component of these visions, but organic mulching is constrained by the large sums of money and labour needed for manufacturing, transportation, and execution.

Though nitrogen fertilizer and mulch are still widely used to increase crop yield, excessive nitrogen fertilizer input is a major source of pollution. However, as the world's population and living standards rise, demand for fibre and food will rise as well, necessitating an increase in crop production. Even if field management practices were improved, NUE in crops would increase. Increasing nitrogen absorption and intake, for example, would result in higher yields if more favorable conditions are created for plants. Nitrogen management techniques for optimal seed absorption include correct application rates, reliable nitrogen application procedures, adequate nitrogen sources, and nitrogen application scheduling.

Nitrogen regulation and mulching quality selection are influenced by crop types, crop management practices, and climatic conditions. Rodent, mosquito, and weed protection are crucial in crop production, and NUE can be increased even further. Improved plant development, WUE, NUE, and yields would benefit from a deeper understanding of the interactions between nitrogen and other nutrients, as well as mulching.

Mulch conserves soil moisture, avoids evaporation, reduces weed formation, encourages soil microbes, regulates soil composition and temperature, and may be attractive.

References

Abd El-Mageed, T. A., Semida, W. M., & Abd El-Wahed, M. H. (2016). Effect of mulching on plant water status, soil salinity and yield of squash under summer-fall deficit irrigation in salt affected soil. *Agricultural Water Management, 173*, 1–12.

Abrantes, J. R. C. B., Prats, S. A., Keizer, J. J., & de Lima, J. L. M. P. (2018). Effectiveness of the application of rice straw mulching strips in reducing runoff and soil loss: Laboratory soil flume experiments under simulated rainfall. *Soil Tillage Res, 180*, 238–249.

Adekiya, A. O. (2018). Legume mulch materials and poultry manure affect soil properties, and growth and fruit yield of tomato. *Agriculturae Conspectus Scientificus, 83*, 161–167.

Agbede, T. M., & Afolabi, L. A. (2014). Soil fertility improvement potentials of mexican sunflower (Tithonia diversifolia) and Siam weed (Chromolaena odorata) using okra as test crop. *Archives of Applied Science Research, 6*, 42–47.

Akhtar, K., Wang, W., Ren, G., Khan, A., Feng, Y., & Yang, G. (2018). Changes in soil enzymes, soil properties, and maize crop productivity under wheat straw mulching in Guanzhong, China. *Soil and Tillage Research, 182*, 94–102.

Al-Kaisi, M. M., & Yin, X. (2003). Effects of nitrogen rate, irrigation rate, and plant population on corn yield and water use efficiency. *Agronomy Journal, 95*, 1475–1482.

Arora, V. K., Singh, C. B., Sidhu, A. S., & Thind, S. S. (2011). Irrigation, tillage and mulching effects on soybean yield and water productivity in relation to soil texture. *Agricultural Water Management, 98*, 563–568.

Augier, J., Oréade, B., Juliane, P., Alice, D., & Vincent, A. (2020). Evaluation support study on the impact of the CAP on sustainable management of the soil. Technical Report number: KF-02-20-617-EN-N. https://doi.org/10.2762/799605.

Berglund, R., Svensson, B., & Gertsson, U. (2006). Impact of plastic mulch and poultry manure on plant establishment in organic strawberry production. *Journal of Plant Nutrition, 29*, 103–112.

Bronick, C. J., & Lal, R. (2005). Soil structure and management: A review. *Geoderma, 124*, 3–22.

Brown, M. W., & Tworkoski, T. (2004). Pest management benefits of compost mulch in apple orchards. *Agriculture, Ecosystems & Environment, 103*, 465–472.

Bu, L., Liu, J., Zhu, L., Luo, S., Chen, X., Li, S., Lee Hill, R., & Zhao, Y. (2013). The effects of mulching on maize growth, yield and water use in a semi-arid region. *Agricultural Water Management, 123*, 71–78.

Chakraborty, D., Garg, R. N., Tomar, R. K., Singh, R., Sharma, S. K., Singh, R. K., Trivedi, S. M., Mittal, R. B., Sharma, P. K., & Kamble, K. H. (2010). Synthetic and organic mulching and nitrogen effect on winter wheat (Triticum aestivum L.) in a semi-arid environment. *Agricultural Water Management, 97*, 738–748.

Chiu, S. B., & Bisad, M. (2006). Mucuna bracteata-biomass, litter and nutrient production. *The Planter, 82*, 247–254.

Chu, G. X., Shen, Q. R., & Cao, J. L. (2004). Nitrogen fixation and N transfer from peanut to rice cultivated in aerobic soil in an intercropping system and its effect on soil N fertility. *Plant and Soil, 263*, 17–27.

Duda, G. P., Guerra, J. G. M., Monteiro, M. T., De-Polli, H., & Teixeira, M. G. (2003). Perennial herbaceous legumes as live soil mulches and their effects on C, N and P of the microbial biomass. *Science in Agriculture, 60*, 139–147.

Fan, M., Jiang, R., Liu, X., Zhang, F., Lu, S., Zeng, X., & Christie, P. (2005a). Interactions between non-flooded mulching cultivation and varying nitrogen inputs in rice–wheat rotations. *Field Crops Research, 91*, 307–318.

Fan, T., Stewart, B. A., Yong, W., Junjie, L., & Guangye, Z. (2005b). Long-term fertilization effects on grain yield, water-use efficiency and soil fertility in the dryland of Loess Plateau in China. *Agriculture, Ecosystems & Environment, 106*, 313–329.

Fang, H., Li, Y., Xiaobo, G., Meng, Y., Chen, P., Li, Y., & Liu, F. (2022). Optimizing the impact of film mulching pattern and nitrogen application rate on maize production, gaseous N emissions, and utilization of water and nitrogen in northwest China. *Agricultural Water Management, 261*, 107350.

Gao, Y., Li, Y., Zhang, J., Liu, W., Dang, Z., Cao, W., & Qiang, Q. (2009). Effects of mulch, N fertilizer, and plant density on wheat yield, wheat nitrogen uptake, and residual soil nitrate in a dryland area of China. *Nutrient Cycling in Agroecosystems, 85*, 109–121.

Gill, H. K., & Goyal, G. (2014). Organic mulches: An innovative pest management strategy. *Popular Kheti, 2*, 118–123.

Gill, H. K., McSorley, R., & Branham, M. (2011). Effect of organic mulches on soil surface insects and other arthropods. *The Florida Entomologist, 94*, 226–232.

Groen, A. H., & Woods, S. W. (2008). Effectiveness of aerial seeding and straw mulch for reducing postwildfire erosion, north-western Montana, USA. *International Journal of Wildland Fire, 17*, 559–571.

Hai, L., Li, X. G., Liu, X.-E., Jiang, X. J., Guo, R. Y., Jing, G. B., Rengel, Z., & Li, F.-M. (2015). Plastic mulch stimulates nitrogen mineralization in urea-amended soils in a semiarid environment. *Agronomy Journal, 107*, 921–930.

Han, M., Okamoto, M., Beatty, P. H., Rothstein, S. J., & Good, A. G. (2015). The genetics of nitrogen use efficiency in crop plants. *Annual Review of Genetics, 49*, 269–289.

He, G., Wang, Z., Li, F., Dai, J., Li, Q., Xue, C., Cao, H., Wang, S., & Malhi, S. S. (2016). Soil water storage and winter wheat productivity affected by soil surface management and precipitation in dryland of the Loess Plateau. *China Agricultural Water Management, 171*, 1–9.

Henschke, M., & Politycka, B. (2016). Application of wood chips for soil mulching in the cultivation of ornamental grasses. *Folia Horticulturae, 28*, 187–194.

Hoagland, L., Carpenter-Boggs, L., Granatstein, D., Mazzola, M., Smith, J., Peryea, F., & Reganold, J. P. (2008). Orchard floor management effects on nitrogen fertility and soil biological activity in a newly established organic apple orchard. *Biology and Fertility of Soils, 45*, 11–18.

http://www.agritech.tnau.ac.in/agricultural_engineering/plastic_mulching.pdf. Accessed 04 Apr. 2022.

https://ensia.com/features/soil-health/. Accessed 04 Apr. 2022.

https://www.fix.com/blog/gardening-with-mulch/. Accessed 04 Apr. 2022.

Huang, Y., Chen, L., Fu, B., Huang, Z., & Gong, J. (2005). The wheat yields and water-use efficiency in the Loess Plateau: Straw mulch and irrigation effects. *Agricultural Water Management, 72*, 209–222.

Huang, Z., Xu, Z., & Chen, C. (2008). Effect of mulching on labile soil organic matter pools, microbial community functional diversity and nitrogen transformations in two hard wood plantations of subtropical Australia. *Applied Soil Ecology, 40*, 229–239.

Ingman, M., Santelmann, M., & Tilt, B. (2015). Agricultural water conservation in China: Plastic mulch and traditional irrigation. *Ecosyst Health Sustain, 1*, 12.

Jama, B., Palm, C. A., Buresh, R. J., Niang, A., Gachengo, C., & Nziguheba, G. (2000). Tithonia diversifolia as a green manure for soil fertility improvement in Western Kenya: A review. *Agroforestry Systems, 49*, 201–221.

Jia, Y., Li, F.-M., Wang, X.-L., & Yang, S.-M. (2006). Soil water and alfalfa yields as affected by alternating ridges and furrows in rainfall harvest in a semiarid environment. *Field Crops Research, 97*, 167–175.

Jimenez, M. N., Pinto, J. R., Ripoll, M. A., Sanchez-Miranda, A., & Navarro, F. B. (2017). Impact of straw and rock-fragment mulches on soil moisture and early growth of holm oaks in a semiarid area. *CATENA, 152*, 198–206.

Jordan, A., Zavala, L. M., & Gil, J. (2010). Effects of mulching on soil physical properties and runoff under semi-arid conditions in southern Spain. *Catena, 81*, 77–85.

Kader, M. A., Senge, M., Mojid, M. A., & Ito, K. (2017). Recent advances in mulching materials and methods for modifying soil environment. *Soil and Tillage Research, 168*, 155–166.

Kar, G., & Kumar, A. (2007). Effects of irrigation and straw mulch on water use and tuber yield of potato in eastern India. *Agricultural Water Management, 94*, 109–116.

Kolota, E., & Sowinska, K. A. (2013). Living mulches in vegetable crops production: Perspectives and limitations. *Acta Scientiarum Polonorum, Hortorum Cultus, 12*, 127–142.
Krapp, A. (2015). Plant nitrogen assimilation and its regulation: A complex puzzle with missing pieces. *Current Opinion in Plant Biology, 25*, 115–122.
Lakaria, B. L., Dotaniya, M. L., Meena, B. P., Wanjari, R. H., & Biswas, A. K. (2019). Soil health: Concept, components, management and opportunities. In *Advances in compost production technology* (pp. 95–103). IARI.
Lele, U. (2010). Food security for a billion poor. *Science, 327*, 1554.
Li, F., Wang, J., Xu, J., & Xu, H. (2004a). Productivity and soil response to plastic film mulching-durations for spring wheat on entisols in the semiarid Loess Plateau of China. *Soil and Tillage Research, 78*, 9–20.
Li, F.-M., Song, Q.-H., Jjemba, P. K., & Shi, Y.-C. (2004b). Dynamics of soil microbial biomass C and soil fertility in cropland mulched with plastic film in a semiarid agro-ecosystem. *Soil Biology & Biochemistry, 36*, 1893–1902.
Li, R., Hou, X., Jia, Z., Han, Q., Ren, X., & Yang, B. (2013). Effects on soil temperature, moisture, and maize yield of cultivation with ridge and furrow mulching in the rainfed area of the Loess Plateau China. *Agricultural Water Management, 116*, 101–109.
Liang, Y., Zhang, C. E., & Guo, D. W. (2002). Mulch types and their benefit in cropland ecosystems on the loess plateau in China. *Journal of Plant Nutrition, 25*, 945–955.
Lin, J., Zhu, G., Wei, J., Jiang, F., Wang, M., & Huang, Y. (2018). Mulching effects on erosion from steep slopes and sediment particle size distributions of gully colluvial deposits. *Catena, 160*, 57–67.
Linnell, E., Burney, J. R., Richter, G., & MacRae, A. H. (2000). Evaluation of compost and straw mulching on soil-loss characteristics in erosion plots of potatoes in Prince Edward Island, Canada. *Agriculture, Ecosystems & Environment, 81*, 217–222.
Liu, X. J., Wang, J. C., Lu, S. H., Zhang, F. S., Zeng, X. Z., Ai, Y. W., Peng, S. B., & Christie, P. (2003). Effects of non-flooded mulching cultivation on crop yield, nutrient uptake and nutrient balance in rice–wheat cropping systems. *Field Crops Research, 83*, 297–311.
Liu, J., Zhan, A., Bu, L., Zhu, L., Luo, S., Chen, X., Cui, Z., Li, S., Lee Hill, R., & Zhao, Y. (2014). Understanding dry matter and nitrogen accumulation for high-yielding film mulched maize. *Agronomy Journal, 106*, 390–396.
Liu, C., Cutforth, H., Chai, Q., & Gan, Y. (2016). Farming tactics to reduce the carbon foot print of crop cultivation in semiarid areas. *A Review Agronomy for Sustainable Devlopment, 36*, 69.
Martin, D. L., Watts, D. G., Mielke, L. N., Frank, K. D., & Eisenhauer, D. E. (1982). Evaluation of nitrogen and irrigation management for corn production using water high in nitrate. *Soil Science Society of America Journal, 46*, 1056–1062.
Matkovic, A., Bozic, D., Filipovic, V., Radanovic, D., Vrbnicanin, S., & Markovic, T. (2015). Mulching as a physical weed control method applicable in medicinal plants cultivation. *Lekovite Sirovine, 25*, 37–51.
Mochiach, M. B., Baidoo, P. K., & Acheampong, G. (2012). Effects of mulching materials on agronomic characteristics, pests of pepper (Capsicum annum) and their natural enemies population. *Agriculture and Biology Journal of North America, 3*, 253–261.
Montenegro, A. A. A., Abrantes, J. R. C. B., de Lima, J. L. M. P., Singh, V. P., & Santos, T. E. M. (2013). Impact of mulching on soil and water dynamics under intermittent simulated rainfall. *Catena, 109*, 139–149.
Murungu, F. S., Chiduza, C., Muchaonyerwa, P., & Mnkeni, P. N. S. (2011). Mulch effects on soil moisture and nitrogen, weed growth and irrigated maize productivity in a warm temperate climate of South Africa. *Soil and Tillage Research, 112*, 58–65.
Ngosong, C., Tanyi, C. B., Njume, C. A., Mfombep, P. M., Okolle, J. N., Njock, T. E, Nkongho, R. N., & Tening, A. S. (2017). Potential of dual-purpose organic amendment for enhancing tomato (Lycopersicon esculentum M.) performance and mitigating seedling damage by mole cricket (Gryllotalpa africana spp.). *International Journal of Plant & Soil Science, 20*(1–12109), 139–149.

Oelofse, M., Markussen, B., Knudsen, L., Schelde, K., Olesen, J. E., Jensen, L. S., & Bruun, S. (2015). Do soil organic carbon levels affect potential yields and nitrogen use efficiency? An analysis of winter wheat and spring barley field trials. *European Journal of Agronomy, 66*, 62–73.

Olabode, O. S., Sola, O., Akanbi, W. B., Adesina, G. O., & Babajide, P. A. (2007). Evaluation of Tithonia diversifolia (Hemsl.) a gray for soil improvement. *World Journal of Agricultural Science, 3*, 503–507.

Oroka, F. O., & Omovbude, S. (2016). Effect of mulching and period of weed interference on the growth, flowering and yield parameters of okra (Abelmoschus esculentus). *IOSR Journal of Agricultural Veterinary Science, 9*, 52–56.

Passioura, J. (2006). Increasing crop productivity when water is scarce—From breeding to field management. *Agricultural Water Management, 80*, 176–196.

Payam, P., Tehranifar, A., Nemati, H., Llakzian, A., & Kharrazi, M. (2013). Effect of different mulching materials on soil properties under semi-arid conditions in Northeastern Iran. *Wudpecker Journal of Agricultural Research, 2*, 80–85.

Pinamonti, F. (1998). Compost mulch effects on soil fertility, nutritional status and perfor mance of grapevine. *Nutrient Cycling in Agroecosystems, 51*, 239–248.

Pompili, L., Mellina, A. S., & Benedetti, A. (2006). Microbial indicators for evaluating soil quality in differently managed soils. *Geophysical Research Abstracts, 8*, 06991.

Prats, S. A., Wagenbrenner, J., Malvar, M. C., Martins, M. A. S., & Keizer, J. J. (2016). Hydrological implications of post-fire mulching across different spatial scales. *Land Degradation and Development, 27*, 1440–2145.

Prosdocimi, M., Jordan, A., Tarolli, P., Keesstra, S., Novara, A., & Cerda, A. (2016). The immediate effectiveness of barley straw mulch in reducing soil erodibility and surface runoff generation in Mediterranean vineyards. *Science of the Total Environment, 547*, 323–3302.

Qin, J., Hu, F., Zhang, B., Wei, Z., & Li, H. (2006). Role of straw mulching in non continuously flooded rice cultivation. *Agricultural Water Management, 83*, 252–260.

Rahma, A. E., Wang, W., Tang, Z., Lei, T., Warrington, D. N., & Zhao, J. (2017). Straw mulch can induce greater soil losses from loess slopes than no mulch under extreme rainfall conditions. *Agricultural and Forest Meteorology, 232*, 141–151.

Ram, D., Ram, M., & Singh, R. (2006). Optimization of water and nitrogen application to menthol mint (Mentha arvensis L.) through sugarcane trash mulch in a sandy loam soilof semi-arid subtropical climate. *Bioresource Technology, 97*, 886–893.

Rasmussen, P. E., Douglas, C. L., Jr., Collins, H. P., & Albrecht, S. L. (1998). Long-term croppingsystem effects on mineralizable nitrogen in soil. *Soil Biology & Biochemistry, 30*, 1829–1837.

Robichaud, P. R., Jordan, P., Lewis, S. A., Ashmun, L. E., Covert, S. A., & Brown, R. E. (2013). Evaluating the effectiveness of wood shred and agricultural straw mulches as a treatment to reduce postwildfire hill slope erosion in southern British Columbia. *Geomorphology, 197*, 21–33.

Schimel, D. S. (2010). Drylands in the earth system. *Science, 327*, 418–419.

Sharma, A. R., Singh, R., Dhyani, S. K., & Dube, R. K. (2010). Moisture conservation and nitro gen recycling through legume mulching in rainfed maize (Zea mays)–wheat (Triticum aestivum) cropping system. *Nutrient Cycling in Agroecosystems, 87*, 187–197.

Shcherbak, I., Millar, N., & Robertson, G. P. (2014). Global metaanalysis of the nonlinear response of soil nitrous oxide (N_2O) emissions to fertilizer nitrogen. *Proceeding of the National Academy of Sciences of the United States of America, 111*, 9199–9204.

Shi, Z. H., Yue, B. J., Wang, L., Fang, N. F., Wang, D., & Wu, F. Z. (2013). Effects of mulch cover rate on interrill erosion processes and the size selectivity of eroded sediment on steep slopes. *Soil Science Society of America Journal, 77*, 257–267.

Siczek, A., & Lipiec, J. (2011). Soybean nodulation and nitrogen fixation in response to soil compaction and surface straw mulching. *Soil and Tillage Research, 114*, 50–56.

Steiner, C., Teixeira, W. G., Lehmann, J., Nehls, T., de Macedo, J. L. V., Blum, W. E. H., & Zech, W. (2007). Long term effects of manure, charcoal and mineral fertilization on crop production and fertility on a highly weathered central amazonian upland soil. *Plant and Soil, 291*, 275–290.

Su, Y.-Z., Wang, F., Suo, D.-R., Zhang, Z.-H., & Du, M.-W. (2006). Long-term effect of fertilizer and manure application on soil-carbon sequestration and soil fertility under the wheat–wheat-maize cropping system in northwest China. *Nutrient Cycling in Agroecosystems, 75*, 285–295.

Thy, S., & Buntha, P. (2005). Evaluation of fertilizer of fresh solid manure, composted manure or biodigester effluent for growing Chinese cabbage (Brassica pekinen-sis). *Livestock Research for Rural Development, 17*, 149–154.

Tiquia, S. M., Lloyd, J., Herms, D. A., Hoitink, H. A. J., & Michel, F. C., Jr. (2002). Effects of mulching and fertilization on soil nutrients, microbial activity and rhizosphere bacterial community structure determined by analysis of TRFLPs of PCR-amplified 16S rRNA genes.

Teame, G., Tsegay, A., & Abrha, B. (2017). Effect of organic mulching on soil moisture, yield, and yield contributing components of sesame (Sesamum indicum L.). *International Journal of Agronomy,*. ID 4767509, 1–6.

Vohland, K., & Barry, B. (2009). A review of in situ rainwater harvesting (RWH) practices modifying landscape functions in African drylands. *Agriculture, Ecosystems & Environment, 131*, 119–127.

Walsh, O., Raun, W., Klatt, A., & Solie, J. (2012). Effect of delayed nitrogen fertilization on maize (Zea mays L.) grain yields and nitrogen use efficiency. *Journal of Plant Nutrition, 35*, 538–555.

Wan, X., Wei, W., & Liao, Y. (2021). Mitigating ammonia volatilization and increasing nitrogen use efficiency through appropriate nitrogen management under supplemental irrigation and rainfed condition in winter wheat. *Agricultural Water Management, 255*, 107050.

Wang, X.-L., Li, F.-M., Jia, Y., & Shi, W.-Q. (2005). Increasing potato yields with additional water and increased soil temperature. *Agricultural Water Management, 78*, 181–194.

Wang, Y., Xie, Z., Malhi, S. S., Vera, C. L., Zhang, Y., & Wang, J. (2009). Effects of rainfall harvesting and mulching technologies on water use efficiency and crop yield in the semi-arid Loess Plateau China. *Agricultural Water Management, 96*, 374–382.

Wang, Y., Xie, Z., Malhi, S. S., Vera, C. L., Zhang, Y., & Guo, Z. (2011). Effects of gravel–sand mulch, plastic mulch and ridge and furrow rainfall harvesting system combinations on water use efficiency, soil temperature and watermelon yield in a semi-arid Loess Plateau of northwestern China. *Agricultural Water Management, 101*, 88–92.

Wang, Z., Chen, J., Mao, S., Han, Y., Chen, F., Zhang, L., Li, Y., & Li, C. (2017). Comparison of greenhouse gas emissions of chemical fertilizer types in China's crop production. *Journal of Cleaner Production, 141*, 1267–1274.

Xu, X., Pei, J., Xu, Y., & Wang, J. (2020). Soil organic carbon depletion in global Mollisols regions and restoration by management practices: A review. *Journal of Soils Sediment, 20*, 1173–1181.

Yi, L., Shenjiao, Y., Shiqing, L., Xinping, C., & Fang, C. (2010). Growth and development of maize (Zea mays L.) in response to different field water management practices: Resource capture and use efficiency. *Agricultural and Forest Meteorology, 150*, 606–613.

Youkhana, A., & Idol, T. (2009). Tree pruning mulch increases soil C and N in a shaded coffee agroecosystem in Hawaii. *Soil Biology & Biochemistry, 41*, 2527–2534.

Zhang, Y., Kendy, E., Qiang, Y., Changming, L., Yanjun, S., & Hongyong, S. (2004). Effect of soil water deficit on evapotranspiration, crop yield, and water use efficiency in the north China Plain. *Agricultural Water Management, 64*, 107–122.

Zhang, S., Lovdahl, L., Grip, H., Tong, Y., Yang, X., & Wang, Q. (2009). Effects of mulching and catch cropping on soil temperature, soil moisture and wheat yield on the Loess Pla teau of China. *Soil and Tillage Research, 102*, 78–86.

Zhang, S., Li, P., Yang, X., Wang, Z., & Chen, X. (2011). Effects of tillage and plastic mulch on soil water, growth and yield of spring-sown maize. *Soil and Tillage Research, 112*, 92–97.

Zhao, Y., Pang, H., Wang, J., Huo, L., & Li, Y. (2014). Effects of straw mulch and buried straw on soil moisture and salinity in relation to sunflower growth and yield. *Field Crops Research, 161*, 16–25.

Zhou, L.-M., Li, F.-M., Jin, S.-L., & Song, Y. (2009). How two ridges and the furrow mulched with plastic film affect soil water, soil temperature and yield of maize on the semiarid Loess Plateau of China. *Field Crops Research, 113*, 41–47.

Zotarelli, L., Dukes, M. D., Scholberg, J. M., Hanselman, T., Le Femminella, K., & Munoz Carpena, R. (2008). Nitrogen and water use efficiency of zucchini squash for a plastic mulch bed system on a sandy soil. *Scientia Horticulturae, 116*, 8–16.

Zribi, W., Aragues, R., Medina, E., & Faci, J. M. (2015). Efficiency of inorganic and organic mulching materials for soil evaporation control. *Soil and Tillage Research, 148*, 40–45.

Implications of Mulching on Weed Management in Crops and Vegetable

Bilal Ahmad Khan, Aneela Nijabat, Muhammad Ishfaq Khan, Imtiaz Khan, Saima Hashim, Muhammad Athar Nadeem, and Muhammad Ikram

Abstract Global efforts are being made to reduce the world's high reliance on synthetic herbicides for weed control to protect human health and the environment and avoid the outbreak of weeds in various crops. Concerning the adverse effects on humans and the environment of the use of herbicides, a fair and cautious approach to restricting or even stopping the use of agrochemical products must be envisaged. Several methods provide information on agroecological activities in this context, such as mulching, which can contribute to the sustainable management of weeds in various field crops worldwide. In organic farming, mulching, by providing a barrier to sun, heat, or moisture exchange, is helpful as one strategy for integrated weed control. Evaporation is minimized, moisture is retained, and structure and temperature are controlled. On the other hand, the Mulching practice has many benefits, such as improving soil structure and texture by increasing infiltration and water retention and providing many insects and earthworms with a refuge. In addition, it promotes root penetration and growth and thus can also minimize erosion through nutrient uptake from deeper soil layers. Mulches control weeds by keeping the surface of the soil from receiving sunlight as light is needed for some weeds to germinate, and even necessary for all green plants to grow. Therefore, Mulches can be the best choice to control weeds in the field and reduce dependence on synthetic herbicides for weed control to prevent weeds' germination in various agronomic crops.

Keywords Crops · Mulches · Weeds · Management

B. A. Khan · M. A. Nadeem
Department of Agronomy College of Agriculture, University of Sargodha, Sargodha 40100, Pakistan

A. Nijabat
Department of Botany, Ghazi University, Dera Ghazi Khan 32200, Pakistan

M. I. Khan (✉) · I. Khan · S. Hashim
Department of Weed Science, University of Agriculture, Peshawar 25130, Pakistan
e-mail: mishfaq@aup.edu.pk

M. Ikram
Department of Botany, Pir Mehr Ali Shah Arid Agriculture University, Rawalpindi, Punjab, Pakistan

K. Akhtar et al. (eds.), *Mulching in Agroecosystems*,
https://doi.org/10.1007/978-981-19-6410-7_13

1 Introduction

1.1 Mulches Definition and Concepts

Mulch, also known as molsch, is from German language that means (easy to decay). It is evident that different kinds of mulches have been used in agronomic crops and vegetable since long time ago (Lightfoot, 1994). Mulches are different materials used to cover the soil surface to reduce moisture loss and weed population to increase agricultural production (Kader et al., 2019; Nalayini, 2007). Mulches can also help preserve evapotranspiration by reducing irrigation water runoff, improving soil permeation capacity, limiting weed multiplication, and reducing weed multiplication (Rathore et al., 1998). In addition to these mulches have numerous other ecological benefits, including regulation of soil temperature that helps plant root to grow vigoursly, decrease nutrient losses, reduced erosion and compactness of the soil (Lamont, 2005; Ngouajio & McGiffen, 2004). It was noted that mulches drew the interest of growers in 1930s because they would change the conditions surrounding rural, woodland, and horticultural lands. In previous research studies, mulches have been shown to have various adverse effects (Bedford & Pickering, 1919). For the plantation of trees and shrubs deep mulches were also used in 1941 (Pirone, 1941) as they provide protection against different environmental threats like drought stress, cold and other related threats that damage the plantation.

Singh et al. (1991) revealed that more moisture contents were preserved when the similar quantity mulching material was used as mulch relative to the instant soil in which they were applied. Mulches were made from agricultural and forestry by-products in the mid-nineteenth century (Clifford & Massello, 1965). Other materials, such as tree and shrub trimmings, animal waste, stubbles, and crop plant residues, have also been used as mulch. Landscape mulches were also seen in 1957, but no scientific research was conducted on them. Mulches can reduce crop plant irrigation requirements and, in some cases, completely eliminate the need for irrigation (Ahmad et al., 2015, 2020; Iqbal et al., 2019; Kader et al., 2019). Some organic mulch acts as a sponge, absorbing runoff and providing water when crops need it, thus conserving rainfall and irrigation water. Also, the water runoff was declined by 43% using straws mulches as surface cover (Borst & Woodburn, 1942). Plastic mulches were first used commercially in the early 1960s, primarily for the cultivation of vegetables. Mulches reduce the need for additional irrigation by storing water, reducing runoff in the soil profile simultaneously (Smith, 2000). Additionally, mulching material protects vulnerable crop from different environmental condition that are triggered by extreme weather, weeds and insect. These concepts make mulches as the most important aspects of weeds under field condition and thus contribute in uplift of modern agriculture.

Table 1 Use of various kinds of plastic mulches in different crops for sustainable production

Mulch Type	Crop	References
Transparent plastic mulch	Maize	Zhang et al.
	Potato	Zaho et al.
	Potato	Wang et al. (2009)
Black plastic mulch	Cucumber	Torres-Olivar et al.
	Maize	Xiukang et al.
	Maize	Li et al.
Silver plastic mulch	Cucumber	Torres-Olivar et al.
Degradable plastic mulch	Maize	Li et al.
Plastic and straw mulch	Maize and wheat	Yin et al.

1.2 Plastic Mulch Application and Response of Soil

The types of plastics including ethylene vinyl acetate (EVA), polyethylene (PE) and polyvinyl chloride (PVC), the majority of these are PE polymers made of petroleum, and they are usually non-biodegradable (Hussain & Hamid, 2003). The global use of plastic films for greenhouse and mulching applications is expected to rise by 69% from 4.4 million tons in 2012 to 7.4 million tons in 2019 (Sintim & Flury, 2017) Unfortunately, the deterioration of polyethylene mulch in the soil can lead to the damaging of aldehydes and ketones that are threats to the biodiversity (Hakkarainen & Albertsson, 2004). Polyethylene takes about 100 years to decompose fully due to its chemical stability (Blick et al., 2010). To overcome this issue, the best altherna-tive should be a biodegradable plastic that are entirely mineralized at the end of the process. Biodegradable plastic mulch has been created using modern plastic processing technology (Zhang et al., 2017). Use of various plastic mulches reported in previous studies has improved the crop yield by regulating the soil temperature (Table 1).

1.3 Concept of Biodegradable Mulches

Biodegradable plastic mulch is made from several biopolymers, including polyhydroxyalkanoates and polylactic acid. Biodegradable polymers can be produced by adding using additives to enhance the physical and mechanical properties of the resultant film (Zhang et al., 2017). According to a study, crop yield can be improved by incorporating bioactive films using mineral herbicides and mineral nutrients (Minuto et al., 2008). Nitrogen in combination with biodegradable film has resulted in a significant increase in nitrogen uptake. The use of biodegradable film as rice mulch allowed this improved fertilization efficiency (Zhang et al., 2017). Legumes crops have the potential to fix nitrogen in a symbiotic relationship with other food crops, benefiting agricultural and environmental sustainability around the world.

1.4 Global Crop Production Interests and an Integrated Concept of Weed Management

The invention of novel approaches for weed control in legumes crops has reignited scientific interest in cropping systems worldwide (Avola et al., 2008). A key principle of conservation agriculture is to cover the soil with mulch, especially when growing a vegetable crop for green pods, to increase yield (Kumar & Angadi, 2016). The key biotic restriction in agriculture is weed infestation (Chauhan & Abugho, 2013) reported that if mulch and herbicides were used together, they resulted long term and more effective weed management that gave enhanced yield in direct seeded rice system. The invention of agrochemicals, such as herbicides, was responsible for many of the agricultural advancements (Zimdahl, 2015).

One of the groundbreaking methods recently explored was the use of controlled releases. Although maintaining high crop productivity, these controlled release carrier systems can reduce negative environmental impacts. Researchers have changed the formulations and preparations for the herbicide chloridazone's sustained release with biodegradable polymers e.g. ethylcellulose and lignin (Grillo et al., 2011). The usage of the instant biopolymers under controlled release formulations has improved herbicide delivery performance. Various methods of integrated weed management (IWM) in different are enlisted in Table 2.

Similarly, summarizing the herbicide amertyn in the shape of micro particles by using the polymers e.g. poly (3-hydroxybutyrate-co-3-hydroxyvalerate) and poly (3-hydroxybutyrate) enhanced the herbicide amertyn's release profile, with delivery being slower and longer than when the herbicide was applied freely. The herbicide ametryn was successfully compressed in polymeric micro particles, implying that these methods could be helpful in decreasing the herbicide's adverse ecological

Table 2 Direct methods of integrated weed management in crops and vegetables

Crop	Method	Integration type	References
Maize	Physical–Mechanical	Banded flaming–Cultivator	Stepanovic et al. (2016)
Maize	Biological–Chemical	Bioherbicide–Herbicide	Ehlert et al. (2014)
Sunflower	Chemical–Mechanical	Herbicides–Hoeing	Pannacci and Tei (2014)
Hemp	Chemical–Mechanical	Herbicides–Ploughing	Vasileiadis et al. (2012)
Okra	Mechaniqcal–Biological	Reduced tillage–Bioherbicides	Wagner and Mitschunas (2008)
Onion	Mechanical–Physical	Hoeing–Brush weeding	Melander and Rasmussen (2001)

Table 3 Successful weed and water management by mulch use in various vegetable crops

Mulch nature	Mulch type	Yield increased (%)
Organic mulches	Bark	30% increase in crop vegetation yield
	Grass	73% increase in legumes yield
	Dry leaves	20% increase in groundnuts yield
	Saw dust	79% increase in berries yield
	Newspaper	16% increase in pot vegetables yield
	Alfalfa	36% increase in sweet corn yield
	Seaweeds	47% increase in kitchen garden yield
	Ash	35% increase in garlic yield
	Stubble	78% increase in stevia and mint yield
	Straw	27% increase in tomato yield
	Sunhemp	90% increase in maize yield
	Leaf litter	8% increase in home garden yield
	Peat	7.3% increase in *Beta vulgaris* yield
	Grass	30% increase in onion yield
Synthetic mulches	Plastic	17% increase in cauliflower yield
	Black plastic	25% increase in muskmelon yield
	Clear plastic	45% increase in muskmelon yield
	Red plastic	70% increase in tomato and zucchini yield
	Yellow plastic	46% increase in black pepper and zucchini yield
	Gravels, pebbles	20% increase in kitchen garden yield
	Chips	23% increase in potted vegetables yield
	Rubber	8% increase in kitchen garden yield

effects. These results indicate that MCPA-PHBV conjugates may have a longer duration of association between herbicides and weed species in agricultural systems. This could make it easier to target weeds more accurately while also reducing environmental and non-target crop plant impacts. As a result, using these mulches biofilm formulation and the consequent release of active herbicide ingredients could help in emission reduction by requiring fewer agrochemicals to achieve the same results as traditional applications (Grillo et al., 2011). Previous studies reported that various organic and synthetic mulch types has significantly increased the vegetable crop yield by successful weed and water management (Table 3) (Chopra & Koul, 2020).

1.5 Cover Crops as an Instrumental Mulching

To avoid weed germination in the spring, organic farmers may mow or crimp fall cover crops as mulches instead of tillage (Robb et al., 2018; Ward et al., 2011). This works by physically impeding seedling emergence by simulating conditions found more profound in the soil (lower light, lower temperatures (Upadhyaya & Blackshaw, 2007). Even though crop mulch can also help to suppress the early season weeds (Blanco-Canqui et al., 2015). The research studies compared the combined effects of chemical and cultural weed control by comparing the combinations of three cover crop mulches (crimson clover, *Trifolium incarnatum* L. and *Secale cereal* L. and fallow control) commonly used in organic vegetable production and three organic herbicide treatments (capric and caprylic acid [CCA], corn gluten meal [CGM], and herbal gluten meal [CGM] can also be used for raising of organic vegetable production. They aimed to see whether herbicides that had recently been approved for use in organic vegetable production could be a valuable tools for IPM and look for non-target effects on beneficial insects. They measured weed pressure and assessed the operation densities of seed predators and arthropod predators, as well as the biological control services of weed seeds, in each of the combined treatments. They predicted that combining cover crops and organic herbicides would minimize weed pressure more effectively than either method alone, but that since CCA is acidic (DiTomaso et al., 2017), it would reduce the effectiveness of natural ground-active enemies and the biological control services they provide.

1.6 Control of Direct and Indirect Weeds Through Mulches

Weed control can be achieved in a variety of ways, both direct and indirect. Thermal weed control, mulching, and biological weed control are examples of direct weed control methods. In contrast, indirect weed control methods include, but are not limited to, selecting crop cultivars that tolerate or suppress weeds, using intercropping or mulching materials, crop rotation, and cultivation (Bond & Grundy, 2001; Wei et al., 2010). Weeds are influenced by management strategies such as tillage or lack of tillage combined with mulching and viability. Crop viability was lower in mulched soil samples than in tilled parcels for some species (*Amaranthus* spp., *Cuscuta* spp. L.) (Moonen & Bàrberi, 2004). Consider the germination of *Amaranthus retroflexus* L. Seeds grew slower in rye mulch than in control plots.

On the other hand, mulching with poplar (*Populus deltoides* Bartr.) did not affect *Echinochloa* Crus-galli as it does not leach any phytotoxic compounds from poplar mulch because it is chemically inert (Moonen & Bàrberi, 2006). Rye (*Secale cereale* L.) straw, on the other hand, can release allochemicals or phytotoxic acids that inhibit weed seed germination (Creamer et al., 1996; Blum, 1997; Burgos & Talbert, 2000). Rice straw mulch was almost as effective as polyethylene mulch (Ramakrishna et al., 2006). Straws from other plants, such as rye or buckwheat (Fagopyrum esculentum Moench), have been shown to suppress weeds (Zaniewicz-Bajkowka et al., 2009; Kosterna, 2014) due to allelopathic effects on seed germination and growth (Creamer et al., 1996). Perennial weeds can be minimized using various mulch materials, including straw, peat, and sawdust, but other mulch materials, such as grass clippings, decompose more quickly, resulting in a decrease in ecacy in the second year

(Pupalienė et al., 2015). Compost, weed, and alfalfa clippings are not as essential as paper, black plastic coating, or rye straw, according to a study on the e-effect of mulching in weed control (Radics & Bognár, 2004). Depending on the origin of the composted materials, acetic acid and organic acid can leach from the compost. Weeds and certain crops can be harmed by these compounds (Ozores-Hampton, 1998). Some recent studies reported combined use of various organic and synthetic mulches as most appropriate and effective control measure of weed management in several vegetable crops including carrot (Jaysawal et al., 2018), garlic (Najafabadi et al., 2012), onion (Larentzaki et al., 2008), potato, tomato (Tu et al., 2006), brinjal (Kumar et al., 2019) and cauliflower (Bhoutekar et al., 2017).

1.7 Impact of Plastic Mulches on Weed Control and Soil Resources

In both conventional and organic fresh vegetable production systems, plastic mulch film is commonly used (Lamont, 2005; Tarara, 2000). Temperature management, reduced nutrient leaching, increased moisture retention,, reduced soil splash and effective weed suppression in the crop rows are only a few of the advantages of plastic mulch, all of which will help to reduce the incidence of many soil-borne diseases. In the plastics industry, these advantages mostly account for marketable vegetable and higher yields (Kasirajan & Ngouajio, 2012; Tarara, 2000). In addition to these benefits, soil erosion in plastic culture systems can be greatly increased, as the impermeable plastic-covered beds surface can cause runoff between beds in the uncovered areas (Steinmetz et al., 2016). This happened when weeds are controlled with herbicides and planting in between-bed areas, leaving the soil bare (Steinmetz et al., 2016; Wan & El-Swaify, 1999). When combined with bare soil, the accumulation of water after rain events within the beds area will increasing the possibility of pesticide and nutrient leaching and runoff from plastic systems, posing a danger to local water sources (Arnold et al., 2004; Rice et al., 2004). Soil erosion can be minimized if cover crop can be grown as living mulch between the plastic-mulched beds for crop productivity.

In addition to other projects, these soil and nutrient management concerns include organic matter and weed suppression contributions in rigorous vegetable cropping system or cool climates where it may be difficult to incorporate a productive cover crop into the rotation (Paine & Harrison, 1993; Rice et al., 2004; Sarrantonio, 1992; Snapp et al., 2005). Species selection is likely to be a key factor in maximizing potential benefits while mitigating output risk from living mulches in plastic cultivation systems. A good living mulch can keep weeds at bay while also limiting cash crop interference (Hartwig & Ammon, 2002). The following plant characteristics are likely to balance weed suppression and cash crop interference: (1) rapid germination and establishment to exclude weeds, (2) thorough soil coverage, (3) resistance to drought and low-nutrient environments, (4) a low and efficiently managed growth habit (Adamczewska-Sowinska et al., 2009; Hartwig & Ammon, 2002; Brennan & Smith, 2018).

1.8 Mechanism of Integrated Weed Management (IWM)

For integrated weed management the cover crops (Fall-to-spring) are considered to be the potential source in temperate climate cropping pattern (IWM). A general mechanism of integrated weed management is shown in Fig. 1. In comparison to organic mulches, cover crops (fall-to-spring) are occasionally separated from the leading crops in the area. Moreover, 6 weeks cycle with favorable environmental conditions is required for good stand of cover cropping pattern. In regions with temperate climate, cover crops can be planted between the harvesting of oil-seed rape, cereals, and legumes, and the cultivation of spring season crops like sugar beet, maize, soybeans, potatoes, and spring cereals will grow profitably in such conditions. Cover crops such as (winter-kill fall-to-spring) cover crops include *Phacelia tanacetifolia*, *Sinapis alba,* and *Raphanus sativus* var. Oleiformis, as well as some grasses and clover. In addition to this the other cover crops that recently mentioned by researchers with enough tolerance to warm and dry weather conditions include; common buckwheat, *Vicia sativa, Camelina sativa, Avena strigosa,, linseed, Guizotia abyssinica* and sunflower (Blanco-Canqui et al., 2015; Teasdale, 1996). This analysis focuses on recent research on using cover crops to improve weed suppression by optimizing competition, allelopathy, and mixture efficacy. The suppression of weeds by cover crops is most likely due to a combination of factors. Due to canopy closure, rapid shading is perceived the important factor for effective weed control (Brust et al, 2014). Common cover crops such as buckwheat, sunflower, linsead and *A. strigosa* are considered to be the Western Europe fastest-growing cover crops. They grow very quickly by covering 40–80%t of the soil surface in 4–8 weeks and produce 200 gm^{-2} of its vegetative biomass (Brust et al., 2014). This trait allows them to compete with weeds effectively. Consequently they suppress even the nearby perennial weeds that include; *Cyperus esculentus, Cirsium arvense,* and *Elymus repens* (Brust et al., 2011; Kunz et al., 2016; Lawley et al., 2011; Osipitan et al., 2018; Kruidof et al., 2008; Bezuidenhout et al., 2012).

1.9 Mulches and Allelopathy

It is worth mentioning to quantify the qualities of cover crops like their competitive, physical and biochemical abilities on weed management. Plant breeders and the large scale growers sometime may combine many weed suppressive capabilities of cover crops in a wide range of crop types. The presence of competing organisms slows the early growth of weeds. Allelopathic plants stop weeds from germinating by releasing biochemicals and thereby affect the neighbouring plants; this process is

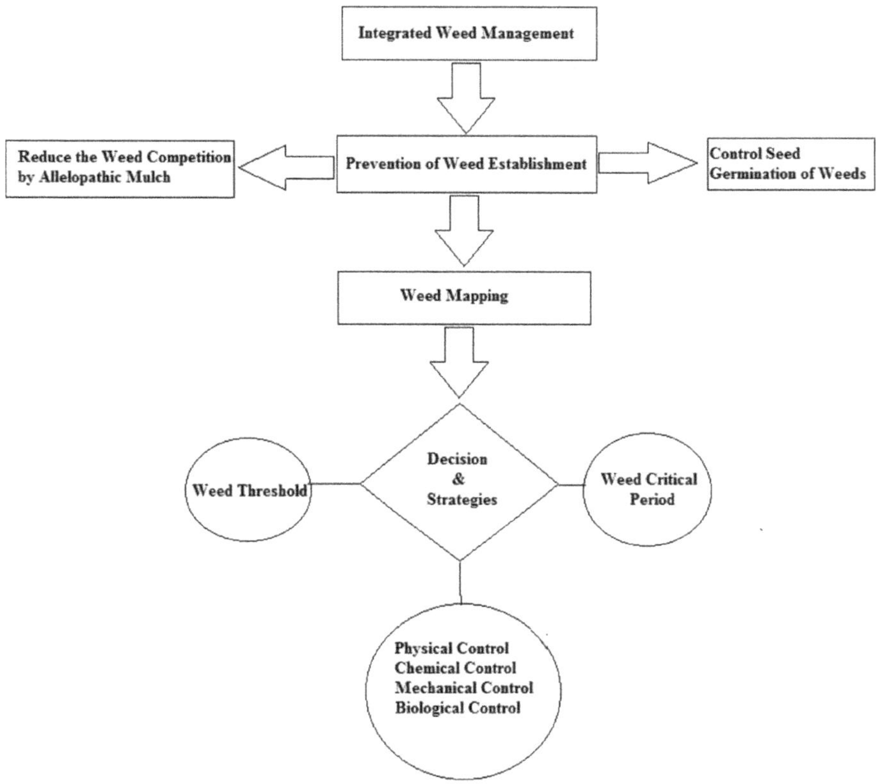

Fig. 1 A schematic view of mechanism of integrated weed management (IWM)

also called biological warfare among the plants for their survival. It is very important to mention that integrated weed management including cover crops and allelopathy is the alternative to the extensive use of herbicides and tillage operation to achieve economical weed suppression. General mechanism of allelopathic repression from competition has been exhibited in Fig. 2.

A three years research studies under field condition were conducted to assess the canopy and aboveground dry biomass of *Phacelia tanacetifolia*, *Sinapis alba*, and *Raphanus sativus* var. Oleiformis. It was observed that the density of Matricaria chamomilla¸ Chenopodium album, and Stellaria media was suppressed significantly that was dominant weeds in the field. These cover crops inhibited 60% of weeds growth on average as compared to control plots without cover crops, and light competition played only a minor role in overall weed suppression. Allelopathic effects accounted for more than half of the difference in cannabis density. These allelopathic properties were assessed using aqueous extracts of leaves from the instant cover crops in a Petri dish bioassay (125 mg mL^{-1}). Root length was measured of the all the weed species to determine allelopathic behavior in a Petri dish test. Land trials with the five most common weed species were used in petri dish experiments

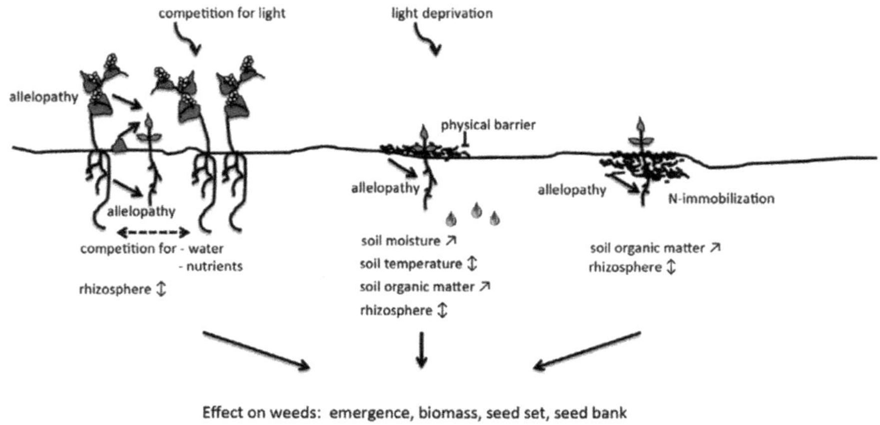

Fig. 2 A general mechanism of allelopathic mulches on weeds (Falquet et al., 2015)

(Kunz et al., 2016). The possibility of seed germination test in cover crops aqueous extracts on the other hand, is debatable (Sturm et al., 2017). In field conditions, most cover crop extracts suppress seed germination of the weeds at higher concentration (Rueda-Ayala et al., 2015), which is impractical. Unlike plant extracts, secondary metabolites are not release at a time by the living cover crops into the soil. Cover crop extract germination bioassays, on the other hand, can be used to determine if a cover crop has a low, moderate or high potential of allelopathic nature (Kunz et al., 2016; Sturm et al., 2016). To distinguish allelopathic and competitive impacts, an experiment was undertaken on weeds and cover crops in a separate system with or without active carbon mixed into the soil (Sánchez-Moreiras et al., 2003). It was observed that allelochemicals found in the soil, especially the one released from cover crop roots, were adsorbted and inactivated using active carbon. The word "S. biomass" refers to the amount of biomass generated by S. The media was reduced by 60–90% in pots containing activated carbon, and voluntary wheat by 40–70%. Except for linseed, all cover crops are suppressed. It is generally accepted that allelopathy had a major consequences on weed management, but cover crop rivalry was still the most important factor in weed suppression, according to this report (Gfeller et al., 2018). In another study it was found that cover crops roots exudates had a greater allelopathic effects on weed management (Falquet et al., 2015; Gfeller et al., 2018). These facts highlight the significance of using or investigating the potential of cover crop allelopathy and its consequences on weed suppression.

2 Conclusion

Finally, cover cropping has been shown in recent studies to be a successful method for IWM because it puts a range of pressures on weeds and voluntary crops. It integrates well with current cropping systems. During the fall-to-spring season, cover crops will reduce negative impacts while also offering additional benefits if they emerge quickly after the previous cash crop has been harvested and their residues provide adequate biochemical and physical weed control before the next cash crop emerges. Recognizing competitive and allelopathic species and combining them correctly will increase the effectiveness of cover crop mixtures, and we now have several methods for doing so. It may be possible to increase allelopathy via stress or injury, resulting in even better weed suppression when crops are covered. Mulching with various materials can help to preserve soil moisture, reduce evaporation, and control weed growth. Mulching techniques had a big influence on crop growth, yield, and quality. There are some uncertainties in mulching's results, as numerous scientists have documented the negative effects of mulching. While mulches' benefits are most dominant in these inconsistencies, these drawbacks stated by various scientists are not as harmful in natural field conditions. However, it can be concluded from the literature that mulches are a cheap source for the reduction of weed populations and for the substantial preservation of soil moisture content. Therefore in water deficit/drought conditions, properly managed mulching strategies could compensate for the water requirements of crops. In addition, integrating the partial root zone drying (PRD) mulching system (wheat straw, cotton sticks, black plastic, maize straw) could serve as an efficient technique to improve overall crop growth, development, and yield.

References

Adamczewska-Sowinska, K., Kolota, E., & Winiarska, S. (2009). Living mulches in field cultivation of vegetables. *Vegetable Crops Research Bulletin, 70*(1), 19–29.

Ahmad, S., Raza, M. A. S., Saleem, M. F., Zaheer, M. S., Iqbal, R., Haider, I., Aslam, M. U., Ali, M., & Khan, I. H. (2020). Significance of partial root zone drying and mulches for water saving and weed suppression in wheat. *Journal of Animal and Plant Sciences, 30*, 154–162.

Ahmad, S., Raza, M. A. S., Saleem, M. F., Zahra, S. S., Khan, I. H., & Ali, M. (2015). Mulching strategies for weeds control and water conservation in cotton. *Journal of Agricultural and Biological Sciences, 8*, 299–306.

Arnold, G. L., Luckenbach, M. W., & Unger, M. A. (2004). Runoff from tomato cultivation in the estuarine environment: Biological effects of farm management practices. *Journal of Experimental Marine Biology and Ecology, 298*(2), 323–346.

Avola, G., Tuttobene, R., Gresta, F., & Abbate, V. (2008). Weed control strategies for grain legumes. *Agronomy for Sustainable Development, 28*, 389–395.

Bedford, H. A. R., & Pickering, P. S. U. (1919). Science and fruit growing: Being an account of the results obtained at the woburn experimental fruit farm since its foundation in 1894 Macmillan.

Bezuidenhout, S. R., Reinhardt, C. F., & Whitwell, M. I. (2012). Cover crops of oats stooling rye and three annual ryegrass cultivars influence maize and *Cyperus esculentus* growth. *Weed Research, 52*, 153–160.

Bhoutekar, S., Luchon, S., Bonti, G., & Sonbeer, C. (2017). Fertigation level and mulching in Cauliflower (*Brassica oleraceae* L. var. botrytis) cv. Snowball White. *International Journal of Agriculture Sciences, 9*, 4226–4228.

Bilck, A. D., Grossmann, M. V. E., & Yamashita, F. (2010). Biodegradable mulch films for strawberry production. *Polymer Testing, 29*, 471–476.

Blanco-Canqui, H., Shaver, T. M., Lindquist, J., Shapiro, C. A., & Elmore, R. W. (2015). Cover crops and ecosystem services: Insights from studies in temperate soils. *Agronomy & Horticultural Faculty Publication, 844*, 1–15.

Blum, U. (1997). Benefits of citrate over EDTA for extracting phenolic acids from soils and plant debris. *Journal of Chemical Ecology, 23*, 347–362.

Bond, W., & Grundy, A. C. (2001). Non-chemical weed management in organic farming systems. *Weed Research, 41*, 383–405.

Borst, H. L., & Woodburn, R. (1942). The effect of mulching and methods of cultivation on runoff and erosion from Muskingham silt loam. *Journal of Agricultural Engineering, 23*, 19–22.

Brennan, E. B., & Smith, R. F. (2018). Mustard cover crop growth and weed suppression in organic strawberry furrows in California. *HortScience, 53*, 432–440.

Brust, J., Claupein, W., & Gerhards, R. (2014). Growth and weed suppression ability of common and new cover crops in Germany. *Crop Protect, 63*, 1–8.

Brust, J., Gerhards, R., Karanisa, T., Ruff, L., & Kipp, A. (2011). Why under sown and cover crops become important again for weed suppression in European cropping systems. *Gesunde Pflanzen, 63*, 191–198.

Burgos, N. R., & Talbert, R. E. (2000). Differential activity of allelochemicals from *Secale cereale* in seedling bioassays. *Weed Science, 48*, 302–310.

Chauhan, B. S., & Abugho, S. B. (2013). Integrated use of herbicide and crop mulch in suppressing weed growth in a dry-seeded rice system. *American Journal of Plant Sciences, 4*, 1611–1616.

Chopra, M., & Koul, B. (2020). Comparative assessment of different types of mulching in various crops: A review. *Plant Archives, 20*, 1620–1626.

Clifford, E. D., & Massello, J. W. (1965). Mulching materials for nursery seedbeds. *Tree Planters Notes, 72*, 18–22.

Creamer, N. G., Bennett, M. A., Stinner, B. R., Cardina, J., & Regnier, E. E. (1996). Mechanisms of weed suppression in cover crop-based production systems. *Hortscience, 31*, 410–413.

DiTomaso, J., Kyser, G., Lewis, D., & Roncoroni, J. (2017). Conventional and organic options for the control of woolly distaff thistle (*Cathamus lana- tus*). *Invasive Plant Science and Management, 10*, 72–79.

Ehlert, K. A., Mangold, J. M., & Engel, R. E. (2014). Integrating the herbicide imazapic and the fungal pathogen Pyrenophora semeniperda to control Bromus tectorum. *Weed Research, 54*, 418–424.

Falquet, B., Gfeller, A., Pourcelot, M., Tschuy, F., & Wirth, J. (2015). Weed suppression by common buckwheat: A review. *Environmental Control in Biology, 53*(1), 1–6.

Gfeller, A., Herrera, J. M., Tschuy, F., & Wirth, J. (2018). Explanations for *Amaranthus retroflexus* growth suppression by cover crops. *Crop Protection, 104*, 11–20.

Grillo, R., SPereira, A. E., de Melo, N. F. S., Porto, R. M., Feitosa, L. O., Tonello, P. S., Filho, N. D., Rosa, A. H., Lima, R., & Fraceto, L. F. (2011). Controlled release system for ametryn using polymer microspheres: Preparation characterization and release kinetics in water. *The Journal of Hazardous Materials, 186*, 1645−1651.

Hakkarainen, M., & Albertsson, A. C. (2004). Environmental degradation of polyethylene. *Advances in Polymer Science, 169*, 177–200.

Hartwig, N. L., & Ammon, H. U. (2002). Cover crops and living mulches. *Weed Science, 50*(6), 688–699.

Hussain, I., & Hamid, H. (2003). Plastics in agriculture. In A. L. Andrady (Ed.), *Plastic and the Environment* (pp. 185–206). Wiley.

Iqbal, R., Muhammad, A. S. R., Muhammad, F. S., Imran, H. K., Salman, A., Muhammad, S. Z., Muhammad, U., & Imran, H. (2019). Physiological and biochemical appraisal for mulching and partial rhizosphere drying of cotton. *Journal of Arid Land, 11*, 785–794.

Iqbal, R., Raza, M. A. S., Valipour, M., Saleem, M. F., Zaheer, M. S., Ahmad, S., Toleikiene, M., Haider, I., Aslam, M. U., & Nazar, M. A. (2020). Potential agricultural and environmental benefits of mulches—a review. *Bulletin of the National Research Centre, 44*(1), 1–16.

Jaysawal, N., Singh, G., Kanojia, A., & Debbarma, B. (2018). Effect of different mulches on growth and yield of carrot (Daucus carota L.). *International Journal of Chemical Studies, 6*(4), 381–384.

Kader, M. A., Singha, A., Begum, M. A., Jewel, A., Khan, F. H., & Khan, N. I. (2019). Mulching as water-saving technique in dry land agriculture. *Bulletin of the National Research Centre, 43*, 1–6.

Kasirajan, S., & Ngouajio, M. (2012). Polyethylene and biodegradable mulches for agricultural applications: A review. *Agronomy for Sustainable Development, 32*(2), 501–529.

Kosterna, E. (2014). The effect of soil mulching with organic mulches on weed infestation in broccoli and tomato cultivated under polypropylene fibre and without a cover. *Journal of Plant Protection Research, 54*, 188–198.

Kruidof, H. M., Bastiaans, L., & Kropff, M. J. (2008). Ecological weed management by cover cropping: Effects on weed growth in autumn and weed establishment in spring. *Weed Research, 48*, 492–502.

Kumar, B. R. M., & Angadi, S. S. (2016). Effect of tillage mulching and weed management practices on the performance and economics of chickpea. *Legume Research, 39*, 786–791.

Kumar, P., Kumar, S., Kumari, M., & Kumar, V. (2019). Effect of mulching on brinjal cultivation. *International Journal of Science, Environment and Technology, 8*(3), 624–629.

Kunz, C. D., Sturm, J., Varnholt, D., Walker, F., & Gerhards, R. (2016). Allelopathic effects and weed suppressive ability of cover crops. *Plant, Soil and Environment, 62*, 60–66.

Lamont, W. J. (2005). Plastics: Modifying the microclimate for the production of vegetable crops. *Horticultural Technology, 15*, 477–481.

Larentzaki, E., Plate, J., Nault, B. A., & Shelton, A. M. (2008). Impact of straw mulch on populations of onion thrips (Thysanoptera: Thripidae) in onion. *Journal of Economic Entomology, 101*(4), 1317–1324.

Lawley, Y. E., Weil, R. R., & Teasdale, J. R. (2011). Forage radish cover crop suppresses winter annual weeds in fall and before corn planting. *Agronomy Journal, 103*, 137–144.

Lightfoot, D. R. (1994). Morphology and ecology of lithic-mulch agriculture. *Geography Review, 25*, 172–185.

Mehetre, Y. D., Pawar, R. D., Mangave, K. K., Sonawane, P. N., & Dhumal, S. S. (2018). Effects of different mulches on the yield and productivity of drip irrigated chilli cv. Phule Jyoti. *Trends in Biosciences Dheerpura Society for Advancement of Science and Rural Development An International Journal, 2726*.

Melander, B., & Rasmussen, G. (2001). Effects of cultural methods and physical weed control on intrarow weed numbers manual weeding and marketable yield in direct-sown leek and bulb onion. *Weed Research, 41*, 491–508.

Minuto, G., Pisi, L., Tinivella, F., Bruzzone, C., Guerrini, S., Versari, M., Pini, S., & Capurro, M. (2008). Weed control with biodegradable mulch in vegetable crops. *Acta Horticulture, 801*, 291–298.

Moonen, A. C., & Bàrberi, P. (2004). Size and composition of the weed seedbank after 7 years of different cover crop maize management systems. *Weed Research, 44*, 163–177.

Moonen, A. C., & Bàrberi, P. (2006). An ecological approach to study the physical and chemical effects of rye cover crop residues on *Amaranthus retroflexus Echinochloa crus-galli* and maize. *The Annals of Applied Biology, 148*, 73–89.

Najafabadi, M., Peyvast, G. H., Hassanpour, A. M., Olfati, J. A., & Rabiee, M. (2012). Mulching effects on the yield and quality of garlic as second crop in rice fields.

Nalayini, P. (2007). Poly-mulching a case study to increase cotton productivity. Senior scientist *Central Institute for Cotton Research* Regional Station Coimbatore.

Ngouajio, M., & McGiffen, M. E. (2004). Sustainable vegetable production: Effects of cropping systems on weed and insect population dynamics. *Acta Horticulture, 638*, 77–83.

Osipitan, O. A., Dille, J. A., Assefa, Y., & Knezevic, S. Z. (2018). Cover crop for early season weed suppression in crops: Systematic review and meta-analysis. *Agronomy Journal, 110*, 2211–2221.

Ozores-Hampton, M. (1998). Compost as an alternative weed control method. *HortScience, 33*, 938–940.

Paine, L. K., & Harrison, H. (1993). The historical roots of living mulch and related practices. *HortTechnology, 3*, 137–143.

Pannacci, E., & Tei, F. (2014). Effects of mechanical and chemical methods on weed control weed seed rain and crop yield in maize sunflower and soybean. *Crop Protection, 64*, 51–59.

Pirone, P. P. (1941). Freak weather damages trees and shrubs. *New Jersey Agriculture, 23*, 3.

Pupalienė, R., Sinkevičienė, A., Jodaugienė, D., & Bajorienė, K. (2015). Weed control by organic mulch in organic farming system. In V. Pilipavicius (Ed.), *Weed biology and control* (pp. 65–86). InTech.

Radics, L. E., & Bognár, S. (2004). Comparison of different mulching methods for weed control in organic green bean and tomato. *Acta Horticulturae, 638*, 189–196.

Ramakrishna, A., Tam, H. M., Wani, S. P., & Long, T. D. (2006). Effect of mulch on soil temperature moisture weed infestation and yield of groundnut in northern Vietnam. *Field Crops Research, 95*, 115–125.

Rathore, A. L., Pal, A. R., & Sahu, K. K. (1998). Tillage and mulching effects on water use root growth and yield of rain-fed mustard and chickpea grown after lowland rice. *Journal of the Science of Food and Agriculture, 78*, 149–161.

Rice, P. J., Harman-Fetcho, J. A., Teasdale, J. R., Sadeghi, A. M., McConnell, L. L., Coffman, C. B., Herbert, R. R., Heighton, L. P., & Hape-man, C. J. (2004). Use of vegetative furrows to mitigate copper loads and soil loss in runoff from polyethylene (plastic) mulch vegetable production systems. *Environmental Toxicology and Chemistry, 23*(3), 719–725.

Robb, D., Zehnder, G., Kloot, R., & Bridges, W. (2018). Weeds nitrogen and yield: Measuring the effectiveness of an organic cover cropped vegetable no-till system. *Renewable Agricultural and Food Systems, 34*, 439–446.

Rueda-Ayala, V., Jäck, O., & Gerhards, R. (2015). Investigation of biochemical and competitive effects of cover crops on crops and weeds. *Crop Protection, 72*, 79–87.

Sánchez-Moreiras, A. M., Weiss, O. A., Reigosa-Roger, M. J. (2003). Allelopathic evidence in the Poaceae. *Botanical Review, 69*, 300–319.

Sarrantonio, M. (1992). Opportunities and challenges for the inclusion of soil-improving crops in vegetable production systems. *HortScience, 27*, 754–758.

Singh, S. B., Pramod, K., Prasad, K. G., & Kumar, P. (1991). Response of *Eucalyptus* to organic manure mulch and fertilizer sources of nitrogen and phosphorus. *VanVig, 29*, 200–207.

Sintim, H. Y., & Flury, M. (2017). Is biodegradable plastic mulch the solution to agriculture's plastic problem? *Environmental Science and Technology, 51*, 1068–1069.

Smith, M. W. (2000). Cultivar and mulch affect cold injury of young pecan trees. *Journal of American Pomological Society, 54*, 29–33.

Snapp, S. S., Swinton, S. M., Labarta, R., Mutch, D., Black, J. R., Leep, R., Nyiraneza, J., & O'Neil, K. (2005). Evaluating cover crops for benefits costs and performance within cropping system niches. *Agronomy Journal, 97*(1), 322–332.

Steinmetz,Z., Wollmann, C., Schaefer, M., Buchmann, C., David, J., Tröger, J., Muñoz, K., Frör, O., Schaumann, G. E. (2016). Plastic mulching in agriculture Trading short-term agronomic benefits for long-term soil degradation? *Science of the Total Environment, 550*, 690–705.

Stepanovic, S., Datta, A., Neilson, B., Bruening, C., Shapiro, C. A., Gogos, G., & Knezevic, S. Z. (2016). Effectiveness of flame weeding and cultivation for weed control in organic maize. *Biological Agriculture & Horticulture, 32*, 47–62.

Sturm, D., Peteinatos, G., & Gerhards, R. (2017). Contribution of allelopathic effects to the overall weed suppression by different cover crops. *Weed Research, 58*, 331–337.

Sturm, D. J., Kunz, C., & Gerhards, R. (2016). Inhibitory effects of cover crop mulch on germination and growth of *Stellaria media* (L) Vill *Chenopodium album* L and *Matricaria chamomilla* L. *Crop Protection, 90*, 125–131.

Tarara, J. M. (2000). Microclimate modification with plastic mulch. *HortScience, 35*, 169–180.

Teasdale, J. R. (1996). Contribution of cover crops to weed management in sustainable agricultural systems. *Journal of Production Agriculture, 9*, 475–479.

Tu, C., Ristaino, J. B., & Hu, S. (2006). Soil microbial biomass and activity in organic tomato farming systems: Effects of organic inputs and straw mulching. *Soil Biology and Biochemistry, 38*(2), 247–255.
Upadhyaya, M., & Blackshaw, R. (2007). Non-chemical weed management: Principles concepts and technology. CABI Publishing Cambridge.
Vasileiadis, V. P., Froud-Williams, R. J., & Eleftherohorinos, I. G. (2012). Tillage and herbicide treatments with inter-row cultivation influence weed densities and yield of three industrial crops. *Weed Biology and Management, 12*, 84–90.
Wagner, M., & Mitschunas, N. (2008). Fungal effects on seed bank persistence and potential applications in weed biocontrol: A review. *Basic and Applied Ecology, 9*, 191–203.
Wan, Y., & El-Swaify, S. A. (1999). Runoff and soil erosion as affected by plastic mulch in a Hawaiian pineapple field. *Soil and Tillage Research, 52*(1–2), 29–35.
Wang, F. X., Feng, S. Y., Hou, X. Y., Kang, S. Z., & Han, J. J. (2009). Potato growth with and without plastic mulch in two typical regions of Northern China. *Field Crops Research, 110*(2), 123–129.
Ward, M., Ryan, M., Curran, W., Barbercheck, M., & Mortensen, D. (2011). Cover crops and disturbance influence activity-density of weed seed pred- ators *Amara aenea* and *Harpalus pensylvanicus* (Coleoptera: Carabidae). *Weed Science, 59*, 76–81.
Wei, D., Liping, C., Zhijun, M., Guangwei, W., & Ruirui, Z. (2010). Review of non-chemical weed management for green agriculture. *International Journal of Agricultural Biology Engineering, 3*, 52–60.
Zaniewicz-Bajkowska, A., Franczuk, J., & Kosterna, E. (2009). Direct and secondary effects of soil mulching with straw on fresh mass and number of weeds vegetable yield. *Polish Journal of Environmental Studies, 18*, 1185–1190.
Zhang, Y., Liu, M., Dannenmann, M., Tao, Y., Yao, Z., Jing, R., Zheng, X., Butterbach-Bahl, K., & Lin, S. (2017). Benefit of using biodegradable film on rice grain yield and N use efficiency in ground cover rice production system. *Field Crops Research, 201*, 52–59.
Zimdahl, R. (2015). *Six Chemicals That Changed Agriculture* (p. 216), First edn. Academic Press.

Effects of Mulching on Crop Growth, Productivity and Yield

Adeel Ahmad, Muhammad Yaseen, Hammad Hussain, Muhammad Naveed Tahir, Aqarab Husnain Gondal, Muhammad Iqbal, Amir Aziz, Muhammad Irfan, and Zahoor Ahmad

Abstract Mulching refers to the covering of bare land with organic or inorganic material for the betterment of soil and plants. Mulching not only improves the soil properties but also improves the growth and yield of many crops. Mulching improves the moisture status, temperature and nutrient status of the soil that are necessities for the better growth and yield of the crops. Ultimately mulching enhances the yield of many crops. This chapter will discuss the importance of mulching in agroecosystems-plants and soils. This chapter will highlight important aspects related to soil characteristics, the role of mulching in soil health and quality. It also includes important facters affecting the crop yield and impact of mulching for controlling these factors. It will also highlight the importance of mulching for vegetables, orchards and cereal crops relating to their yield characteristics.

A. Ahmad (✉) · M. Yaseen · M. N. Tahir · A. H. Gondal
Institute of Soil and Environmental Sciences, University of Agriculture Faisalabad, Faisalabad, Pakistan
e-mail: Adeelgujjar046@yahoo.com

H. Hussain · M. Iqbal
Institute of Horticultural Sciences, University of Agriculture, Faisalabad, Pakistan

A. Aziz
Department of Soil and Environmental Sciences, College of Agriculture, University of Sargodha, Sargodha, Pakistan

M. Irfan
School of Environmental Science and Engineering, Tianjin University, Tianjin 300354, PR China

Z. Ahmad
Department of Botany, University of Central Punjab, Punjab Group of College, Bahawalpur, Pakistan

K. Akhtar et al. (eds.), *Mulching in Agroecosystems*,
https://doi.org/10.1007/978-981-19-6410-7_14

1 Introduction

Rainfed agriculture is responsible for 80% of the world's planted land and 60% of global grain production (Rockstrom et al., 2007; UNESCO, 2009). The inadequate water supply, degraded soil fertility and nutrient supplies are common causes of low productivity in the semi-arid and arid rainfed agricultural system; and crop yield can be improved by a range of methods, including plastic mulching and straw mulching (Li et al., 2013; Gan et al., 2013). Mulching is a process used by farmers and horticulturists to improve the quality of agricultural soils by coating the soil surface with various materials. This system prevents not only evaporation but also wind erosion and soil runoff from lands (Tan et al., 2015). Soil mulching management techniques can minimize degradation, evaporation, adjust soil temperature and reduce weed infestation, resulting in increased yield and possibly nitrogen use efficiency and water use efficiency as well (Qin et al., 2013).

Mulch is most likely derived from the German word "molsch," which means "soft to decompose," which refers to the use of straw and leaves spread around a field by a gardener (Jack & Diaconis, 1955). Mulching prevents soil erosion by avoiding soil degradation, runoff, water evaporation, and weed infestation, and as a result, it aids in the regulation of temperature variations, preservation of soil moisture and improvement in the physical, biological and chemical properties of soil, it also subsidizes nutrients to the soil, and eventually improves crop growth and yield (Akhtar et al., 2018; Nawaz et al., 2017; Nzeyimana et al., 2017). According to an estimate, mulching increases yield by 50–60% in rainfed conditions compared to no mulched area (Dilip et al., 1990).

Plant performance increases, as the physical condition of the soil changes (Chakraborty et al., 2008; Van der Putten et al., 2013). Mulch may help to introduce the organic matter to the soil, control weed growth, and mitigate or prevent erosion (Bot & Benites, 2005). Several types of organic mulches are often used in landscaping to combat weeds and improve plant quality (Ranjan et al., 2017). Though organic matter mulching was used in ancient agriculture, lithic (stone) mulches have a long tradition. Mulching can be done with a variety of materials in today's crop processing. Plant residues, various types of biodegradable films, plastics, and various types of paper mulches non-coated or coated with biodegradable films or plastic are all options available to farmers and horticulturists (Haapala et al., 2014). Due to differences in production methods, growing conditions, and crop types, the impact of mulch type on crop growth can vary (Ashworth & Harrison, 1983). Various factors that affect the growth and yield of the crop are given below.

2 Factors Affecting Plant Growth and Yield

Plant development is influenced by four key factors: temperature, water, light, and nutrients (Gondal et al., 2021; Gondal & Tayyiba, 2022). These four factors influence the plant's growth hormones, and other growth-related processes, causing it to develop slower or faster (Lauridsen et al., 2020). For instance, plants that are exposed to light deficiencies or obtain inadequate blue light have a number of effects on their growth and yield. According to Rhoades (2021) reduction in light leads to stem elongated or leggy, leaves become small enough, leaves with brown margins or tips, lower leaves tend to dry out and variegation on variegated leaves disappears. Similarly, many plants can survive in the majority of soil environments. There are several causes of nutrient deficiency that necessitate fertilization and/or soil amendment. Certain crops or plants have the potential to deplete the soil. Plants, like all living organisms, need nutrients and minerals to survive (Ann et al., 2011).

Most plant cycles, such as transpiration, photosynthesis, germination, respiration, and flowering, is affected by temperature by triggering chemical reactions within the plant's cells (Wahid et al., 2007). Temperature impacts the conversion from vegetative to reproductive development when combined with day length (Dorais, 2003). For example, cool-season crops like spinach can bloom if the weather is hot and the days are long and warm-season crops like tomatoes, on the other hand, cannot set fruit if temperatures are too cold (Abou-Hussein, 2012; Sohail et al., 2021). Water supports a plant's growth by carrying vital nutrients. The plant takes nutrients from the soil and uses them. Plants droop if there isn't enough water in their cells, so water makes them stand. The dissolved sugar and other nutrients are transported through the plant by water. The plant is not only malnourished without the correct water balance. Different plants necessitate various quantities of water (Armstrong, 2021).

3 Role of Mulching in Reducing Stress Factors

Due to all the above factors, possibly growth and yield of the various crops are affected and thus; mulching is the best way to cope with all these factors to maximize the growth.

3.1 *Temperature*

Mulching lowers the temperature of the soil in the summer, increases it in the winter, which avoids temperature extremes. In the peak winter season, wheat straw mulch increased soil temperature by 2–30 °C. When opposed to bare earth, the temperature of the soil under transparent mulch can be up to 7 °C higher (Lamont, 1993). Park et al. (1996) found a 2.4 °C rise in average soil temperature at 15 cm depth when the clear film was used as mulch and a 0.8 °C increase was noticed with the use of black film. Condensation on the underside of the mulch absorbs the longwave radiation released by the soil at night, causing the soil to cool more slowly (Lamont, 2005).

3.2 Water

Mulching prevents unproductive evaporation from the soil surface, allowing more water to be required for transpiration (Chakraborty et al., 2008). This is beneficial in water-limited environments, as plant water status is sustained. Organic mulch also prevents nitrogen depletion by drainage and leaching by covering the soil surface (Erenstein, 2002). When compared to the monitor, straw mulch conserved 55% more soil moisture (Rajput & Singh, 1970). Mulching wheat residue @ 6730 kg/ha greatly improved usable soil moisture deposited up to 1.5 m depth of soil as compared to bare soil (Black, 1973). Over the control, okra production was substantially higher under straw mulch, followed by dust mulch (Batra et al., 1985).

3.3 Nutrients

Chilli leaf N, P, and K content improved after coconut fronds were mulched. Previous studies results revealed that plants grew quicker, ripened earlier, and had lower P and higher N concentrations in their leaves and fruits. Besides, they observed that rice straw mulch increased K-content and decreased P concentration in bell pepper leaves as compared to no-mulch leaves. In tomato, Hundal et al. (2000) observed that mulched crops had slightly higher nitrogen and phosphorus concentrations and nutrient absorption than un mulched plots. Mulch prevents the soil crust from erosion, reduces nitrogen leaching and defends against adverse influences leading to improved crop growing conditions.

4 Role of Mulching in the Improvement of Plant and Soil Health

Mulching is considered useful in moisture conservation and in improving crop productivity, Sharma et al. (2010) reported growth promotion trend in maize, wheat (Chakraborty et al., 2008), vegetables (Mahadeen, 2014) and other crops (Farrukh & Safdar, 2004). Weeds can be controlled through mulching (Erenstein, 2002) and soil moisture could also be retained. Combination of irrigation with mulching technology is advocated for better uptake of water by the spring wheat to reduce the number of irrigations. All the results showed that conserved moisture through mulching is effective during stress for plants.

In sustainable agriculture crop rotations, the use of cover crops, mulching and good crop husbandry are very useful measures to suppress weeds (Erenstein, 2003). Therefore, environment-friendly weed control methods are required to be used for weeding management in crops to avoid the incidence of undesirable effects. Mulching also helps in the retention of soil moisture contents and suppresses weeds without herbicide application (Asif et al., 2020).

Mulch is a material that may be organic or inorganic which spread on the surface of the soil and provide shelter from raindrop damage, evaporation, and solar radiations. Mulches help to preserve moisture, suppress weeds, and improve soil stability and avoid insect pest attack. Organic mulches help to moderate soil temperature, provide efficient control of weeds, decrease the rate of evaporation, and add nutrients and humus to the soil (Iqbal et al., 2020). Mulching prevents soil erosion and can reduce soil-borne diseases. Chemical mulch offers a slow release of humic acids, nitrogen, phosphorus, and potassium in the soil which facilitate to increase their uptake and utilization. Biological mulch is the component of integrated management of pest which provides control to phytophthora root rot, against dual competitive and aggressive microbes (Ghorbani et al., 2009).

5 Role of Mulching in the Improvement of Plant Nutrition

Soil biota can be increased through mulching because it provides a compound of carbon, nitrogen, and other nutrients and this is the key role in the cycling of nutrients, these are also the prolonged resources to get healthy crop (Bot & Benites, 2005). Soil macro and microbiota get nutrients from organic mulches and provide suitable environmental conditions to improve crop growth (Lal Bhardwaj, 2013). Plastic mulch increased the population of actinobacteria and proteobacteria in comparison to the control treatment (Farmer et al., 2017), however, the population of the invertebrates decreased (Bandopadhyay et al., 2018). Microbial activity depends on the temperature, under low-temperature mulches brings soil temperature closer to microbial optima and vice versa. Under high temperature, it resulted in a reduced microbial population (Brodhagen et al., 2015).

6 Impact of Mulching on Plant Growth Mechanisms

It is well-known, mulches can improve plant growth through different aspects i.e. by conserving soil moisture and temperature. Nutrient availability increased because of weed and pest population reduction (Thakur & Kumar, 2020). The increased height of the tea olive (*Osmanthus fragrans* Lour.) was observed by using gravels and wood chips mulch and the trunk diameter and also increased amount of chlorophyll, rhizospheric nutrients, development of roots and soluble sugar (Ni et al., 2016). Qayyum et al. (2020) noticed that paddy straw mulch improves the spike length of gladiolus. Rice straw mulch also increases the number of branches, root length, number of leaves, and plant height in patchouli (*Pogostemon cablin* Benth) as compared to black plastic silver mulch and without mulch. Aromatic plants like rosemary (*Rosmarinus officinalis* L.), lavender (*Lavandula officinalis* L.), thyme (*Thymus vulgaris* L.), and damask rose (*R. damascene* Mill.) produced maximum plant height and diameter with the applications of mulches (Hussain et al., 2019).

Xianchen et al. (2020) reported that black polythene mulch, increased the temperature of the soil, resulted in poor water and nutrients absorption, root growth, and consequently low yield returns. The Colour of the mulch also affects plant yield, as in sweet basil (*O. Basilicum* L.), the significantly higher yield was recorded under red colour mulch. Higher biomass was reported under organic mulch as compared to bare soil in basil (*O. Basilicum* L.), citronella (*Cymbopogon citratus* L.), and geranium (*P. graveolens* L.) (Mahadeen, 2014).

7 Mulching and Water Productivity

The most limiting source is water for the farming system among all-natural resources. Biomass produced per unit of water used is called water use efficiency (Farmer et al., 2017). The requirement of water is different depending upon the species of plants (Qayyum et al., 2020). The overall yield of crop plant depending upon rainfall, transpiration, drainage system and rate of evaporation (Bot & Benites, 2005). According to Qayyum et al. (2020), the mulching technique has proved to enhance the yield and WUE (Lal Bhardwaj, 2013). Plastic mulch enhanced WUE (20–60%) by decreasing evaporation rate (Bandopadhyay et al., 2020) and ultimately improves the soil water retention and infiltration and provide a favourable environment to root propagation and seed germination (Folino et al., 2020). Irrigation water requirement in bell pepper was decreased by 14–29% by covering with the plastic film due to limited moisture losses. Improved water use efficiency and yield potential of tomato were achieved in polyethylene mulched soil under all (surface and drip) levels of irrigation. In brinjal crop soil moisture of 29–56% and 22–107% conserved by using black plastic mulches over straw mulches and control, respectively (Sharma et al., 2010). It has been demonstrated that black polyethylene mulch is found to be useful in achieving the early harvest and yield of muskmelon. The yield of brinjal increased by 3.5–5.2 folds by white and black polyethylene over control probably because of slow water percolation and restricted nutrient loss from the top 15 cm of soil (Sharma et al., 2010).

Black polyethylene mulch boosted the soil moisture, reduced soil evaporation, altered microbial population, and hence produced higher quality and yield of production, which enhanced the economic value for farmers (Thakur & Kumar, 2020). There is mixed mulch which is a combination of organic and inorganic mulch. Phyto degradable and biodegradable mulches are a new type of mulch for easy use and versatility (Folino et al., 2020). It has been developed to reduce the accumulation of low-density polyethylene (LDPE) and environmental pollution produced from plastic wastes (Kader et al., 2017).

8 Influence of Mulching on Crop Productivity and Yield

8.1 Mulching for Vegetable Crops

Ashrafuzza man et al. (2011) reported that mulch increased the yield and quality of fruit in chillies when mature, the tallest plant (78.45 cm) was observed in transparent mulch, followed by black (77.58 cm) and blue (77.03 cm) plastic mulch. The smallest observation was in the control plot (61.15 cm). Moreover, un-mulched chillies had fewer branches as compared to mulched chillies. The highest number of branches was observed in black plastic mulch as compared to transparent mulch and blue mulch. Similarly, mulched had good effects on the root elongation. This is due to better soil water use efficiency and the most suitable soil temperature. Rajablariani et al. (2012) performed an experiment on tomato grown on bare soil and polyethylene mulch film and reported that the number of leaves and branches in tomato plants observed better in the plastic film as compared to bare soil. The early production was attained from transparent plastic mulch due to light entrance and increasing soil temperature.

As the plants grown on silver/black plastic mulch suggested a 65% improvement in marketable mulch relative to control treatment, mulching improved marketable yield compared to the bare soil. The production of silver and black increased by 65% respectively, followed by black (50%), blue (40%), red (26%) and transparent plastic cover (24%). The increase in yield in the covered area may be related to the preservation of soil surface and surface soil moisture, the improvement of the microclimate and the control of a large number of weeds, especially in the silver and black plastic coverings. Singh et al. (2009) found that the use of black polyethylene shields and drip irrigation further increased tomato yield by 57.87 tons/ha.

The mulch helps prevent fruits such as tomatoes from touching the ground. This reduces decay and helps keep the product clean. In many cases, its cracking and flower end rot is reduced on fruits. It tends to be smoother and have fewer scars on fruits. The correct installation of plastic mulch helps prevent soil from splashing on the plants when it rains, which can reduce the grading time.

8.1.1 Plant Growth and Yield

It has been proven that black polyethylene mulch can be used to achieve early harvest and yield of melons. After applying white polyethylene and black polyethylene, the yield of eggplant increased by 3.5–5.2 times compared with the control, which may be due to slow water leakage and limiting the nutrient loss of 15 cm from the soil surface (Singh et al., 2006).

Compared with white and reflective plastics, black plastic mulches were effective in promoting early tomato yields because of their high temperature inducing properties. Early tomato yields were often recorded as black and transparent plastic mulches because they prioritize the allocation of carbon to fruits rather than leaves. Conversely, when using plastic mulch in summer, high ambient temperature and high solar radiation often lead to poor growth and low yields.

8.1.2 Germination, Seedling Establishment and Growth

The mulch film produced some phytotoxic allelochemicals, which reduced the germination rate and seedling raising rate. Researchers founded that the effect of rice straw on the field to raise seedlings was better.

8.1.3 Effects on Plant Microclimate

The microclimate of the plant can be changed by changing the balance of energy in soil and by control the soil water evaporation. Root zone temperature (RZT) is one of the main advantages associated with the use of plastic coverings. Under controlled conditions, the growth of the root system increases with the increase of temperature until it reaches the optimum. Further increase in the temperature of the root zone may adversely affect the growth of the root system and stem. Under controlled conditions, the temperature and maximum root zone suitable for plant growth are considered to fluctuate under the conditions of air and root zone temperature field.

8.2 *Mulching for Orchards*

8.2.1 Banana

Mulch works as a 'lid,' allowing water to percolate into the soil while reducing evaporation. Because of this, your weekly watering will go a long way. Ensure that the mulch circle is well-watered so that the roots have plenty of room to grow. As a result, plants yield and growth increased. For instance, in banana orchards, the use of wheat straw and banana straw as a mulching material is very helpful to conserve the soil moisture and increasing the bunch weight of the banana plant and this mulching material applied at the start of the summer session, especially at February month. Stewart et al. (1926) evaluated the effect of asphalt-impregnated paper on pineapple weed control. In the experiment, the mulching film was unfolded with the mulch layer fixed by the tractor, much like the mulch layer currently used. In addition to reducing weed pressure on crops, paper mulch usually raises the soil temperature by several degrees Fahrenheit on sunny days, with little effect on cloudy or rainy days. Moisture and nitrate levels are also usually higher under mulch than bare ground, leading to increased pineapple production.

The citrus fruit quality improves by the use of mulching covering but on fruit appearance has a negative influence. Zhang and Xie (2014) indicated that the use of mulching in mandarin during the early stages of plant and in cell division increased the reduced sugar and total sugars content. It has been reported that when transparent plastics are used in combination with soil fumigation of methyl bromide and clopyralid, the production of strawberries (*Fragaria* sp.) increases (Johnson & Fennimore, 2005). The de Araújo et al. (2022) and Wang et al. (1998) showed that different mulch types have a significant effect on the concentration of ellagic acid in strawberry fruits. Ali and Gaur (2013) used the rice straw mulching method to record the maximum number of strawberry runners per plant, the number of platelets per planter, and the number of runners per planter. Singh et al. (2005) found that the growth, fruit weight, yield and quality of the black polyethylene film in strawberries were the best. Cover the strawberry to protect the flower bud temperature below 15°F (Tyagi et al., 2015). Similarly mulching improve the soil properties, and yield of various horticultural crops as shown in Table 1.

8.3 Effect of Mulching on the Production of Cereal Crops

Maize and wheat are globally primary crop due to their importance in food security and food production. Seventy percent of the global cereal crop production are maize and wheat due to lean availability of nutrients and water significantly affected their yield especially in the semi-arid and arid area of the world (Rockstrom et al., 2010). Rice is mostly grown in heavily irrigated or in paddy field and mulching has not been generally practiced in rice crop therefore I excluded rice crop. The actual obtainable grain yield is just 30–80%. In a region where nutrient and water are sufficiently available, 8–10% losses of maize and wheat are observed due to suboptimal field practices (Zwart et al., 2004; Vitousek et al., 2009). In a dry environment where soil organic matter is generally less than two percent in this situation the availability of water and temperature are dominant factors and play a vital role in determining the yield of crops.

Mulching effects for maize and wheat crop are different, in maize absorbed more positive effect on yield as compared to wheat. Maize can use sunlight more efficiently for photosynthesis because maize is a C4 plant and wheat is a C3 plant (Long et al., 2006). During the growing season of maize crop evaporation and temperature are higher than wheat because maize grows in the summer season and wheat is a winter season crop that grows when evaporation and temperature are low as compared to maize. With the use of mulching 28% in wheat and 40% in maize temperature and evaporation can reduce (Zhang et al., 2013). The water requirement for the wheat

Table 1 Role of mulching in horticultural crops and yield

Mulching source	Orchard or plants	Crop growth and yield	Soil properties	References
Pruning mulching	Pear	Improved	–	Moniruzzaman et al. (2007)
Black plastic mulch	Mango	Improved	Increased	
mulch	Nectarine	Improved	–	Andreotti et al. (2009)
Black plastic mulch	Kiw	Improved	Increased	Pratima et al. (2016)
Black plastic sheeting and weed barrier grids	Olive	Improved		Camposeo and Vivaldi (2011)
Black polythene sheet mulch	Guava	Improved	Increased	Das et al. (2010)
Organic mulch	Guava	Improved	Increased	Das et al. (2010)
Straw mulching	Raspberry	Improved	–	Trinka and Pritts (1992)
Mulching	Peach	Improved	Increased	Lordan et al. (2015), Neri et al. (2022)
Black plastic mulch	Currants	Improved	Increased	Dale (2000)
Organic mulches	High bush blueberry	Improved	Increased	Mercik and Smolarz (1995), Spiers (1986)

crop is less than maize. Delta of water for wheat is 25–1000 mm and for maize 150–2000 mm. The nitrogen requirement for wheat ranged 20–200 kg/ha and for maize ranged 30–400 kg/ha. The efficiency to utilize of nitrogen and water increased in both crop maize and wheat with the use of mulching as compared to no mulching.

The plastic mulch effect in wheat crop vary with respect to the availability of water, 15% yield increased under the low water condition and 35% positive response on yield observed when water is available in sufficient quantity. In maize crop mulching of straw show 20% increase in yield and not affected by the input level of water. Maize crop perform better under the low water application with plastic mulch compared to plastic mulch with high water. When used plastic mulch 60 and 40% maize yield increase under the low and high water application respectively. The temperature of soil increase with the use of plastic mulch and mulching of straw decrease the temperature of soil. Germination of seed is effected by soil temperature, when we used straw mulch in wheat crop 5–7 decreased in yield observed due to decreased in soil temperature. Mulching with Plastic sheet favor the early seed germination and better growth of roots of maize crop and give positive response on yield (Li et al., 2013; Gan et al., 2013). However the use of plastic mulch in winter wheat crop increased soil temperature and favor the seed germination. Use of straw in tropical region help in maintain the soil temperature and increase the yield of crops. Wheat grow at low temperature and has more growing time compared to maize crop. Therefore, the use of plastic mulch in wheat crop contribute in the increase of soil temperature than straw mulch. Conclude that mulching in wheat and maize crop prevent the losses of water and evaporation of nitrogen from field hence increase 60 and 20% yield of maize and wheat, respectively.

9 Conclusion and Remarks

From the above discussion, it is clear that mulching induces positive effects on soil quality as well as crop yield. Soil physical, chemical and biological characteristics are under the strong influence of mulching. As plants need proper temperature, moisture and nutrients for their survival and growth that is a necessity for better yield of the crops. Mulching induces these characters in the soil very efficiently. Mulching makes the soil a suitable medium for the proper growth and yield of many crops by improving the moisture and nutrient status of the soil. It also improves the structure of soil by increasing the organic matter concentration of the soil. Thus, all these characteristics bring maximum yield in vegetables, fruits and cereals.

References

Abou-Hussein, S. D. (2012). Climate change and its impact on the productivity and quality of vegetable crops. *Journal of Applied Sciences Research, 8*, 4359–4383.

Akhtar, K., Wang, W., Ren, G., Khan, A., Feng, Y., & Yang, G. (2018). Changes in soil enzymes, soil properties, and maize crop productivity under wheat straw mulching in Guanzhong, China. *Soil and Tillage Research, 182*, 94–102.

Ali, A., & Gaur, G. S. (2013). Effect of organic mulches on runner production of strawberry (Fragaria × ananassa Duch.). *Asian Journal of Bio Science, 8*(2), 175–179.

Andreotti, C., Ravaglia D., & Costa, G. (2009). Innovative light management to improve production sustainability, overall quality, and the phenolics composition of nectarine (Prunus persica cv. Stark Red Gold). *Journal of Horticultural Science & Biotechnology. ISAFRUIT Special Issue,* 145–149.

Ann, M., Clain, J., & Jeff, J. (2011). Plant Nutrient Functions and Deficiency and Toxicity Symptoms. *Nutrient Management Module, 9.*

Armstrong. S. (2021). How does water affect plant growth? https://www.gardeningknowhow.com/special/children/how-does-water-affect-plantgrowth.htm#:~:text=Water%20helps%20a%20plant%20by,other%20nutrients%20throuh%20the%20plant.

Ashrafuzza Man, M., Halim, A., Ismail, M., Shahidullah, S., & Hossain, A. (2011). Effect of plastic mulch on growth and yield at chilli. *Brazilian Archives of Biology and Technology, 54*(2), 321–330.

Ashworth, S., & Harrison, H. (1983). Evaluation of mulches for use in the home garden. *HortScience, 18*(2), 180–182.

Asif, M., Nadeem, M. A., Aziz, A., Safdar, M. E., Adnan, M., Ali, A., Ullah, N., Akhtar, N., & Abbas, B. (2020). Mulching improves weeds management, soil carbon and productivity of spring planted maize (Zea mays L.). *International Journal of Botany Studies, 5*, 57–61.

Bandopadhyay, S., Martin-Closas, L., Pelacho, A. M., & DeBruyn, J. M. (2018). Biodegradable plastic mulch films: Impacts on soil microbial communities and ecosystem functions. *Frontiers in Microbiology, 9*, 819.

Bandopadhyay, S., Sintim, H. Y., & DeBruyn, J. M. (2020). Effects of biodegradable plastic film mulching on soil microbial communities in two agroecosystems. *PeerJ, 8*, e9015.

Batra, S. K. (1985). Other long vegetable fibres: Abaca, banana, sisal, henequen, flax, ramie, hemp, sunn, and coir. *Handbook of fiber science and technology, 4*, 727–807.

Black Jr, C. C. (1973). Photosynthetic carbon fixation in relation to net CO2 uptake. *Annual Review of Plant Physiology, 24*(1), 253–286.

Bot, A., & Benites, J. (2005). The importance of soil organic matter. Key to drought-resistant soil and sustained food production. *FAO Soils Bulletin, 80.*

Brodhagen, M., Peyron, M., Miles, C., & Inglis, D. A. (2015). Biodegradable plastic agricultural mulches and key features of microbial degradation. *Applied Microbiology and Biotechnology, 99*(3), 1039–1056.

Camposeo, S., & Vivaldi, G. A. (2011). Short-term effects of de-oiled olive pomace mulching application on a young super high-density olive orchard. *Scientia Horticulturae, 129*, 613–621.

Chakraborty, D., Nagarajan, S., Aggarwal, P., Gupta, V. K., Tomar, R. K., Garg, R. N., Sahoo, R. N., Sarkar, A., Chopra, U. K., Sarma, K. S., & Kalra, N. (2008). Effect of mulching on soil and plant water status, and the growth and yield of wheat (Triticum aestivum L.) in a semi-arid environment. *Agricultural Water Management, 95*(12), 1323–1334.

Dale, A. (2000). Black plastic mulch and between-row cultivation increase black currant yields. *Hort Technology, 10*(2), 307–308.

Das B. C., Mahi S., & Mulieh, S. R. (2010). Response of soil covers on guava cv. L-49. *Journal of Crop and Weed, 6*(2), 10–14.

de Araújo, D. L., de Luna Souto, A. G., Cavalcante, A. G., Cavalcante, L. F., Pereira, W. E., & de Melo, A. S. (2022). Physiological aspects of yellow passion fruit with use of hydrogel and mulching. *Revista Caatinga, 35*(2), 382.

Dilip, K. G., Sachin, S. S., & Rajesh, K. (1990). Importance of mulch in crop production. *Indian Journal of Soil Conservation, 18*, 20–26.

Dorais, M. (2003). The use of supplemental lighting for vegetable crop production: Light intensity, crop response, nutrition, crop management, cultural practices. In *Canadian Greenhouse Conference* (Vol. 9).

Erenstein, O. (2002). Crop residue mulching in tropical and semi-tropical countries: An evaluation of residue availability and other technological implications. *Soil and Tillage Research, 67*(2), 115–133.

Erenstein, O. (2003). Smallholder conservation farming in the tropics and sub-tropics: A guide to the development and dissemination of mulching with crop residues and cover crops. *Agriculture, Ecosystems & Environment, 100*(1), 17–37.

Farmer, J., Zhang, B., Jin, X., Zhang, P., & Wang, J. (2017). Long-term effect of plastic film mulching and fertilization on bacterial communities in a brown soil revealed by high through-put sequencing. *Archives of Agronomy and Soil Science, 63*(2), 230–241.

Farrukh, I., & Safdar, A. (2004). Impact of different types of mulches on soil moisture. *Sarhad Journal of Agriculture, 20*(4), 571–573.

Folino, A., Karageorgiou, A., Calabrò, P. S., & Komilis, D. (2020). Biodegradation of wasted bioplastics in natural and industrial environments: A review. *Sustainability, 12*(15), 6030.

Gan, Y. T., et al. (2013). Ridge-furrow mulching systems-an innovative technique for boosting crop productivity in semiarid rain-fed environments. *Advances in Agronomy, 118*, 429–476. https://doi.org/10.1016/B978-0-12-405942-9.00007-4.

Ghorbani, R., Wilcockson, S., Koocheki, A., & Leifert, C. (2009). Soil management for sustainable crop disease control: a review. In *Organic farming, pest control and remediation of soil pollutants* (pp.177–201).

Gondal, A. H., & Tayyiba, L. (2022). Prospects of using nanotechnology in agricultural growth, environment and industrial food products. *Reviews in Agricultural Science, 10*, 68–81.

Gondal, A. H., Tampubolon, K., Toor, M. D., & Ali, M. (2021). Pragmatic and fragile effects of wastewater on a soil-plant-air continuum and its remediation measures: A perspective. *Reviews in Agricultural Science, 9*, 249–259.

Haapala, T., Palonen, P., Korpela, A., & Ahokas, J. (2014). Feasibility of paper mulches in crop production—A review. *Agricultural and Food Science, 23*(1), 60–79.

Hundal, R. S., Krssak, M., Dufour, S., Laurent, D., Lebon, V., Chandramouli, V., Inzucchi, S. E., Schumann, W. C., Petersen, K. F., Landau, B. R. & Shulman, G. I. (2000). Mechanism by which metformin reduces glucose production in type 2 diabetes. *Diabetes, 49*(12), 2063–2069.

Hussain, A., Illahi, B. A., Iqbal, A. M., Jehaangir, I. A., & Hussain, S. T. (2019). Agronomic management of saffron (Crocus sativus)-A review. *Indian Journal of Agronomy, 64*(2), 147–164.

Iqbal, R., Raza, M. A. S., Valipour, M., Saleem, M. F., Zaheer, M. S., Ahmad, S., Toleikiene, M., Haider, I., Aslam, M. U., & Nazar, M. A. (2020). Potential agricultural and environmental benefits of mulches-a review. *Bulletin of the National Research Centre, 44*, 1–16.

Jack, J. R., & Diaconis, N. S. (1955). *Variation of boundary-layer transition with heat transfer on two bodies of revolution at a Mach number of 3.12* (No. NACA-TN-3562).

Johnson, M. S., & Fennimore, S. A. (2005). Weed and crop response to colored plastic mulches in strawberry production. *Hort Science, 40*, 1371–1375.

Kader, M. A., Senge, M., Mojid, M. A., & Ito, K. (2017). Recent advances in mulching materials and methods for modifying soil environment. *Soil and Tillage Research, 168*, 155–166.

Lal Bhardwaj, R. (2013). Effect of mulching on crop production under rainfed condition-a review. *Agricultural Reviews, 34*(3).

Lamont, W. J. (2005). Plastics: Modifying the microclimate for the production of vegetable crops. *HortTechnology, 15*(3), 477–481.

Lamont, B. B., Klinkhamer, P. G., & Witkowski, E. T. F. (1993). Population fragmentation may reduce fertility to zero in Banksia goodii—A demonstration of the Allee effect. *Oecologia, 94*(3), 446–450.

Lauridsen, T. L., Mønster, T., Raundrup, K., Nymand, J., & Olesen, B. (2020). Macrophyte performance in a low arctic lake: Effects of temperature, light and nutrients on growth and depth distribution. *Aquatic Sciences, 82*(1), 1–14.

Li, S. X., Wang, Z. H., Li, S. Q., Gao, Y., & Tian, X. H. (2013). Effect of plastic sheet mulch, wheat straw mulch and maize growth on water loss by evaporation in dryland areas of China. *Agricultural Water Management, 116*, 39–49. https://doi.org/10.1016/j.agwat.2012.10.004

Long, S. P., Zhu, X. G., Naidu, S. L., & Ort, D. R. (2006). Can improvement in photosynthesis increase crop yields?. *Plant, cell & environment, 29*(3), 315–330.

Lordan, J., Pascual, M., Villar, J. M., Fonseca, F., Papió, J., Montilla, V., & Rufat, J. (2015). Use of organic mulch to enhance water-use efficiency and peach production under limiting soil conditions in a three-year-old orchard. *Spanish Journal of Agricultural Research, 13*(4), e0904.

Mahadeen, A. Y. (2014). Effect of polyethylene black plastic mulch on growth and yield of two summer vegetable crops under rainfed conditions under semi-arid region conditions. *American Journal of Agricultural and Biological Sciences, 9*(2), 202–207.

Mercik, S., & Smolarz, K. (1995). Influence of fertilization and mulching on the growth, fruiting and chemical composition of soil and leaves of high bush blueberry. *Acta Horticulture, 383*, 323–329.

Moniruzzaman, M., Mozumde, S. N., & Islam, M. R. (2007). Effect of mulching and pruning on yield and quality of pear. *Bangladesh Journal of Agriculture Research, 32*(2), 225–233.

Nawaz, A., Lal, R., Shrestha, R. K., & Farooq, M. (2017). Mulching affects soil properties and greenhouse gas emissions under long-term no-till and plough-till systems in Alfisol of central Ohio. *Land Degradation & Development, 28*(2), 673–681.

Neri, D., Zikeli, S., Lepp, B., Malusa, E., Fernandez, M. M., Boutry, C., et al. (2022). Dynamic sod mulching and use of recycled amendments to increase biodiversity, resilience and sustainability of intensive organic apple orchards and vineyards (DOMINO).

Ni, X., Song, W., Zhang, H., Yang, X., & Wang, L. (2016). Effects of mulching on soil properties and growth of tea olive (Osmanthus fragrans). *PLoS ONE, 11*(8), e0158228.

Nzeyimana, I., Hartemink, A. E., Ritsema, C., Stroosnijder, L., Lwanga, E. H., & Geissen, V. (2017). Mulching as a strategy to improve soil properties and reduce soil erodibility in coffee farming systems of Rwanda. *CATENA, 149*, 43–51.

Park, C. L., Cohen, L. H., & Murch, R. L. (1996). Assessment and prediction of stress-related growth. *Journal of personality, 64*(1), 71–105.

Pratima, P., Sharma, N., & Sharma, D. P. (2016). Canopy temperature and water relations of kiwifruit cultivar Allison in response to deficit irrigation and in situ moisture conservation. *Current Science, 111*(2), 375–379.

Qayyum, M. M., Hassan, I., Abbasi, N. A., & Khalid, A. (2020). Mitigation of low temperature stress by polythene for quality production of Gladiolus (Gladiolus hortulanus L.) during winter. *Applied Ecology and Environmental Research, 18*(3), 4469–4486.

Qin, W., Chi, B. L., Oenema, O. (2013). Long-term monitoring of rainfed wheat yield and soil water at the loess plateau reveals low water use efficiency. *Plos One, 8*, ARTN e78828, https://doi.org/10.1371/journal.pone.0078828.

Rajablariani, H., Khan, F. H., & Rafezi, R. (2012). Effect at colored plastic mulches on yield at tomato and weed biomass. *International Journal of Environmental Science and Development, 3*(6), 590–593.

Rajput, R. K., & Singh, M. (1970). Efficacy of different mulches in conserving soil moisture in cotton. *Indian Journal of Agronomy, 15*(1), 41–5.

Ranjan, P., Patle, G. T., Prem, M., & Solanke, K. R. (2017). Organic mulching-A water saving technique to increase the production of fruits and vegetables. *Current Agriculture Research Journal, 5*(3), 371–380.

Rhoades, H. (2021). How light affects the growth of a plant & problems with too little light. https://www.gardeningknowhow.com/plant-problems/environmental/how-lightaffects-the-growth-of-a-plant-problems-with-too-little-light.htm.

Rockstrom, J., Hatlbu, N., Owels, T. Y., & Wani, S. P. (2007). Managing water in rainfed agriculture.

Rockström, J., Karlberg, L., Wani, S. P., Barron, J., Hatibu, N., Oweis, T., Bruggeman, A., Farahani, J. and Qiang, Z. (2010). Managing water in rainfed agriculture—The need for a paradigm shift. *Agricultural Water Management, 97*(4), 543–550.

Sharma, A. R., Singh, R., Dhyani, S. K., & Dube, R. K. (2010). Moisture conservation and nitrogen recycling through legume mulching in rainfed maize (*Zea mays*)–wheat (Triticum aestivum) cropping system. *Nutrient Cycling in Agroecosystems, 87*(2), 187–197.

Singh, R., Sharma, R. R., & Jain, R. K. (2005). Planting time and mulching influenced vegetative and reproductive traits in strawberry in India. *Fruits, 60*, 395–403.

Singh, R., Jhorar, R. K., Van Dam, J. C., & Feddes, R. A. (2006). Distributed ecohydrological modelling to evaluate irrigation system performance in Sirsa district, India II: Impact of viable water management scenarios. *Journal of Hydrology, 329*(3-4), 714–723.

Singh, R., Kumar, S., Nangare, D. D., & Meena, M. S. (2009). Drip irrigation and black polyethylene mulch influence on growth, yield and water-use efficiency of tomato. *African Journal of Agricultural Research, 4*(12), 1427–1430.

Sohail, S., Gondal, A. H., Farooq, Q., Tayyaba, L., Zainab, D. E., Ahmad, I. A., et al. (2021). Organic vegetable farming; a valuable way to ensure sustainability and profitability. In *Vegetable growing*. IntechOpen.

Spiers, J. M. (1986). Root distribution of 'Tifblue' rabbiteye blueberry as influenced by irrigation, incorporated peatmoss, and mulch. *Journal of American Society for Horticultural Science, 111*, 877–880.

Stewart, G., Thomas, E., & Horner, J. (1926). Some effects of mulching paper on Hawaiian soils. *Soil Science, 22*, 35–39.

Tan, C., Cao, X., Yuan, S., Wang, W., Feng, Y., & Qiao, B. (2015). Effects of long-term conservation tillage on soil nutrients in sloping fields in regions characterized by water and wind erosion. *Scientific Reports, 5*(1), 1–8.

Thakur, M., & Kumar, R. (2020). Mulching: Boosting crop productivity and improving soil environment in herbal plants. *Journal of Applied Research on Medicinal and Aromatic Plants,* 100287.
Trinka, D. L., & Pritts, M. P. (1992). Micropropagated raspberry plant establishment responds to weed control practice, row cover use, and fertilizer placement. *Journal of American Society for Horticultural Science, 117*(6), 874–880.
Tyagi, S., Ahmad, M., Sahay, S., Nanher, A. H., & Nandan, B. (2015). Strawberry: A potential cash crop in India. *Rashtriya krishi, 10*(2), 57–59.
UNESCO. (2009). The United Nations World Water Development Report 3: Water in a Changing World. Earthscan.
Van der Putten, W. H., Bardgett, R. D., Bever, J. D., Bezemer, T. M., Casper, B. B., Fukami, T., et al. (2013). Plant–soil feedbacks: The past, the present and future challenges. *Journal of Ecology, 101*(2), 265–276.
Vitousek, P. M., Naylor, R., Crews, T., David, M. B., Drinkwater, L. E., Holland, E., Johnes, P. J., Katzenberger, J., Martinelli, L. A., Matson, P. A. & Nziguheba, G. (2009). Nutrient imbalances in agricultural development. *Science, 324*(5934), 1519–1520.
Wahid, A., Gelani, S., Ashraf, M., & Foolad, M. R. (2007). Heat tolerance in plants: An overview. *Environmental and Experimental Botany, 61*(3), 199–223.
Wang, S. Y., Galletta, G. J., Camp, M. J., & Kasperbauer, M. J. (1998). Mulch types affect fruit quality and composition of two strawberry genotypes. *Hort Science, 33*, 636–640.
Xianchen, Z., Huiguang, J., Xiaochun, W., & Yeyun, L. (2020). The effects of different types of mulch on soil properties and tea production and quality. *Journal of the Science of Food and Agriculture, 100*(14), 5292–5300.
Zhang, G., & Xie, S. (2014). Effects of plastic film mulching on quality and appearance of Statuma mandarin fruit. *American Journal of Plant Sciences, 5*, 3829–3835.
Zhang, X., Wang, H., He, L., Lu, K., Sarmah, A., Li, J., Bolan, N. S., Pei, J., & Huang, H. (2013). Using biochar for remediation of soils contaminated with heavy metals and organic pollutants. *Environmental Science and Pollution Research, 20*(12), 8472–8483.
Zwart, P. J., Gerber, A. G., & Belamri, T. (2004, May). A two-phase flow model for predicting cavitation dynamics. In *Fifth international conference on multiphase flow, Yokohama, Japan* (Vol. 152).

Comparative Effects of Living and Non-living Mulches on Insect Pest Management in Agroecosystems

Abrar Muhammad, Muhammad Ali, Muhammad Shakeel, Supaporn Buajan, and Habib Ali

Abstract Organic farming and sustainable agriculture call for non-chemical, economic, and eco-friendly pest management techniques. Scientific efforts are underway to develop new or optimize the existing techniques. In this regard, habitat modification with mulching has been widely investigated on the suppression of insect pest abundance. In general, vegetationally diverse cropping systems impairs the herbivore's ability to locate the host plant by creating physical barriers, disrupting the visual and olfactory cues, and enhancing plant defenses that lead to the reduction of pest abundance and disease incidences. Also, mulching increases natural enemy population density because of greater habitat diversity and food resources, thereby decreasing herbivore abundance indirectly by improving biological control activity. However, in some cases, mulching has negatively affected the crop yield by competing for available resources (water, nutrients, light, space, etc.), impacted natural enemy abundance and efficiency, or provided alternate hosts/refuge for the pest insects. Besides, mulching can also play a pivotal role in conserving and support the declining pollinator population by providing nectar and pollen, nesting sites, and

A. Muhammad (✉)
Max Planck Partner Group, Institute of Sericulture and Apiculture, College of Animal Sciences, Zhejiang University, Hangzhou, China
e-mail: abrar_ento334@zju.edu.cn

M. Ali
Department of Horticulture, College of Agriculture and Biotechnology, Zhejiang University, Hangzhou, China

M. Shakeel
Laboratory of Bio-Pesticide Innovation and Application of Guangdong Province, College of Plant Protection, South China Agricultural University, Guangzhou 510642, China

S. Buajan
Faculty of Environment and Resource Studies, Mahidol University, Nakhon Pathom 73170, Thailand

H. Ali (✉)
Department of Agricultural Engineering, Khwaja Fareed University of Engineering and Information Technology, Rahim Yar Khan 64200, Pakistan
e-mail: habib_ali1417@yahoo.com

Department of Entomology, University of Agriculture Faisalabad, Faisalabad, Pakistan

K. Akhtar et al. (eds.), *Mulching in Agroecosystems*,
https://doi.org/10.1007/978-981-19-6410-7_15

refuge from predators. Overall, if appropriately planned, mulching might contribute significantly to insect pests' non-chemical control and promote the diversity and abundance of natural enemies and pollinators.

Keywords Integrated pest management · Living mulches · Synthetic mulches · Insect pests · Natural enemy · Pollinators

1 Introduction

Agricultural and forest plants are under constant threat from various biotic and abiotic factors limiting their growth and production. Among the biotic factors, insect pests are alone responsible for most of the crop production losses as herbivores or vectors of plant diseases (Oerke, 2006). In agriculture, insect pests are referred to as all kinds of insects that directly or indirectly negatively affect a crop or tree. The chronicle of pests competing with agricultural products has existed since the birth of farming activities. Meanwhile, the practice of pest control aimed to reduce the damage to cultivated crops can be traced back to ancient civilizations as well (Rapisarda & Cocuzza, 2017). For a long time, inorganic chemicals, e.g., Sulphur, remained in practice for pest control until world war II. After the war, a radical shift occurred towards synthetic organic insecticides that witnessed an unprecedented boom of chemical pressure in agroecosystems (Köhler & Triebskorn, 2013; Lin et al., 2013; Rapisarda & Cocuzza, 2017). The adverse effects inflicted on public and environmental health using chemical control galvanized global awareness that led to the concept of integrated pest management (IPM). During the mid of 20th century, IMP was developed as an alternative to chemical control by combining different tactics that would keep the pest population below reaching the economic injury level (Kogan, 1998; Stern et al., 1959). The philosophy of IPM is aimed to achieve long-term prevention of pest damage through a combination of cultural control, biological control, mechanical/physical control, genetic control, and chemical control, being the last resource only after monitoring (Fig. 1). Scientific efforts are always continued in search of new environmentally friendly or optimization of the existing techniques for pest control (Lewis et al., 1997).

Physical control involved the use of physical barriers such as trenches, fences, mulches, etc., that modify the physical environment and render it least favorable for the pest population (Vincent et al., 2003). Mulch is the passive method of physical control where any living or non-living materials are used to discourage pest movement or their establishment in the agroecosystems (Vincent et al., 2003). In agriculture, mulch refers to any material (organic or inorganic, dead or alive) that is applied on the soil surface as a protective cover to regulate soil temperature, prevent water loss through evaporation, provide pest control (weeds, insects, and disease), protect the young seedlings and newly planted trees, and improve soil fertility, etc. (Alyokhin et al., 2019; Brown & Tworkoski, 2004; Jabran, 2019; Liebman et al., 2001).

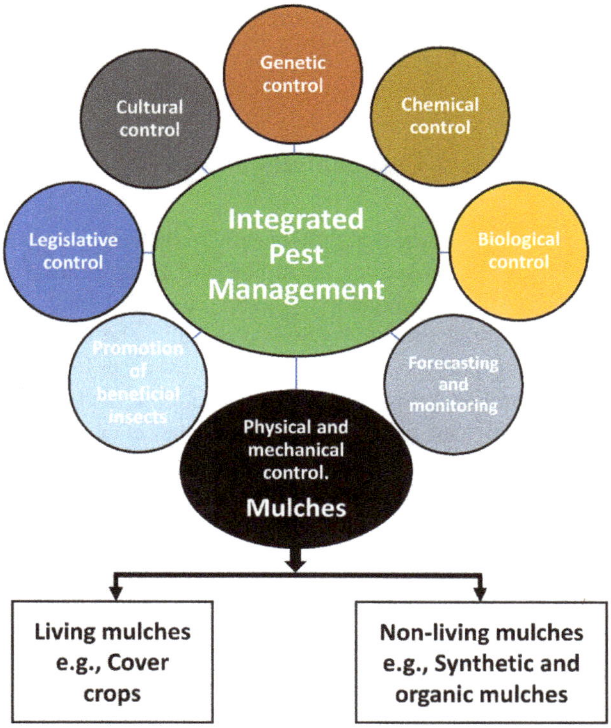

Fig. 1 Integrated pest management (IPM) combines various techniques to solve pest problems at minimum expense to public and environmental health. Mulching is one of the components of IPM used to suppress pest abundance and support beneficial insects, including natural enemies and pollinators

Fundamentally, mulching creates a micro-climate suitable for plant growth and soil activities. They can also reduce soil erosion and runoff and enhance soil quality (Alyokhin et al., 2019). These positive effects of mulching are attributed to increasing soil organic content, aggregates stability, bulk density, water holding capacity, microbial activity, and earthworm density. Although less appreciated, an added benefit of mulching is its critical role in organic farming to achieve non-chemical pest control (Alyokhin et al., 2019). Many studies have pointed out the potential of mulching to achieve non-chemical and integrated pest control (Brown & Tworkoski, 2004; Bryant et al., 2013; Burgio et al., 2014; Depalo et al., 2016; Frank & Liburd, 2005; Gill et al., 2010; Jabran, 2019; Kogan, 1998; Othim et al., 2018a; Schmidt et al., 2004; Teasdale et al., 2002; Vincent et al., 2003). However, some studies have reported negative effects on crop yield by increasing insect herbivory or competition for available resources with the crop plants. In this chapter, we reviewed the current literature on the use of mulching, specifically exploited in different cropping systems for suppressing insect pests and promoting beneficial insects including natural enemies (predators and parasitoids) and pollinators. Given the multifaceted benefits of mulching, this practice certainly offers a viable alternative to synthetic pesticides in organic farming and sustainable agriculture.

2 Mulching for Insect Pest Management

Evidence suggests that often pest populations tend to be least abundant by adopting habitat modification practices such as mulching and other cultural practices. The effects of mulching on the abundance of insect pests and their natural enemies in some studies are summarized in Table 1. Generally, the diversity of non-crop species in a variety of cropping systems in fruit and field crops has shown to increase the abundance of natural enemies, reduce pest population, and decrease herbicides and fertilizers inputs (Bryant et al., 2013; Letourneau et al., 2011; Rieux et al., 1999). Habitat modification practices may enhance plant defenses and increase the density and biological activity of natural enemies (Alyokhin et al., 2019). However, there have been some instances (15.3%) where the mulching (especially the living mulches) favored the pest herbivory and increased crop losses, whereas about 20% responded variably in polyculture (Andow, 1991). In addition, the living mulch may

Table 1 Effects of mulching on pest control and natural enemy populations

Mulch type	Crop	Insect pests/natural enemies	Impact	References
Living mulches *Trifolium repens* L., *Trifolium fraguferum* L., *Trifolium praetense* L., and *Lotus corniculatus* L	Broccoli	*Myzus persicae* and the cabbage aphid, *Brevicoryne brassicae*	Reduced pest population	Costello and Altieri (1995)
Living/dead mulches *Avena sativa* L	Cabbage *Brassica oleraceae* L	Cabbageworm Cabbage looper Cabbage aphid Chalcidoidea Propylea quatuordecimpunctata Lady beetle larvae *Diadegma insulare*	Lowered abundance of several key cabbage pests Greater abundance of important natural enemies' species	Bryant et al. (2013)
Living mulches *Medicago polimorpha* L. var anglona	Cauliflower *Brassica oleracea* L. var. botrytis	*Pieris brassicae* *Pieris rapae* *Phyllotreta spp.*	Increased natural enemies (rove beetle and spider) activity	Burgio et al. (2014)

(continued)

Table 1 (continued)

Mulch type	Crop	Insect pests/natural enemies	Impact	References
Organic/living mulches *Vigna unguculata* L Crotalaria juncea L *Sorghum bicolor* × *S. sudanese*	Beans *Phaseolus vulgaris* L	Aphididae Thripidae Aleyrodidae	Increased density of Dipteran	Gill et al. (2011)
Living mulches *Fagopyrum esculentum* *Trifolium repens* L	Zucchini *Cucurbita pepo* L	*Bemisia argentifolii* Aphids	Increased natural enemies' population	Frank and Liburd (2005)
Synthetic mulch Reflective and white	Zucchini *Cucurbita pepo* L	*Bemisia argentifolii* Aphids	Decreased pest population	Frank and Liburd (2005)
Living mulches *Medicago sativa* L *Trifolium ambiguum*	*Zea mays* L *Glycine max* L	*Ostrinia nubilalis* Hubner *Poecilus chalcites* *Scarites quadriceps* Chaudior	Increased natural enemies' population	Prasifka et al. (2006)
Living mulch *Medicago polimorpha* L	Cauliflower *Brassica oleracea* L. var. botrytis	Aphids *Pieris brassicae*	Decreased pest population Increased parasitization	Depalo et al. (2016)
Living mulches *Fagopyrum esculentum* *Sinapis alba* L	Zucchini *Cucurbita pepo* L	*Bemisia argentifolii* *Aphis gossypii*	Decreased pest incidence	Hooks et al. (1998)
Living mulches *Agrostis palustris* *Festuca rubra L* *Trifolium repens L* *T. repens*	Cabbage	*Phyllotreta cruciferae* *Brevicoryne brassicae*	Lowered pest density	Andow et al. (1986)
Leaf mulch	Amaranths	*Disonycha glabrata*	Resulted in higher leaf damage due to increased pest population	Vorsah et al. (2020)
Compost mulch (Poultry manure)	Apple orchard	*Phyllonorycter blancardella* *Eriosoma lanigerum*	More predators Fewer herbivores	Brown and Tworkoski (2004)
Living mulches: *Capsicum annuum* *Melilotus officinalis*	Broccoli *Brassica oleracea*	*Trichoplusia ni* *Hellula undalis* *Artogeia rapae*	Lowered pest densities	Hooks and Johnson (2006)

(continued)

Table 1 (continued)

Mulch type	Crop	Insect pests/natural enemies	Impact	References
Living mulch: Red clover	Corn Tomato Cauliflower	Aphids Lygaeids Leafhoppers Carabidae Staphylinidae Spiders	Higher abundance of natural enemies' populations	Altieri et al. (1985)
Living mulches: *Trifolium fragiferum L* *Trifolium repens* *Melilotus officinalis*	Broccoli	Lepidopteran pests Spiders	Decreased lepidopteran eggs and larval densities Increased natural enemy abundance	Hooks and Johnson (2004)
Straw mulch	Sweet potato *Ipomoea batatas* L	Rove beetles, fire ants, and carabid beetles	Increased predator abundance	Jackson and Harrison (2008)
Hay mulch	Beans *Phaseolus vulgaris*	Cornstalk borer *Elasmopalpus lignosellus*	Decreased pest damages	Gill et al. (2010)
White and reflective synthetic mulches	Zucchini	Aphids and white flies	Decreased pest damages	Frank and Liburd (2005)

also compete with the main crop for resources such as light, nutrients, water, and space (Andow, 1991). This variation of mulching effects on the pest population can be partly attributed to the mulch type, natural enemy feeding range, and herbivore response to diverse plantings in polyculture (Costello & Altieri, 1995). According to the enemy hypothesis, 'the natural enemies tend to be more abundant in vegetational diverse agroecosystems due to broader host ranges and food resources (Costello & Altieri, 1995; Root, 1973). On the other hand, vegetationally diverse cropping systems can also interfere with natural enemies' abilities to locate their host due to plant structural complexity and mixed chemical cues emanating from diverse plantings use in host finding (Andow, 1991).

3 Types of Mulches Used to Manage Insect Populations

Mulches can be broadly classified into two major categories: living and nonliving, with the latter further subdivided into organic and inorganic/synthetic mulches (Fig. 1). Living mulches refer to the plants that are grown in the main crop to suppress weeds or pests, reduce soil erosion, prevent water loss, or increase soil fertility (Hartwig & Ammon, 2002; Hooks et al., 1998; Lanini et al., 1989; Prasifka

Fig. 2 Types of mulches used in different cropping systems to minimize pest populations and increase crop yield. **a** Living mulches, **b** organic mulches, and **c** synthetic mulches

et al., 2006) (Fig. 2a). Organic mulches are derived from plant-like sources such as grass, hay, compost, clippings, etc. (Fig. 2b). On the contrary, inorganic/synthetic mulches come from plastic or other equivalent materials, e.g., plastic sheets (transparent mulch, black mulch, and double colored mulch), aluminized films, and other photo biodegradable materials (Vincent et al., 2003) (Fig. 2c). Effects of living and non-living mulches on the population abundance of insect pests and their natural enemies are discussed following in more detail.

4 Instances of Pest Suppression in Living Mulches

In agriculture, growing cover crops and intercrops as living mulch is a common practice for its benefits of enhancing soil quality and control of weeds, insects, and disease (Gill et al., 2011; Prasifka et al., 2006). As indicated by their name, cover crops are plants that are planted to cover the soil to increase soil fertility, manage soil erosion, preserve soil moisture, and control weeds and insect pests in an agroecosystem (Gill et al., 2011). It has been pointed out that cover crops can change vegetational complexity in agricultural fields that may affect arthropods density and enhance natural enemy abundance (Bryant et al., 2013; Burgio et al., 2014; Sans & Altieri, 2005). Cover crop mulches have exhibited suppressive effects on weeds, thereby decreasing herbicide inputs in the fields as well (Bryant et al., 2013). Furthermore, cover crops in -strip-tillage systems reduced runoff of agrichemicals and groundwater contamination (Luna et al., 2012). If killed before sowing the main crop, cover crops can act as 'non-living mulch' but, if retained during part or all of the crop life cycle, act as 'living much' (Bryant et al., 2013). Using cover crops as mulching showed beneficial effects by reducing insect pests in vegetable crops (Altieri & Schmidt, 1987; Altieri et al., 1985). The reduction of the major insect pests in the cereal crops was attributed to higher predation rates in the cover crop mulches (Lundgren & Fergen, 2010; Schmidt et al., 2004). Bryant et al., (2013) investigated the influence of cover crop mulch on weed management and arthropod communities in strip-tilled cabbage. Natural enemy communities were more responsive and influenced by changes in habitat complexity. The oat row mulch positively impacted the predatory lady beetles *Propylaea quatuordecimpunctata* and predatory thrips. Also, it increased the abundance of a specialist imported cabbageworm (Bryant et al., 2013). However, the abundance of a specialist natural enemy *Diadegma insulare* was negatively affected by habitat complexity in the cabbage field (Bryant et al., 2013). On the contrary, the *D. insulare* abundance and parasitism rate were higher in non-crop flowering plants, which apparently provided the type of resources that exerted a positive effect (Lee & Heimpel, 2005). Overall, the oat cover crop mediated habitat complexity in the cabbage field enhanced predator populations and reduced some pests. But it also decreased the cabbage yield and declined the abundance of some specialist parasitoids (Bryant et al., 2013). Sun hemp *Crotalaria juncea* L. living mulch in zucchini plants significantly lowered the incidence of striped cucumber beetles *Acalymma vittatum* F. as compared with non-mulch plants (Hinds & Hooks, 2013). Earlier, Hooks et al. (1998) shown that the population densities of melon aphid, *Aphis gossypii* Glover and silver leaf whitefly, *Bemisia argentifolii* Bellows, and perrings grown in zucchini crop were significantly lower in living mulches of buckwheat, *Fagopyrum esculentum* Moench, and yellow mustard, *Sinapis alba* L. Besides, the percentage incidence of papaya ringspot virus-watermelon strain (PRSV-W) transmitted by the pests as mentioned earlier was significantly lower in living mulches as compared to bare-ground (Hooks et al., 1998). Decreased pest abundance and disease incidence in living mulches resulted in higher and good quality marketable yield.

Alfalfa living mulch grown with soyabean increased the abundance of natural enemies and delayed the pest *Aphis glycines*' establishment (Schmidt et al., 2007). Similarly, the predator rove beetle and spider activity density were higher in the living mulch of Medicago (*Medicago polimorpha* L. var anglona) in cauliflower plot (Burgio et al., 2014). Broccoli grown in living mulches with three leguminous cover crops included strawberry clover *Trifolium fragiferum* L., white clover *T. repens* L., and a mixture of red clover *Trifolium praetense* and birdsfoot trefoil *Lotus corniculatus* L. significantly reduced the populations of the green peach aphid *Myzus persicae* and cabbage aphid *Brevicoryne brassicae* (Costello & Altieri, 1995). In another study, Gill et al. (2011) evaluated the effects of four different types of organic mulches: sunn hemp *Crotalaria jumcea* L., cowpea *Vigna unguiculata* L., sorghum-sudangrass (*Sorghum bicolor* × *S. sudanese*), and pine bark nuggets on the populations of soil surface insects and other related arthropods. Data collected from the pitfall traps revealed that these mulches impacted a wide range of insects. For instance, Dipterans were more abundant in pine bark mulch plots, whereas small plant-feeding insects (Thripidae, Aphididae, and Aleyrodidae) were more densely populated in the cowpea and control plots.

On the other hand, some predatory arthropods such as predatory beetles, spiders, Collembolans, and Orthopterans (Gryllidae and Acrididae) abundances were not affected significantly by mulches (Gill et al., 2011). Predation activity of carabid ground beetles against European corn borer *Ostrinia nubilalis* increased in kura clover *Trifolium ambiguum* M. and Alfalfa *Medicago sativa* L. living mulches grown in corn *Zea mays* L. and soyabean *Glycine max* L. crops (Prasifka et al., 2006). In Italy and Denmark, the living mulch of *Medicago polimorpha* L. on cauliflower *Brassica oleracea* L. var. *botrytis* decreased aphids population, increased larval parasitization, and positively affected the activity of carabid beetles (Depalo et al., 2016).

Nevertheless, there have been instances where mulching did not exert any positive or negative effects on pest abundance. For example, in Italy, the living mulch of *Medicago polimorpha* L. on cauliflower *Brassica oleracea* L. var. *botrytis* did not affect the cabbage caterpillar *Pieris* spp., infestation (Depalo et al., 2016). Similarly, three different types of mulches including living mulch of cowpea plant, straw mulch, and plastic mulch used to study their effects on pests abundance and natural enemies population of pepper, *Capsicum annum* L. had no significant effect on the pest *A. gosypii* colonization (Mochiah & Baidoo, 2012), suggesting that the used mulching materials did not exhibit pest repellent properties. Incongruent with this, Saucke et al. (2009) also found no suppressing effect of mulching on aphids' population (Saucke et al., 2009).

5 Instances of Pest Suppression in Non-living Mulches

Non-living or synthetic mulches are composed of organic or inorganic natural materials or synthetic materials manufactured specifically for mulching purposes (Grundy & Bond, 2007). Examples include grass, hay, compost, fine particles of woods, grass clippings, crop wastes, gravel mulch, plastic sheets (transparent mulch, black mulch, and double colored mulch), aluminized films, and other photo biodegradable materials (Grundy & Bond, 2007; Vincent et al., 2003). The above section made it clear how the living mulches may provide resources to support natural enemies and suppress pest populations. Non-living mulches derived from plant sources (hay, straw, composted wastes, etc.) may also offer beneficial effects in this regard.

Sweet potato *Ipomoea batatas* L. grown in killed cover crop mulch had a higher abundance of predatory insects, including rove beetles, fire ants, and carabid beetles, captured using pitfall traps (Jackson & Harrison, 2008). Moreover, the killed cover crop mulch also reduced the injury level to the sweet potato roots from the soil insect pests (Jackson & Harrison, 2008). Damages caused by cornstalk borer *Elasmopalpus lignosellus* to the beans *Phaseolus vulgaris* were minimized substantially using the hay mulch from sunn hemp (Gill et al., 2010). In various vegetable crops, aluminum-colored mulches have been used successfully to control insect pests (Porter et al., 1982; Schalk et al., 1979). Frank and Liburd, (2005) evaluated the effects of synthetic and living mulches to control aphids and whiteflies (*Bemisia argentifolii* Bellows and Perring) in zucchini plantings. They used two synthetic (white and reflective) and two living mulches (white clover *Trifolium repens* L. and buckwheat *Fagopyrum esculentum* Moench). They recorded the incidence of pests and natural enemies' diversity (Frank & Liburd, 2005). Whiteflies and aphids' populations varied in white and reflective synthetic mulches, with a higher number of adult flies and aphids in white mulch treatment than in reflective mulch type. On the other hand, natural enemy populations were consistently higher in the two living mulches than in synthetic mulches or bare-ground treatment (Frank & Liburd, 2005). Several studies have reported the use of silver or gray reflective mulches that successfully reduced the colonization of aphids or the incidences of aphids borne diseases in various crops (Brown et al., 1993; Stapleton & Summers, 1997, 2002; Stapleton et al., 1993). Reduction of aphids' colonization in reflective mulches can be attributed to their property of 'short wave light reflection' which acts as repellent for alate aphids (Harpaz, 1982; Kring, 1972; Loebenstein & Raccah, 1980). Biodegradable silver pigmented, synthetic latex spray mulches are postulated to enhance plants' photosynthetic activity that significantly increased (42–237%) the production of eggplant cv. Millionaire (*Solanum melongena* L.) (Mahmoudpour & Stapleton, 1997). An increase in the crop yield is due to the retention of soil moisture and heat during the nighttime (Stapleton & Summers, 2002). Reflective mulch reduced the incidence of aphids transmitted diseases and colonization by the whiteflies *Bemisia argentifolii* as well (Stapleton & Summers, 2002). Aphid numbers on the leaves of cucurbitaceous crops were consistently lower on reflective polyethylene and biodegradable synthetic latex mulches than on bare soil (Stapleton & Summers, 2002). The synthetic mulches delayed the onset of aphids transmitted diseases such as watermelon mosaic potyvirus, cucumber mosaic cucumovirus, and zucchini yellow mosaic potyviruses. Resultantly, the mulching practice increased the marketable yield to 9.5- and 2.5-fold for polyethylene and spray mulch, respectively (Stapleton & Summers, 2002). Given the suppressive nature of reflective mulches against the whiteflies and aphids in selected vegetable crops, they have been suggested as alternative to conventional insecticides for organic pest control (Frank & Liburd, 2005). However, despite the benefits of synthetic mulches in increasing crop yield and managing weeds and insect pests, they always come at a price. For example, the cost and difficulty of disposing synthetic mulches sometimes discourage the adoption of such control measures (Stapleton & Summers, 2002).

6 Instances of Mulching Increased Insect Herbivory and Crop Losses

It should be noted that mulching sometimes negatively affects the crop yields by competing for resources (water, nutrients, space, etc.), impacting natural enemy abundance and efficiency, or providing alternate hosts or refuge for the pest insects (Andow & Risch, 1985; Costello & Altieri, 1995; Legrand & Barbosa, 2003; Smid et al., 2002). For instance, the abundance of a coccinellid predator (*Coleomegilla maculate*) in the maize crops was reduced in the polyculture due to the difficulty of visually searching for prey (Andow & Risch, 1985). Also, the predatory lady beetles, *Coccinella septempunctata* L. (Coleoptera: Coccinellidae) ability to visually search for prey was negatively impacted in structurally complex agroecosystems (Legrand & Barbosa, 2003). Some studies have indicated that vegetationally diverse cropping systems can decrease the rate of parasitism or parasitoid abundance because of interference with chemical or visual signals from prey hosts in brussels sprout and broccoli crops (Costello & Altieri, 1995; Smid et al., 2002). In a recent study, Vorsah et al. (2020) found that the mulch treatment significantly increased the herbivory of *Disonycha glabrata* on different Amaranthus varieties. Both larval and adult stages of *D. glabrata* cause significant defoliation and are considered a major pest on Amaranthus in the United States (Garcia et al., 2011; Stegmaier, 1950; Vorsah et al., 2020). In addition, the treatment also increased the abundance of insects including Chrysomelidae and Curculionidae families in the order Coleoptera, family Agromyzidae in the order Diptera, families Miridae, Blissidae, Membracidae, Cercopidae, Cicadellidae, Pentatomidae, Acanaloniidae, Coreidae, and Aphididae in the order Hemiptera, families Hespiridae and Crambidae in order Lepidoptera, and family Acrididae in order Orthoptera (Vorsah et al., 2020). All these insects have been reported on Amaranthus spp. and cause significant damage (Othim et al., 2018a, 2018b; Smith et al., 2018; Vorsah et al., 2020).

Increased herbivory of pest insects occurs when mulching alters the microenvironment in their favor, resulting in increased growth and development of pest arthropods. In contrast, some studies have suggested that the dark color of the mulch, humidity profile, and odor may have attracted these insects (Vorsah et al., 2020). Higher moisture content in mulched plots had a higher density of *D. glabrata* suggesting that a suitable microhabitat was created lacking repellence in the specific leaf compost of the used mulch (Brown & Tworkoski, 2004; Vorsah et al., 2020). Moreover, the higher C:N ratio in mulched plots increase the availability of nitrogen element to plants enhanced their vegetative growth (leaf tissues and biomass), and rendered the plant attracted to pest herbivory (Leghari et al., 2016). That being said, habitat modification with mulching for pest management should be carefully articulated to avoid pest herbivory and other unwanted ramifications.

7 Mulching Supports Insect Pollinators

Indiscriminate use of broad-spectrum synthetic insecticides has not only threatened the natural enemies' population but also resulted in the decline of insect pollinators such as honeybees, bumblebees, butterflies, and pollinating flies (Goulson et al., 2015; Hannon & Sisk, 2009; Majewska & Altizer, 2020; Pywell et al., 2011). Farmers undertake various maintenance practices such as planting crop margins and hedgerows to restore the natural landscape and recover declining pollinators populations (Hannon & Sisk, 2009; Menz et al., 2011; Pywell et al., 2011). In addition, mulching, weeding, and cleaning practices can also contribute to pollinators' promotion (Clayton, 2007; Goddard et al., 2013). Besides the benefits mentioned above of mulching in pests suppression and promotion of natural enemies, mulching can also play a pivotal role in conserving and support of the declining pollinators population (Majewska & Altizer, 2020). Mulching provides greater habitat diversity and hence a substrate for nesting for andrenid bees and other pollinators as well as hiding sites from predators (Cane, 2015; Fortel et al., 2016; Majewska & Altizer, 2020).

Furthermore, the practice of weeding and mulching reduces competition for the main crop by eliminating the competitors and investing more resources in growth and reproduction that leading to an increase in nectar and pollen of cultivated crops, thereby positively affecting the diversity and abundance of pollinators (Johnson, 1971; Majewska & Altizer, 2020). The ground-nesting squash bees, *Peponapis pruinosa* is a major pollinator for high pollination demanding zucchini squash, *Cucurbita pepo*. Conventionally, the tillage operation or other cultivation practices would disturb their nests in crop fields. However, using different mulch materials such as municipal wastes, woodchips, grass clippings, and shredded newspapers for weed control may spare the tillage and other management practices, thus enhancing the squash bees' population pollination services (Splawski et al., 2014). Since mulching can suppress weed emergence and reduce herbicides, application and tillage operation may directly benefit pollinators that make their nests in agroecosystems.

Apart from arthropod pollinators, mulch materials may also provide substrates for nesting for other wildlife such as birds, rodents, and reptiles that potentially participate in pollination (Majewska & Altizer, 2020). Given that mulches may provide nectar and pollen, nesting sites, and other essential resources for insect pollinators, studies on the effects of mulching on pollinators' diversity and population abundance are extremely limited. Therefore, studying the overall effects of mulching should also include crop pollinators to conserve their diversity and improve the declining populations.

8 Mulching May also Suppress Weeds Emergence and Decrease Herbicides Application

An added benefit of mulching is to limit the need for tillage for weed control in fruits and vegetable crops. In this regard, different mulch materials, including living or dead plant residues, wood fiber, plastic film, and paper, have been used to suppress weeds and improve crop yield (Splawski et al., 2014; Teasdale & Mohler, 2000; Teasdale et al., 2002). Living and dead mulches have been shown to decrease the emergence of small-seeded annual weeds and reduced the crop loss yield caused by weeds and soil-inhabiting herbivores (Teasdale et al., 2002). Growing a mixture of cover crops such as legume and cereal were particularly effective against the suppression of weeds. Plant residues maintained on the soil's surface in a no-tillage farming system hold several opportunities to regulate pest population and suppress weed emergence (Teasdale et al., 2002). Suppression of weeds emergence mediated by mulching may vary in magnitude according to the type of mulch (living/dead), crop species, weed species, and residues biomass (Liebman et al., 2001; Teasdale et al., 2002). Mulch area index (area of mulch material per area of soil) is one of the defining factors directly related to weed suppression (Teasdale & Mohler, 2000). In addition, mulches with a high area to weight ratio may provide more efficient weed suppression than mulches with a small area to weight ratio (Teasdale et al., 2002). Also, weed suppression is closely related to seed size, i.e., the smaller the seed, the more sensitive is the weed species to suppression and vice versa (Teasdale & Mohler, 2000; Teasdale et al., 2002). Although cover crops may provide a degree of weed suppression in the early season of a no/reduced-tillage system, yet for achieving optimum weed control, other control tactics such as herbicides application may be needed as well. But for sure, the cover crops would allow a reduction of herbicide input. Rice straw mulch suppress weeds in wheat crop effectively up to 69% (Nawaz et al., 2017). Similarly, the use of straw mulch in organic production of potato crops increased total and marketable yield as well as effectively control broadleaf weeds (Dvorak et al., 2015; Genger et al., 2018). Different types of mulches include plastic, straw, paper, agricultural wastes, and manures have been used to control weeds in agroecosystems (Cirujeda et al., 2012; Dvorak et al., 2015; Jabran, 2019; Nawaz et al., 2017; Steinmetz et al., 2016).

Cover crop residues or inorganic mulches mediated weed suppression is mainly achieved by creating the soil conditions unfavorable for weeds' emergence, physically obstructing their emergence/dispersal, and releasing allelochemicals (Teasdale & Mohler, 2000; Teasdale et al., 2002). In addition, the crops residues add organic matter to soil that increases soil fertility that can improves plant growth and vigor and reduce weeds damage to crop plants. Interception and reflection of incoming radiation by mulches modify the quantity of light required for weed species to activate the phytochrome-mediated germination process, thus inhibiting weed emergence through light extinction (Teasdale et al., 2002). Allelopathic compounds released from cover crops inhibit the germination and growth of weed plants (Liebman et al., 2001; Teasdale et al., 2002). On the other hand, allelopathic compounds may also have stimulatory effects on some weed species' germination, such as nitrates released from legumes residues facilitate the emergence of certain weed species (Teasdale et al., 2002). These attributes may contribute to mulching-derived weed suppression and integrated weed management.

9 Conclusion

Integrated pest management combines various management tools to achieve effective and sustainable pest control. The potential of mulching in the control of insect pests as well as the promotion of beneficial insects such as natural enemies and pollinators has been documented here. Different mulch materials, including dead or living, organic or synthetic, have been employed in different fruit and vegetable crops against a wide range of pest insects. In most cases, the mulching practice decreased pest populations and increased crop yield. However, in some instances, the approach backfired and increased insect herbivory. Mulching creates a complex ecosystem that impairs the herbivore's ability to locate the host plant and create diversions to non-crop plants by imposing physical barriers, disrupting visual and olfactory cues, and enhancing plant defenses, subsequently suppressing pests' abundance. In addition, habitat modification through mulching may increase natural enemies' population densities due to greater and more diverse habitats and food resources, thereby indirectly reducing herbivore abundance by improving the activity of biological control agents. Also, mulching can play an essential role to conserve and support the declining pollinators population.

Conflict of Interest The authors declare no conflict of interest.

Funding This work received no external funding.

References

Altieri, M., & Schmidt, L. J. C. A. (1987). Mixing broccoli cultivars reduces cabbage aphid numbers. *California Agriculture, 41*, 24–26.

Altieri, M. A., Wilson, R. C., & Schmidt, L. L. J. C. P. (1985). The effects of living mulches and weed cover on the dynamics of foliage-and soil-arthropod communities in three crop systems. *Crop Protection, 4*, 201–213.

Alyokhin, A., Nault, B., & Brown, B. (2019). Soil conservation practices for insect pest management in highly disturbed agroecosystems—A review. *Entomologia Experimentalis Et Applicata, 168*, 7–27.

Andow, D. A. (1991). Vegetational diversity and arthropod population response. *Annual Review of Entomology, 36*, 561–586.

Andow, D. A., Nicholson, A. G., Wien, H. C., & Willson, H. R. (1986). Insect populations on cabbage grown with living mulches. *Environmental Entomology, 15*, 293–299.

Andow, D. A., & Risch, S. J. (1985). Predation in diversified agroecosystems—Relations between a coccinellid predator Coleomegilla-maculata and its food. *Journal of Applied Ecology, 22*, 357–372.

Brown, J. E., Dangler, J. M., Woods, F. M., Tilt, K. M., Henshaw, M. D., Griffey, W. A., & West, M. S. (1993). Delay in mosaic-virus onset and aphid vector reduction in summer squash grown on reflective mulches. *HortScience, 28*, 895–896.

Brown, M. W., & Tworkoski, T. (2004a). Pest management benefits of compost mulch in apple orchards. *Agriculture Ecosystems & Environment, 103*, 465–472.

Bryant, A., Brainard, D. C., Haramoto, E. R., & Szendrei, Z. (2013). Cover crop mulch and weed management influence arthropod communities in strip-tilled cabbage. *Environmental Entomology, 42*, 293–306.

Burgio, G., Kristensen, H. L., Campanelli, G., Bavec, F., Bavec, M., von Fragstein und Niemsdorff, P., Depalo, L., Lanzoni, A., & Canali, S. J. B. O. B. (2014). Effect of living mulch on pest/beneficial interaction. *Building Organic Bridges, 3*, 741–744.

Cane, J. H. (2015). Landscaping pebbles attract nesting by the native ground-nesting bee Halictus rubicundus (Hymenoptera: Halictidae). *Apidologie, 46*, 728–734.

Cirujeda, A., Aibar, J., Anzalone, A., Martin-Closas, L., Meco, R., Moreno, M. M., Pardo, A., Pelacho, A. M., Rojo, F., Royo-Esnal, A., Suso, M. L., & Zaragoza, C. (2012). Biodegradable mulch instead of polyethylene for weed control of processing tomato production. *Agronomy for Sustainable Development, 32*, 889–897.

Clayton, S. (2007). Domesticated nature: Motivations for gardening and perceptions of environmental impact. *Journal of Environmental Psychology, 27*, 215–224.

Costello, M. J., & Altieri, M. A. (1995). Abundance, growth-rate and parasitism of brevicoryne-brassicae and Myzus-Persicae (Homoptera, Aphididae) on Broccoli grown in living mulches. *Agriculture Ecosystems & Environment, 52*, 187–196.

Depalo, L., Burgio, G., von Fragstein, P., Kristensen, H. L., Bavec, M., Robačer, M., Campanelli, G., & Canali, S. (2016). Impact of living mulch on arthropod fauna: Analysis of pest and beneficial dynamics on organic cauliflower (Brassica oleracea L. var. botrytis) in different European scenarios. *Renewable Agriculture and Food Systems, 32*, 240–247.

Dvorak, P., Tomasek, J., Hamouz, K., & Kuchtova, P. (2015). Reply of mulch systems on weeds and yield components in potatoes. *Plant Soil and Environment, 61*, 322–327.

Fortel, L., Henry, M., Guilbaud, L., Mouret, H., & Vaissiere, B. E. (2016). Use of human-made nesting structures by wild bees in an urban environment. *Journal of Insect Conservation, 20*, 239–253.

Frank, D. L., & Liburd, O. E. (2005). Effects of living and synthetic mulch on the population dynamics of whiteflies and aphids, their associated natural enemies, and insect-transmitted plant diseases in zucchini. *Environmental Entomology, 34*, 857–865.

Garcia, A. A., Huato, M. A. D., Lara, M. H., Saenz-de-Cabezon, F. J., Perez-Moreno, I., Marco-Mancebon, V., & Lopez-Olguin, J. F. (2011). Insect occurrence and losses due to phytophagous species in the amaranth Amaranthus hypocondriacus L. crop in Puebla Mexico. *African Journal of Agricultural Research, 6*, 5924–5929.

Genger, R. K., Rouse, D. I., & Charkowski, A. O. (2018). Straw mulch increases potato yield and suppresses weeds in an organic production system. *Biological Agriculture & Horticulture, 34*, 53–69.

Gill, H. K., McSorley, R., & Branham, M. (2011). Effect of organic mulches on soil surface insects and other arthropods. *Florida Entomologist, 94*, 226–232.

Gill, H. K., McSorley, R., Goyal, G., & Webb, S. E. (2010). Mulch as a potential management strategy for lesser cornstalk borer, elasmopalpus lignosellus (Insecta: Lepidoptera: Pyralidae), in Bush Bean (Phaseolus Vulgaris). *Florida Entomologist, 93*, 183–190.

Goddard, M. A., Dougill, A. J., & Benton, T. G. (2013). Why garden for wildlife? Social and ecological drivers, motivations and barriers for biodiversity management in residential landscapes. *Ecological Economics, 86*, 258–273.

Goulson, D., Nicholls, E., Botias, C., & Rotheray, E. L. (2015). Bee declines driven by combined stress from parasites, pesticides, and lack of flowers. *Science, 347*, 1255957.
Grundy, A., & Bond, B. (2007). Use of non-living mulches for weed control. *Non-Chemical Weed Management*, 135–153.
Hannon, L. E., & Sisk, T. D. (2009). Hedgerows in an agri-natural landscape: Potential habitat value for native bees. *Biological Conservation, 142*, 2140–2154.
Harpaz, I. (1982). Nonpesticidal control of vector-borne viruses.
Hartwig, N. L., & Ammon, H. U. (2002). 50th Anniversary—Invited article—Cover crops and living mulches. *Weed Science, 50*, 688–699.
Hinds, J., & Hooks, C. R. R. (2013). Population dynamics of arthropods in a sunn-hemp zucchini interplanting system. *Crop Protection, 53*, 6–12.
Hooks, C. R. R., & Johnson, M. W. (2004). Using undersown clovers as living mulches: Effects on yields, lepidopterous pest infestations, and spider densities in a Hawaiian broccoli agroecosystem. *International Journal of Pest Management, 50*, 115–120.
Hooks, C. R. R., & Johnson, M. W. (2006). Population densities of herbivorous lepidopterans in diverse cruciferous cropping habitats: Effects of mixed cropping and using a living mulch. *BioControl, 51*, 485–506.
Hooks, C. R. R., Valenzuela, H. R., & Defrank, J. (1998). Incidence of pests and arthropod natural enemies in zucchini grown with living mulches. *Agriculture Ecosystems & Environment, 69*, 217–231.
Jabran, K. (2019). *Role of mulching in pest management and agricultural sustainability*. Springer.
Jackson, D. M., & Harrison, H. F., Jr. (2008). Effects of a killed-cover crop mulching system on sweetpotato production, soil pests, and insect predators in South Carolina. *Journal of Economic Entomology, 101*, 1871–1880.
Johnson, B. J. (1971). Effect of Weed Competition on Sunflowers. *Weed Science, 19*, 378–380.
Kogan, M. (1998). Integrated pest management: Historical perspectives and contemporary developments. *Annual Review of Entomology, 43*, 243–270.
Köhler, H.-R., & Triebskorn, R. (2013). Wildlife ecotoxicology of pesticides: Can we track effects to the population level and beyond? *Science, 341*, 759–765.
Kring, J. B. (1972). Flight behavior of aphids. *Annual Review of Entomology, 17*, 461.
Lanini, W., Pittenger, D., Graves, W., Munoz, F., & Agamalian, H. (1989). Subclovers as living mulches for managing weeds in vegetables. *California Agriculture, 43*, 25–27.
Lee, J. C., & Heimpel, G. E. (2005). Impact of flowering buckwheat on Lepidopteran cabbage pests and their parasitoids at two spatial scales. *Biological Control, 34*, 290–301.
Leghari, S. J., Wahocho, N. A., Laghari, G. M., HafeezLaghari, A., MustafaBhabhan, G., Hussain-Talpur, K., Bhutto, T. A., Wahocho, S. A., & Lashari, A. A. (2016). Role of nitrogen for plant growth and development: A review. *Advances in Environmental Biology, 10*, 209–219.
Legrand, A., & Barbosa, P. (2003). Plant morphological complexity impacts foraging efficiency of adult Coccinella septempunctata L. (Coleoptera : Coccinellidae). *Environmental Entomology, 32*, 1219–1226.
Letourneau, D. K., Armbrecht, I., Rivera, B. S., Lerma, J. M., Carmona, E. J., Daza, M. C., Escobar, S., Galindo, V., Gutiérrez, C., & López, S. D. (2011). Does plant diversity benefit agroecosystems? A synthetic review. *Ecological Applications, 21*, 9–21.
Lewis, W. J., van Lenteren, J. C., Phatak, S. C., & Tumlinson, J. H. (1997). A total system approach to sustainable pest management. *Proceedings of the National Acadamy Science United States of America, 94*, 12243–12248 (3rd edn).
Liebman, M., Mohler, C. L., & Staver, C. P. (2001). *Ecological management of agricultural weeds*. Cambridge University Press.
Lin, P. C., Lin, H. J., Liao, Y. Y., Guo, H. R., & Chen, K. T. (2013). Acute poisoning with neonicotinoid insecticides: A case report and literature review. *Basic & Clinical Pharmacology & Toxicology, 112*, 282–286.
Loebenstein, G., & Raccah, B. (1980). Control of non-persistently transmitted aphid-borne viruses. *Phytoparasitica, 8*, 221–235.

Luna, J. M., Mitchell, J. P., & Shrestha, A. (2012). Conservation tillage for organic agriculture: Evolution toward hybrid systems in the western USA. *Renewable Agriculture and Food Systems, 27*, 21–30.

Lundgren, J. G., & Fergen, J. K. (2010). The effects of a winter cover crop on Diabrotica virgifera (Coleoptera: Chrysomelidae) populations and beneficial arthropod communities in no-till maize. *Environmental Entomology, 39*, 1816–1828.

Mahmoudpour, M. A., & Stapleton, J. J. (1997). Influence of sprayable mulch colour on yield of eggplant (Solanum melongena L. cv. Millionaire). *Scientia Horticulturae, 70*, 331–338.

Majewska, A. A., & Altizer, S. (2020). Planting gardens to support insect pollinators. *Conservation Biology, 34*, 15–25.

Menz, M. H., Phillips, R. D., Winfree, R., Kremen, C., Aizen, M. A., Johnson, S. D., & Dixon, K. W. (2011). Reconnecting plants and pollinators: Challenges in the restoration of pollination mutualisms. *Trends in Plant Science, 16*, 4–12.

Mochiah, M., & Baidoo, P. (2012). Effects of mulching materials on agronomic characteristics, pests of pepper (Capsicum annuum L.) and their natural enemies population.

Nawaz, A., Farooq, M., Lal, R., Rehman, A., Hussain, T., Nadeem, A., & Development. (2017). Influence of sesbania brown manuring and rice residue mulch on soil health, weeds and system productivity of conservation rice–wheat systems. *Land Degradation, 28*, 1078–1090.

Oerke, E. C. (2006). Crop losses to pests. *Journal of Agricultural Science, 144*, 31–43.

Othim, S. T. O., Kahuthia-Gathu, R., Akutse, K. S., Foba, C. N., & Fiaboe, K. K. M. (2018a). Seasonal occurrence of amaranth Lepidopteran defoliators and effect of attractants and amaranth lines in their management. *Journal of Applied Entomology, 142*, 637–645.

Othim, S. T. O., Ramasamy, S., Kahuthia-Gathu, R., Dubois, T., Ekesi, S., & Fiaboe, K. K. M. (2018b). Expression of resistance in amaranthus spp. (Caryophyllales: Amaranthaceae): Effects of selected accessions on the behaviour and biology of the amaranth leaf-webber, spoladea recurvalis (Lepidoptera: Crambidae). *Insects, 9*, 62.

Porter, W., WC, P., & WW, E. (1982). Effects of aluminium-painted and black polyethylene mulches on bell pepper, capsicum annum L.

Prasifka, J. R., Schmidt, N. P., Kohler, K. A., O'neal, M. E., Hellmich, R. L., & Singer, J. W. (2006). Effects of living mulches on predator abundance and sentinel prey in a corn–soybean–forage rotation. *Environmental Entomology, 35*, 1423–1431.

Pywell, R. F., Meek, W. R., Hulmes, L., Hulmes, S., James, K. L., Nowakowski, M., & Carvell, C. (2011). Management to enhance pollen and nectar resources for bumblebees and butterflies within intensively farmed landscapes. *Journal of Insect Conservation, 15*, 853–864.

Rapisarda, C., & Cocuzza, G. E. M. (2017). *Integrated pest management in tropical regions*. CABI.

Rieux, R., Simon, S., & Defrance, H. (1999). Role of hedgerows and ground cover management on arthropod populations in pear orchards. *Agriculture Ecosystems & Environment, 73*, 119–127.

Root, R. B. (1973). Organization of a plant-arthropod association in simple and diverse habitats: The fauna of collards (Brassica oleracea). *Ecological Monographs, 43*, 95–124.

Sans, F., & Altieri, M. (2005). Effects of intercropping and fertilization on weed abundance, diversity and resistance to invasion. *International Society of Organic Agricultural Research, Bonn.*

Saucke, H., Juergens, M., Doring, T. F., Fittje, S., Lesemann, D. E., & Vetten, H. J. (2009). Effect of sowing date and straw mulch on virus incidence and aphid infestation in organically grown faba beans (Vicia faba). *Annals of Applied Biology, 154*, 239–250.

Schalk, J. M., Creighton, C. S., Fery, R. L., Sitterly, W. R., Davis, B. W., Mcfadden, T. L., & Day, A. (1979). Reflective film mulches influences insect control and yield in vegetables. *Journal of the American Society for Horticultural Science, 104*, 759–762.

Schmidt, M. H., Thewes, U., Thies, C., & Tscharntke, T. (2004). Aphid suppression by natural enemies in mulched cereals. *Entomologia Experimentalis et Applicata, 113*, 87–93.

Schmidt, N. P., O'Neal, M. E., & Singer, J. W. (2007). Alfalfa living mulch advances biological control of soybean aphid. *Environ Entomol, 36*, 416–424.

Smid, H. M., van Loon, J. J. A., Posthumus, M. A., & Vet, L. E. M. (2002). GC-EAG-analysis of volatiles from Brussels sprouts plants damaged by two species of Pieris caterpillars: Olfactory receptive range of a specialist and a generalist parasitoid wasp species. *Chemoecology, 12*, 169–176.

Smith, J. D., Dinssa, F. F., Anderson, R. S., Su, F. C., & Srinivasan, R. (2018). Identification of major insect pests of Amaranthus spp. and germplasm screening for insect resistance in Tanzania. *International Journal of Tropical Insect Science, 38*, 261–273.

Splawski, C. E., Regnier, E. E., Harrison, S. K., Goodell, K., Bennett, M. A., & Metzger, J. D. (2014). Mulch effects on floral resources and fruit production of squash, and on pollination and nesting by squash bees. *HortTechnology, 24*, 535–545.

Stapleton, J., & Summers, C. (1997). Reflective mulch for managing aphids, aphid-borne viruses, and silverleaf whitely: 1996 season review. *UC Plant Protection Quarterly, 7*, 13–15.

Stapleton, J., Summers, C., Turini, T., & Duncan, R. (1993). Effect of reflective polyethylene and spray mulches on aphid populations and vegetative growth of bell pepper. In *Proceedings of the National Agricultural Plastics Congress* (Vol. 24, pp. 117–122).

Stapleton, J. J., & Summers, C. G. (2002). Reflective mulches for management of aphids and aphid-borne virus diseases in late-season cantaloupe (Cucumis melo L. var. cantalupensis). *Crop Protection, 21*, 891–898.

Stegmaier, C. E. (1950). Insects associated with the rough pigweed, Amaranthus retroflexus L. (Amaranthaceae).

Steinmetz, Z., Wollmann, C., Schaefer, M., Buchmann, C., David, J., Troger, J., Munoz, K., Fror, O., & Schaumann, G. E. (2016). Plastic mulching in agriculture. Trading short-term agronomic benefits for long-term soil degradation? *Science of the Total Environment, 550*, 690–705.

Stern, V., Smith, R., van den Bosch, R., & Hagen, K. (1959). The integration of chemical and biological control of the spotted alfalfa aphid: The integrated control concept. *Hilgardia, 29*, 81–101.

Teasdale, J. R., Abdul-Baki, A. A., Mill, D., & Thorpe, K. W. (2002). Enhanced pest management with cover crop mulches. In *XXVI International Horticultural Congress: Sustainability of Horticultural Systems in the 21st Century* (Vol. 638, pp. 135–140).

Teasdale, J. R., & Mohler, C. L. (2000). The quantitative relationship between weed emergence and the physical properties of mulches. *Weed Science, 48*, 385–392.

Vincent, C., Hallman, G., Panneton, B., & Fleurat-Lessard, F. (2003). Management of agricultural insects with physical control methods. *Annual Review of Entomology, 48*, 261–281.

Vorsah, R. V., Dingha, B. N., Gyawaly, S., Fremah, S. A., Sharma, H., Bhowmik, A., Worku, M., & Jackai, L. E. (2020). Organic mulch increases insect herbivory by the flea beetle species, disonycha glabrata, on amaranthus spp. *Insects, 11*.

Environment Section

Mulching Effects on Soil Greenhouse Gas Emission in Agricultural Systems

Xiaolin Liao, Saadatullah Malghani, Ahmad Ali, and Ghulam Haider

Abstract Agricultural activities are among the most important sources of atmospheric GHGs that are the primary causes of adverse climate changes. Therefore, efforts must be made to reduce GHGs source capacities of different agricultural activities, including agronomic management practices. Due to associated benefits such as weed control, reduced irrigation and fertilizer demands, and improved nutrient and water use efficiency, soil surface mulching is gaining popularity especially plastic film mulch. Biomass residues and plastic film, the two most common materials used in soil mulching, may have similar benefits as mulch but differ significantly in their effects on soil physicochemical properties and GHGs-producing biochemical processes. This chapter discussed the impacts of plastic film and crop residue mulching approaches on GHGs emissions in upland and paddy cropping systems under different management strategies. The literature suggests that surface mulches shift GHGs emission rates by altering soil physicochemical properties, including soil moisture, temperature, pH, redox potential, and nutrients. A few studies included changes in the microbial community responsible for the producing and consuming of GHGs in soil, suggesting the complex role of surface mulching.

Keywords Greenhouse gases emission · Cropping system · Mulch · Agriculture activities

X. Liao (✉) · S. Malghani (✉)
College of Biology and the Environment, Nanjing Forestry University, Nanjing, China
e-mail: liaoxiaolin@njfu.edu.cn

S. Malghani
e-mail: saadat.malghani@njfu.edu.cn; sum@plen.ku.dk

A. Ali · G. Haider
Department of Plant Biotechnology, Atta-Ur-Rahman School of Applied Biosciences, NUST, Islamabad, Pakistan

S. Malghani
Department of Plant and Environmental Sciences, University of Copenhagen, Copenhagen, Denmark

K. Akhtar et al. (eds.), *Mulching in Agroecosystems*,
https://doi.org/10.1007/978-981-19-6410-7_16

1 Introduction

High emissions of atmospheric greenhouse gases (GHGs) are the leading cause of global warming and various associated environmental problems (Stock et al., 2013; IPCC, 2014). Among different GHGs, carbon dioxide (CO_2), methane (CH_4), and nitrous oxide (N_2O) are considered the most important. Agriculture is the second-largest source of GHGs, accounting for about 14% of CO_2 (Vermeulen et al., 2012), 60% of N_2O, and 50% of CH_4 emissions (Smith et al., 2018), and total, about 13.5% of global anthropogenic emissions (IPCC, 2013). Specifically, rice paddies have been considered significant contributors to CH_4 and N_2O emissions, accounting for 50% of cropland CH_4 emissions and 10% of cropland N_2O emissions (Carlson et al., 2017; Kritee et al., 2018).

Agricultural practices, such as soil management, fertilization, and water management, have been identified as the main drivers behind the significant emissions of agricultural GHGs. Alarmingly, the synthetic fertilizer is applied in considerably large quantities beyond the crop demand, providing sufficient substrate for microbial communities to produce GHGs. While, continuous use of conventional farming practices based on extensive tillage, especially when combined with in situ burning of crop residues, can magnify the problem by triggering the release of carbon stocks otherwise stored in soil and crop biomass (Govaerts et al., 2009; Tanveer et al., 2013). Although the agricultural sector is one of the significant contributors to atmospheric GHGs concentrations, it is also considered the most vulnerable sector, facing significant challenges in providing sufficient food for the growing world population while attenuating the adverse effects of climate change (Zhang et al., 2020). Therefore, it is challenging to develop a low-carbon or carbon-neutral future for global crop production without considering sustainable agricultural management practices, such as optimized fertilization, less tillage or zero tillage, irrigation, cover crop, and mulching, to reduce total GHG emissions but without compromising the crop productivity (Lehner & Rosenberg, 2017; Gao et al., 2018).

Various types of mulches such as thin plastic film, gravel and sand, rock fragments, crop residue (straw and stubble), concrete, volcanic ash, paper pellets, and livestock manures are commonly applied at the soil surface in both dryland and irrigated areas (Chap. 1 in Sect. 1 of this book). However, plastic film and crop straw residues are the most popular.

Using crop straw residue as agricultural soil surface mulch is one of the pioneer approaches in mulching. Traditionally, it is a residue management approach, especially in conservation agriculture, to improve soil quality, reduce soil erosion, conserve water, and regulate soil temperature (Gan et al., 2013). The importance of crop residues as mulch is evident from expansion in areas under conservation agriculture worldwide (FAO, 2001; Govaerts et al., 2009; Hobbs, 2007). Currently, about 11% of global cropland (157 M ha) is under conservation agriculture, of which South America has the most significant area with 66.4 M ha (49.5% of the global total), followed by North America with 54.0 M ha (40.3%) and Asia with 10.3 M ha (7.7%) (Kassam et al., 2015).

Plastic film mulching was invented as an advanced agriculture cultivation technology for saving water and improving crop yield in the 1950s and has been widely applied worldwide since then (Kasirajan & Ngouajio, 2012). Approximately 19% of the total arable land in China in 2014, while 0.85% of the arable land around the world was cultivated using plastic film mulch (He et al., 2018; Ming et al., 2018). Soil surface mulching via plastic film can reduce evaporation by physically blocking the movement of water vapour from the soil surface to the air (Liu et al., 2016a, 2016b), as well as increase soil water storage and plant transpiration (He et al., 2018), and the amount of water available to plants during periods of peak water demand. Plastic film mulching can be in different colours (black or white) and materials (polyethylene or biodegradable materials). Previous studies have mainly discussed the effects of the most widely used transparent plastic film mulching, and a few studies used black or any other colour plastic film (Cuello et al., 2015).

Both crop straw and plastic film mulching are expected to benefit soil quality and crop growth, yet there are differences between their effects on soil physico-chemical properties. For example, plastic film mulching increases soil temperature almost continuously, while residue mulching generally has a cooling effect during hot summer days (Liu et al., 2014; Fan et al., 2019). Moreover, crop straw mulch also acts as a nutrient source for crop plants upon decomposition and can directly sequester carbon in soils (Singh et al., 2008; Yadvinder-Singh et al., 2008; Lenka & Lal, 2013; Chen et al., 2017). It was reported that the return to the soil of 1 Mg/ha of straw (rice, wheat, or maize) each year could sequester about 130 kg C/ha/year (Lu et al., 2009). Therefore, it will not be surprising that the two types of mulching would exhibit a contrasting effect on soil GHG emissions.

2 Mulching in Dryland and Rice Paddies Cropping Systems

Agricultural soil surface mulching is most prevalent in dryland cropping systems where water shortage and low winter temperature could lead to crop failure. Conventionally, crops are cultivated on a flat surface covered with mulching material (Fig. 1). However, flat surface cultivation can be replaced by the ridge and furrow system (RFMS) which, an innovative water-saving, high-yield and low-input farming practice (Zhou et al., 2009; Mo et al., 2016; Yao et al., 2017; Luo et al., 2018). Alternate ridges and furrows are two functional areas for producing and collecting surface run-off (Li et al., 2007). In practice, crops are planted in the furrow to maximize water uptake, while ridges are covered with mulch to reduce water loss via evaporation (Ren et al., 2008). There are several ridge-furrow mulching configurations, including (i) surface mulching (commonly plastic film) only on ridges, (ii) ridges with plastic film mulching and furrows with residue mulching, (iii) ridges and furrows all covered with plastic mulch, sometimes called "fully-mulch" (Zhou et al., 2009; Ye and Liu, 2012; Liu et al., 2014); and recently (iv) alternating small and large ridges with complete film mulching (Mo et al., 2018) (Fig. 1). Several studies compared different RFMS configurations on soil temperature, crop yields, and water use efficiency (Zhou et al.,

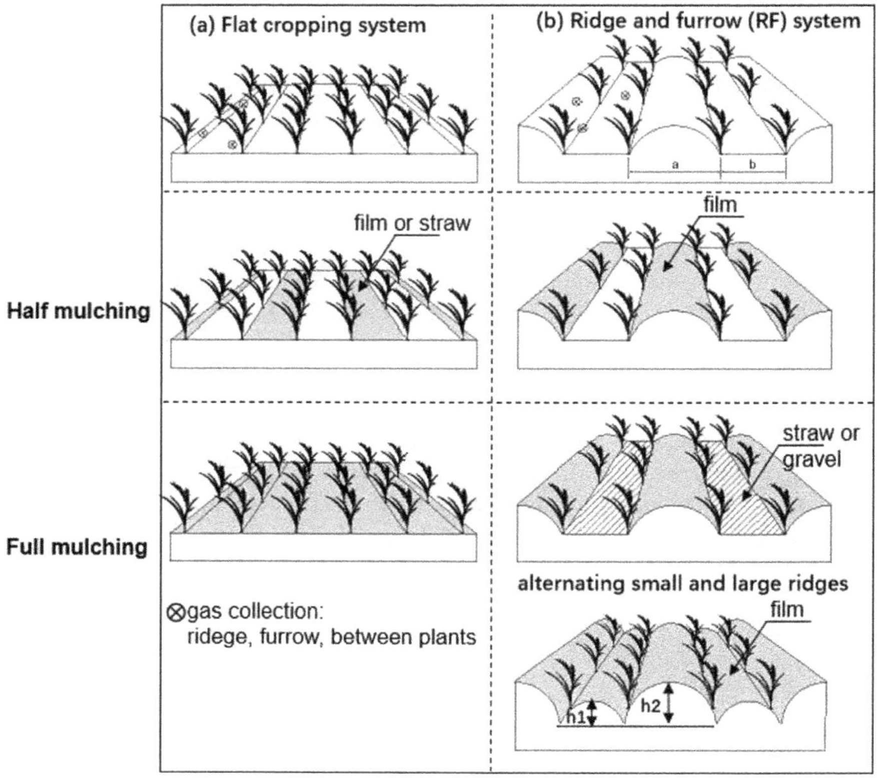

Fig. 1 Schematic diagram of the field layout showing different planting practices and mulching materials. **a** Flatting cropping systems, **b** ridge and furrow systems. Modified from Mo et al. (2017, 2018), Li et al. (2019a, 2019b), Jia et al. (2021)

2009; Mo et al., 2016, 2017; Luo et al., 2018), but a few discussed its role on GHGs emissions (Zhang et al., 2015; Li et al., 2019a, 2019b; Jia et al., 2021).

Agronomic practices for rice cropping can be divided into two basic systems based on water management: upland and lowland rice (Dossou-Yovo et al., 2016). For the upland rice (aerobic rice) cultivation system, cultivation is done in non-flooded and well-drained soils (Lal et al., 2013). While, for lowland rice (paddy rice) cultivation, flooding conditions are maintained throughout the cropping season with occasional drainage to maximize crop yield and reduce GHG emissions (Andriesse & Fresco, 1991). Considering the constant threat of decreasing fresh water supply for paddy cultivation, a promising water-saving ground cover rice production (GCRP) system for lowland rice production has gained popularity. Under the GCRP system, water levels are significantly reduced by maintaining field soil at 80–90% of the water-holding capacity instead of flooding. To reduce water evaporation, soil surface is covered with rice straw or plastic film mulching materials. Although significant research paid attention to the effect of the ground covering approach for lowland paddy on rice productivity, few studies indicate its promising role in the mitigation of GHGs emissions.

Agricultural soil surface mulching can influence soil microbial community activities related to carbon and nitrogen cycling by affecting the soil's critical parameters, including soil temperature, moisture, water-filled porosity, and aeration (Sect. 1 in this book), leading to changes in GHG emissions rates from agricultural soils (Zhang et al., 2015). In addition, agricultural soil mulching practice may trigger or mitigate GHG emissions by altering the agronomic inputs, i.e., a significant increase in organic matter supply either via direct residue application or via enhanced crop growth, reduction in fertilizer application rates, etc. In literature, several reports have quantified the effect of agricultural soil surface mulching on GHGs emissions and pointed out the significant role of mulching type (residue vs. plastic), combined use of mulch, and other conservation agriculture practices in different cultivation systems. Therefore, this book chapter will elaborate on the potential effects of residue and plastic film mulching under dryland and paddy cultivation systems with different management practices, such as tillage, fertilization, irrigation, and cropping, on CO_2, CH_4, and N_2O emissions.

3 Mulching Effects on Soil CO_2 Emissions

3.1 Atmospheric CO_2 Cycling Through the Agricultural Crop Production System

The source capacity of the agriculture ecosystem for atmospheric CO_2 depends on the net difference between the supply of CO_2 via crop plants and emission rates via soil respiration. On the one hand, crop plants uptake the atmosphere CO_2 via natural carboxylation processes and add it as soil organic carbon via below-ground biomass, rhizodeposition, and plant litter, commonly called net primary productivity (NPP) (Schlesinger & Andrews, 2000). On the other hand, soils emit CO_2 back into the atmosphere via microbial respiration. Soil respiration or CO_2 emissions involve microorganisms metabolizing substrates within the soil matrix (heterotrophic respiration). However, CO_2 emitted from the soil surface represents the cumulative emission rates from autotrophic (plant roots) and heterotrophic respiration (Fig. 2).

Soil CO_2 emission from the soil surface is a two-step process, including CO_2 production (biological process) and its diffusion from the soil-water matrix to the atmosphere (physical process), driven by both biotic and abiotic factors (Liu et al., 2014). Temperature and soil water content are the two most important abiotic factors with the potential to influence both steps of soil CO_2 emission (Zhang et al., 2015; Wang et al., 2019; Peter et al., 2020). Increasing soil temperature promotes CO_2 production rates by accelerating the soil organic matter decomposition process (Steinmetz et al., 2016). Meanwhile, an optimum moisture content range is also crucial for soil microbial activities as low and high moisture contents influence the supply of

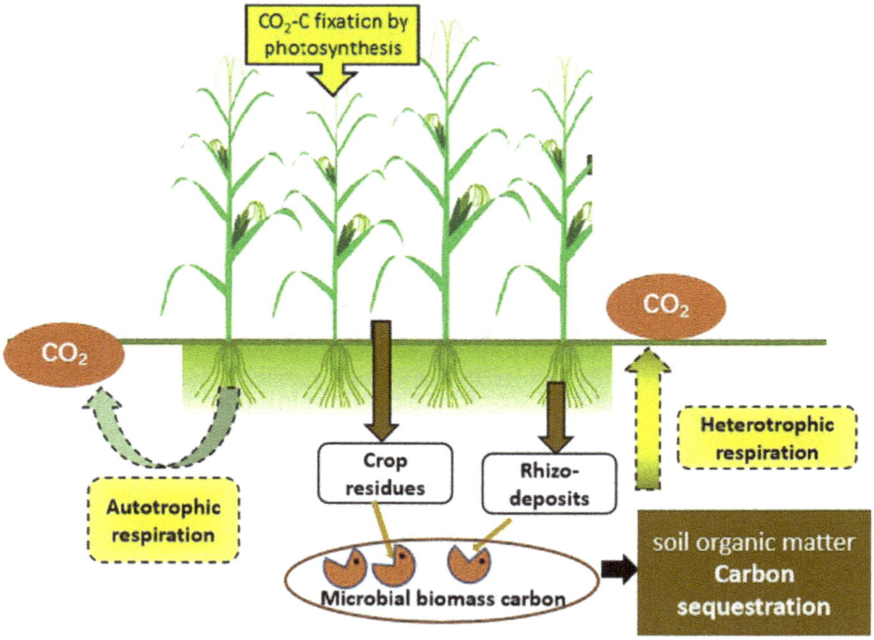

Fig. 2 A general sketch of atmospheric CO_2 cycling through an agricultural crop cultivation system

C source and oxygen (Falloon et al., 2011; Trugman et al., 2018). Agricultural soil surface mulching could promote the supply of organic matter to microbial communities and directly affect the soil temperature and moisture. Thus, it possesses the ability to alter CO_2 emission rates.

3.2 *Effects of Crop Straw Mulch on Soil CO_2 Emission*

Most field experiments reported higher CO_2 emission rates with residue mulch (Table 1) and attributed it to the supply of extra C to soil microbial communities by crop residue (Bhattacharyya et al., 2012; Dossou-Yovo et al., 2016). However, large variabilities in an absolute increase in CO_2 emission rates were associated with the decomposition rates of straw, governed by several factors related to the quality and quantity of input crop straw as well as soil properties (e.g., soil moisture, soil temperature, and soil nitrogen content) (Abro et al., 2011; Liu et al., 2014; Akhtar et al., 2020). For instance, Jacinthe and Lal (2003) recorded a positive linear relationship between annual CO_2 flux and residue-C input from a central Ohio Luvisol influenced by wheat residue management. Liu et al. (2019) reported that when two cultivars of maize were surface applied to black soil in northeast China, their decomposition rates differed, and soil CO_2 emission varied. In contrast, some studies reported lower soil

Table 1 A comprehensive review of field-based studies investigating the effect of residue mulch under different cultivation systems. The $\Delta CO_2\%$ represents the differences in C means the changes in CO_2 emission from soils with mulching compared to those without mulching [a, b] represents the two numbers in the two years (in order), and a ~ b represents the ranges conducted in more than two years

References	Location	Crop type	Year	Mulch type	Other treatments	$\Delta CO_2\%$
Akhtar et al. (2020)	34°20′N, 108°24′E Shaanxi, China	Maize-wheat	2015–2016	Maize straw	Straw amount: 4500, 9000 kg/ha Fertilizer: 0, 192, 240 kg N/ha Tillage: conventional tillage	1.62 ~ 13.6
Chen et al. (2017)			2013–2015	Wheat straw		[21.9, 15.9]
Tanveer et al. (2013)		Corn-wheat	2010–2012	Corn residue	Fertilizer: 0, 80, 160, 240, 320 kg N/ha Tillage: 4 tillage methods	−11.2 ~ 13.9
Wang et al. (2019)	34°12′N, 108°07′E Shaanxi, China	Wheat-soybean	2013–2014	Soybean straw + wheat straw	Straw amount: 700 + 3000 kg/ha; 1400 + 6000 kg/ha	0.33 ~ 7.44
Liu et al. (2010)	35.2°N, 107.8°E Shaanxi, China	Spring maize	2008	Corn straw		46.1
Jia et al. (2021)	35°75′N, 107°63′E Gansu, China	Maize	2017–2018	Maize straw		[30.4, 23.7]
Fan et al. (2019)	37°96′N, 102°64′E Gansu, China	Maize	2014	Straw	430 kg N/ha	−15.9 ~ 0.58
Ren et al. (2017)	36°10′9′N, 117°9203203″E Shandong China	Winter wheat	2013–2015	Maize straw	Conventional versus wide precision planting	−28.0 ~ −16.0
Liu et al. (2014)		Summer maize	2012	Wheat residue	3 planting density	−39.8 ~ −10.8

(continued)

Table 1 (continued)

References	Location	Crop type	Year	Mulch type	Other treatments	ΔCO_2%
Li et al. (2016)	34°31′N, 110°59′E Henan, China	Corn	2013	Corn straw	Tillage: convention versus no tillage 3 locations	−17.7 ~ 20.9
Zhang et al. (2014)	29°55′N, 115°30′ E Hubei, China	Oilseed rape-rice	2010	Oilseed rape residue	Straw amount: 3000, 4000, 6000 kg/ha; no-tillage	73.0 ~ 186
Liang et al. (2007)	47°26’N, 126°38’E Heilong Jiang, China	Rice	2003	Rice straw	No-tillage	28.9
Langeroodi et al. (2019)	36°53′N, 54°24′E Iran	Wheat-soybean	2014–2015	Wheat residue	Fertilizer: 0, 40, 80, 120 kgN/ha No tillage	9.52 ~ 10.1
Dossou-Yovo et al. (2016)	1°01’–1°14’E 10°42’–10°57’N Benin, Africa	Upland rice field	2014–2015	Rice straw	Fertilizer: 0, 60, 120 kg N/ha Tillage: no-till, manual tillage	−9.58 ~ 57.5
Nawaz et al. (2017)	40°00′N, 83°01′W Ohio, USA	No crop	2015	Wheat residue	No-tillage versus plough tillage	[45.0, −5.69]
Lenka and Lal (2013)			2011	Wheat residue	Straw amount: 8, 16 Mg/ha/y Fertilizer: 0, 244 kg N/ha/y No-till	7.39 ~ 49.6
Jacinthe et al. (2002)			2000			[49.0,76.9]

CO_2 flux from mulching treatments than non-mulched treatments and attributed it to a decrease in root respiration due to low N supply (Ren et al., 2017; Fan et al., 2019; Liu et al., 2014). In addition, others pointed out the significant effects of residue mulch on seasonal CO_2 fluxes through its heat-blocking effect. For instance, Jacinthe and Lal (2003) found that the emissions of CO_2 from the non-mulched plots responded quicker to temperature change, while, in the mulch treatments, CO_2 fluxes showed delays before reaching their late winter peaks and their lowest levels in autumn.

3.3 *Effects of Plastic Film Mulch on* CO_2 *Emissions*

Most studies reported increased soil respiration or CO_2 emission under plastic film mulching (Table 2). In a recent meta-analysis study, Mo et al. (2020a) integrated the effects of plastic mulching on soil CO_2 emissions in different cultivation systems of China's drylands. They pointed out that the induced CO_2 emissions by plastic film mulch were primarily due to a significant increase in plant net primary productivity (NPP). Specifically, they recorded 1.41 Mg C/ha/y more aboveground NPP and 437 kg C/ha/y more CO_2 emission rates under plastic mulching. The positive effect of plastic mulching on soil CO_2 emissions could also be related to the increase in soil temperature primarily driven by the greenhouse effect of transparent plastics. In a recent study, Liu et al. (2016a, 2016b) compared the effects of transparent and black plastic film's contrasting radiative properties in a spring maize field. Despite relatively low crop yield, they found significantly-high CO_2 emissions rates under transparent plastic film mulching than black plastic.

Plastic mulching is an integral part of the ridge and furrow mulching system (RFMS), where ridges are usually covered with plastic film among different cultivation approaches in dryland regions. Different ridge/furrow rows are commonly adopted among farmers depending on the crop type and regional precipitation patterns, and they showed different effects on CO_2 emissions. For example, Li et al. (2019a) found that in comparison with conventional flat planting systems, RFMS with ridge/furrow ratios of 40:70 cm, 55:55 cm, and 70:40 cm emitted 10.6%, 19.6%, and 20.4% more CO_2 respectively. However, differences in the range of ridge/furrows ratios were positively related to increased crop yield, suggesting a higher proportion of plastic mulched ridges could promote crop yield and CO_2 emission rates. Although most reports accounted for CO_2 emission rates from both ridges and furrows, a few exceptional cases only measured gas emissions from furrows (Liu et al., 2016a, 2016b), which may lead to underestimating of soil flux. We recommend that the field average CO_2 flux from the RFMS system be interpreted as the sum of ridge and furrow soil CO_2 emissions multiplied by their cover area ratios. The pathways for gas emissions in a RFMS may include ridges, furrows, and planting holes; therefore, a careful representative sampling is necessary (Ming et al., 2018).

Apart from crop productivity and soil temperature on soil CO_2 emission rates from RFMS, Ming et al. (2018) pointed out the significant role of precipitation intensity in arid regions. The plastic film reduces rainwater infiltration in ridges

Table 2 Review of field-based studies investigating the effect of plastic film mulching (PFM) on soil CO_2 emissions rates in different cropping systems. $\Delta CO_2\%$ means the percentage changes of CO_2 emission with mulching compared to without mulching. RFMS represents the ridge and furrow mulching system

References	Location	Crop	Year	Treatment	$\Delta CO_2\%$
Ming et al. (2018)	41°36'N, 86°12'E Xinjiang, China	Cotton	2014–2016	RFMS	Furrow: 2.88 ~ 31.4 Ridge: 8.08 ~ 57.5
Yu et al. (2016)	40°37'N, 80°45'E Xinjiang, China		2014–2015	PFM	Ridge: 3.2 Furrow: 21 Total: 8
Li et al. (2012)	44°17′N, 87°56′E Xinjiang, China		2010	Traditional cultivation system with no mulching versus PFM with drip irrigation	−99.1
Li et al. (2011)			2009	PFM	−23.9
Jia et al. (2021)	35°75′N, 107°63′E Gansu, China	Maize	2017–2018	conventional tillage + flat, ridge tillage + RFMS	18.3 ~ 61.7
Fan et al. (2019)	37°96′N, 102°64′E Gansu, China		2014–2015	PFM: common, biodegradable	0.8 ~ 59.2
Zhang et al. (2018)	36.03°N, 104.42°E Gansu, China		2011–2014	RFMS	6.83
Zhang et al. (2015)			2012	RFMS	27.6
Xu et al. (2020)	34°20′N, 108°24′E Shaanxi, China	Wheat-maize rotation	2016–2018	Irrigation: 150, 75, 37.5, 0 mm; Rainfall: 275, 200, 125 mm	5.93–33.7
Chen et al. (2018)			2013–2015	PFM: half and full	11.3 ~ 39.8
Luo et al. (2018)			2014–2015	PFM: half, full Irrigation: rainfed, irrigated	−12.3 ~ 40.1

(continued)

Table 2 (continued)

References	Location	Crop	Year	Treatment	$\Delta CO_2\%$
Liu et al. (2016a, 2016b)	35°12′N, 107°40′E Shaanxi, China	Maize	2013–2014	PFM: transparent and black Fertilizer: Urea, Control released fertilizer	Urea: 16.3 ~ 24.4 CRF: – 1.87–3.85
Liu et al. (2010)	35.2°N, 107.8°E Shaanxi, China	Maize	2008	PFM	67.9
Li et al. (2020a, 2020b)	34°32′N, 112°16′E Henan, China	Wheat	2015–2018	Conventional flat tillage RFMS	2.56 ~ 20.0
Chen et al. (2019)	30°26′N, 106°26′E Chongqing, China	Rice-rapeseed rotation	2015–2016	PFM	−0.13
Lee et al. (2019)	36°50′N, 128°26′E South Korea	Maize	2014–2015	Fertilizer: chemical fertilizer, organic fertilizer	Fallow: −3.33 ~ 8.91 Growing: 34.6 ~ 57.9

inhibiting moisture-induced CO_2 emissions. Similarly, plastic mulch may promote or reduce the coupling effect of agricultural management practices, including irrigation, fertilization, and tillage, on soil microbial activities and CO_2 emission rates.

4 Effects of Mulching on Soil CH_4 Emission or Uptake

4.1 The Mechanism Behind the Soil CH_4 Emissions or Uptake

In general, CH_4 emitted from the soil surface of any ecosystem represents the net results of three biophysical steps, including CH_4 production, oxidation, and transportation (Fig. 2). Methane is usually produced under highly reduced conditions without O_2 and other electron acceptors by a group of microbial communities commonly known as methanogens (Schulz & Conrad, 1996). A large proportion of CH_4 produced is immediately consumed by oxic and anoxic microbial communities known as methanotrophs, leading to a significant decrease in CH_4 concentrations within the soil matrix before leaving the soil (Le Mer & Roger, 2001). During the exchange of CH_4 between soil and atmosphere, the vertical transportation of CH_4 to the atmosphere occurs mainly through diffusion between soil-atmosphere interfaces for dryland systems. In contrast, transportation includes two significant pathways for a rice paddy, i.e., loss through ebullition (the release of gas bubbles) and release from plants (Hussain et al., 2015; Le Mer & Roger, 2001). Overall, CH_4 emission rates depend on the balance between methanogenic and methanotrophic activities and the transportation of CH_4 from the soil matrix to the atmosphere. Which is closely related to soil and environmental factors such as soil C composition and quality, soil temperature and moisture, microbial activity, soil N availability, and soil gas diffusivity (Le Mer & Roger, 2001).

As agricultural soil surface mulching can affect CH_4 production and oxidation rates, the net effect may vary depending on the mulch type and agricultural management systems. Since there is a remarkable distinction in source capacity of CH_4 emission between drylands (mainly as a sink) and rice fields (prominent as a source), we will separately discuss the effects of mulching on soil CH_4 emission rates under two agricultural systems.

4.2 Mulching Effects on CH_4 Emission in Dryland Cropping Systems

Dryland ecosystems act as a sink for atmospheric CH_4, with occasional CH_4 emissions triggered by heavy rainfall that promote methanogenic activities. Therefore, agricultural soil surface mulching generally depends on the interaction of mulching material with heavy rainfall events under dryland cropping systems. Crop residues could promote CH_4 production rates through enhanced substrate supply to heterotrophic methanogenic communities. Residue mulch directly affects soil moisture and reduces evaporation rates. Such effects have been commonly observed in field studies under humid environments and are the primary reason for discouraging the use of residue mulching. Waterlogged conditions promoted by residue mulch after heavy precipitation often inhibit seedling emergence at the initial stages of crop plants. Although CH_4 emissions in a dry climate are strictly linked with precipitation events, the positive impact of residue mulching can significantly induce the intensity of CH_4 flux rates. Nevertheless, some studies also have reported a significant decrease in CH_4 emission rates under residue mulching, indicating complexity in the role of residue mulch in driving CH_4 emission (Chen et al., 2017).

In contrast, plastic mulch reduces rain water infiltration and promotes soil physical structure, thus may promote atmospheric CH_4 uptake in dry cropping systems (Table 3). However, there is ambiguity regarding the gaseous permeability of plastic film used as surface mulch. Some studies observed low GHG emission rates in plastic mulching plots and argued that the low GHGs emission rates were associated with the

Table 3 Mulching effect on soil CH_4 and N_2O emission on dryland cropping systems. ΔCH_4% and ΔN_2O % mean the percentage changes of CH_4 and N_2O emission with mulching compared to without mulching. PFM: plastic film mulching; RFMS: represent ridge and furrow mulching system. $\Delta(CH_4, N_2O)$ % means the percentage changes of CH_4/N_2O emission compared with non-mulching treatment. [a, b] denotes the value in the two growing seasons in order; a ~ b denotes the values in a range

Author	Location	Crop	Time period	Mulch type	Other treatments	ΔCH_4%	ΔN_2O%
Wang et al. (2021)	38°23′N, 109°43′E Shaanxi, China	Potato	2019	Polyethylene mulch, transparent and black degradable	RFMS: drip irrigation		−29.2 ~ −11.2
Jia et al. (2021)	35°75′N, 107°63′E Gansu, China	Maize	2017–2018	Straw	Flat croppying system + conventional tillage	[−37.5, 7.5]	[3.92, 4.65]
				PFM		[−75, 15]	[7.84, 9.3]
				PFM	RFMS + ridge tillage	[94.4, 3.85]	[4.46, 9.76]
				RFMS + straw furrow		[189, 7.69]	[8.93, 19.5]
Jia et al. (2021)				PFM	Fertilizer: 0, 120, 240 kg N/ha Cropping system: RFFM versus flatting	132 ~ 192	−5.01 ~ 11.6
Meng et al. (2020)	N37°52′, E102°50′ Gansu, China	Potato	2016–2018	Black PFM	Fertilizer: chemical NPK, cattle manure, combined	8.33 ~ 40	
Akhtar et al. (2020)	34°20′N, 108°24′E Shaanxi, China	Maize-wheat	2015–2017	Maize straw	Straw amount: 4500, 9000 kg/ha Fertilizer: 0, 192, 240 kg N/ha		12.5 ~ 100
Xu et al. (2020)			2016–2018	PFM	Irrigation: 150, 75, 37.5 mm Rainfall 275, 200,125 mm	−17.4 ~ 171	−52.3 ~ 13.5

(continued)

Table 3 (continued)

Author	Location	Crop	Time period	Mulch type	Other treatments	ΔCH_4%	ΔN_2O%
Luo et al. (2018)			2014–2015	PFM	PFM: half and full Irrigation: rainfed versus irrigated	6.44 ~ 76.9	−53.3 ~ −9.34
Chen et al. (2017)			2013–2015	Wheat straw		[−733, −67.4]	[−32.0, 12.2]
				PFM	PFM: half and full	−244 ~ 45.8	−23.1 ~ 16.7
Ye et al. (2020)	41°49′22″N, 123°33′55″E Liaoning, China	Tomato	2015	PFM	Irrigation: mulched drip irrigation versus conventional irrigation		−5.76
Wu et al. (2018)	28°55′N, 111°27′E Hunan, China	Maize	2015	Rice straw	Amount: 5000, 10,000 kg/ha		[−24.1, −34.4]
Pinheiro et al. (2019)	29°42′54″S, 53°42′23″W, Brazil		2015–2016	sugarcane straw	Amount: 4, 8, 12 Mg/ha		11.8 ~ 107
Schmatz et al. (2020)		Winter-summer crops	2016	Vetch straw Wheat straw	Amount: 3, 6, 9 Mg/ha		145 ~ 767
Lee et al. (2019)	36°50′N, 128°26′E Republic of Korea	Maize	2014–2015	PFM	Fertilizer: chemical NPK versus organic fertilization	130 ~ 260	5 ~ 10
Kim et al. (2010)			2013–2014	PFM	4 mixing ratios of cover crop residues		15.8 ~ 30.7

(continued)

Table 3 (continued)

Author	Location	Crop	Time period	Mulch type	Other treatments	$\Delta CH_4\%$	$\Delta N_2O\%$
Cuello et al. (2015)	35°146′N, 128°096′E South Korea	Maize	2012–2013	Black PFM	Fertilizer: NPK versus two green manure	50.3 ~ 200	12.5 ~ 65.6
Berger et al. (2013)	38°16′26.211″N, 128°8′45.354″E South Korea	Soybean	2011	Ridges covered with impervious black PFM			50
Yu et al. (2018)	40°37′N, 80°45′E, China	Cotton	2014–2015	PFM			[−28.5, −19.0]
Li et al. (2014)	87°56′E, 44°17′N Xinjiang, China	Cotton	2009–2010	PFM	Irrigation: drip irrigation versus tradition mulch-free flood-irrigated	[150, 120]	[−40, −28]
Nawaz et al. (2017)	40°00′N, 83°01′W Ohio, USA	No crop	2015	wheat straw	Tillage: no-tillage versus plough tillage	[400, −12.7]	[470, 162]
Lenka and Lal (2013)			2011		Mulch amount: 8, 16 Mg/ha/yr	104 ~ 213	12.1 ~ 59.9
Jacinthe and Lal (2003)			2000		Fertilizer: 0, 244 kg N/ha/yr, no tillage	−4.78 ~ 132	~ 128

impermeability of plastic film. At the same time, others disagreed with it and considered plastic films to be permeable for gaseous exchange. A recent meta-analysis study suggested that plastic film mulch induces atmospheric CH_4 oxidation approximately by 0.25 kg C/ha/y in agricultural systems of China (Mo et al., 2020a, 2020b). However, their conclusions were based on a meta-analysis of data from only three published reports in the same area, thus demanding further investigations.

4.3 Mulching Effects on CH_4 Emission in Rice-Based Systems

Rice fields significantly contribute to atmospheric CH_4, accounting for about 10% and 1.5% of the total anthropogenic CH_4 budget and global anthropogenic GHG emissions (IPCC, 2013). Simultaneously, paddy cultivation is also the major consumer of water among different cultivation methods. Several water management approaches are replacing traditional paddy systems, including alternate wetting and drying, mid-season drainage, unsaturated soil (upland) rice farming, and saturated soils condition without standing water using ground covering (GCRP) (IRRI, 2009; Kreye et al., 2007). The efficiency of these innovative rice production systems is intensively studied for their potential to save water and improve rice yield. The water-saving cultivation approaches can potentially reduce CH_4 emission rates by inhibiting methanogenic activities and inducing methanotrophy.

The use of soil surface mulching material is an integral part of an innovative water-saving GCRP system, thus demanding a separate investigation regarding its effect on GHG emissions from rice cultivation. Under the GCRP system, the soil is no longer submerged but kept near saturation (70–90% of water holding capacity) over the rice-growing season. The soil surface is covered with rice straw or plastic film to reduce water evaporation. Due to the distinguished characteristics of residue and plastic film mulch, the GCRP system may exhibit contrasting effects on CH_4 emission rates regarding traditional paddy cultivation.

Adopting the GCRP system and crop straws as mulching materials may not mitigate CH_4 emissions as residues introduce a labile carbon pool that could trigger methanogenic activities (Ma et al., 2008, 2009; Xu et al., 2000; Zhang et al., 2020). However, the intensity of induced methanogenesis may depend on various factors, including quality and quantity (different C:N ratio), the application time of the crop straws, and the application method (e.g., incorporation, mulching, compost, and burning in situ) (Sanchis et al., 2012)., In this regard, literature on the use of residue in traditional paddy systems pointed out that CH_4 emission rates may positively respond to increasing rates of residue application following either linear or quadratic fashion (Wang et al., 1992; Watanabe et al., 2005). In contrast, a negative relationship between CH_4 emission rates and the quantity of residue mulch has also been reported. For instance, Zhang et al. (2014) recorded a decrease in cumulative CH_4 emissions by 34, 52, and 75% at straw mulch rates of 3, 4 and 6 Mt/ha with non-mulching treatment. Additionally, several studies pointed out that the straw application time and method may affect CH_4 emission rates (Watanabe et al., 1995; Yan et al., 2005; Zou et al., 2005). The use of residue as surface mulch during the winter fallow season and later incorporation is considered promising for CH_4 emission mitigation (Lu et al., 2000; Xu et al., 2000). When the surface soil was drier than puddled soil, much less CH_4 was emitted from the residue surface mulching than residue incorporation (Singh et al., 2008).

In comparing the GCRP system with straw mulching, plastic film mulching does not introduce organic carbon or nutrients to the rice fields. It mainly affects CH_4 emission physically by changing soil temperature and soil moisture. Therefore, GCRP systems with plastic film mulching usually release less CH_4 than those with straw mulching (Chen et al., 2021; Liu et al., 2021). Like CO_2 emission, modifying tillage permutations and fertilization in the GCRP system, besides water management, also affects CH_4 emissions (Hussain et al., 2015).

5 Effects of Mulching on N_2O Emission

5.1 Soil N_2O Emission in the Agriculture System

Agricultural soils are the primary source of N_2O emissions, contributing about 87.2% of N_2O emissions, mainly through the intensive use of nitrogenous synthetic and organic fertilizers (e.g., urea and manure) (Shcherbak et al., 2014). Unlike CH_4, N_2O is produced in soil via several pathways driven by distinct microbial communities. Among different pathways, nitrification and denitrification are the most critical processes due to the ubiquitous nature of microbial communities that carry out these processes and their potential to produce a large quantity of N_2O (Butterbach-Bahl et al., 2013). The nitrification process occurs under aerobic conditions by autotrophic microbial communities, known as nitrifiers. Nitrifiers oxidize NH_{4+} into NO_{2}- and NO_{3}- and release N_2O as a byproduct. Whereas denitrification is adopted by heterotrophic anaerobic microbes that sequentially reduce the oxide of mineral N (NO_3-, NO_2-, NO, N_2O) into N_2 in the absence of O_2 (Kuypers et al., 2018). Denitrification is not only the source of N_2O but also acts as a natural sink depending on the conditions suitable for N_2O reducing communities and microbial community composition. In general, continuously reduced conditions favour N_2 as the end product of the denitrification process. Otherwise, N_2O acts as the final product and is released into the atmosphere. It is commonly accepted that the nitrification process is dominant process under upland soil conditions, providing substrates for denitrification to release N_2O or N_2 once conditions are achieved. This combo effect is considered the primary cause behind large spatial and temporal variability in N_2O emissions from soil, making N_2O measurements challenging. Some researchers used terms like "hot spot" and "hot moment" for significant soil N_2O emission events that

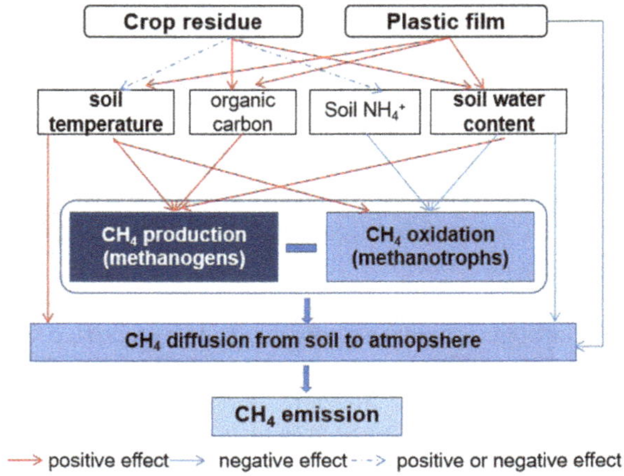

Fig. 3 A conceptual sketch of potential mechanisms underlying the effects of agricultural soil surface mulching on CH_4 emissions

are commonly observed immediately after irrigation or rainfall events in arable soils (Flessa et al., 1995).

The role of agricultural soil surface mulching on soil N_2O emission rates is complex due to its effect on a range of soil properties, including soil temperature, soil moisture, and substrate supply that can either induce or reduce the nitrifiers and denitrifier activities (Kim et al., 2010; Knorr et al., 2005) (Fig. 3). Several mechanisms have been proposed to explain the mulching effects on soil N_2O emission. For example, mulching may accelerate nitrification rates, leading to an increase in N_2O emissions, triggering the organic matter mineralization via high temperature and supply of labile carbon pool (Berger et al., 2013; Nan et al., 2016; Yu et al., 2017). Similarly, soil moisture conservation promotes the formation of microsites suitable for denitrifying microbial communities. In contrast, promoting nitrogen use efficiency in crop plants via mulch could significantly reduce the supply of mineral N–N_2O producing microbial communities (Liu et al., 2014). Surface mulching improves soil aggregation and structure, leading to high aeration that reduces denitrification. Additionally, organic mulching with a high C:N ratio would enhance the immobilization of fertilizer N with straw, resulting in less NH_{4+} available for nitrification-denitrification. Some researchers also pointed out that the physical impermeability of mulching material may also reduce the N_2O exchange between the soil surface and atmosphere (Yang et al., 2015; Yu et al., 2018).

5.2 *Mulching Effects on Soil N_2O Emission in Dryland Cultivation Systems*

In dryland systems, particularly in semi-arid and arid regions, soil N_2O emissions are driven by fertilizer applications, heavy rainfall events and/or water management practices (Berger et al., 2013). A thorough review of the literature investigating the effect of mulching on N_2O emission rates suggests that mulching may induce or reduce or have no effect on N_2O emissions depending on the type of mulching materials and water management. Plastic mulch generally reduces the triggering effects of heavy rainfall events by controlling the infiltration of rainfall water. Mo et al. (2020b) concluded that the mitigation effect of plastic mulching on N_2O emissions from arid and semi-arid cropping regions was as small as 7%. Many studies did not find a mitigating effect of plastic mulch on N fertilizer-induced N_2O emission rates (Tables 3 and 4) (Fig. 4).

Although the role of organic mulch on soil N_2O emissions rates is rather complex, a relatively large number of studies agreed that organic mulch increases fertilizer-induced N_2O emissions by favouring the denitrifying communities. With high moisture contents, organic mulch materials promote the formation of suitable micro-sites and provide labile organic substrates for the activities of denitrifying communities (Garcia-Ruiz & Baggs, 2007). Additionally, the quantity and quality of the organic mulch affect the N_2O production process via variable decomposition rates (Kim et al., 2010). Crop residues, upon degradation, act as a source of the substrate of the N_2O-producing process; however, the relative proportion of residue N acting as a substrate is generally below 2%, depending on the quality of the residue. In this regard, Garcia-Ruiz and Baggs (2007) reported 0.9%, 2.5%, and 0.01% of N contents of rye, wheat and organic olive crop weed residues, released as N_2O, respectively, depending on their different C:N ratios. Regarding the effect of the quantity of residue as surface mulch, Pinheiro et al. (2019) recorded a quasi-linear relationship between N_2O emissions and the amount of straw on the soil surface.

5.3 *Mulching Effects on Soil N_2O Emission in Rice-Based Systems*

In the traditional waterlogged rice paddy, soil N_2O emissions are generally negligible as consistent flooding conditions halt the nitrification process and reduce substrate supply to denitrifying microbial communities. Moreover, consistently reduced conditions promote the complete sequential reduction of NO_3- into N_2, reducing the chances of N_2O release as a final product of denitrification pathways (Kreye et al., 2007; Xu et al., 2004). Approximately 60% of the total N_2O emission from traditional paddy cultivation systems occurs during the fallow winter when the fields are under drainage (Xing, 1998). However, a significant increase in N_2O emission during rice growing has been commonly observed under water-saving approaches, as drainage generally promotes N_2O production (Kreye et al., 2007). Several studies reported that alternative irrigation practices during rice production effectively reduced CH_4 emissions but exacerbated N_2O emissions, thus may not mitigate net GHG emissions (Fawibe et al., 2019; Ma et al., 2009; Yao et al., 2014).

Unlike in dryland cropping systems, it is hard to discuss the mulching effect alone because mulching is often combined with specific water regimes or irrigation

Table 4 Literature review of studies investigating the effect of surface mulching in rice-based cultivation systems on CH_4 and N_2O emission rates. $\Delta CH_4\%$ and ΔN_2O % represent the percentage difference in CH_4, and N_2O emissions, respectively, between treatments with and without surface mulching. PFM: plastic film mulching. [a, b] denotes the value in the two growing seasons in order; a ~ b denotes the values in a range

Author	Location	Ecosystem	Time period	Mulch type	Irrigation	Growing stage	$\Delta CH_4\%$	$\Delta N_2O\%$
Liu et al. (2021)	31.13°N, 104.37°E Sichuan, China	Rice-winter wheat	2015–2016	PFM	Initial flooding and intermittent irrigation	Growing season	[−86.6, −19.2]	[41.2, 26.7]
				Wheat straw	Initial flooding and intermittent irrigation		[−78.2, 14.7]	[21.6, 24.4]
Zhang et al. (2020)	30°05′N, 104°34′E Ziyang, Sichuan	Winter flooded paddy fields	2013–2016	PFM	GPRS-waterlogged in ditches surrounding the raised beds of rice growth VS rainfed (TC1 treatment)	Winter season	−84.1 ~ −57.8	75 ~ 580
						Rice season	−11.6 ~ 32.0	−2.33 ~ 155
						Annual	−35.0 ~ −5.8	10 ~ 152
					GPRS-waterlogged in ditches surrounding the raised beds of rice growth VS traditional irrigation (TC2 treatment)	Winter season	−84.7 ~ −58.8	250 ~ 750
						Rice season	−69.9 ~ −60.2	160 ~ 965
						Annual	−70.1 ~ −60.3	194 ~ 908
Zhang et al. (2018)	30°05′N, 104°34′E Sichuan, China	Rice	2010–2014	PFM	GPRS-waterlogged in ditches surrounding the raised beds of rice growth vs. rainfed	Growing season	−20.5 ~ 8.72	2.44 ~ 144

(continued)

Table 4 (continued)

Author	Location	Ecosystem	Time period	Mulch type	Irrigation	Growing stage	$\Delta CH_4\%$	$\Delta N_2O\%$
					GPRS versus traditional irrigation Fertilizer: urea or controlled-release fertilizer, urease and nitrification inhibitors		−85.5 ~ −46.3	206 ~ 1153
Chen et al. (2019)	E106°26′, N30°26′ Chongqing, China	Rice-rapeseed rotation	2015–2016	PFM		All the year	148	−18.2
						Rice season	79.4	39.4
						Fallow season	567	119
						Rapeseed season	147	−21.9
Chen et al. (2021)	33°27′N, 120°11′E Jiangsu, China	Rice-barley rotation	2015–2016	Rice straw	GPRS-waterlogged in ditches surrounding the raised beds of rice growth vs. conventional waterlogged	Growing season	[−27.4, −23.1]	
				PFM			[−40.1, −60.8]	
Ma et al. (2009)	31°17′N, 119°54′E Jiangsu, China	Rice-winter wheat	2004–2006	Wheat straw incorporation	Intermittent irrigation pattern Mulching amount: 3.75, 4.8 t/ha	Growing season	394 ~ 948	−77.8 ~ −0.91
			2004–2006	Wheat straw burying			[307, 336]	[−7.08, −30.3]
			2004–2006	Wheat straw ditch mulching			[204, 610]	[−42.6, −18.7]

(continued)

Table 4 (continued)

Author	Location	Ecosystem	Time period	Mulch type	Irrigation	Growing stage	$\Delta CH_4\%$	$\Delta N_2O\%$
			2006	Wheat straw strip mulching	Intermittent irrigation pattern Mulching amount: 4.8 t/ha		168	13.9
Xu et al. (2004)	33°27′N, 120°11′E Jiangsu, China	Lowland rice field	2001	PFM	Mulching with unsaturated soil (direct seeding) vs. puddled waterlogged soil (traditional transplanting)	Growing season	−87.3	486
				Rice straw			−94.6	416
Wang et al. (2020a, 2020b)	29°51′N, 115°33′E Hubei, China	Rice-oilseed rape	2018–2019	Oilseed rape straw mulch	Rice fields were flooded at 3–5 cm water levels except for the drainage about 1–2 weeks before the tillering and harvest stages	Growing season	[21.9, 20.6]	[1.44, 4.72]
				Straw incorporated			[40.6, 34.3]	[0, 4.72]
Liu et al. (2016a, 2016b)	30°05′N, 113° 45′ E Hubei, China	Mono-rice	2010–2013	Rice straw mulching	During the winter fallow season, the fields were naturally drained. Before land preparation (at the end of April), the rainwater was conserved in the field. During the rice-growing season, the water regime was under the flooding-mid-season drainage-reflooding-moist by intermittent irrigation but without waterlogged conditions (F-D-F-M)	Land preparation period before rice transplanting	[36.4, 496]	[10.7,183]
						Growing season	[186, 448]	[22.2, 41.8]
						Winter fallow	[713,103]	[104, 60]
						Total	[153,354]	[39.5, 69.8]

(continued)

Table 4 (continued)

Author	Location	Ecosystem	Time period	Mulch type	Irrigation	Growing stage	ΔCH_4%	ΔN_2O%
Zhang et al. (2014)	29°55′ N, 115°30′ E Hubei, China	Oilseed rape-rice	2010	Oilseed rape residue mulching	Conventional irrigation-drainage practices Mulching amount: 3000, 4000, 6000 kg/ha	Growing season	−74.6 ~ −34.5	44.5 ~ 128
Yao et al. (2017)	32°07′13″N, 110°43′04″E Hubei China	Rice	2012–2014	PFM and degradable PFM	Conventional, GCRP-saturated: soil water content was held nearly saturated, GCRP-low: with near saturation until the rice-regreening stage and at approximately 80% of the GCRPSsat management for the remainder of the season Fertilizer: 0, 150 kg N/ha	Growing season	−90.7 ~ −57.2	−42.1 ~ 217
						Fallow	−69.1 ~ 18.6	−69 ~ 71.4
						Annual	−91.2 ~ −57.7	−50.5 ~ 92.7
Wu et al. (2019)	28°55′N, 111°27′E Hunan, China	Rice field	2015	Straw incorporated	Continuous flooding (CF), flooding—mid-season drying—flooding (FDF), flooding for transplanting—rainfed (RF)	Growing season	51.7 ~ 310	
				Straw mulching	Continuous flooding (CF)		25 ~ 46.5	

(continued)

Table 4 (continued)

Author	Location	Ecosystem	Time period	Mulch type	Irrigation	Growing stage	$\Delta CH_4\%$	$\Delta N_2O\%$
Kreye et al. (2007)	Beijing, China	Lowland single rice cropping	2001–2002	PFM	GCRP: dry land preparation and direct dry seedling, aerobic, irrigation stop in mid-September; Traditional: transplanting, flooded-drainage-irrigation stop six weeks before harvest	Growing season	[−96.5, −103]	[1685, 50]
				Rice straw mulch			[−127, −100]	[1439, 18.9]
Liang et al. (2007)	47°26'N, 126°38' E Heilong Jiang, China	Single rice cropping	2003	Rice straw mulch	Flooded 40–50 days before transplanting to the tillering stage. Depending on the weather, a drainage period lasting 7–10 days was practiced, and then the fields were re-flooded until about one month before harvesting. During the fallow season, the fields wee drained	Fallow season	272	30.8
				Rice straw incorporation		Fallow season	38.8	91.2
Fawibe et al. (2019)	35°42.7′N, 140°2.6′E Japan	Rice	2016–2017	PFM	Drip irrigation with plastic-film mulch system DP vs. continuous flooded rice cultivation system (CF)	Growing season	[−194, −93.6]	[−97.3, −20.5]

Fig. 4 A conceptual sketch to describe the role of agricultural soil surface mulching in regulating N_2O production and emissions

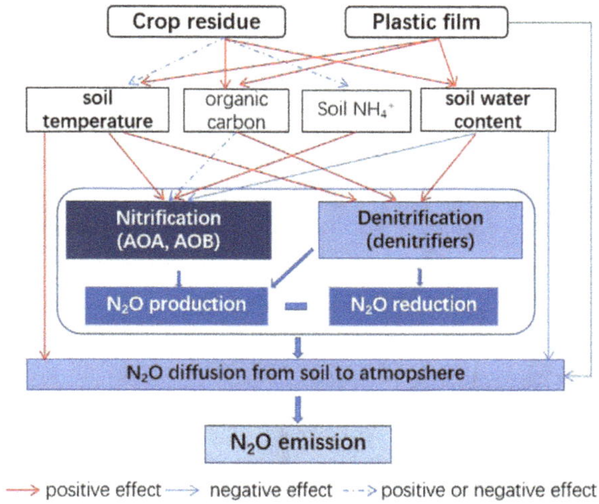

patterns in the rice paddy. The comparison is thus mostly made between water-saving systems with mulching and conventional irrigation systems without mulching, with two variables involved. In general, rice fields under the GCRP system with mulching promote soil N_2O emission compared to the traditional rice cropping systems (Table 4), mainly because the soil is no longer submerged but kept near saturation (70–90% of water holding capacity) (Yao et al., 2017). Additionally, the type of crop straw rather than the timing and application method is more critical in determining N_2O emission that crop residues with a high C:N ratio (e.g., wheat straw) usually decreased N_2O emission through N immobilization (Singh et al., 2008; Ma et al., 2009), while a low C:N residue (e.g., rapeseed cake) increased N_2O emission (Zou et al., 2005; Wang et al., 2020a, 2020b).

6 Combined Effects of Agricultural Soil Surface Mulching with Other Agricultural Management on GHGs Emissions

Agricultural management is a systematic package, and mulching involves covering the soil surface with different materials to conserve soil moisture and minimize temperature fluctuation. Mulching will inevitably be combined with other agronomic practices, such as optimizing fertilizer and irrigation application, proper residue incorporation, and reduced and minimum tillage. All of those practices influence on GHGs emissions, and after combing with mulching, their interaction effects on GHGs emission would be intensified, weakened, or overturned. The following discussion will focus on the combined effects of mulching with tillage, fertilization, and irrigation on GHGs emissions (Fig. 5).

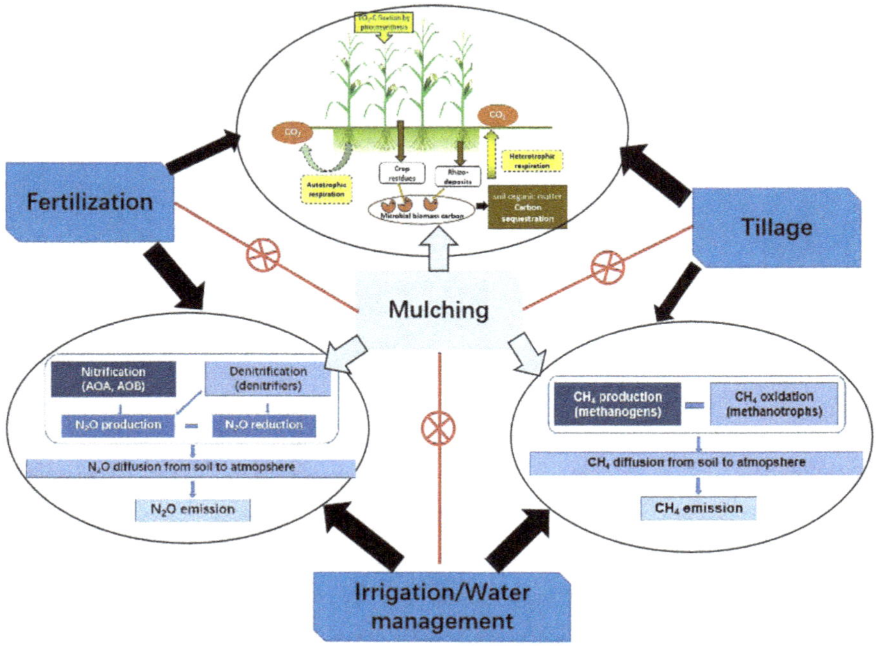

Fig. 5 Combination effects of mulching and other agricultural management on soil greenhouse gases (CO_2, CH_4, and N_2O) emissions. The thicker the arrow, the greater the effect

6.1 Mulching with Tillage

Mulching is part of residue management and conservation agriculture, aiming to conserve water and reduce soil erosion by tillage (Kassam et al., 2015). Tillage affects soil aggregation directly by physical disruption of the macro-aggregates and indirectly by altering of biochemical factors (Barto et al., 2010). It is recommended to adopt residue mulching before shifting conservation tillage (e.g., no-tillage and ridge tillage) to enhance soil fertility and increase crop yield. Conservation tillage could substantially reduce soil CO_2 and CH_4 emissions by improving soil aggregation, rainwater infiltration, and root growth (Huang et al., 2019; Yagioka et al., 2015). However, the effects on N_2O emission remain controversial (Van Kessel et al., 2013; Zhao et al., 2016). As previously mentioned, most studies reported increased CO_2 emission when applying either crop straw or plastic film mulching (Table 1). Incorporating conservation tillage into mulching management would help reduce CO_2 emissions (Tanveer et al., 2013; Zhang et al., 2015; Dossou-Yovo et al., 2016; Nawaz et al., 2017; Abbas et al., 2020). However, their combining effects on CH_4 and N_2O emission would vary with cropping systems (e.g., dryland vs. rice paddies). Regardless, it is still reasonable to go for conservation tillage under mulching because of more benefits for soil carbon sequestration than the long-term adverse effects on GHGs emissions.

6.2 Mulching with Fertilization

Fertilization is necessary to improve crop growth but can increase GHG emissions (Al-Kaisi et al., 2008; Zhang et al., 2013). N fertilizer application is widely accepted to substantially increases N_2O emission (Shcherbak et al., 2014), which generally outweighs the mulching effect on N_2O emission. On the other hand, it is generally believed that N fertilizer has a complicated influence on soil CH_4 oxidation (Bodelier & Laanbroek, 2004). Some research indicated that N fertilization reduces CH_4 oxidation due to the interference of NH_4^+ with CH_4 oxidation, especially at high rates (Chen et al., 2021; Lenka & Lal, 2013). In contrast, others reported positive effects of N fertilizer application on CH_4 uptake (Bodelier & Laanbroek, 2004; Liu et al., 2014). Given that mulching could affect soil conditions (e.g., water content) and N cycling in the soil-plant system, N fertilization on CH_4 uptake may become more complicated under mulching conditions.

Specifically, crop straw can release nutrients via decomposition when returned to the soils. Thus, adding crop residues potentially reduces the additional fertilization demand and, therefore, reduces GHGs emissions associated with chemical fertilizers (Akhtar et al., 2020). However, most studies comparing GHGs emissions from agricultural systems with and without surface mulching used the same amount of chemical fertilizer. Only a few reports applied different rates of N fertilizers together with different amounts of mulching materials to test the hypothesis, trying to find an optimum combination to balance the GHGs emission and grain yields (Ma et al., 2009; Lenka & Lal, 2013; Dossou-Yovo et al., 2016; Wang et al., 2018; Langeroodi et al., 2019; Li et al., 2019a, 2019b; Akhtar et al., 2020). Some others also hypothesized that different types of fertilizers (e.g., organic fertilizer, inorganic fertilizer, and control releasing fertilizer) affect the mulching effects on GHGs emissions (Rabenarivo et al., 2014; Zhang et al., 2018; Kyulavski et al., 2019; Lee et al., 2019). For instance, Yao et al. (2013) compared CH_4 and N_2O fluxes and heterotrophic soil respiration (CO_2 emission) between six fertilizer treatments for conventional paddy and GCRP systems. They concluded that using organic fertilizers for the GCRP system considerably reduced annual emissions of CH_4 and N_2O and increased soil carbon sequestration. Accordingly, a water-saving GCRP system with organic fertilizer amendments was considered the most promising management regime for simultaneously achieving relatively high grain yield and reducing greenhouse gas emissions simultaneously.

6.3 Mulching with Irrigation

Besides fertilizer, irrigation in agriculture is another key to plant growth and GHGs emission, especially in semi-arid and arid areas. Moreover, water requirements and irrigation modes vary with crop types and climate zones. Soil CO_2 emissions in agricultural systems in the arid region of China are approximately 2–5 times greater than in natural ecosystems (Lal et al., 2013) because irrigation overcomes the limitations imposed by drought and fertilizer application, increases the growth of crops and stimulates soil respiration rates (Li et al., 2014, 2018). Therefore, agricultural water-saving irrigation technology has become more critical, which has essential and strategic significance for reducing greenhouse effects (Ali et al., 2013; Chai et al., 2011; Ye et al., 2020).

In the dryland system, Xu et al. (2020) demonstrated that RFMS remarkably reduced GHG emissions regardless of rainfall compared to traditional flat planting under the same amount of irrigation. While the absorption of CH_4 in the winter wheat field increased as the supplementary irrigation decreased, indicating the predominant role of irrigation in deciding the GHG emissions. Li et al. (2011) showed that mulched drip irrigation reduced soil respiration by 25% compared to non-mulched furrow irrigation in an oasis cotton field. However, Zong et al. (2020) reported that drip irrigation with plastic mulching significantly increased soil CO_2 emissions compared to drip irrigation with no mulch in oasis cotton fields in Xinjiang, China. The difference is likely mainly driven by different irrigation in non-mulched sites. Li et al. (2011) compared mulched drip irrigation with non-mulched furrow irrigation, while Zong et al. (2020) compared mulched and mulched non-mulched under the same drip irrigation.

For rice paddy, the share of rice in CO_2 emission is less than that of CH_4 and N_2O emission, as the anaerobic condition in flooded soil is not conducive for C oxidation (Hussain et al., 2015). We highlight the importance of water regimes in controlling soil N_2O and CH_4 emissions. The issue is that most of the time, innovative irrigation significantly reduces CH_4 emissions at the expense of increasing N_2O emissions (Yadvinder-Singh et al., 2008). The incorporation of mulch into the rice paddy under water-saving irrigation (e.g., GCRP system) has an unclear effect on N_2O depending on different water regimes (Table 4). Zou et al. (2005) suggested that water management by flooding with mid-season drainage and frequent waterlogging without organic amendments (crop straw incorporation or surface mulching) is an effective option for mitigating the combined climatic impacts from CH_4 and N_2O in paddy rice production. When crop straw mulching is replaced by plastic film mulch, a better N fertilizer management such as applying controlled-release fertilizer and organic fertilizer or the addition of nitrification inhibitors (e.g., nitrapyrin) to N fertilizers would be the possible solution for reducing N_2O fluxes and CH_4 in rice paddy (Zhang et al., 2018).

Overall, the mulching effect on GHG emissions is not that remarkable and predominant when combining with tillage, fertilization and irrigation practices. However, mulching still needs to be applied, considering other benefits of mulching to the agricultural systems, e.g., weed control, water conservation, reduced soil erosion, and enhanced crop yields.

7 Conclusion and Future Perspectives

This chapter focused on the two primary mulching materials, i.e., crop residues/straw and plastic film, to investigate their effects on the three major GHGs emissions in different cropping systems. Overall, agricultural soil surface mulching may trigger, reduce, or show no effect on atmospheric GHGs emissions depending on mulching, cultivation systems, and agronomic management practices. Therefore, a holistic approach is needed to evaluate the role of agricultural soil surface mulching in achieving sustainable crop yield with minimum detrimental environmental effects during the realization of the carbon-neutral concept. Specifically, we have the following highlights for future research:

(1) Based on our review, it is clear that different GHGs could respond differently to the same mulching systems, e.g., mulching could substantially decrease CH_4 emission from rice fields but at the cost of increasing N_2O emission. In this regard, some integrative parameters may work better, such as global warming potential (GWP) calculated as CO_2 equivalents (CO_2-eq) over a 100-year time zone using the radioactive forcing potential of 34 for CH_4 and 298 for N_2O (IPCC, 2013). While greenhouse gas emission intensity (GHGI) is calculated as the GWP divided by grain yield and net ecosystem economic budget (NEEB) is the balance between economic benefits and environmental costs (NEEB = Yield gains − Input costs − GWP costs, Zhang et al., 2018). These indicators will serve as a comprehensive evaluation of the mulching effects on global warming by incorporating grain yield, the primary concern for the farmers and economic benefits, which are the public concerns for achieving sustainable agriculture.
(2) Several studies have reported the effect of mulching materials on GHGs emissions in various dryland and wetland (rice paddies) cropping systems; however, the underlying mechanisms behind contrasting responses to GHGs emissions are unclear. So far, most studies suggest soil surface mulching shifts GHGs emission rates by altering soil physicochemical properties, including soil moisture, temperature, pH, redox potential, and nutrients. Few studies included changes in the microbial community responsible for producing and consuming of GHGs in soil, suggesting the complex role of surface mulching rather. For instance, Chen et al. (2021) recorded a direct correlation between the CH_4 flux rates and the differences in abundances of microbial functional genes (*mcrA, pmoA*) involved in CH_4 production and consumption under the GCRP system for rice cultivation. In comparison, Wang et al. (2021) found a significant increase in the abundance and diversity of ammonia-oxidizing bacteria and N_2O emission rates under biodegradable film mulching. We recommend further revealing the underlying mechanisms driving the surface mulching effects on GHGs emissions from contrasting agricultural systems.
(3) The ultimate goal for field observations or mechanism exploration is to guide agronomic practices and even projects into the future. Modeling is such a powerful tool to fill in this gap. Until now, little effort has been made to incorporate the mulching effect into the DNDC model for predicting N_2O emission

rates and the results were quite promising (Han et al., 2014; Yang et al., 2012). For instance, Yu et al. (2017, 2018) tested a modified DNDC model to simulate changes in soil N_2O emissions, respiration, and crop growth under plastic film mulching in an oasis cotton field. Chen et al. (2019) applied the DNDC model to predict the impacts of different mulching practices (i.e., no-mulching, straw mulching, and plastic film mulching) on wheat (*Triticum aestivum* L.) and maize (*Zea mays* L.) yields and N_2O emissions under future climate scenarios in the South Loess Plateau of China. Meanwhile, Lu and Zuo (2018) introduced a new land-surface process model, Common Land Model (CoLM), to predict changes in the surface albedo of agricultural soil surface mulching. The updated model elements are based on observations from field experiments determining the land-atmosphere interaction under plastic film mulching. Initial validations of CoLM suggested that the improved CoLM-mulch could reasonably simulate the diurnal variations of soil temperature and moisture, together with radiation, water, heat, and carbon dioxide (CO_2) fluxes, on the cropland underlying a surface with a plastic film covering. However, the land surface models did not contain key physical processes driven by plastic film mulching, such as inhibition of evaporation by plastic material. It is thus urgent to develop or modify the current process-based models to well project the possible climate effects on a large temporal and spatial scale.

References

Abbas, F., Hammad, H. M., Ishaq, W., Farooque, A. A., Bakhat, H. F., Zia, Z., Fahad, S., Farhad, W., & Cerda, A. (2020). A review of soil carbon dynamics resulting from agricultural practices. *Journal of Environmental Management, 268*.

Akhtar, K., Wang, W., Ren, G., Khan, A., Enguang, N., Khan, A., Feng, Y., Yang, G., & Wang, H. (2020). Straw mulching with inorganic nitrogen fertilizer reduces soil CO_2 and N_2O emissions and improves wheat yield. *Science of the Total Environment, 741*.

Ali, M. A., Hoque, M. A., & Kim, P. J. (2013). Mitigating global warming potentials of methane and nitrous oxide gases from rice paddies under different irrigation regimes. *Ambio, 42*, 357–368.

Al-Kaisi, M. M., Kruse, M. L., & Sawyer, J. E. (2008). Effect of nitrogen fertilizer application on growing season soil carbon dioxide emission in a corn-soybean rotation. *Journal of Environmental Quality, 37*, 325–332.

Andriesse, W., & Fresco, L. O. (1991). Characterization of rice-growing environments in West Africa. *Agriculture Ecosystems & Environment, 33*(4):377–395. https://doi.org/10.1016/0167-8809(91)90059-7.

Barto, E. K., Alt, F., Oelmann, Y., Wilcke, W., & Rillig, M. C. (2010). Contributions of biotic and abiotic factors to soil aggregation across a land use gradient. *Soil Biology and Biochemistry, 42*, 2316–2324.

Berger, S., Kim, Y., Kettering, J., & Gebauer, G. (2013). Plastic mulching in agriculture—Friend or foe of N_2O emissions? *Agriculture, Ecosystems & Environment, 167*, 43–51.

Bhattacharyya, P., Roy, K., & Neogi, S. (2012). Effects of rice straw and nitrogen fertilization on greenhouse gas emissions and carbon storage in tropical flooded soil planted with rice. *Soil and Tillage Research, 124*, 119–130.

Bodelier, P. L. E., & Laanbroek, H. J. (2014). Nitrogen as a regulatory factor of methane oxidation in soils and sediments. *FEMS Microbiology Ecology, 47*(3), 265–277.

Butterbach-Bahl, K., Baggs, E. M., Dannenmann, M., Kiese, R., & Zechmeister-Boltenstern, S. (2013). Nitrous oxide emissions from soils: how well do we understand the processes and their controls? *Philosophical Transactions of the Royal Society B, 368*. https://doi.org/10.1098/rstb.2013.0122.

Carlson, K. M., Gerber, J. S., Mueller, N. D., Herrero, M., MacDonald, G. K., Brauman, K. A., Havlik, P., O'Connell, C. S., Johnson, J. A., Saatchi, S., & West, P. C. (2017). Greenhouse gas emissions intensity of global croplands. *Nature Climate Change, 7*(1), 63–68. https://doi.org/10.1038/nclimate3158

Chai, T. Y., et al. (2011). Recent progress of research on comprehensive efects of diferent rates of straw mulch on rained farming areas in China II. Problems and prospects of study on efects of diferent rates of straw mulch on physiological ecology of crop. *Agricultural Research in the Arid Areas, 29*, 108–114.

Chen, H., Liu, J., Zhang, A., Chen, J., Cheng, G., Sun, B., Pi, X., Dyck, M., Si, B., Zhao, Y., & Feng, H. (2017). Effects of straw and plastic film mulching on greenhouse gas, emissions in Loess Plateau, China: A field study of 2 consecutive wheat-maize rotation cycles. *Science of the Total Environment, 579*, 814–824.

Chen, Z., Lin, S., Yao, Z., Zheng, X., Gschwendtner, S., Schloter, M., Liu, M., Zhang, Y., Butterbach-Bahl, K., & Dannenmann, M. (2018). Enhanced nitrogen cycling and N_2O loss in water-saving ground cover rice production systems (GCRPS). *Soil Biology & Biochemistry, 121*, 77–86.

Chen, S.-J., Jiang, C.-S., Ni, X., Li, X.-X., & Hao, Q.-J. (2019). Effect of plastic film mulching on greenhouse gas emissions from rice-rapeseed rotation in cropland. *Huan jing ke xue = Huanjing kexue, 40*(9), 4213–4220.

Chen, J., Chen, L., Kong, D., Geng, Y., Wang, H., Fang, X., Li, S., Hu, Z., Liu, S., & Zou, J. (2021). Incorporating DNA-level microbial constraints helps decipher methane emissions from Chinese water-saving ground cover rice production systems. *Field Crops Research, 260*, 107992.

Cuello, J. P., Hwang, H. Y., Gutierrez, J., Kim, S. Y., & Kim, P. J. (2015). Impact of plastic film mulching on increasing greenhouse gas emissions in temperate upland soil during maize cultivation. *Applied Soil Ecology, 91*, 48–57.

Dossou-Yovo, E. R., Brueggemann, N., Jesse, N., Huat, J., Ago, E. E., & Agbossou, E. K. (2016). Reducing soil CO_2 emission and improving upland rice yield with no-tillage, straw mulch and nitrogen fertilization in northern Benin. *Soil & Tillage Research, 156*, 44–53.

Falloon, P., Jones, C. D., Ades, M., & Paul, K. (2011). Direct soil moisture controls of future global soil carbon changes: An important source of uncertainty. *Global Biogeochemical Cycles, 25*(3).

Fan, M., Li, Q., Zhang, E., Liu, Q. & Wang, Q. (2019). Effects of mulching on soil CO_2 fluxes, hay yield and nutritional yield in a forage maize field in Northwest China. *Scientific Reports, 9*.

FAO. (2001). Conservation agriculture case studies in Latin America and Africa. In *Introduction FAO Soils Bulletin* (No. 78). FAO.

Fawibe, O. O., Honda, K., Taguchi, Y., Park, S., & Isoda, A. (2019). Greenhouse gas emissions from rice field cultivation with drip irrigation and plastic film mulch. *Nutrient Cycling in Agroecosystems, 113*(1), 51–62.

Flessa, H., Dörsch, P., & Beese F. (1995). Seasonal variation of N_2O and CH_4 fluxes in differently managed arable soils in southern Germany. *Journal of Geophysical Research, 100*, 23115–23124.

Gan, Y., Siddique, K. H. M., Turner, N. C., Li, X.-G., Niu, J.-Y., Yang, C., Liu, L., & Chai, Q. (2013). Ridge-furrow mulching systems—An innovative technique for boosting crop productivity in semi-arid rain-fed environments. In D. L. Sparks (Ed.), *Advances in agronomy* (pp. 429–476). Academic Press.

Gao, B., Huang, T., Ju, X. T., Gu, B. J., Huang, W., Xu, L. L., Rees, R. M., Powlson, D. S., Smith, P., & Cui, S. H. (2018). Chinese cropping systems are a net source of greenhouse gases despite soil carbon sequestration. *Global Change Biology, 24*(12), 5590–5606.

Garcia-Ruiz, R., & Baggs, E. M. (2007). N_2O emission from soil following combined application of fertiliser-N and ground weed residues. *Plant and Soil, 299*, 263–274.

Govaerts, B., Verhulst, N., Castellanos-Navarrete, A., Sayre, K. D., Dixon, J., & Dendooven, L. (2009). Conservation agriculture and soil carbon sequestration: Between myth and farmer reality. *Critical Reviews in Plant Sciences, 28*, 97–122.

Han, J., Jia, Z., Wu, W., Li, C., Han, Q., & Zhang, J. (2014). Modeling impacts of film mulching on rainfed crop yield in Northern China with DNDC. *Field Crops Research, 155*, 202–212.

He, G., Wang, Z., Li, S., & Malhi, S. S. (2018). Plastic mulch: Tradeoffs between productivity and greenhouse gas emissions. *Journal of Cleaner Production, 172*, 1311–1318.

Hobbs, P. R. (2007). Conservation agriculture: What is it and why is it important for future sustainable food production? *Journal of Agricultural Science, 145*, 127–137.

Huang, D. D., Cao, G. J., Geng, Y. H., Wang, L. C., Chen, X. W., & Liang, A. Z. (2019). Impact of agricultural waste return on soil greenhouse gas emissions. *Applied Ecology and Environmental Research, 17*(1), 1321–1335.

Hussain, S., Peng, S., Fahad, S., Khaliq, A., Huang, J., Cui, K., & Nie, L. (2015). Rice management interventions to mitigate greenhouse gas emissions: A review. *Environmental Science and Pollution Research, 22*, 3342–3360.

IPCC. (2013). Contribution of working group I to the fifth assessment report of the intergovernmental panel on climate change. Cambridge University Press.

IPCC. (2014). Climate Change 2014: Synthesis report. contribution of working groups I, II and III to the fifth assessment report of the Intergovernmental Panel on Climate Change. In R. K. Pachauri, L. A. Meyer (Eds.).

IRRI. (2009). Saving water: Alternate wetting and drying (AWD). IRRI Rice Fact Sheet. Los Baños, Philippines: International Rice Research Institute (IRRI). http://www.knowledgebank.irri.org/factsheetsPDFs/watermanagement_FSAWD3.pdf.

Jacinthe, P. A., & Lal, R. (2003). Nitrogen fertilization of wheat residue affecting nitrous oxide and methane emission from a central Ohio Luvisol. *Biology and Fertility of Soils, 37*(6), 338–347.

Jacinthe, P. A., Lal, R., & Kimble, J. M. (2002). Annual carbon budget and seasonal carbon dioxide emission from mulch-covered soils. *Soil & Tillage Research, 67*, 147–57.

Jia, Q. M., Zhang, H. X., Wang, J., Xiao, X. M., Chang, S. H., Zhang, C., Liu, Y. J., & Hou, F. J. (2021). Planting practices and mulching materials improve maize net ecosystem C budget, global warming potential and production in semi-arid regions. *Soil & Tillage Research, 207*.

Kasirajan, S., & Ngouajio, M. (2012). Polyethylene and biodegradable mulches for agricultural applications: A review. *Agronomy for Sustainable Development, 32*, 501–529.

Kassam A., Friedrich T., Derpsch R., & Kienzle J. (2015). Overview of the worldwide spread of conservation agriculture. *Field Actions Science Reports, 8*. http://journals.openedition.org/factsreports/3966.

Kim, W.-S., Rom, C., 최현석, 최경주., & 이연. (2010). Effect of organic fertilizer and mulch sources on growth and CO_2 assimilation in MM.106 apple trees. *Korea Journal of Organic Agriculture, 18*(2), 245–255.

Knorr, W., Prentice, I., House, J., et al. (2005). Long-term sensitivity of soil carbon turnover to warming. *Nature, 433*, 298–301. https://doi.org/10.1038/nature03226

Kreye, C., Dittert, K., Zheng, X. H., Zhang, X., Lin, S., Tao, H. B., & Sattelmacher, B. (2007). Fluxes of methane and nitrous oxide in water-saving rice production in North China. *Nutrient Cycling in Agroecosystems, 77*(3), 293–304.

Kritee, K., Nair, D., Zavala-Araiza, D., Proville, J., Rudek, J., Adhya, T. K., Loecke, T., Esteves, T., Balireddygari, S., Dava, O., Ram, K. S. R. A., Madasamy, M., Dokka, R. V., Anandaraj, D., Athiyaman, D., Reddy, M., Ahuja, R., & Hamburg, S. P. (2018). High nitrous oxide fluxes from rice indicate the need to manage water for both long- and short-term climate impacts. *Proceedings of the National Academy of Sciences of the United States of America, 115*(39), 9720–9725. https://doi.org/10.1073/pnas.1809276115

Kuypers, M. M., Marchant, H. K., & Kartal, B. (2018). The microbial nitrogen-cycling network. *Nature Reviews Microbiology, 16*, 263–276. https://doi.org/10.1038/nrmicro.2018.9.

Kyulavski, V., Recous, S., Garnier, P., Paillat, J.-M., & Thuries, L. (2019). Application of N fertilizer to sugarcane mulches: Consequences for the dynamics of mulch decomposition and CO_2 and N_2O fluxes. *Bioenergy Research, 12*(3), 484–496.
Lal, R. (2013). Soil carbon management and climate change. *Carbon Management, 4*(4), 439–462. https://doi.org/10.4155/cmt.13.31
Langeroodi, A. R. S., Osipitan, O. A., & Radicetti, E. (2019). Benefits of sustainable management practices on mitigating greenhouse gas emissions in soybean crop (Glycine max). *Science of the Total Environment, 660*, 1593–1601.
Le Mer, J., & Roger, P. (2001). Production, oxidation, emission and consumption of methane by soils: A review. *European Journal of Soil Biology, 37*, 25–50. https://doi.org/10.1016/S1164-5563(01)01067-6.
Lee, J. G., Cho, S. R., Jeong, S. T., Hwang, H. Y., & Kim, P. J. (2019). Different response of plastic film mulching on greenhouse gas intensity (GHGI) between chemical and organic fertilization in maize upland soil. *Science of the Total Environment, 696*.
Lehner, P., & Rosenberg, N. (2017). Legal pathways to carbon-neutral agriculture. In M. B. Gerrard & J. C. Dernbach (Eds.), *Legal pathways to deep decarbonization in the United States* (2018 Forthcoming) (Vol. 47, p. 10845). Environmental Law Reporter. https://ssrn.com/abstract=3040919.
Lenka, N. K., & Lal, R. (2013). Soil aggregation and greenhouse gas flux after 15 years of wheat straw and fertilizer management in a no-till system. *Soil & Tillage Research, 126*, 78–89.
Li, X. L., Su, D. R., & Yuan, Q. H. (2007). Ridge-furrow planting of alfalfa (Medicago sativa L.) for improved rainwater harvest in rainfed semiarid areas in Northwest China. *Soil Research, 93*(2007), 117–125.
Li, Z. G., Zhang, R. H., Wang, X. J., Wang, J. P., Zhang, C. P., & Tian, C. Y. (2011). Carbon dioxide fluxes and concentrations in a cotton field in Northwestern China: Effects of plastic mulching and drip irrigation. *Pedosphere, 21*(2), 178–185.
Li, Z.-G., Zhang, R.-H., Lai, D.-M., Yan, Z.-Y., Jiang, L., & Tian, C.-Y. (2012). Effects of drip irrigation with plastic mulching on the net primary productivity, soil heterotrophic respiration, and net CO_2 exchange flux of cotton field ecosystem in Xinjiang, Northwest China. *Ying yong sheng tai xue bao = The Journal of Applied Ecology, 23*(4), 1018–1024.
Li, Z., Zhang, R., Wang, X., Chen, F., Lai, D., & Tian, C. (2014). Effects of plastic film mulching with drip irrigation on N_2O and CH_4 emissions from cotton fields in arid land. *Journal of Agricultural Science, 152*(4), 534–542.
Li, X. Y., Liu, L. J., Yang, H. J., & Li, Y. (2018). Relationships between carbon fluxes and environmental factors in a drip-irrigated, film-mulched cotton field in arid region. *Plos One, 13*(2).
Li, J. H., Li, H., Zhang, Q., Shao, H. B., Gao, C. H., & Zhang, X. Z. (2019a). Effects of fertilization and straw return methods on the soil carbon pool and CO_2 emission in a reclaimed mine spoil in Shanxi Province, China. *Soil & Tillage Research, 195*.
Li, W. W., Zhuang, Q. L., Wu, W., Wen, X. X., Han, J., & Liao, Y. C. (2019b). Effects of ridge-furrow mulching on soil CO_2 efflux in a maize field in the Chinese Loess Plateau. *Agricultural and Forest Meteorology, 264*, 200–212.
Liu, Y., Li, S. Q., Yang, S. J, Hu, W., & Chen, X. P. (2010). Diurnal and seasonal soil CO_2 flux patterns in spring maize fields on the Loess Plateau, China. *Acta Agriculturae Scandinavica, Section B-Soil & Plant Science, 60*(3), 245–255. https://doi.org/10.1080/09064710902878121.
Liang, W., Shi, Y., Zhang, H., Yue, J., & Huang, G. H. (2007). Greenhouse gas emissions from northeast China rice fields in fallow season. *Pedosphere, 17*(5), 630–638.
Liu, Q., Liu, X., Bian, C., Ma, C., Lang, K., Han, H., & Li, Q. (2014). Response of Soil CO_2 Emission and summer maize yield to plant density and straw mulching in the North China plain. *Scientific World Journal*.
Liu, Q., Chen, Y., Li, W., Liu, Y., Han, J., Wen, X., & Liao, Y. (2016a). Plastic-film mulching and urea types affect soil CO_2 emissions and grain yield in spring maize on the Loess Plateau, China. *Scientific Reports, 6*.

Liu, Q., Chen, Y., Li, W., Liu, Y., Han, J., Wen, X., & Liao, Y. (2016b). Plastic-film mulching and urea types affect soil CO2 emissions and grain yield in spring maize on the Loess Plateau. *China Science Reports, 6*, 28150. https://doi.org/10.1038/srep28150.

Liu, S., Fan, R. Q., Yang, X. M., Zhang, Z. H., Zhang, X. P., & Liang, A. Z. (2019). Decomposition of maize stover varies with maize type and stover management strategies: A microcosm study on a Black soil (Mollisol) in northeast China. *Journal of Environmental Management, 234*, 226–236.

Liu, J. L., Huang, X. Y., Jiang, H. B., & Chen, H. (2021). Sustaining yield and mitigating methane emissions from rice production with plastic film mulching technique. *Agricultural Water Management, 245*.

Lu, S., & Zuo, H. C. (2018). Improvement and validation of the common land model on cropland covered by plastic film in the arid Region of China. *Journal of Applied Meteorology and Climatology, 57*(9), 2071–2089.

Luo, X.-Q., Zhang, A. F., Chen, H.-X., & Feng, H. (2018). Effects of plastic film mulching patterns and irrigation on yield of summer maize and greenhouse gas emissions intensity of field. *Huan jing ke xue = Huanjing kexue, 39*(11), 5246–5256.

Ma, J., Xu, H., Yagi, K., & Cai, Z. C. (2008). Methane emission from paddy soils as affected by wheat straw returning mode. *Plant and Soil, 313*(1–2), 167–174.

Ma, J., Ma, E. D., Xu, H., Yagi, K., & Cai, Z. C. (2009). Wheat straw management affects CH_4 and N_2O emissions from rice fields. *Soil Biology & Biochemistry, 41*(5), 1022–1028.

Meng, C. B., Wang, F. X., Yang, K. J., Shock, C. C., Engel, B. A., Zhang, Y. L., Tao, L. J., & Gu, X. X. (2020). Small wetted proportion of drip irrigation and non-mulched treatment with manure application enhanced methane uptake in upland field. *Agricultural and Forest Meteorology, 281*.

Ming, G. H., Hu, H. C., Tian, F. Q., Peng, Z. Y., Yang, P. J., & Luo, Y. Q. (2018). Precipitation alters plastic film mulching impacts on soil respiration in an arid area of northwest China. *Hydrology and Earth System Sciences, 22*(5), 3075–3086.

Mo, F., Wang, J. Y., Xiong, Y. C., Nguluu, S. N., & Li, F. M. (2016). Ridge-furrow mulching system in semi-arid Kenya: A promising solution to improve soil water availability and maize productivity. *European Journal of Agronomy, 80*, 124–136.

Mo, F., Wang, J. Y., Zhou, H., Luo, C. L., Zhang, X. F., Li, X. Y., et al. (2017). Ridge-furrow plastic-mulching with balanced fertilization in rainfed maize (Zea mays L.): An adaptive management in east African Plateau. *Agricultural and Forest Meteorology, 236*, 100–112.

Mo, F., Li, X. Y., Niu, F. J., Zhang, C. R., Li, S. K., Zhang, L., & Xiong, Y. C. (2018). Alternating small and large ridges with full film mulching increase linseed (Linumusitatissimum L.) productivity and economic benefit in a rainfed semi-arid environment. *Field Crops Research, 219*, 120–130.

Mo, F., Yu, K. L., Crowther, T. W., Wang, J. Y., Zhao, H., Xiong, Y. C., & Liao, Y. C. (2020a). How plastic mulching affects net primary productivity, soil C fluxes and organic carbon balance in dry agroecosystems in China. *Journal of Cleaner Production, 263*.

Mo, F., Han, J., Wen, X. X., Wang, X. K., Li, P. F., Vinay, N., Jia, Z. K., Xiong, Y. C., & Liao, Y. C. (2020b). Quantifying regional effects of plastic mulch on soil nitrogen pools, cycles, and fluxes in rain-fed agroecosystems of the Loess Plateau. *Land Degradation & Development, 31*(13), 1675–1687.

Nan, W.-G., Yue, S.-C., Huang, H.-Z., Li, S.-Q., & Shen, Y.-F. (2016). Effects of plastic film mulching on soil greenhouse gases (CO_2, CH_4 and N_2O) concentration within soil profiles in maize fields on the Loess Plateau China. *Journal of Integrative Agriculture, 15*(2), 451–464.

Nawaz, A., Lal, R., Shrestha, R. K., & Farooq, M. (2017). Mulching affects soil properties and greenhouse gas emissions under long-term no-till and plough-till systems in Alfisol of central Ohio. *Land Degradation & Development, 28*(2), 673–681.

Peters, S. J. W., Saikawa, E., Markewitz, D., Sutter, L., Avramov, A., Sanders, Z. P., Yosen, B., Wakabayashi, K., Martin, G., Andrews, J. S., & Hill, N. S. (2020). Soil trace gas fluxes in living mulch and conventional agricultural systems. *Journal of Environmental Quality, 49*(2), 268–280.

Pinheiro, P. L., Recous, S., Dietrich, G., Weiler, D. A., Schu, A. L., Bazzo, H. L. S., & Giacomini, S. J. (2019). N_2O emission increases with mulch mass in a fertilized sugarcane cropping system. Biology and Fertility of Soils, 5, 511–523.

Rabenarivo, M., Wrage-Moennig, N., Chotte, J.-L., Rabeharisoa, L., Razafimbelo, T. M., & Chapuis-Lardy, L. (2014). Emissions of CO_2 and N_2O from a pasture soil from madagascar-simulating conversion to direct-seeding mulch-based cropping in incubations with organic and inorganic inputs. *Journal of Plant Nutrition and Soil Science, 177*(3), 360–368.

Ren, X., Chen, X., Cai, T., Wei, T., Wu, Y., Ali, S., Zhang, P., & Jia, Z. K. (2017). Effects of ridge furrow system combined with different degradable mulching materials on soil water conservation and crop production in semi-humid areas of China. *Frontiers in Plant Science, 8*.

Sanchis, E., Ferrer, M., Torres, A. G., Cambra-López, M., & Calvet, S. (2012). Effect of water and straw management practices on methane emissions from rice fields: A review through a meta-analysis. *Environmental Engineering Science, 29*, 1053–1062

Schlesinger, W. H., & Andrews, J. A. (2000). Soil respiration and the global carbon cycle. *Biogeochemistry, 48*(1), 7–20.

Schmatz, R., Recous, S., Weiler, D. A., Pilecco, G. E., Schu, A. L., Giovelli, R. L., & Giacomini, S. J. (2020). How the mass and quality of wheat and vetch mulches affect drivers of soil N_2O emissions. *Geoderma, 372*.

Schulz, S., & Conrad, R. (1996). Influence of temperature on pathways to methane production in the permanently cold profundal sediment of Lake Constance. *FEMS Microbiology Ecology, 20*(1), 1–14.

Shcherbak, I., Millar, N., Robertson, G. P. (2014). Global meta-analysis of the nonlinear response of soil nitrous oxide (N_2O) emissions to fertilizer nitrogen. *Proceedings of The National Academy of Sciences of the United States of America, 111*(25), 9199–9204.; Xu, H., Cai, Z.C., Li, X. P., & Tsuruta, H. (2000). Effect of antecedent soil water regime and rice straw application time on CH_4 emission from rice cultivation. *Australian Journal of Soil Research, 38*, 1–12.

Singh, B., Shan, Y. H., Johnson-Beebout, S. E., Singh, Y., & Buresh, R. J. (2008). Crop residue management for lowland rice-based cropping systems in Asia. *Advances in Agronomy,* (98), 117–199.

Smith, P., Martino, D., Cai, Z., Gwary, D., Janzen, H., Kumar, P., McCarl, B., Ogle, S., O'Mara, F., Rice, C., Scholes, B., Sirotenko, O., Howden, M., McAllister, T., Pan, G., Romanenkov, V., Schneider, U., & Towprayoon, S. (2007). Policy and technological constraints to implementation of greenhouse gas mitigation options in agriculture. *Agriculture, Ecosystems & Environment, 118*(1–4), 6–28. https://doi.org/10.1016/j.agee.2006.06.006.

Smith, K. A., Ball, T., Conen, F., Dobbie, K. E., Massheder, J., Rey, A. (2018). Exchange of greenhouse gases between soil and atmosphere: Interactions of soil physical factors and biological processes. *European Journal of Soil Science, 69*, 10–20.

Steinmetz, Z., Wollmann, C., Schaefer, M., Buchmann, C., David, J., Tröger, J., Muñoz, K., Frör, O., & Schaumann, G. E. (2016). Plastic mulching in agriculture. Trading short-term agronomic benefits for long-term soil degradation? *Science of the Total Environment, 550*, 690–705.

Stocker, T. F., Qin, D., Plattner, G.-K., Tignor, M., Allen, S. K., Boschung, J., Nauels, A., Xia, Y., Bex, B., & Midgley, B. (2013). IPCC, 2013: Climate change 2013: The physical science basis. In *Contribution of working group I to the fifth assessment report of the intergovernmental panel on climate change*. Cambridge University Press.

Tanveer, S. K., Wen, X., Lu, X. L., Zhang, J., & Liao, Y. (2013). Tillage, Mulch and N fertilizer affect emissions of CO_2 under the rain fed condition. *Plos One, 8*(9).

Trugman, A. T. T., Medvigy, D., Mankin, J. S., & Anderegg, W. R. L. (2018). Soil moisture stress as a major driver of carbon cycle uncertainty. *Geophysical Research Letters, 45*(13), 6495–6503.

van Kessel, C., Venterea, R., Six, J., Adviento-Borbe, M., Linquist, B., & van Groenigen, K. (2013). Climate, duration, and N placement determine N_2O emissions in reduced tillage systems: A meta-analysis. *Global Change Biology Bioenergy, 19*, 33–44.

Vermeulen, S. J., Campbell, B. M., & Ingram, J. S. I. (2012). Climate change and food systems. *Annual Review of Environment and Resources, 37*, 195–222.

Wang, Z., Delaune, R. D., Lindau, C. W. et al. (1992). Methane production from anaerobic soil amended with rice straw and nitrogen fertilizers. *Fertilizer Research, 33*, 115–121. https://doi.org/10.1007/BF01051166.

Wang, D. L., Feng, H., Liu, X. Q., Li, Y., Zhou, L. F., Zhang, A. F., & Dyck, M. (2018). Effects of gravel mulching on yield and multilevel water use efficiency of wheat-maize cropping system in semi-arid region of Northwest China. *Field Crops Research, 218*, 201–212.

Wang, W. Y., Akhtar, K., Ren, G. X., Yang, G. H., Feng, Y. Z., & Yuan, L. Y. (2019). Impact of straw management on seasonal soil carbon dioxide emissions, soil water content, and temperature in a semi-arid region of China. *Science of the Total Environment, 652*, 471–482.

Wang, D., Li, Y., Zhang, T., Zhou, L., Ge, J., Zhang, L., Dyck, M., & Feng, H. (2020a). Greenhouse gas emissions and carbon footprint under gravel mulching on China's Loess Plateau. *Agronomy Journal, 112*(2), 733–747.

Wang, N. J., Ding, D. Y., Malone, R. W., Chen, H. X., Wei, Y. S., Zhang, T. B., Luo, X. Q., Li, C., Chu, X. S., & Feng, H. (2020b). When does plastic-film mulching yield more for dryland maize in the Loess Plateau of China? A meta-analysis. *Agricultural Water Management, 240*.

Watanabe, A., Satoh, Y., & Kimura, M. (1995). Estimation of the increase in CH_4 emission from paddy soils by rice straw application. *Plant and Soil, 173*, 225–231.

Watanabe, A., Yamada, H., & Kimura, M. (2005). Analysis of temperature effects on seasonal and interannual variation in CH4 emission from rice-planted pots. *Agriculture, Ecosystems & Environment, 105*(1–2), 439–443.

World Meteorological Organization. (2017). Manual on the WMO integrated global observing system (WMO-No. 1160), 2015 edn, updated in 2017.

Wu, J. P., Xiao, K., Zhao, C., Yu, A. Z., Feng, F. X., Li, L., & Chai, Q. (2018). Ridge-furrow cropping of maize reduces soil carbon emissions and enhances carbon use efficiency. *Agriculture Ecosystems & Environment, 256*, 153–162.

Wu, X. H., Wang, W., Xie, K. J., Yin, C. M., Hou, H. J., & Xie, X. L. (2019). Combined effects of straw and water management on CH_4 emissions from rice fields. *Journal of Environmental Management, 231*, 1257–1262.

Xing, G. X. (1998). N_2O emission from cropland in China. *Nutrient Cycling in Agroecosystems, 52*, 249–254.

Xu, H., Cai, Z. C., Li, X. P., & Tsuruta, H. (2000). Effect of antecedent soil water regime and rice straw application time on CH_4 emission from rice cultivation. *Australian Journal of Soil Research, 38*, 1–12.

Xu, Y. C., Shen, Q. R., Li, M. L., Dittert, K., & Sattelmacher, B. (2004). Effect of soil water status and mulching on N_2O and CH_4 emission from lowland rice field in China. *Biology and Fertility of Soils, 39*(3), 215–217.

Xu, Y., Wang, Y., Ma, X., Liu, X., Zhang, P., Cai, T., & Jia, Z. (2020). Ridge-furrow mulching system and supplementary irrigation can reduce the greenhouse gas emission intensity. *Science of the Total Environment, 717*.

Yadvinder-Singh, B. N., Humphreys, E., Singh, B., & Timsina, J. (2008). Yield and nitrogen use efficiency of permanent bed rice–wheat systems in northwest India: Effect of N fertilization, mulching and crop establishment method. In E. Humphreys & C. H. Roth (Eds.), *Permanent beds and rice-residue management for rice–wheat systems in the Indo-Gangetic Plain. Proceedings of a workshop held in Ludhiana* (pp. 62–78). ACIAR Proceedings No. 127, Australian Centre for International Agricultural Research. http://www.aciar.gov.au/publication/term/18. Accessed 17 Nov. 08.

Yagioka, A., Komatsuzaki, M., Kaneko, N., & Ueno, H. (2015). Effect of no-tillage with weed cover mulching versus conventional tillage on global warming potential and nitrate leaching. *Agriculture Ecosystems & Environment, 200*, 42–53.

Yan, X. Y., Yagi, K., Akiyama, H., & Akimoto, H. (2005). Statistical analysis of the major variables controlling methane emission from rice fields. *Global Change Biology, 11*, 1131–1141.

Yang, Q., Zuo, H., Xiao, X., Wang, S., Chen, B., & Chen, J. (2012). Modelling the effects of plastic mulch on water, heat and CO_2 fluxes over cropland in an arid region. *Journal of Hydrology, 452*, 102–118.

Yang, N., Sun, Z., Feng, L., Zheng, M., & Chi, D. (2015). Plastic film mulching for water efficient agricultural applications and degradable films materials development research. *Materials and Manufacturing Process*, 37–41.

Yao, Z., et al. (2013). Greenhouse gas fluxes and NO release from a Chinese subtropical rice-winter wheat rotation system under nitrogen fertilizer management. *Journal of Geophysical Research: Biogeosciences, 118*, 623–638.

Yao, Z., Du, Y., Tao, Y., Zheng, X., Liu, C., Lin, S., & Butterbach-Bahl, K. (2014). Water-saving ground cover rice production system reduces net greenhouse gas fluxes in an annual rice-based cropping system. *Biogeosciences, 11*(22), 6221–6236.

Yao, Z. S., Zheng, X. H., Liu, C. Y., Lin, S., Zuo, Q., & Butterbach-Bahl, K. (2017). Improving rice production sustainability by reducing water demand and greenhouse gas emissions with biodegradable films. *Scientific Reports, 7*.

Ye, J., & Liu, C.-A. (2012) Suitability of mulch and ridge-furrow techniques for maize across the precipitation gradient on the Chinese Loess Plateau. *Journal of Agricultural Science, 4*(10). https://doi.org/10.5539/jas.v4n10p182.

Ye, X. H., Liu, H. D., Zhang, X. H., Ma, J. H., Han, B., Li, W., Zou, H. T., Zhang, Y. L., & Lin, X. G. (2020). Impacts of irrigation methods on greenhouse gas emissions/absorptions from vegetable soils. *Journal of Soils and Sediments, 20*(2), 723–733.

Yu, Y., Zhao, C., Stahr, K., Zhao, X., & Jia, H. (2016). Plastic mulching increased soil CO_2 concentration and emissions from an oasis cotton field in Central Asia. *Soil Use and Management, 32*(2), 230–239.

Yu, Y., Tao, H., Jia, H., & Zhao, C. (2017). Impact of plastic mulching on nitrous oxide emissions in China's arid agricultural region under climate change conditions. *Atmospheric Environment, 158*, 76–84.

Yu, Y. X., Tao, H., Yao, H. Y., & Zhao, C. Y. (2018). Assessment of the effect of plastic mulching on soil respiration in the arid agricultural region of China under future climate scenarios. *Agricultural and Forest Meteorology, 256*, 1–9.

Zhang, X. B., Xu, M. G., Sun, N., Wang, X. J., Wu, L., Wang, B. R., & Li, D. C. (2013). How do environmental factors and different fertilizer strategies affect soil CO_2 emission and carbon sequestration in the upland soils of southern China? *Applied Soil Ecology, 72*, 109–118.

Zhang, Z.-S., Cao, C.-G., Guo, L.-J., & Li, C.-F. (2014). The effects of rape residue mulching on net global warming potential and greenhouse gas intensity from no-tillage paddy fields. *Scientific World Journal.*

Zhang, F., Li, M., Qi, J. H., Li, F. M., & Sun, G. J. (2015). Plastic film mulching increases soil respiration in ridge-furrow maize management. *Arid Land Research and Management, 29*(4), 432–453.

Zhang, G. B., Ma, J., Yang, Y. T., Yu, H. Y., Song, K. F., Dong, Y. J., Lv, S. H., & Xu, H. (2018). Achieving low methane and nitrous oxide emissions with high economic incomes in a rice-based cropping system. *Agricultural and Forest Meteorology, 259*, 95–106.

Zhang, G. B., Yang, Y. T., Huang, Q., Ma, J., Yu, H. Y., Song, K. F., Dong, Y. J., Lv, S. H., & Xu, H. (2020). Reducing yield-scaled global warming potential and water use by rice plastic film mulching in a winter flooded paddy field. *European Journal of Agronomy, 114*.

Zhao, X., Liu, S., Pu, C., et al. (2016). Methane and nitrous oxide emissions under no-till farming in China: A meta-analysis. *Glob. Change Biol. Bioenergy, 22*, 1372–1384.

Zhou, L.-M., Li, F.-M., Jin, S.-L., & Song, Y. (2009). How two ridges and the furrow mulched with plastic film affect soil water, soil temperature and yield of maize on the semiarid Loess Plateau of China. *Field Crops Research, 113*, 41–47. https://doi.org/10.1016/j.fcr.2009.04.005

Zong, R., Wang, Z. H., Wu, Q., Guo, L., & Lin, H. (2020). Characteristics of carbon emissions in cotton fields under mulched drip irrigation. *Agricultural Water Management, 231*.

Zou, J., Huang, Y., Jiang, J., Zheng, X., & Sass, R. L. (2005). A 3-year field measurement of methane and nitrous oxide emissions from rice paddies in China: Effects of water regime, crop residue, and fertilizer application. *Global Biogeochemical Cycles, 19*, GB2021.

Mulching: A New Concept for Climate Smart Agriculture

Izhar Ullah, Syed Tanveer Shah, Abdul Basit, Muhammad Sajid, Muhammad Arif, Noman Ahmad, and Fahad Noor

Abstract Climate change is a state of change in climatic condition driven by human activity directly or indirectly causing alteration of global temperature which becomes a drastic challenge to sustainable food security leads to a major loss to agricultural produce. This alteration of global temperature produce long term change in weather distribution patterns (CO_2 availability, global temperature, intensity and rate of harsh weather events, up surging sea level and weather variability etc.). In order to reduce these damages, a new approach of Climate smart agriculture must be promoted for achieving sustainable agricultural productivity and incomes and also to mitigate the adverse effect of climate change and reduce greenhouse gases emissions. The need for providing large access of quality food and environment to individuals has encouraged the use of organic or inorganic materials as mulch that can maintain the physical and chemical condition of soil. Mulching is a covering material of soil surface with organic or inorganic material for improving soil structure, conserving soil moisture condition, soil temperature and reducing nutrient loss, salinity and erosion problems. Therefore, the present chapter gives a clear overview about the importance of mulching as organic fertilizer, soil regulator, water, nutrient and residue manager, improver in crop yield and productivity. The potential role of mulching in different climatic zones of Pakistan and also the role of mulch in minimizing different environmental stresses is also highlighted in this chapter.

Keywords Climate change · Climate smart agriculture · Environmental stresses · Mulching sustainable agriculture

---

S. T. Shah (✉) · A. Basit · M. Sajid · N. Ahmad · F. Noor
Department of Horticulture, Faculty of Crop Production Sciences, The University of Agriculture Peshawar, Peshawar 3125, Pakistan
e-mail: syedtanveer07@yahoo.com; syedtanveer07@aup.edu.pk

I. Ullah
Department of Horticulture, Ondokuz Mayis University, 55200 Samsun, Turkey

M. Arif
Department of Agronomy, Faculty of Crop Production Sciences, The University of Agriculture Peshawar, Peshawar 3125, Pakistan

K. Akhtar et al. (eds.), *Mulching in Agroecosystems*,
https://doi.org/10.1007/978-981-19-6410-7_17

1 Introduction

Mulching is a covering material of soil surface with organic or inorganic material for improving soil structure, conserving soil moisture condition, soil temperature and reducing nutrient loss, salinity and erosion problems. However due to modernization in agriculture sector these practiced decreased largely but in the framework of sustainable agriculture once again these practices got more importance. Climate changes have caused enormous damages like high temperature, landslides, flashfloods etc. Different types of mulches had exhibited significant performance as compared to agricultural soil in term of reducing soil erosion more than 90% (Mostaghimi et al., 1994). The need for providing large access of quality food and environment to individuals has encourage the use of organic or inorganic materials as mulch that can maintain the physical and chemical condition of soil (Armbrust & Jackson, 1977). Climate change is a state of change in climatic condition driven by human activity directly or indirectly causing alteration of global temperature which becomes a drastic challenge to sustainable food security leads to a major loss to agricultural produce (FAO, 2010a, 2010b). This alteration of global temperature produce long term change in weather distribution patterns (CO_2 availability, global temperature, intensity and rate of harsh weather events, up surging sea level and weather variability etc.). In addition this alteration also promote other environmental events like water scarcity, storms, fire incidence, pest infestation etc. which adversely affect the health of ecosystem but also cause a major damage to agricultural productivity and food security. Though, the fundamental agriculture practices to climate smart systems must be adopted to minimize the ill effect due to the change in climatic conditions. Substantial transformation in agriculture is needed in developing countries to achieve sustainable food security and response to climate change. Climate change show drastic reduction in agricultural production, its stability and incomes in some areas that also face food insecurity. In order to reduce these damages, a new approach of Climate smart agriculture must be promoted for achieving sustainable agricultural productivity and incomes and also to mitigate the adverse effect of climate change and reduce greenhouse gases emissions. The objectives of climate smart agriculture could be achieved by following some strategies like use of renewable energy technologies (solar panels, pyrolysis unit, bio-operated energy units and wind mills) and resource conserving technologies like zero tillage for avoiding the detrimental effect of heat on grain filling of wheat after planting of rice or cotton harvest. The development of heat, salinity and drought tolerant varieties also contribute in achieving the objective of climate smart agriculture. The threats of climate change could be minimized by getting advantage of weather forecast system and early warning system. It is necessary to differentiate the climate change proned regions so that appropriate strategies could be adopted. The idea of climate smart agriculture can be successfully integrated by overcoming the three interlinked issues like improve in productivity of agriculture produce in a manner of sustainability and at the same time to adopt climate change condition and reduced greenhouse gases emission in the atmosphere. This approach is developed by linking the agricultural production and food security to the strategies mitigated

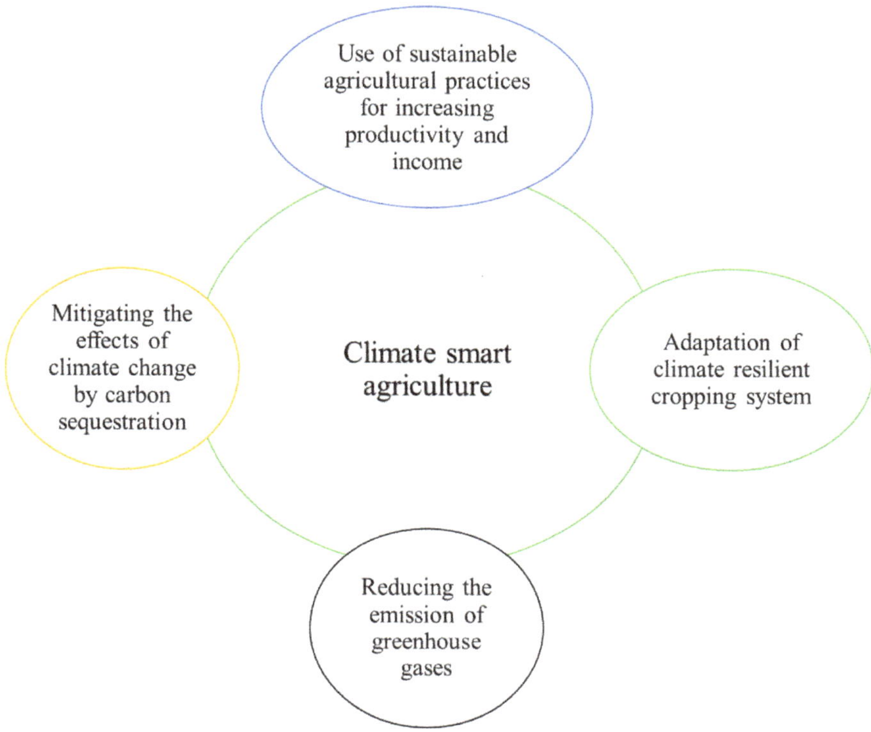

Fig. 1 Schematic representation of climate smart agriculture

and adopted to climate change, promote the agriculture and food system management at various levels (Fig. 1).

2 Mulching: An Advance Tool in Climate Change

Mulching is an organic or inorganic covering material of soil which help in regulating moisture content, soil temperature, weeds suppression, increase soil fertility, prevent soil erosion and reduce pest infestation (He et al., 2016).In addition, mulching has an effective influence on reducing water flow on the surface and increase the soil surface hydraulic roughness (Prats et al., 2016). Organic mulch is helpful in enhancing health condition of the soil and conserve soil moisture and regulate soil temperature by reducing the loss of nutrients and evaporation from the surface of the earth (Montenegro et al., 2013). Furthermore, living mulch (ground covers) is also useful to promote symbiosis (plants and main crop) and plant capability to fix nitrogen. It also helps to improve status nitrogen due to biological fixation and also increase the pool of nitrogen, phosphorus and carbon in the microbial biomass. Similarly the use of straw as a mulch observed reduction in the compaction of soil and improve nodule formation and nitrogenase activity and yield related attributes like protein and seed content of grain legumes and alleviate the hostile influence of saline stress in the rhizosphere (Abd El-Mageed et al., 2016). Mulching is an approach of sustainable agriculture of promoting the production and navigating the adverse influence of environmental stress. The population of the world are increasing day by day and to fulfill the demand of food for 30% increase in population globally under availability of scarce land, water and energy resources, promotion of a new sustainable agriculture approach Climate smart agriculture are needed to secure food security under climate resilient cropping system and also the navigate the adverse impact of climate change by carbon sequestration and NO tillage or reduced tillage system.

2.1 Potential Effects (Adoption of Sustainable Land Management Practices)

An approach to adapt practices and measures at socio-economic and biophysical conditions for the conservation, protection and manageable water use, soil and biodiversity is known as Sustainable land management (SLM). In addition it also promotes the function of ecosystem and restores the degraded natural resources. The main advantage of adopting practices of sustainable land management is the prevention of vulnerable land conversion, mitigation of land degradation, control of soil erosion, enhancing soil health condition, rehabilitation and sustainable management of dry land environments, improving the crop productivity and soil salinity management in irrigated dry land agriculture (FAO, 2021) and also promote system flexibility leads to enhanced the livelihoods and food security and decreased the risks of agricultural productivity (Woodfine, 2009).

2.1.1 Agronomy

Cover crops are the plants that are cultivated for covering the soil rather for the purpose of being harvested or the intercropping of the crop with the main crop in a region like semi-arid regions of the Sahel that is specified by single and relatively short rainy season. Cultivation of cover crops improve soil fertility, soil quality, conserve moisture, mitigate environmental stress, reduce soil erosion and pest and disease infestation as result increase yield (Olson et al., 2010). Using cover crop as mulch in no till crop production system had improved availability of nitrogen (Smith et al., 1987), improve soil fertility (Cavigelli & Thien, 2003), conserve moisture content of soil, decrease insect and disease infestation (Sustain-able Agriculture Network, 1998), reduce influence of environmental stress and increase crop yield (Triplett et al., 1996). Similarly Altieri (2001) reported that the use of cover crops improved crop productivity of maize by 198–246% in Brazil.

2.1.2 Organic Fertilization

Organic fertilization is the application of naturally produced materials having high carbon content in order to improve the soil fertility and growth of the plant. Wang et al. (2018) reported that mulching had an effective role in improving the soil fertility, promoting enzyme activity, conserving soil moisture content, weed suppression, adjusting soil temperature and microbial activity of soil. The use of organic fertilizer has played an important role in improving the productivity of most crops. Hussain et al. (2020) revealed that application of organic fertilizer had increased yield of cauliflower as compared to inorganic fertilizer. Similarly Hine and Pretty (2008) observed increase in yield of maize in Kenya by 100%; Scialabba and Hattam (2002) evaluated the yield of potato between the early 1980s and 2000s in Bolivia and observed 250–375% increase in yield of potato (from 4 to 10–15 tons/ha). Atijegbe et al. (2014) reported that poultry manure supplemented with NPK application improve crop yield and quality of okra in Nigeria.

2.1.3 Soil Regulation

A wide range of terminologies and literature are available to determine different practices that can play an effective role to reduce soil disturbances. Amongst these practices is the structural regulation of the topmost layer of soil (Stavi & Lal, 2012) which can be meet with the most effective methods like permanent covering of the soil with organic or inorganic residue (mulching) and minimum use of tillage in preparation of seedbed. Minimum tillage practices (zero, zonal and ridge tillage) coupled with mulching practices had a significant effect on soil structure and also on crop yield as compared to conventional tillage practices where ploughing of soil are done with animal or mechanically. In case of zero tillage the seeds are sown through driller and the preparation of seedbed are not done mechanically without disturbing soil through tillage. Similarly in strip or zonal tillage systems, the seed bedding area are divided into different zones and seed planting area are ploughed mechanically or by hand hoe and the zones are leave as undisturbed and are often mulched (Ibraimo & Munguambe, 2007). The regulation of soil profile through minimum disturbance of soil not only maintains soil structure but also improve soil microbial activity which promotes health of the environment (Huggins & Reganold, 2008). The use of mulching had an effective role in maintaining soil health by decrease soil erosion, limiting nutrient leaching, conserve moisture content and improve soil porosity and aeration (Stavi & Lal, 2012), as resulted higher yield are noted (Conant, 2009) water deficit areas (Stavi & Lal, 2012).

2.1.4 Water Management

Water management is the optimum use of water to improve their beneficial use in an efficient way to minimize damage to life. One of the most effective water management techniques like terraces and contour farming had contributed significantly in reducing soil erosion and mitigate runoff velocity and flooding that leads to higher yields. Altieri (2001) reported improvement in upland crops by 150% when Incan terraces were restored. Further, Shively (1999) also observed 15% increase in yields of maize cultivated on contour hedgerows compared to conventional practices used on hillside farms in the Philippines. Global warming had caused drastic damage to agricultural water resources and caused irregular rainfall pattern limiting agricultural productivity in arid and semi-arid regions In dry land farming the conservation of moisture with mulching is one of the most effective approach to save water resources as well as growth and production of plants. Mulching helps in mitigating water stress condition, conserve soil moisture, regulate soil temperature, limit weed growth and reduce soil erosion (Kader et al., 2019). Qin et al. (2015) reported that mulching reduce soil erosion, conserve moisture content in the soil and improve soil structure and fertility due to movement of earthworms into the soil.

2.1.5 Agro Forestry

Agroforestry is the management practices of land use in which perennials plant (trees, shrubs and crops) are integrated to optimize the benefits from the biological interactions. About 46% of all agricultural lands support 30% of all rural populations. The plantation of trees on different ways like contour planting, multiple cropping, intercropping and tree fallows create favorable climatic condition for better land productivity, improve soil health and promote better microbial activity enhance their fertility (Zomer et al., 2009). The drastic effect of change in climatic conditions can be mitigated by planting trees and shrubs which helps in securing sustainable food security, reduce the vulnerability and create an adoptive agricultural systems. In addition it also increase farm productivity and income and reduce risk of market failure or agricultural production (FAO, 2010a, 2010b). It can be widely used as fruits, fodder, building materials and fuel which can further contributes to household risk management by providing a big source of income (Ajayi et al., 2007). Under increasing drastic condition of climate change, crops grown in shade condition had contributed to yield of grooves by 23–38% (Soto-Pinto et al., 2000) and also provide a sustainable growing conditions for better growth of plants (Lin et al., 2008). Hellin and Haigh (2002) observed non-significant results of crop yields in their experimental trials conducted between 1996 and 1998 (Fig. 2).

2.1.6 Mitigation Potential in Relation to Climatic Zones in Pakistan

Climate change has a drastic impact on Pakistan sustainable development and other sectors and ecosystem like energy, food and water, marine and costal environment, forest and biodiversity as well as on the occurrence of environmental stress like drought and flooding. Climate change has an effective potential of seriously harming Pakistan socially, environmentally and economically. Global warming are more

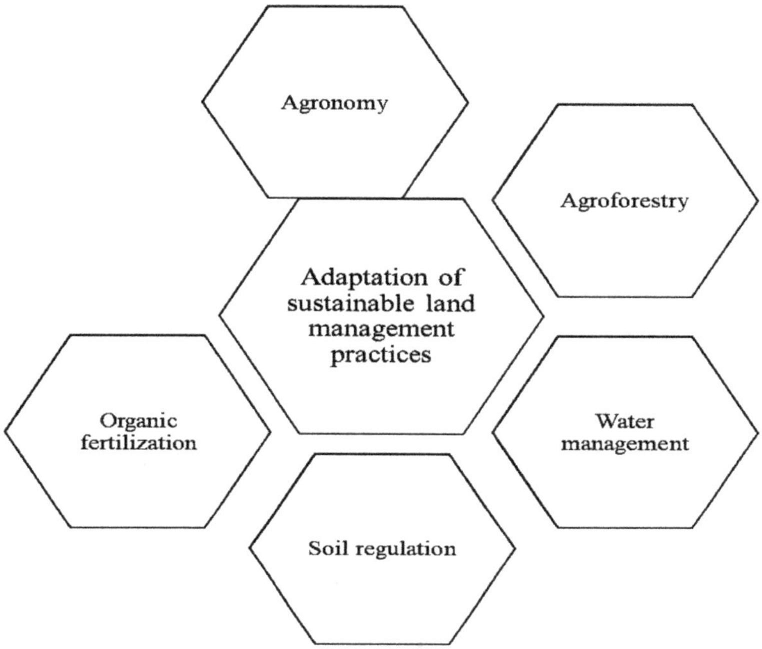

Fig. 2 Schematic representation of adaptation strategies of sustainable land management

drastic in a country where the basic life requirements, growth rate and density are dependent on environment and the community living in that areas which has low adaptation capacity to poverty and impacts related to change in climate. The vulnerability to change in climate has been realized and has given proper attention by the country to develop a national policy to mitigate the negative impact of climate change and also an action plan for the implementation of climate change concerns. According to findings observed by Pakistan Meteorological Department and GCISC (Global Change Impact Studies Centre) reported that climate of Pakistan are changing and during last century about 0.6 °C rise in temperature was recorded. There is also a significant difference in rise of temperature in northern (0.8 °C) over southern regions of Pakistan (0.6 °C). Regional and city-level studies have also revealed rising temperatures in Pakistan due to global warming which has contributed to raise the average annual precipitation over the last century. Over the previous century, it is thought to have increased by 25% (Shakoor et al., 2012). There has also been a rise in the number of wet events across the region. Forty-one meteorological stations out of 54 reported an increased trend in precipitation. In terms of forecasts, today's global circulation models (GCMs) effective at forecasting precipitation. The collective outputs of 13 and 17 GCMs for the A2 and A1B scenario were used to observe the GCISC. The increased and decreased rainfall in summer and winter is likely to occur in both the Northern and Southern parts of Pakistan but there will be no significant increase in the annual rainfall (Sheikh et al., 2009). Pakistan is particularly vulnerable to climate change's negative effects. Out of the 170 countries in the world, Pakistan ranked 16th based on the Vulnerability index to climate change (Maplecroft, 2010a). Pakistan is taking effective global mitigation initiatives in mitigating the negative impact of rising temperature in the region (Cheema et al., 2006). These issues are particularly essential and dire to frame the best possible evolving strategies for the nation, especially under the limited available resources.

Pakistan's freshwater supplies are already in short supply. As the world's population has grown, per capita water supply has decreased from 5600 to 1200 m^3 in 1951–2003, respectively which is now reaching the threshold level of water scarcity i.e. 1000 m^3 (Commission Planning, 2007). Pakistan is already classified as an extremely high-risk country by Maplecroft (2010b) index of water protection risk, ranking 7th out of 165 countries (Maplecroft, 2010b). Rainfall and glacier and snowmelt are the country's two primary sources of water. The contribution of westerly winds and Monsoon is about 50×10^6 acre-feet of rain, or 60×10^9 m^3 of water. Melting of glaciers and snow from the Mountainous regions of Hindukush–Karakoram–Himalayas range gives about 141×10^6 acre feet or 174×10^9 m^3 of water to the Indus River System from snowmelt and glaciers (Commission Planning, 2007). Change in climatic conditions is expected to have an influence on all of these outlets. As a result, given the restricted space for expanding water sources, Pakistan need to focus on increasing the quality of water usage in agriculture. Irrigation from the Indus River System is currently ineffective. Pakistan loses twice as much Indus water in watercourses per year than could be stored at Tarbela dam. According to an estimate, irrigation water of about 60% is lost in the transport process from the source to the field. About 50 and 33% of the losses in watercourse and canal level is due to lack of maintenance and operations in the canal system (Ahmad et al., 2007). Mitigation and development of sustainable land management (SLM) strategies are high in areas with more precipitation except water management. Agronomic, water and integrated nutrient management practice for getting crop yields are more effective in humid areas as compared to dry areas. However, higher average yield and agroforestry systems can be obtained under tillage practices in dry regions which clearly indicate the key role of water to obtain better crop production. These findings also give a clear picture about SLM practices to improve crop production using efficient strategies of water use both in dry and humid regions. In humid areas, terracing and other conservational steps in water and soil are some of the effective water management strategies that can increase nutrient supply and soil organic matter and minimize soil erosion in the root zone of the plant. In contrast, better use of limited amount of water in plants is one of the effective management practices in drier areas. Other strategies (minimized direct evaporation with enhanced hydraulic conductivity of the uppermost layer of the soil as well as porous soil surface) can effectively improve availability of water to plants under low tillage systems (Scopel et al., 2001).

3 Role of Climate Smart Practices in Conservation Agriculture

Climate smart agriculture is the adaptation of climate resilient agricultural practices to navigate the threats of global warming by carbon sequestration, reducing the production of greenhouse gases emissions and increase the productivity and income of agricultural produce. The adaptation of climate smart agriculture strategies must be interlinked with the objective of minimum disturbance of soil structure, maintenance of crop residue, use of crop rotation or intercropping for the diversification of crops (FAO, 2015; Kassam et al., 2009), minimum use of tillage or zero tillage for planting wheat after rice or cotton harvest to mitigate the adverse effect of environmental stress on grain filling and also the use of newly developed heat, drought and salinity tolerant variety. In addition the identification of climate change prone areas are also necessary, so that proper navigating strategies like early warming system and weather forecast could be adopted to diminish the threats of climate change. Climate change has caused a drastic damage to the environment and agricultural productivity throughout the world. In addition they also promote carbon dioxide accumulation and have a negative impact on herbivores or all other member of food chain system. Climate change had a drastic impact on agricultural activities which boost higher accumulation of carbon dioxide, rising the global atmospheric temperature and disturb the sequence of rainfall and overflow of large glaciers all over the world (IPCC, 2007). Agricultural fields are negatively influenced not only by maximum accumulation of carbon dioxide but also by the emission of greenhouse gases like nitrous oxide (N_2O) which contributed to the total greenhouse gases emissions by 58% and methane (CH_4) emission by 40%, released through livestock and cultivation of rice. The importance of agriculture and forestry cannot be ignored and could be effectively managed to navigate the negative impact of climate change (Gitz, 2013).

3.1 Improvement of Crop Yield and Productivity

The adverse impact of climate change on agricultural productivity could be navigated by an effective use of agricultural practices and technologies in a sustainable manner which cam enhance the management of natural resources like genetic resource to cope to climate change (FAO, 2015). The sustainable productivity of agricultural produce and increasing food security and resilience in crop production could be achieved by adopting mitigation practices like development of environmental friendly soil ecosystem by maintenance of soil health, cultivation of a wide range of varieties and species in rotation and sequences, Use of high yielding well adopted varieties to different environmental stress, adaptation of pest management strategies and effective management of water (FAO, 2014). Climate smart agriculture is known to be an effective approach of mitigating the negative effect of changing climate and availing sustainable food security (Tantely et al., 2015). A wide range of sustainable agricultural practices like water conservation, conservation agriculture, irrigation system, agroforestry to increase household income, increase the productivity of agriculture produce, reduce the production of greenhouse gases, effective management of carbon content in soil as well as the development of new varieties to cope the negative impact climate change and to secure food security and higher yield of both legumes and cereals under a varying climatic condition (Thierfelder & Wall, 2010). Population of the world are growing day by day and about 34% increase in population are expected by 2050.The food requirement of these growing population needs about 70% increase in world food production (FAO, 2009). Similarly the production of cereals by 2050 needs a rise of 43% and rice production which feed half population globally need an increase by 0.6–0.9% annually to meet the demand of growing population (Carriger & Vallee, 2007). The increase in production of cereals will be required in a world where variability in rainfall patterns and more frequent weather extremes conditions caused by climate change have adverse impact on the quality productivity of cereals. The improvement in cereal productivity will be achieved by improving the current germplasm for yield potential and enhancing the abiotic stress tolerance of cereal crops to cope the adverse impact of climate change.

3.2 Adoption to Climate Related Stresses

Plants are very delicate and they can acquire the changing environmental condition for achieving important functions of development. The environmental stress and climate variability have negative impact on growth and yield of agricultural produce. Although the developmental of grain yield in cereals depends on the successful reproductive process of the crop in an environment. The size or number of grains of the cereals is determined by the timing of stress stimulus in term of development of reproductive process. Abiotic stress like drought stress cause reduction of yield by impacting crops during the initial stages of reproductive development (Savin & Slafer, 1991). Similarly high temperature stimulus has a drastic impact on normal growth and flower physiology of the plants which directly influence the grain formation and ultimately the crop yield. Plants adopt different responses like molecular, physiological, morphological and biochemical to mitigate the adverse impact of climate change (Johnson et al., 2005). Fluctuation in temperature had significantly reduced the yield of wheat crop by 6% (Reicosky, 2000). Climate smart agriculture have been indorsed a navigating strategies of reducing the emission of greenhouse gases, increasing the productivity of agriculture produce, enhancing the health condition of the soil, conserving soil, moisture, increasing the resilience to climate change and supporting the food security in a sustainable manner (Derpsch et al., 2010).

3.3 Water Infiltration

Water infiltration is the entrance of surface ground water into the soil and is mostly affected by different factors like precipitation, soil characteristics, moisture content of soil, organic contents in soils, slope and land cover. The higher rate of water infiltration could be achieved by adopting the combination of different practices like no tillage, crop rotation and conservation of residues (Thierfelder & Wall, 2009). Increasing water infiltration significantly enhance the conservation of soil moisture which provide a mitigating support to plant in dry spells of the season and benefits the climate agriculture system. Similarly the combine use of no tillage and crop residue provide environmental friendly condition for soil macro fauna to improve the soil structure by moving to the soil surface and enhance the air movement and water infiltration (Kladviko et al., 1986). The use of crop residues provides the growth enhancing and moisture conserving condition for the proliferation of soil organism. Nyamangara et al. (2014a) reported that the adaptation of climate agriculture after years in contrasting soil improved pore volume minimally by 70% in Zimbabwe which contributed to higher soil water infiltration (the conservation of soil moisture and ground water recharge was significantly enhanced by the rotation of legumes with maize which contributed in improving the soil structure and soil porosity) (Rusinamhodzi et al., 2012). The usage of root exudates helps in the formation and stabilization of exudates which enhance the water infiltration and soil porosity (Bronick & Lal, 2005).

3.4 Soil Moisture Conservation

The conservation of moisture content in the soil depends on minimizing the loss of water through evapotranspiration which ultimately have direct impact on irrigation requirement of the crops and also on agricultural productivity. The use of crop residues in agricultural systems enhance the water infiltration and conserve soil moisture by minimizing the loss of water through evaporation (Roth et al., 1988) and provide a mitigating support to plants against environmental stress (Cairns et al., 2013).Climate change has a become serious issue of preventing the damage expected in future. About 1.6–6 °C rise in global temperature are expected by 2050 due to climate change which will directly affect the renewable water resources of the world. The adaptation of climate smart practices will help in mitigating the adverse impact of climate change and management of water reservoirs globally which will promote water security and contribute to development of sustainable agriculture (FAO, 2016). According to Thierfelder and Wall (2010), the adoptative response of climate smart agriculture systems could be achieved under availability of maximum soil moisture content in the system. Furthermore, at a soil depth of 0–60 cm, higher moisture content was observed in Zambia during 3–4 weeks continuous dry spells.

3.5 *Regulation of Cropping Season and High Density Plantation (Labor Savings and Early Planting)*

The ability to react quickly to new circumstances of changing climate condition is an important feature of adaptation. Farmers must take advantage of the first successful rains in a changing environment since this normally extends the growing season and results in higher yields. Labor shortages at the start of the rainy season trigger planting delays and have a significant impact on maize yields (Mazvimavi & Twomlow, 2009). Seeding practices under zero tillage benefits the farmers which permits seeding at a faster rate i.e. ripline or direct system of seeding or/and basin planting system where more labors are required for planting (Sims et al., 2012). Basins, for example, prefers winter season for their preparation (Mazvimavi & Twomlow, 2009), while minimum labor are required with direct planting i.e. no need to wait for field ploughing in ripline or direct seeding system (Thierfelder et al., 2016). Smart climate agriculture practices results in better yield which may be due to plantation on right time which clearly indicates that the system is highly adoptability. Early plantation benefits the farmers in way that it provides larger areas and also allows them to plant another crop. A classic example may be the plantation of beans or cow pea sown after harvesting early maturing maize crop. This adaptation also helps growers to produce crops on marginal lands where limited moisture in the soil can be used easily (Nyamangara et al., 2014b) (Fig. 3).

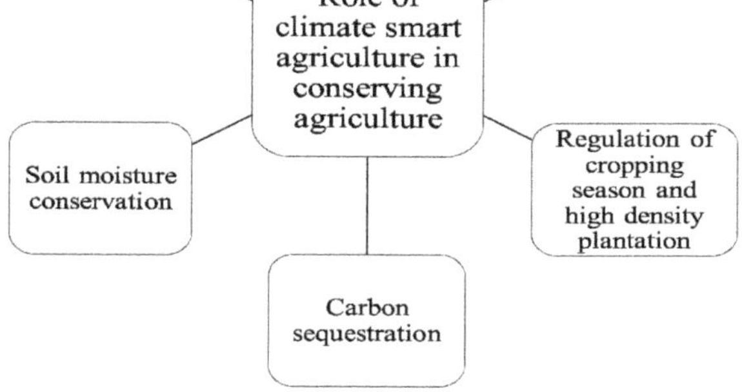

Fig. 3 Schemetric representation of conserving agriculture strategies of climate smart agriculture

3.6 Carbon Sequestrations

Carbon sequestration is an approach of long term removing or sequestrating carbon dioxide from the earth atmosphere and its storage in soil to navigate the adverse impact of global warming. This process is mostly enhanced by an effective process like photosynthesis in which the atmospheric carbon dioxide are used as a raw material to manufacture organic compound that is a vital component of plants growth. After decaying of plants parts, the soil organism bacteria, fungi and earthworm starts decomposition of organic material and convert them into soil organic carbon and prevent its entrance into the atmosphere. Carbon sequestration has the ability to effectively reduce the amount of CO_2 in atmosphere and also to minimize the release of CO_2 from major stationary human sources, including power plants and refineries into atmosphere. Crop land, hedgerows, and semi-natural habitats, which also contain large plants like trees that are more effective at trapping CO_2, also capture a significant amount of carbon. To absorb more carbon, better land management and changing land use can be used to increase soil carbon sequestration (Lal, 2001). It has a significant effect on minimizing the risk of marine and atmospheric accumulation of greenhouse gases. Different strategies of climate smart agriculture like cover crops, tillage conservation and application of biochar have an effective role promoting the carbon sequestration of soil and also navigate the adverse effect of greenhouse gases on crop productivity. Soil organic carbon (SOC) is a key indicator of soil quality, with implications for food production, greenhouse gas balance, and climate change mitigation and adaptation (Lorenz & Lal, 2016). Under long-term constant environmental and management conditions, the equilibrium between carbon inputs (e.g., crop residues and organic fertilizers) and outputs (e.g., decomposition and erosion) controls the dynamic of agricultural SOC. Climate change, on the other hand, is expected to increase SOC decomposition while weakening soil's ability to sequester carbon (Wiesmeier et al., 2016). Climate smart agriculture (CSA) is effective approach of ensuring long-term food security under climate change condition (FAO, 2013), Reduced greenhouse gas emissions and improved soil carbon sequestration and soil quality (Lipper et al., 2014). Growing carbon inputs while lowering carbon outputs is the secret to sequestering more carbon in soils. Adding cover crops to the crop rotation, applying biochar to soils, and minimizing soil tillage are all commonly recommended approaches for SOC sequestration (i.e., conservation tillage). These management methods have been used in major agricultural regions around the world in recent decades, resulting in a significant number of observations/measurements (Clark et al., 2017). Cover crops increase carbon and nitrogen inputs, increase biodiversity in agroecosystems, and provide additional biomass inputs from above and belowground (Blanco-Canqui et al., 2011). Cover crops can also improve soil aggregation and structure (Sainju et al., 2003), reducing carbon loss from soil erosion indirectly (De Baets et al., 2011) (Fig. 4).

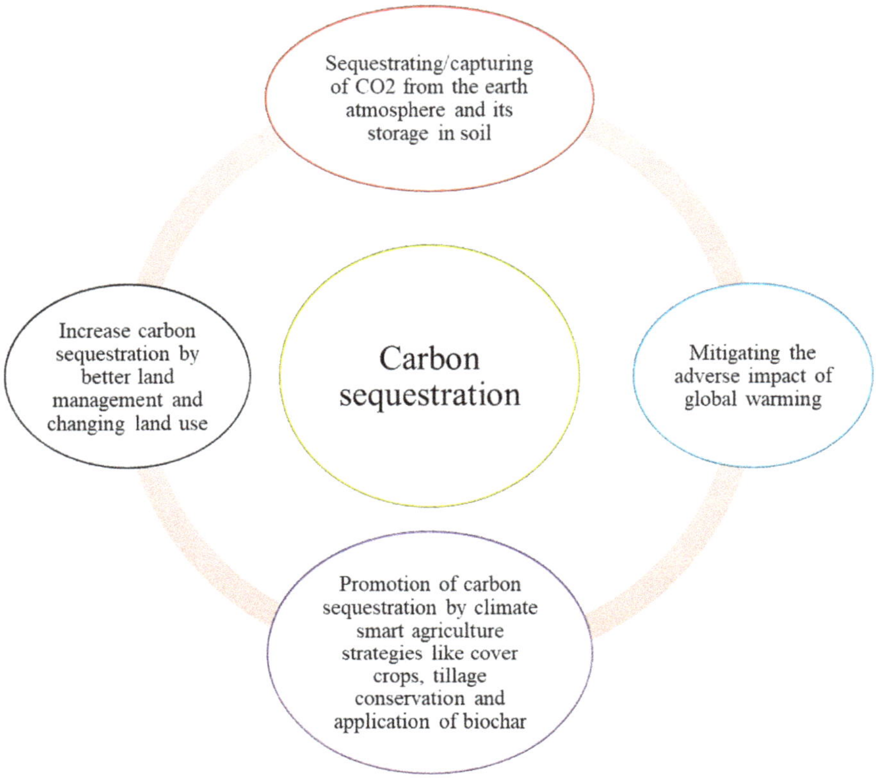

Fig. 4 Schematic description of Carbon sequestration process

4 Crop Residue Management

Crop deposits (residues) had a significant influence on soil water flow, runoff, and infiltration by incorporating a maximum amount of nutrients in the soil to produce the crop in a better way. Efficient crop residue management is an important part of a conservation agriculture (CA) scheme, and in-situ management is the only way to get the most output of conservation agriculture (Jat et al., 2019a, 2019b, 2019c). There is a positive as well negative effect of the decomposition of crop residue on the crop, which greatly depends on the researcher to maximize the beneficial effect of crop residues (Lu, 2020). Crop residues can also improve the soil health in many ways which may include prevention of soil erosion, availability of recycled nutrients to the plants, decomposition of residues, control of pest and weeds and other tillage practice which is used to maximize crop production (Yadvinder-Singh & Timsina, 2005). Recycling of plants nutrients annually is critical in the plant-soil environment for maintaining a sustainable agricultural system and improving nutrient mobilization (Liu et al., 2020). The positive interaction of soil, water and air with plants has an effective role in enhancing the availability of nutrient for better growth and production

(Zhao et al., 2020). Furthermore, crop residues enriched with carbon is a major source of food for microbes which assists to commence framework for biological nutrient recycling. The living organisms and plants utilize the different chemicals produced during decomposition of crop residues in the soil (Meena & Lal, 2018). Crop residues greatly affect availability of macronutrients especially nitrogen, phosphorous and potash using various biological, physical and chemical processes. The amount of nutrients recycled and its availability to the plant from crop residues depends on leftover crop residues from the cropping area and fertilization. The promotion of soil organic matter could be achieved by an effective management of residue of crop for a better sequestration of C by the combine use of nitrogen fertilizer with crop residues (Rathod et al., 2019). The prosperity of human are linked to Agriculture and related industries, are the backbone of many developing and underdeveloped countries' economies and directly or indirectly affect nearly 82% of the world's population. This necessitates require the adoption, growth, and implementation of improved manufacturing technologies (Kar et al., 2021). The most important keys to ensuring food production sustainability are agricultural technologies that conserve energy. Soil nutrients are often depleted as a result of such input-intensive agricultural practices. Heavy machinery like sum cum fertilizer drill, rotavator and combine harvester, nowadays are becoming popular (Mondal et al., 2020a, 2020b). Crop residue production was also increased linearly in modern input-intensive agricultural practices. Onsite burning of crop residues has significant health and environmental consequences (Maneepitak et al., 2019) and accounts for the burning of more than 1/3rd overall biomass in Asian countries (Chen et al., 2019). Environmental pollution is also caused by particulates (PM) emitted by such burning, such as greenhouse gases (GHGs), $PM_{2.5}$ and PM_{10} (Zhao et al., 2020).

4.1 Soil Organic Matter

The drastic decrease in soil organic matter pool as crop residues are incorporated into the soil (Wang et al., 2004). There is a constant increase in soil's light carbon fraction due to uninterrupted addition of residue incorporation for 3 years adds a large amount to overall soil organic C (Conteh et al., 1998). Changes in heavy-fraction carbon (>1.6 gcm^3), based on residue management methods had significantly influence on organic carbon over time (Wang et al., 2004). Intensive farming practices encourage the degradation of organic matter over time adversely affect carbon balance and quality of the soil worldwide. Residues obtained from legumes in comparison with cereals give much more biomass production and net gain of carbon in the soil (Tiemann et al., 2015). Beyond 30 cm soil depth, legume stubbles produce 49% more biomass, 133% more nitrogen, and 60% high soil organic carbon (SOC) than untreated plots with crop residues. Crop residues obtained from legumes minutely affect carbon storage in soil which might be because of tapered C:N ratio and least lignin content (Conteh et al., 1998), a reason for quick decomposition of residues. Residue retention of sun hemp resulted in 0.92% more SOC and 0.64% less soil inorganic carbon (SIC) compared to control plots (no residue incorporation) (Conteh et al., 1998). The difference in SOC differs area to area due change in management practices for crop residues along with soil and environmental factors. For example, it takes round about 10–20 years to bring the soil carbon content a new equipoise using better management of crop residues as compared to Asia and Australia takes about 20 years (Yadvinder-Singh & Timsina, 2005).

4.2 Soil Nutrients (Status and Its Availability)

Crop residues contain nutrients that are organically bound and must be mineralized before they are accessible to plants (Bhupinderpal-Singh & Bowden, 2006). Following absorption, the mineralization and decomposition of residues are brought by a number of microbes residing with them resulting in transformation of residues into monomers. These monomers are then transformed into organic matter by the use of chemical, mechanical and biological assimilation process (Salas et al., 2003). In addition the presence of some nutrients in the crop residues like K^+ and $SO4^{-2}$ are in the soluble organic form or linked with the mineralization of organic materials. For example phosphate Easter or protein bound S (Bhupinderpal-Singh & Bowden, 2006). One of the key benefits of crop residue is to momentarily immobilize the nutrients for the plants by storing the available form of nutrients in the soil. This results in the availability of nutrients for longer period of time to plants as nutrients are not readily available to the plants thus utilize nutrients efficiency and minimize volatilization and leaching of nutrients. Piccoli et al. (2020) found a significant improvement in nutrient usage efficiency as a result of residue retention. Repeated residue decomposition greatly increases nutrient distribution, and previous research has shown that phosphorous accumulation (both organic and inorganic) in the soil is more with modern tillage practices as compared to traditional ones (Du Preez et al., 2001). Higher organic matter deposition in conservation practices is thought to increase the amount of P availability by drenching adsorption sites of P on soil colloids. Similar results were also obtained in case of N availability (Salinas-Garcia et al., 2001). Losses in soil nutrients are much more in leaving the crop residue on soil surface as compared to conservation tillage (Sarkar et al., 2020). Surface residues limit soil erosion and subsequently surface nutrient losses (Lal, 2005). Heavy losses of nitrogen (due to leaching, uptake by weeds and denitrificaiton) might be due high decomposition rate of resides. Removal of residues increases the risk of K deficiency in the soil as they contains high amount of K (Whitbread et al., 2003). Plant tissues do not contain K in their organic structure, so K release is not dependent on residue decomposition which is lost in different ways as if there is no crop demand, rainwater or irrigation can wash it out of the residues. This emphasizes the significance of residue management in terms of nutrient release timing and crop demand. The decomposition of nutrients is greatly influenced by crop residue's quality (presence of polyphenols, nitrogen and lignin) (Whitbread et al., 2003).

4.3 Soil Productivity

Soil productivity is directly related with the amount of soil fertility and agronomic management. The different dimensions of soil fertility like physical, biological and chemical characteristics of soil contributed positively on crop yield (Poeplau et al., 2017) which is correlated with presence of organic matter in the soil (Hijbeek et al., 2017), and residue preservation increases soil organic matter storage. In comparison to other crops, crop residues have the greatest impact on cereal production (Schjonning et al., 2018). More residues are expected to increase soil C and enhance soil properties such as microbial activity, water retention, nutrient mobilization and temperature (Wei et al., 2016). Piccoli et al. (2020) found that integrating crop residues resulted more production in sugar beet and maize by 16 and 12%, respectively, than other sources. However, Hijbeek et al. (2017) found that residue retention enhanced maize yield by 4%. Mandal et al. (2004) also discovered that leaving straw and rice residue (leftover without burning for 13 years) on the field increased wheat yield by 12.3 and 53.25% respectively. They also reported that overall characteristics of soil, especially soil organic matter had significantly improved the yield of maize in a similar pattern due to large contribution of improved activities of microbes and storage of Carbon.

4.4 Improving Activities of Microorganism in Soil

Managing crop residues have a substantial effect on soil microbial biomass regulation. Mondal et al. (2020a, 2020b) have observed significant effect of crop residue as mulch on higher activities of microorganisms in the uppermost soil layer which contributed to effective plant–soil microclimate, soil temperature regulation and increased water and nutrient availability. Chatterjee et al. (2018) found similar results and found that increase in soil microbial biomass carbon (MBC) was higher with the incorporation of wheat residues. The integration of leguminous crop residues like cluster bean, into the soil followed by 2nd crop sowing has been contributed to promote the dehydrogenase activity (DHA) and biomass of soil microbes compared to no crop residue treatment (Smitha et al., 2019). In a wheat–soybean cropping scheme, residue retention under CA has also been shown to help reduce soil nematode populations (Escalante et al., 2020).

4.5 Reduce Soil Degradation

The beneficial microbes of soil uptake nutrients and get shelter from the tropical region cereal crops like wheat, maize and rice which is a major source of carbon (40%), nitrogen (0.8%), phosphorus (0.1%) and potassium (Adimassu et al., 2019).

Furthermore, removal of residues for industrial use or cattle feed results in high removal of nutrients from croplands which may adversely affect soil (problems related to soil erosion, lowering the quality of water, soil and air) As a result, crop residues left over after harvest will effectively help to preserve soil resources and maintain crop productivity.

4.5.1 Minimize Soil Erosion

It has been stated that efficient use of crop residue reduces problems of runoff, sediment transport or losses, and conserves moisture content in the soil (Adimassu et al., 2019). It has been stated that using residue mulching can minimize soil loss extent up to forty three percent when compared to unproductive land. Moreover use of mulching contributes by reducing the runoff, nutrient depletion, and sediment presence in runoff water. Similarly using crop residue as a top layer of soil minimize the topsoil losses by 30%. Legumes are the most effective cover crop because they symbiotically add atmospheric N and thereby enhance soil quality. Increase in mulching and number of plants significantly decrease runoff (Ghosh et al., 2018). The cropping method chosen has a significant impact on soil erosion and removal of the top fertile layer of soil. Consecutive implementation of mono-cropping system with erosion-tolerant crops resulted in increased soil and water loss. Turmel et al. (2015) reported that use of soya bean residue positively contributed by minimizing soil loss by 50% as compared to unregulated residual soil.

4.5.2 Lower Soil Salinity

Surface mulching has a significant and beneficial effect on soil salinity management by lowering evapotranspiration. Brahmachari et al. (2020) observed that salinity was significantly reduced in potato field by using zero tilled mulched as compared to rice fallow fields. Fan et al. (1993) reported that the consecutive use of straw mulching for two years resulted reduction in the soil salinity from 0.44 to 0.07%. Yang et al. (2006) also observed positive impact of mulching of soil minimizing the adverse impact of salinity stress.

4.5.3 Decrease Soil Aridity

Soil moisture is an important factor growth and development of the crop as well as in nutrient availability. The use of plant residue is very helpful for dryland and rain fed regions areas to conserve soil moisture. Residue mulching is considered as economical and effective measure produce quality crop with maximum yield and also regulate the environment (reduce evaporation, regulate soil temperature, improve soil moisture) for the crop (Brahmachari et al., 2020). With increased ground coverage, increased mulch coverage greatly decreases splash erosion. Mulching treatments, on average, store more soil moisture than untreated soil (bare soil) (Jat et al., 2019a, 2019b, 2019c). Surface residue preservation is a possible alternative for conserving soil moisture by minimizing capillary loss in the dry tact (Brahmachari et al., 2020), where rainfall is much less than the average evaporation.

4.5.4 Maintenance of Soil Temperature

Application of crop residue as a thick mulch in the field have an effectively potential of regulating soil temperature by controlling sunlight penetration and retaining heat for better growth during the cropping season. (Mondal et al., 2020a, 2020b; Samui et al., 2020). The potential of Surface crop residues in regulating the moderate temperature are so effective in dry tropical climate where the soil temperature rises too high that have an adverse impact on plant growth (Su et al., 2007). Similarly crop residue mulching has a significant effect in maintaining the soil temperature in cool climates and provide a better environment for plant growth (Shen et al., 2018). In addition, the combine use of residue and various tillage practices in crop fields serve as defensive materials, bringing a consistency. Conservation activities change bulk density, inter-aggregate interaction, moisture and organic matter content of the soil, according to numerous studies which resulted influence soil's heat power, thermal conductivity, and thermal diffusivity (Shukla et al., 2003).

5 Conclusions

It is essential to understand those factors which decreased framework of sustainable agriculture. This will help to reduce soil erosion, improve soil health and regain fertility. One of the techniques to improve soil properties is Mulching with organic and inorganic material. Mulching also helps to cope up the problems related to ever changing environment due to climate change. Mulching promote symbiosis, nitrogen fixation, pool of N, P and C in microbial biomass. The potential effect of mulching includes adaptation of sustainable land management practice, management of water, soil, organic fertilization that not only improves crop productivity but also add fertility to the soil. Climate change has drastically affected most part of world especially Pakistan which is a serious alarming situation which need to be managed properly. Hence, sustainable land management strategies especially tillage system & mulching and climate smart practices in conservation agriculture need to be adopted to overcome such an alarming situation in Pakistan. Climate smart agriculture practices include adaptation responses of plant (molecular, physiological, morphological and biochemical), crop rotation, tillage, conservation and management of crop residues, conservation of soil moisture, change in cropping season, high density plantation, carbon sequestration, increase in soil organic matter, nutrient status and activity of microorganism in soil, decrease soil aridity, soil salinity, soil erosion, soil degradation and regulation of soil temperature.

References

Abd El-Mageed, T. A., Semida, W. M., & Abd El-Wahed, M. H. (2016). Effect of mulching on plant water status, soil salinity and yield of squash under summer-fall deficit irrigation in salt affected soil. *Agricultural Water Management, 173*, 1–12.

Adimassu, Z., Alemu, G., & Tamene, L. (2019). Effects of tillage and crop residue management on runoff, soil loss and crop yield in the Humid Highlands of Ethiopia. *Agricultural Systems, 168*, 11–18.

Ahmad, S., Mohyuddin, J., Siddiqui, S. M., & Bhutta, M. N. (2007). Tree plantation for intercepting canal seepage and controlling water table. *Pakistan Journal of Water Resources, 11*(2), 243–247.

Ajayi, O. C., Akinnifesi, F. K., Sileshi, G., & Chakeredza, S. (2007). Adoption of renewable soil fertility replenishment technologies in the southern African region: Lessons learnt and the way forward. *Natural Resources Forum, 31*(4), 306–317 (Blackwell Publishing Ltd.).

Altieri, M. (2001). Applying agroecology to enhance the productivity of peasant farming systems in Latin America. Paper presented at the Reducing Poverty through Sustainable Agriculture. University of Essex.

Armbrust, D. V., & Jackson, J. D. (1977). Temporary wind erosion control: Cost and effectiveness of 34 commercial materials. *Journal of Soil and Water Conservation, 26*, 154–157.

Atijegbe, S. R., Nuga, B. O., Lale, N. E., & Osayi, R. N. (2014). Effect of organic and inorganic fertilizers on Okra (*Abelmoschus esculentus* L. Moench) production and incidence of insect pests in the humid tropics. *IOSR Journal of Agricultural and Veterinary Science, 7*(9), 25–30.

Bhupinderpal-Singh, R. Z., & Bowden, J. W. (2006). Carbon, nitrogen and sulphur cycling following incorporation of canola residue of different sizes into a nutrient-poor sandy soil. *Soil Biology & Biochemistry, 38*, 1591–1597.

Blanco-Canqui, H., Mikha, M. M., Presley, D. R., & Claassen, M. M. (2011). Addition of cover crops enhances no-till potential for improving soil physical properties. *Soil Science Society of America Journal, 75*(4), 1471–1482.

Brahmachari, K., Nanda, M. K., Saha, H., Goswami, R., Ray, K., Sarkar, S., & Ghosh, A. (2020). Final report of the project on Cropping systems intensification in the salt affected coastal zones of Bangladesh and West Bengal, India (CSI4CZ). Bidhan Chandra Krishi Viswavidyalaya, West Bengal India. *Plos ONE, 15*, 1–88.

Bronick, C. J., & Lal, R. (2005). Soil structure and management: a review. *Geoderma, 124*(1), 3–22.

Commission Planning. (2007). Pakistan in the 21st Century: Vision 2030. Government of Pakistan.

Cairns, J. E., Hellin, J., Sonder, K., Araus, J. L., MacRobert, J. F., & Thierfelder, C. (2013). Adapting maize production to climate change in sub-Saharan Africa. *Food Security, 5*(3), 345–360.

Carriger, S., Vallee, D. (2007). More crop per drop. *Rice Today, 6*(2), 10–13.

Cavigelli, M. A., & Thien, S. J. (2003). Phosphorus bioavailability following incorporation of green manure crops. *Soil Science Society of America Journal, 67*, 1186–1194.

Chatterjee, S., Bandyopadhyay, K. K., Pradhan, S., Singh, R., & Datta, S. P. (2018). Effects of irrigation, crop residue mulch and nitrogen management in maize (*Zea mays* L.) on soil carbon pools in a sandy loam soil of Indo-Gangetic plain region. *CATENA, 165*, 207–216.

Cheema, M. A., Farooq, M., Ahmad, R., & Munir, H. (2006). Climatic trends in Faisalabad (Pakistan) over the last 60 years (1945–2004). *Journal of Agriculture and Social Science, 2*(1), 42–46. http://www.fspublishers.org.

Chen, J., Gong, Y., Wang, S., Guan, B., Balkovic, J., & Kraxner, F. (2019). To burn or retain crop residues on croplands? An integrated analysis of crop residue management in China. *Science of the Total Environment, 662*, 141–150.

Clark, R. T., Bett, P. E., Thornton, H. E., & Scaife, A. A. (2017). Skilful seasonal predictions for the European energy industry. *Environmental Research Letters, 12*(2), 024002.

Conant, R. T. (2009). Rebuilding resilience: sustainable land management for climate mitigation and adaptation. Technical Report on Land and Climate Change for the Natural Resources Management and Environment Department. Food and Agriculture Organization of the United Nations.

Conteh, A., Blair, G. J., & Rochester, I. J. (1998). Soil organic carbon fractions in a Vertisol under irrigated cotton production as affected by burning and incorporating cotton stubble. *Soil Research, 36*, 655–667.
De Baets, S., Poesen, J., Meersmans, J., & Serlet, L. (2011). Cover crops and their erosion reducing effects during concentrated flow erosion. *CATENA, 85*(3), 237–244.
Derpsch, R., Friedrich, T., Kassam, A., & Li, H. (2010). Current status of adoption of no-till farming in the world and some of its main benefits. *International Journal of Agriculture and Biological Engineering, 3*(1), 1–25.
Du Preez, C. C., Steyn, J. T., & Kotze, E. (2001). Long-term effects of wheat residue management on some fertility indicators of a semi-arid Plinthosol. *Soil and Tillage Research, 63*, 25–33.
Escalante, L. E., Brye, K. R., & Faske, T. R. (2020). Nematode populations as affected by residue and water management in a long-term wheat-soybean double-crop system in Eastern Arkansas. *Applied Soil Ecology, 157*, 103761.
Fan, X. W., Chi, B. L., Jiao, X. Y., Li, D. W., & Zhang, Z. P. (1993). Soil improvement and yield increment in salt-alkaline fields by straw mulch. *Agricultural Research in the Arid Areas, 11*, 13–18.
FAO. (2009). Climate change implications for fisheries and aquaculture. FAO fisheries and aquaculture Technical Paper. No. 530. http://www.fao.org/docrep/012/i0994e/i0994e00.htm.
FAO. (2010a). "Climate-smart" agriculture. Policies, practices and financing for food security, adaptation and mitigation. Food and Agriculture Organization of the United Nations.
FAO. (2010b). Report of the FAO workshop on climate change and fisheries in the African Great Lakes. Bujumbura. http://www.academia.edu/12458723/Report_of_the_FAO_workshop_on_climate_change_and_fsheries_in_the_African_Great_Lakes.
FAO. (2013). Coping with climate change—The roles of genetic resources for food and agriculture. http://www.fao.org/3/a-i3866e.pdf.
FAO. (2014). Enabling farmers to face climate change. Second cyclè of the benefit sharing fund projects. Secretariat of the International Treaty on Plant Genetic Resources for Food and Agriculture. http://www.planttreaty.org/sites/default/files/BSF_2nd_cycle booklet.pdf.
FAO. (2015). Conservation agriculture. http://www.fao.org/ag/ca/.
FAO. (2016). The state of food and agriculture. Climate Change, Agriculture and Food Security. SOFA.
FAO. (2021). Sustainable land management (SLM) practices. www.fao.org/landwater/land/sustainable-landmanagement/slm-practices/en/.
Ghosh, K., Sarkar, S., Brahmachari, K., & Porel, S. (2018). Standardizing row spacing of Vetiver for river bank stabilization of Lower Ganges. *Current Journal of Applied Science and Technology, 26*, 1–12.
Gitz, V. (2013). Usage des terres et politiques climatiques globales: la physique, l'économie et les politiques de l'usage des puits de carbone pour lutter contre le changement climatique. Presses Académiques Francophones. Saarbrücken.
He, G., Wang, Z., Li, F., Dai, J., Li, Q., Xue, C., Cao, H., Wang, S., & Malhi, S. S. (2016). Soil water storage and winter wheat productivity affected by soil surface management and precipitation in dryland of the Loess Plateau. *China Agricultural Water Management, 171*, 1–9.
Hellin, J., & Haigh, M. J. (2002). Impact of *Vetiverua Zizanioides* (Vetiver Grass) live barriers on maize production in Honduras. In *Paper presented at the 12th Conference of the International Soil Conservation Organization.*
Hijbeek, R., van Ittersum, M. K., ten Berge, H. F. M., Gort, G., Spiegel, H., & Whitmore, A. P. (2017). Do organic inputs matter: A meta-analysis of additional yield effects for arable crops in Europe. *Plant and Soil, 411*, 293–303.
Hine, R., & Pretty, J. (2008). Organic agriculture and food security in Africa. United Nations Conference on Trade and Development and United Nations Environment Programme.
Huggins, D. R., & Reganold, J. P. (2008). No-till: The quiet revolution. *Scientific American, 299*, 70–77. https://doi.org/10.1038/scientificamerican0708-70.

Hussain, Z., Alam, M., Ullah, I., Ahmad, I., Sajid, M., Alam, I., Rehman, A. U., Shah, M. A., & Asif, M. (2020). Effect of organic and inorganic regimes on growth, production and quality characteristics of cauliflower. *Bioscience Research, 17*(2), 1289–1298.

Ibraimo, N., & Munguambe, P. (2007). Rainwater harvesting technologies for small scale rainfed agriculture in arid and semi-arid areas. Department of Rural Engineering, Faculty of Agronomy and Forestry Engineering, University Eduardo Mondlane, Maputo.

IPCC. (2007). Climate change 2007: Mitigation. In *Contribution of working group III to the fourth assessment report of the Intergovernmental Panel on Climate Change, chapter 8, agriculture*. Cambridge University Press.

Jat, L., Rai, S. K., Choudhary, J. R., Bawa, V., Bharti, R., Sharma, M., & Sharma, M. (2019a). Phenotypic evaluation of genetic diversity of diverse Indian mustard (*Brassica Juncea* L. Czern and Coss) genotypes using correlation and path analysis. *International Journal of Bio-Resource Stress Management, 10*(5), 467–471.

Jat, R. K., Singh, R. G., Gupta, R. K., Gill, G., Chauhan, B. S., & Pooniya, V. (2019b). Tillage, crop establishment, residue management and herbicide applications for effective weed control in direct seeded rice of eastern Indo-Gangetic Plains of South Asia. *Crop Protection, 123*, 12–20.

Jat, S. L., Parihar, C. M., Singh, A. K., Nayak, H. S., Meena, B. R., Kumar, B., Parihar, M. D., & Jat, M. L. (2019c). Differential response from nitrogen sources with and without residue management under conservation agriculture on crop yields, water-use and economics in maize-based rotations. *Field Crops Research, 236*, 96–110.

Johnson, J. M. F., Reicosky, D. C., Allmaras, R. R., Sauer, T. J., Venterea, R. T., & Dell, C. J. (2005). Greenhouse gas contributions and mitigation potential of agriculture in the central USA. *Soil and Tillage Research, 83*(1), 73–94.

Kader, M. A., Nakamura, K., Senge, M., Mojid, M. A., & Kawashima, S. (2019). Numerical simulation of water and heat flow regimes of mulched soil in rain-fed soybean field in Central Japan. *Soil and Tillage Research, 191*, 142–155.

Kar, S., Pramanick, B., Brahmachari, K., Saha, G., Mahapatra, B., Saha, A., & Kumar, A. (2021). Exploring the best tillage option in rice based diversified cropping systems in Alluvial Soil of Eastern India. *Soil and Tillage Research, 205*, 104761.

Kassam, A. H., Friedrich, T., Shaxson, F., & Pretty, J. (2009). The spread of conservation agriculture: Justification, sustainability and uptake. *International Journal of Agriculture Sustainability, 7*(4), 1–29.

Kladviko, E. J., Mackay, A. D., & Bradford, J. M. (1986). Earthworms as a factor in the reduction of soil crusting. *Soil Science Society of America Journal, 50*, 191–196.

Lal, R. (2001). World cropland soils as a source or sink for atmospheric carbon. *Advances in Agronomy, 71*, 145–191.

Lal, R. (2005). World crop residues production and implications of its use as a biofuel. *Environment International, 31*, 575–584.

Lin, B. B., Perfecto, I., & Vandermeer, J. (2008). Synergies between agricultural intensification and climate change could create surprising vulnerabilities for crops. *BioScience, 58*(9), 847–854.

Lipper, L., Thornton, P., Campbell, B. M., Baedeker, T., Braimoh, A., Bwalya, M., & Torquebiau, E. F. (2014). Climate-smart agriculture for food security. *Nature Climate Change, 4*(12), 1068–1072.

Liu, Z., Gao, T., Tian, S., Hu, H., Li, G., & Ning, T. (2020). Soil organic carbon increment sources and crop yields under long-term conservation tillage practices in wheat-maize systems. *Land Degradation and Development, 31*, 1138–1150.

Lorenz, K., & Lal, R. (2016). Environmental impact of organic agriculture. *Advances in Agronomy, 139*, 99–152.

Lu, X. (2020). A meta-analysis of the effects of crop residue return on crop yields and water use efficiency. *PLoS ONE, 15*, e0231740.

Mandal, K. G., Misra, A. K., Hati, K. M., Bandyopadhyay, K. K., Ghosh, P. K., & Mohanty, M. (2004). Rice residue management options and effects on soil properties and crop productivity. *Journal of Food Agriculture and Environment, 2*, 224–231.

Maneepitak, S., Ullah, H., Paothong, K., Kachenchart, B., Datta, A., & Shrestha, R. P. (2019). Effect of water and rice straw management practices on yield and water productivity of irrigated lowland rice in the Central Plain of Thailand. *Agricultural Water Management, 211*, 89–97.

Maplecroft-Global Risks Portfolio and services. (2010a). Food security risk index. http://www.maplecroft.com/about/news/food-security.html.

Maplecroft, B. (2010b). Labour standards and environmental report.

Mazvimavi, K., & Twomlow, S. (2009). Socioeconomic and institutional factors influencing adoption of conservation farming by vulnerable households in Zimbabwe. *Agricultural Systems, 101*(1–2), 20–29.

Meena, R. S., & Lal, R. (2018). *Legumes for soil health and sustainable management*. Springer.

Mondal, M., Garai, S., Banerjee, H., Sarkar, S., & Kundu, R. (2020a). Mulching and nitrogen management in peanut cultivation: An evaluation of productivity, energy trade-off, carbon footprint and profitability. *Energy, Ecology and Environment,* 1–15.

Mondal, M., Skalicky, M., Garai, S., Hossain, A., Sarkar, S., Banerjee, H., Kundu, R., Brestic, M., Barutcular, C., & Erman, M. (2020b). Supplementing nitrogen in combination with rhizobium inoculation and soil mulch in peanut (*Arachis hypogaea* L.) production system: Part II. Effect on phenology, growth, yield attributes, pod quality, profitability and nitrogen use efficiency. *Agronomy, 10*, 1513.

Montenegro, A. A. A., Abrantes, J. R. C. B., de Lima, J. L. M. P., Singh, V. P., & Santos, T. E. M. (2013). Impact of mulching on soil and water dynamics under intermittent simulated rainfall. *CATENA, 109*, 139–149.

Mostaghimi, S., Gidley, T. M., Dillaha, T. A., & Cooke, R. A. (1994). Effectiveness of different approaches for controlling sediment and nutrient losses from eroded land. *Journal of Soil and Water Conservation, 49*(6), 615–620.

Nyamangara, J., Marondedze, A., Masvaya, E., Mawodza, T., Nyawasha, R., & Nyengerai, K. (2014a). Influence of basin-based conservation agriculture on selected soil quality parameters under smallholder farming in Zimbabwe. *Soil Use and Management, 30*(4), 550–559.

Nyamangara, J., Nyengerai, K., Masvaya, E., Tirivavi, R., Mashingaidze, N., & Mupangwa, W. (2014b). Effect of conservation agriculture on maize yield in the semi-arid areas of Zimbabwe. *Experimental Agriculture, 50*(02), 159–177.

Olson, K. R., Ebelhar, S. A., & Lang, J. M. (2010). Cover crop effects on crop yields and organic carbon content. *Soil Science, 175*, 89–98.

Piccoli, I., Sartori, F., Polese, R., & Berti, A. (2020). Crop yield after 5 decades of contrasting residue management. *Nutrient Cycling in Agroecosystems, 117*, 231–241.

Poeplau, C., Reiter, L., Berti, A., & Kätterer, T. (2017). Qualitative and quantitative response of soil organic carbon to 40 years of crop residue incorporation under contrasting nitrogen fertilization regimes. *Soil Research, 55*, 1–9.

Prats, S. A., Wagenbrenner, J., Malvar, M. C., Martins, M. A. S., & Keizer, J. J. (2016). Hydrological implications of post-fire mulching across different spatial scales. *Land Degradation and Development, 27*(1440), 1452.

Qin, W., Hu, C., & Oenema, O. (2015). Soil mulching significantly enhances yields and water and nitrogen use efficiencies of maize and wheat: A meta-analysis. *Scientific Reports [Internet], 5*, 16210.

Rathod, P. H., Bhoyar, S. M., Katkar, R. N., Kadu, P. R., Jadhao, S. D., Konde, N. M., Deshmukh, P. W., & Patle, P. N. (2019). Recycling and management of crop residues for sustainable soil health in climate change scenario with farmer's profit as frontline moto. *Journal of Pharmacognosy and Phytochemistry, 5155.*

Reicosky, D. C. (2000). Tillage induced CO_2 emissions from soil. *Nutrient Cycling in Agroecosytems, 49*, 273–285.

Roth, C. H., Meyer, B., Frede, H. G., & Derpsch, R. (1988). Effect of mulch rates and tillage systems on infiltrability and other soil physical properties of an Oxisol in Parafla, Brazil. *Soil and Tillage Research, 11*, 81–91.

Rusinamhodzi, L., Corbeels, M., Nyamangara, J., & Giller, K. E. (2012). Maize–grain legume intercropping is an attractive option for ecological intensification that reduces climatic risk for smallholder farmers in Central Mozambique. *Field Crops Research, 136*, 12–22.

Sainju, U. M., Whitehead, W. F., & Singh, B. P. (2003). Cover crops and nitrogen fertilization effects on soil aggregation and carbon and nitrogen pools. *Canadian Journal of Soil Science, 83*(2), 155–165.

Salas, A. M., Elliott, E. T., Westfall, D. G., Cole, C. V., & Six, J. (2003). The role of particulate organic matter in phosphorus cycling. *Soil Science Society of America Journal, 67*, 181–189.

Salinas-Garcia, J. R., Báez-González, A. D., Tiscareño-López, M., & Rosales-Robles, E. (2001). Residue removal and tillage interaction effects on soil properties under rain-fed corn production in Central Mexico. *Soil and Tillage Research, 59*, 67–79.

Samui, I., Skalicky, M., Sarkar, S., Brahmachari, K., Sau, S., Ray, K., Hossain, A., Ghosh, A., Nanda, MK., & Bell, R. W. (2020). Yield response, nutritional quality and water productivity of tomato (*Solanum Lycopersicum* L.) are influenced by drip irrigation and straw mulch in the coastal saline ecosystem of Ganges Delta, India. *Sustainability, 12*, 6779.

Sarkar, S., Ghosh, A., & Brahmachari, K. (2020). Application of APSIM model for assessing the complexities of rice-based cropping systems of South-Asia. In S. Maitra, B. Pramanick (Eds.) (pp. 212–233). New Delhi Publishers.

Savin, R., & Slafer, G. A. (1991). Shading effects on the yield of an Argentinian wheat cultivar. *Journal of Agriculture Science, 116*(1), 1–7.

Schjonning, P., Jensen, J. L., Bruun, S., Jensen, L. S., Christensen, B. T., Munkholm, L. J., Oelofse, M., Baby, S., & Knudsen, L. (2018). The role of soil organic matter for maintaining crop yields: Evidence for a renewed conceptual basis. *Advances in Agronomy, 150*, 35–79.

Scialabba, N., & Hattam, C. (2002). *Organic agriculture, environment and food security*. Food and Agriculture Organization of the United Nations.

Scopel, E., Tardieu, F., Edmeades, G., & Sebillotte, M. (2001). Effects of conservation tillage on water supply and rainfed maize production in semiarid zones of West-central Mexico, CIMMYT NRG paper 01–01.

Shakoor, M. T., Ayub, S., & Ayub, Z. (2012). Dengue fever: Pakistan's worst nightmare, World Health Organization (WHO) South-East Asia. *Journal of Public Health, 1*(3), 244–247.

Sheikh, M. M., Manzoor, N., Adnan, M., Ashraf, J., & Khan, A. M. (2009). Climate profile and past climate changes in Pakistan, research report No.GCISC-RR-01, Global Change Impact Studies Centre.

Shen, Y., McLaughlin, N., Zhang, X., Xu, M., & Liang, A. (2018). Effect of tillage and crop residue on soil temperature following planting for a black soil in Northeast China. *Science and Reports, 8*, 4500.

Shively, G. E. (1999). Risks and returns from soil conservation: Evidence from low-income farms in the Philippines. *Environmental Monitoring and Assessment, 62*, 55–69. https://doi.org/10.1016/S0169-5150(99)00013-4.

Shukla, M. K., Lal, R., & Ebinger, M. (2003). Tillage effects on physical and hydrological properties of a typic argiaquoll in Central Ohio. *Soil Science, 168*, 802–811.

Sims, B. G., Thierfelder, C., Kienzle, J., Friedrich, T., & Kassam, A. (2012). Development of the conservation agriculture equipment industry in sub-Saharan Africa. *Application of Engineering in Agricultural, 28*(6), 813–823.

Smith, M. S., Frye, W. W., & Varco, J. J. (1987). Legume winter cover crops. *Advances in Soil Science, 7*, 95–139.

Smitha, G. R., Basak, B. B., Thondaiman, V., & Saha, A. (2019). Nutrient management through organics, bio-fertilizers and crop residues improves growth, yield and quality of sacred basil (Ocimum sanctum Linn). *Industrial Crops and Products, 128*, 599–606.

Soto-Pinto, L., Perfecto, I., Castello-Hernandez, J., & Caballero-Nieto, J. (2000). Shade effect on coffee production at the Northern Tzeltal zone in the state of Chiapas Mexico. *Agriculture, Ecosystems & Environment, 80*(1), 61–69. https://doi.org/10.1016/S0167-8809(00)00134-1.

Stavi, I., & Lal, R. (2012). Agroforestry and biochar to offset climate change: A review. *Agronomy for Sustainable Development*. https://doi.org/10.1007/s13593-012-0081-1.

Su, Z., Zhang, J., Wu, W., Cai, D., Lv, J., Jiang, G., Huang, J., Gao, J., Hartmann, R., & Gabriels, D. (2007). Effects of conservation tillage practices on winter wheat water-use efficiency and crop yield on the Loess Plateau, China. *Agricultural Water Management, 87*, 307–314.

Tantely, L. M., Boyer, S., & Fontenille, D. (2015). A review of mosquitoes associated with Rift Valley fever virus in Madagascar. *American Journal of Tropical Medicine and Hygiene, 92*(4), 722–729.

Thierfelder, C., & Wall, P. C. (2009). Effects of conservation agriculture techniques on infiltration and soil water content in Zambia and Zimbabwe. *Soil and Tillage Research, 105*(2), 217–227.

Thierfelder, C., & Wall, P. C. (2010). Investigating conservation agriculture (CA) systems in Zambia and Zimbabwe to mitigate future effects of climate change. *Journal of Crop Improvement, 24*(2), 113–121.

Thierfelder, C., Matemba-Mutasa, R., Bunderson, W. T., Mutenje, M., Nyagumbo, I., & Mupangwa, W. (2016). Evaluating manual conservation agriculture systems in southern Africa. *Agriculture, Ecosystems & Environment, 222*, 112–124.

Tiemann, L. K., Grandy, A. S., Atkinson, E. E., Marin-Spiotta, E., & McDaniel, M. D. (2015). Crop rotational diversity enhances belowground communities and functions in an agroecosystem. *Ecology Letters, 18*, 761–771.

Triplett, G. B., Dabney, S. M., & Siefker, J. H. (1996). Tillage systems for cotton onsilty upland soils. *Agronomy Journal, 88*, 507–512.

Turmel, M. S., Speratti, A., Baudron, F., Verhulst, N., & Govaerts, B. (2015). Crop residue management and soil health: A systems analysis. *Agricultural Systems, 134*, 6–16.

Wang, W. J., Dalal, R. C., & Moody, P. W. (2004). Soil carbon sequestration and density distribution in a Vertosol under different farming practices. *Soil Research, 42*, 875–882.

Wang, X., Wang, N., Xing, Y., & Ben El Caid, M. (2018). Synergetic effects of plastic mulching and nitrogen application rates on grain yield, nitrogen uptake and translocation of maize planted in the loess plateau of China. *Science and Reports, 8*, 14319.

Wei, W., Yan, Y., Cao, J., Christie, P., Zhang, F., & Fan, M. (2016). Effects of combined application of organic amendments and fertilizers on crop yield and soil organic matter: An integrated analysis of long-term experiments. *Agriculture, Ecosystems & Environment, 225*, 86–92.

Whitbread, A., Blair, G., Konboon, Y., Lefroy, R., & Naklang, K. (2003). Managing crop residues, fertilizers and leaf litters to improve soil C, nutrient balances, and the grain yield of rice and wheat cropping systems in Thailand and Australia. *Agriculture, Ecosystems & Environment, 100*, 251–263.

Wiesmeier, M., Poeplau, C., Sierra, C. A., Maier, H., Frühauf, C., Hübner, R., & Knabner, K. I. (2016). Projected loss of soil organic carbon in temperate agricultural soils in the 21st century: Effects of climate change and carbon input trends. *Science and Reports, 6*(1), 1–17.

Woodfine, A. (2009). *The potential of sustainable land management practices for climate change mitigation and adaptation in sub-Saharan Africa*. Food and Agriculture Organization of the United Nations.

Yadvinder-Singh, B. S., & Timsina, J. (2005). Crop residue management for nutrient cycling and improving soil productivity in rice-based cropping systems in the tropics. *Advances in Agronomy, 85*, 269–407.

Yang, Y. M., Liu, X. J., Li, W. Q., & Li, C. Z. (2006). Effect of different mulch materials on winter wheat production in desalinized soil in Heilonggang region of North China. *Journal of Zhejiang University Science B, 7*, 858–867.

Zhao, X., Liu, B. Y., Liu, S. L., Qi, J., Wang, X., Pu, C., Li, S., Zhang, X., Yang, X., & Lal, R. (2020). Sustaining crop production in china's cropland by crop residue retention: A meta analysis. *Land Degradation and Development, 31*, 694–709.

Zomer, R. J., Trabucco, A., Coe, R., & Place, F. (2009). Trees on farm: analysis of global extent and geographical patterns of agroforestry. *ICRAF Working Paper-World Agroforestry Centre*, (89).

Mulches and Microplastic Pollution in the Agroecosystem

Hafeez Ur Rahim, Waqas Ali Akbar, Naheeda Begum, Misbah Uddin, Muhammad Qaswar, and Nawab Khan

Abstract In recent decades, organic (plant and animal residues) and inorganic (plastic polyethylene) mulches have become globally considered an environmentally friendly agricultural practice for their instant benefits, such as preserving soil moisture, reducing soil evaporation, improving water efficiency, soil temperature regulation, shorten growth of weeds, enhancing microbial activity in the soil, higher yields, early harvesting, and improved crop quality. Similarly, the use of plastic mulch as a amendment for the restoration of contaminated soils is becoming increasingly popular, and its application is expanding. In the agroecosystem, microplastics with a particle size of 5 < mm can enter the soil either directly through irrigation water, application of biosolids, and atmospheric deposition or indirectly via the in situ degradation of large pieces of plastic mulch films. The legacy of this is that many soils are now contaminated with large amounts of plastic residues, and it is crucial for evaluating the risk of soil-borne emerging microplastic pollution. Thus, the problem associated with the use of plastic mulch remains poorly understood in the agroecosystem. Therefore, in this chapter, we critically discuss the recent understanding of the use of inorganic mulches related to microplastic pollution in the soil environment. The sources of inorganic mulches in the agroecosystem, distribution, and migration of microplastic in soils, mechanisms of soil microplastic, constraints and

H. U. Rahim (✉) · M. Uddin · M. Qaswar
Key Laboratory of Industrial Ecology and Environmental Engineering (Ministry of Education), School of Environmental Science and Technology, Dalian University of Technology, Dalian 116024, China
e-mail: hafeez.kalpani@aup.edu.pk

W. A. Akbar
Department of Soil and Environmental Sciences, The University of Agriculture, Peshawar, Pakistan

N. Begum
Soybean Research Institution, National Center for Soybean Improvement, Key Laboratory of Biology and Genetics and Breeding for Soybean, Ministry of Agriculture, State Key Laboratory of Crop Genetics and Germplasm Enhancement, Nanjing Agricultural University, Nanjing, China

N. Khan
Livestock Research and Development Station, Surezai Peshawar, Pakistan

Department of Agronomy, The University of Agriculture Peshawar, Peshawar, Pakistan

K. Akhtar et al. (eds.), *Mulching in Agroecosystems*,
https://doi.org/10.1007/978-981-19-6410-7_18

dynamic behavior of microplastics during aging on land, explore the responses of soil fauna to plastic particles at microscales, and mitigation strategies to prevent microplastic pollution is proposed.

Keywords Mulches · Plastic mulch films · Microplastic pollution · Soil environment · Agroecosystem

1 Introduction

The term mulch is derived from the German word "molsch," which means "easy to decay," and has been used widely since ancient times as an agricultural method for vegetable production (Lightfoot, 1994). It refers to the distribution of various kinds of materials on the surface of soil that affects the physicochemical & biological characteristics of the soil (Xu et al., 2020). The history of the application of mulches in agriculture can be traced back to the 1919s, while it got tremendous attention among researchers in the 1930s (Bedford & Pickering, 1919). Mulching materials such as leguminous crop residues and biochar on the soil surface have been documented to minimize moisture losses, water runoff, decrease weed populations, boost soil infiltration capability, regulate soil temperature and thus increase crop yield (Rahim et al., 2019, 2020a, 2020b).

There are two types of origin in the mulching materials, namely organic and inorganic. Organic mulching materials include residues of plants and livestock, such as straws, husks, grasses, cover crops, sawdust, compost, and different forms of manures (Rathore et al., 1998), while the inorganic mulching materials are composed of polyethylene plastic mulches. The use of plastic-mulching polyethylene in agriculture has become prevalent worldwide, and its use is growing with each day. It was reported that, in 1999, plastic mulch was used on more than 22 million hectares of cultivated land worldwide (Miles, 2005). While in China, plastic mulch was distributed over an area of 15 million hectares in 2002 (Xing et al., 2001). On average, 700,000 tons of plastic sheets are used worldwide as mulch annually and 140,000 tons in the USA alone (Shogren, 2000). Plastic mulches are mainly applied to seedlings and shoots by insulation and preventing evaporation, thereby keeping or moderately enhancing the temperature and humidity traits of soil. In addition, weeds and pest pressure are known to be minimized by the application of plastic covers. Minimizing seed and fruit production time, increased yield, preventing soil erosion and weed growth, and, consequently, reducing herbicide and fertilizer use as highlighted in Fig. 1 are commonly reported benefits. These prospects have made plastic mulching in the agroecosystem an upcoming technology. However, adverse effects may arise from plastic additives such as microplastic pollution (MPs) in the soil environment.

MPs refer to contaminants made up of plastic particles less than 5 mm in size, including fiber, fragments, foam, film, and other forms in the atmosphere. The environmentalists were concerned in particular with the source, distribution, emissions,

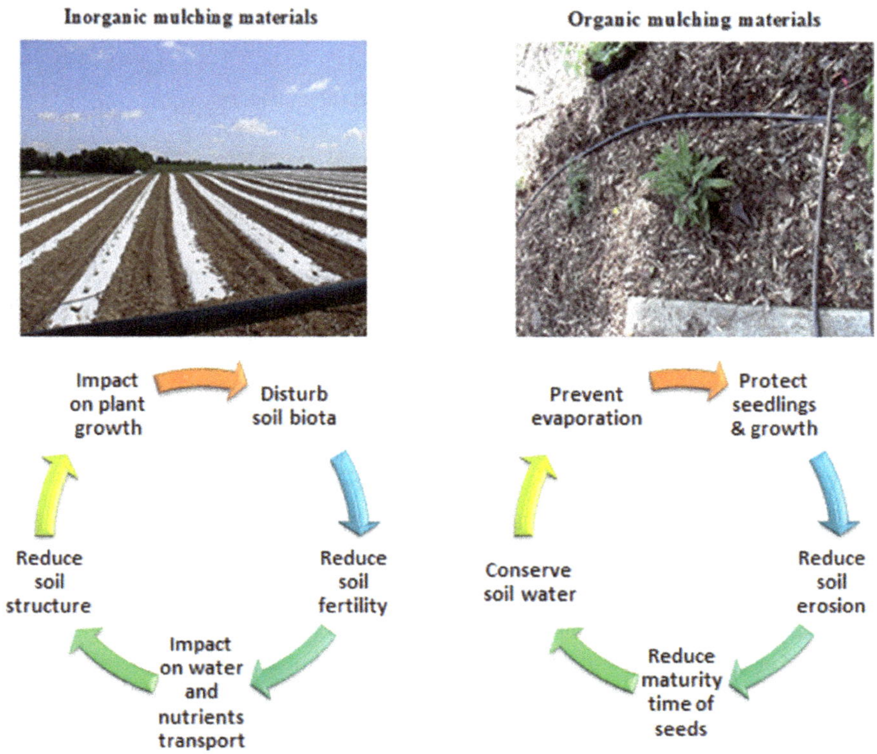

Fig. 1 The beneficial and harmful effects of organic and inorganic mulching materials in agroecosystem

environmental and ecological impacts of MPs in the offshore and tidal beach environments (Jambeck et al., 2015; Primpke et al., 2017). But recently environmental scientists were attracted to the pollution of MPs in the terrestrial ecosystem, particularly in agricultural ecosystems (Rillig, 2012; Wang et al., 2019).

Microplastics (MPs) are emerging pollutants generating a minimum of 300 million tonnes annually, of which a considerable amount ends up in the atmosphere where it remains for decades. To date, the contamination of agricultural soils by MPs is still little documented (O'Kelly et al., 2021; Li et al., 2020b). In this regard, Nizzetto et al. and Qian et al. reported that MPs can affect the physical and chemical properties of soil (Nizzetto et al., 2016a; Qian et al., 2018). Likewise, pH of the soil, soil hydrophobicity, and transport by water and nutrients, and soil carbon sources can be affected by MPs and their additives, but the reports are uncommon. MPs may also influence the growth of plants. For example, various plastic degradation (such as titanate plasticisers) additives may be inserted into mesophyll by respiration of plants, chlorophyll, or chlorophyll formation, thus damaging plant development (Matzek & Carter, 2016; Sun et al., 2015).

This chapter aims at reviewing and analyzing the new literature and knowledge on:

i. The possible sources of inorganic mulches in the agroecosystem.
ii. The micro plastic distribution and migration in soils.
iii. The response of soil properties to micro plastic.
iv. Micro plastic contamination effects on human health.
v. Mitigation strategies to prevent micro plastic pollution.

However, the existing soil contamination-related data is still insufficient. Further study is therefore urgently warranted regarding the prevalence and fate of MPs in agricultural soils. We also propose other opportunities for future studies on microplastic contamination and plastic waste soil ecotoxicity, which will guide such research.

2 Occurrence and Sources of Microplastics in Agroecosystem

MPs in the agroecosystem occur from two main sources namely, primary source and secondary source, which leads to various types of plastics particles. However, the detection of exact sources of MPs in the agroecosystem is still not possible, and the research is in progress to find innovative ways. The primary source of MPs consisted of industrial detergent and cosmetic formulations, while secondary sources include agricultural plastic films, household garbage, sewage and sludge, atmospheric deposition, and vehicle emissions, as shown in Fig. 2. Secondary sources of MPs are projected to be the leading agro-ecosystem MPs. However, the destruction of massive plastic waste into MPs under natural conditions takes hundreds of years. Among those sources, automotive tyre wear is regarded by the rapid growth in the global number of vehicles as a major source of environmental microplastic. There are, however, very few studies on the presence of rubber particles in the atmosphere (An et al., 2020; Guo et al., 2020; Qi et al., 2020).

Agricultural plastic film is another significant source of environmental microplastic. A thin film made up of polyvinyl chloride, polyethylene, and other blow forming additives is one type of agricultural mulch. In agriculture, about 3.4% of plastics in overall global production are used (UN-Environment, 2019). In the early 1950s, the application of plastic mulch films started in agriculture. These films raise the temperature of the soil and reduce contamination of the soil, increase crop production and income, and are very important for food security. However, the high consumption, combined with the short film life cycle of the plastic mulch film, leads to a difficult recovery, poor recycling quality, and easy release into your soil of MPs (Guo et al., 2020; Li et al., 2020b, 2020c).

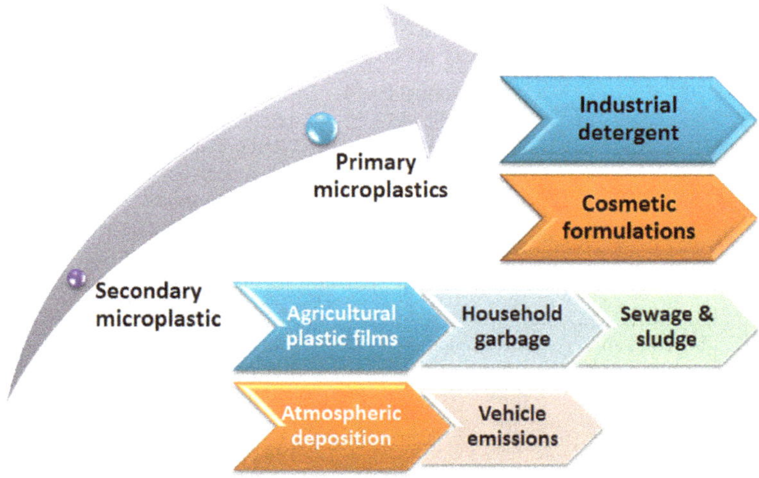

Fig. 2 Sources of microplastic in environmental system

3 The Distribution and Migration Behavior of Microplastics Pollutants in Soils

The migration behavior of MPs in the soil is still unclear due to the very complex system of soil (Li et al., 2020a). It was reported that the MPs on the soil surface tend to be lost as a result of water runoff or air (Nizzetto et al., 2016a). Though, the downward transport of MPs through unknown pathways is evident those MPs in soil could move vertically (Zubris & Richards, 2005).

The migration of MPs in soils is possible due to the porous nature of soil ranging from macro to mesopores, and it was reported in numerous studies that small size particles tend to transport with soil pores via leaching (Bläsing & Amelung, 2018). In this regard, Grayling et al. (2018) extensively investigated that small particles with a size range of between 0.1 and 0.6 μm may transport in the vertical direction. Even so, the influence of external forces, including farming activities and bioturbation, can contribute to the moment of larger MPs in the soil profile. In recent studies, researchers have reported that MPs can be moved by various means which is summarized in Table 1.

4 The Response of Soil Properties to Microplastic

The existence of a huge amount of MPs in soil influences the properties of soil, including soil structure, functions, microbial diversity, soil fertility, and hydraulic conductivity (He et al., 2019; Zhang et al., 2017), which may cause consequent translocation in plants and describe potential concern for food security, ultimately

Table 1 Investigation on MPs movement in soil profile

Location	MPs size	Type of MPs used	Key findings	Refs.
Germany	200–400 μm	Urea-formaldehyde micro-plastic	The experimental findings showed that collembolans species such as *Folsomia candida and Proisotoma* can move Urea-formaldehyde micro-plastic in soil	Maaß et al. (2017)
China	80–250 μm	Commercial PVC	It was evident that mite, i.e., *Hypoaspis aculeifermoved* can translocate and disperse the studied commercial PVC in soil	Zhu et al. (2018a)
China	80–250 μm	Commercial PVC	The results approved that collembolans species *Folsomia candida* can translocate commercial PVC in soil	Zhu et al. (2018b)
Germany	Four types 710–2800 μm	Polyethylene (PE) beads	Four types of Polyethylene (PE) beads with different particle size (710–2800) have been used to investigate the potential of earth worms (*Lumbricus terrestris* L.,) in the movement of PE-beads from soil surface through soil profile. The results suggested that earth worms could be the significant transport agents of MPs in soil	Rillig et al. (2017b)
China	Five MPs were used with different size and densities	Polymers types	The results showed that soil is the feasible entry pathway for MPs from surface to subsurface soil	O'Connor et al. (2019)

noxious to human wellbeing (Murugan et al., 2014). The response of some of the soil properties to MPs will be discussed briefly in this section.

4.1 Soil Structure and Water Transport

Various studies have provided limited information on soil structure in response to MPs. The research studies showed that microplastics influence the soil's characteristics to rely on the microplastic form. Those MPs have forms and dimensions that are nearer to soil particles, and the soil structure and water cycles are less pronounced (Xu et al., 2020). MPs interfere with different soil properties when they come into close contact, and among all properties, soil structure is one of the important properties to understanding the hazard of MPs posed to soil properties (de Souza Machado et al., 2018a). Polyester fibers can substantially increase the ability of water holding and decrease the bulk and water-stable aggregation. The polyethylene and polyacrylic effects, however, were not obvious (de Souza Machado et al., 2018b). Different MPs materials have shown different nature of effects on soil structure. It was reported that the treatment with polyester microfibers treatment didn't change the bulk density of the soil and showed a decline in the ability of the soil to hold water (Zhang et al., 2019). Likewise, MPs change the permeability and retention capacity of water, which resultantly affects the evaporation of water (Zhichao et al., 2015). Collectively, MPs can change the soil water cycles intensify the accumulation of water in the soil, and affect pollutant movement to deep soil (Rillig et al., 2017a). However, more detailed research is required to critically evaluate the positive or negative impacts of MPS on soil structure and water transport.

4.2 Soil Fertility and MPs

MPs can also affect soil fertility because MPs are composed of high carbon polymers, a remarkable source of organic-carbon source (Rillig, 2018). Over 30 days of an experiment, the application of MPs at the rate of 28% (W/W) greatly enhanced the content of dissolved organic carbon, inorganic nitrogen, and total phosphorous in sandy soils as compared to the treatment rate of 7% (W/W) (Liu et al., 2017). However, even this lower amendment rate was higher than environmentally-relevant microplastic concentrations (Xu et al., 2020). The contamination of plastic film residues significantly reduced soil organic matter, nitrogen, and phosphorous content in soil (Hegan et al., 2015). Likewise, the effects on soil structure and water transport, the effect of MPs on soil fertility, and nutrients availability are not well-clear. More research is needed to get in-depth insights into the effects of MPs on soil fertility and nutrients availability.

4.3 Soil Microbes and MPs

Owing to the long residence time of MPs in the soil, it can be ingested by soil microbes which consequently influence the growth, development, and reproduction of soil microbes by destroying the organs of the organism and DNA (Ren et al., 2018). It was reported that the application of MPs greatly intervene in the structure of the microbial community, and the substrate-induced respiration rates were remarkably reduced, resulting in alterations in the functions of soil microbes that were induced by MPs in soil (Judy et al., 2019). Similarly, the application of polystyrene in soil notably reduced microbial biomass, enzyme activities engaged in macronutrients (N, P,) and carbon cycle, and enhanced basal respiration (Awet et al., 2018). Likewise to the soil structure, fertility, and water transport properties, the influence on soil microbial and enzyme activities also depend on the shape, size, treatment rates, compositions of MPs, and the texture of soil (Wang et al., 2016). However, the findings from these studies don't agree with the magnitude of the impact of MPs pollution on these targets and need to be extensively investigated (Xu et al., 2020).

5 Fate of Microplastic Pollution

Polyethylene, polypropylene, and other polymers are used for MPs, which are typically less than 5 mm. Plastics production and utilities have steadily grown over the decades, microplastics often increase in the environment, and these new contaminants are frequently found in rivers (Yonkos et al., 2014), lakes (Free et al., 2014), shorelines (Thompson et al., 2004), and soils (Nizzetto et al., 2016b). From the early discussion, it is clear that soil is not only a sink of MPs but may also represent a source of MPs to groundwater and aquatic environment, as shown in Fig. 3. The effect of microplastics on marine life in our ecosystem is negative. Cells of the blue mustula Mytilus edulis were taken up with microplastics where experimental exposures have harmful effects on the mustula tissue (Von Moos et al., 2012). Zooplankton, normally drifting in salt and fresh water ingests microplastics (Cole, 2013). A group of flame retardant substances commonly used in electronics, polybrominated diphenyl ethers (PBDEs) were demonstrated by marine amphipod to be equated with microplastics(Chua et al., 2014). Owing to their hydrophobic nature, microplastics appear, along with other persistent, organic contaminants in water, to absorb PBDEs, endocrine disrupting compounds (EDC), pharmaceuticals, and personal care products (PPCPs). PBDE, EDC, and PPCP concentrations found in various effluent samples in parts per trillion (Cole, 2013; Nelson et al., 2011), could be adsorbed and microplastic particulate surfaces (MPPs) enriched, these toxic contaminants could eventually reach the food chain of an environment, if fish, aquatic invertebrate and other wildlife eat the polluted plastic residues (do Sul & Costa, 2014).

As stated in the previous section, modified plastic mulch soil conditions are expected to speed up the deterioration of the soil and can cause unwanted changes in

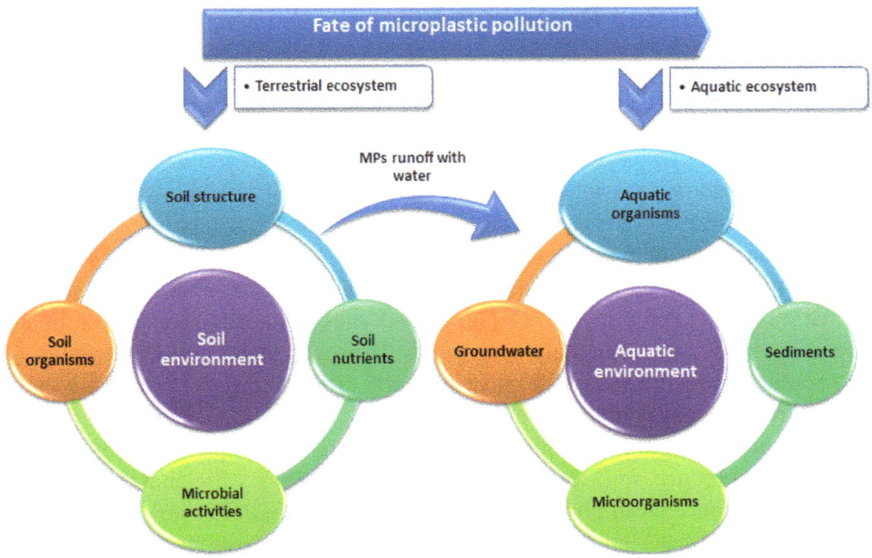

Fig. 3 Fate of microplastic in environmental system

soil organism populations and affect ecosystem engineers, including earthworms and nematodes. The soil's chemical and biological quality and function play an important role in regulating the decomposition, sequestration of carbon, and mineralization of organic matter (Steinmetz et al., 2016).

Despite risks to the agroecosystem, the exposure of human health to microplastics by the ingestion of contaminated food is inevitable and poses a risk to human health (Sharma & Chatterjee, 2017). Potential dangers can lead to chromosome changes, leading to infertility, obesity, and cancer (De-la-Torre, 2020).

6 Mitigation Strategies to Prevent Microplastic Pollution

Quite apart from the global emphasis on plastic contamination and its impacts in recent years, regulations to deal with the effects of secondary MPs are not yet developed. (Karbalaei et al., 2018). Various clean-up measures have been suggested in recent years to mitigate the adverse effects of plastic waste but they are unable to cope with increasing plastic volumes entering the environment. Therefore, a global multidisciplinary strategy would prioritize reducing plastic input to the ecosystem (Prata et al., 2019). These clean-up or mitigation activities include improving the production efficiency of plastic products by using alternative materials such as recycled or biodegradable plastic materials, reducing the consumption of plastic materials, and improving the collection and disposal of waste (Prata et al., 2019).

7 Conclusions and Recommendations

This chapter briefly discusses the recent progress in the microplastic film mulches pollution in the agroecosystem, linked with microplastic pollution in the soil environment. The sources of inorganic mulches in the agroecosystem, the distribution and migration of microplastic in soils, the mechanisms of soil microplastic, constraints and dynamic behavior of micro-plastics during aging on land, explore the responses of soil fauna to plastic particles at microscales, and mitigation strategies to prevent microplastic pollution. From the overall discussion, it was concluded that the applications of mulches in the agroecosystem enhance the physicochemical and biological properties of soil and, ultimately, crop yields. However, their continuous and long-term applications without any life cycle and ecological integrity and risk assessment can disturb the ecosystem services.

Microplastic i.e., both organic and inorganic, have great potential to conserve soil moisture, reduce soil evaporation, enhance water use efficiency, control soil temperature, reduce weed growth, enhance microbial activities in soil, increase crop yield, earlier harvest, and improve crop. However, there is a need to explore the sources of inorganic mulches in the agroecosystem, the distribution and migration of microplastic in soils, the mechanisms of soil microplastic, constraints, and dynamic behavior of micro-plastics during aging on land, explore the responses of soil fauna to plastic particles at microscales. There is a further need to explore mitigation strategies to prevent microplastic pollution.

References

An, L., Liu, Q., Deng, Y., Wu, W., Gao, Y., & Ling, W. (2020). Sources of microplastic in the environment. https://doi.org/10.1007/698_2020_449.

Awet, T., Kohl, Y., Meier, F., Straskraba, S., Grün, A.-L., Ruf, T., Jost, C., Drexel, R., Tunc, E., & Emmerling, C. (2018). Effects of polystyrene nanoparticles on the microbiota and functional diversity of enzymes in soil. *Environmental Sciences Europe, 30*, 1–10. https://doi.org/10.1186/s12302-018-0140-6.

Bedford, H. A. R., & Pickering, P. S. U. (1919). *Science and fruit growing: Being an account of the results obtained at the Woburn experimental fruit Farm Since its Foundation in 1894*. Macmillan.

Bläsing, M., & Amelung, W. (2018). Plastics in soil: Analytical methods and possible sources. *Science of the Total Environment, 612*, 422–435. https://doi.org/10.1016/j.scitotenv.2017.08.086.

Chua, E. M., Shimeta, J., Nugegoda, D., Morrison, P. D., & Clarke, B. O. (2014). Assimilation of polybrominated diphenyl ethers from microplastics by the marine amphipod, Allorchestes compressa. *Environmental Science and Technology, 48*, 8127–8134. https://doi.org/10.1021/es405717z.

Cole, M. (2013). Microplastic Swallowing Zooplankton. *Environmental Science and Technology, 47*, 6646–6655. https://doi.org/10.1021/es400663f.

de Souza Machado, A. A., Kloas, W., Zarfl, C., Hempel, S., & Rillig, M. C. (2018a). Microplastics as an emerging threat to terrestrial ecosystems. *Global Change Biology, 24*, 1405–1416. https://doi.org/10.1111/gcb.14020.

de Souza Machado, A. A., Lau, C. W., Till, J., Kloas, W., Lehmann, A., Becker, R., & Rillig, M. C. (2018b). Impacts of microplastics on the soil biophysical environment. *Environmental Science and Technology, 52*, 9656–9665. https://doi.org/10.1021/acs.est.8b02212.

De-la-Torre, G. E. (2020). Microplastics: An emerging threat to food security and human health. *Journal of Food Science and Technology, 57*, 1601–1608. https://doi.org/10.1007/s13197-019-04138-1.

do Sul, J. A. I., & Costa, M. F. (2014). The present and future of microplastic pollution in the marine environment. *Environmental Pollution, 185*, 352–364. https://doi.org/10.1016/j.envpol.2013.10.036.

Free, C. M., Jensen, O. P., Mason, S. A., Eriksen, M., Williamson, N. J., & Boldgiv, B. (2014). High-levels of microplastic pollution in a large, remote, mountain lake. *Marine Pollution Bulletin, 85*, 156–163. https://doi.org/10.1016/j.marpolbul.2014.06.001.

Grayling, K., Young, S. D., Roberts, C. J., de Heer, M. I., Shirley, I., Sturrock, C., & Mooney, S. J. (2018). The application of X-ray micro Computed Tomography imaging for tracing particle movement in soil. *Geoderma, 321*, 8–14. https://doi.org/10.1016/j.geoderma.2018.01.038.

Guo, J.-J., Huang, X.-P., Xiang, L., Wang, Y.-Z., Li, Y.-W., Li, H., Cai, Q.-Y., Mo, C.-H., & Wong, M.-H. (2020). Source, migration and toxicology of microplastics in soil. *Environment International, 137*, 105263. https://doi.org/10.1016/j.envint.2019.105263.

He, P., Chen, L., Shao, L., Zhang, H., & Lü, F. (2019). Municipal solid waste (MSW) landfill: A source of microplastics?-Evidence of microplastics in landfill leachate. *Water Research, 159*, 38–45. https://doi.org/10.1016/j.watres.2019.04.060.

Hegan, D., Tong, L., Zhiquan, H., Qinming, S., & Ru, L. (2015). Determining time limits of continuous film mulching and examining residual effects on cotton yield and soil properties. *Journal of Environmental Biology, 36*, 677.

Jambeck, J., Geyer, R., Wilcox, C., Siegler, T., Perryman, M., Andrady, A., & Naray, R. (2015). *Science, 347*(6223), 768–771. https://doi.org/10.1126/science.1260352.

Judy, J. D., Williams, M., Gregg, A., Oliver, D., Kumar, A., Kookana, R., & Kirby, J. K. (2019). Microplastics in municipal mixed-waste organic outputs induce minimal short to long-term toxicity in key terrestrial biota. *Environmental Pollution, 252*, 522–531. https://doi.org/10.1016/j.envpol.2019.05.027.

Karbalaei, S., Hanachi, P., Walker, T. R., & Cole, M. (2018). Occurrence, sources, human health impacts and mitigation of microplastic pollution. *Environmental Science and Pollution Research, 25*, 36046–36063. https://doi.org/10.1007/s11356-018-3508-7.

Li, J., Song, Y., & Cai, Y. (2020a). Focus topics on microplastics in soil: Analytical methods, occurrence, transport, and ecological risks. *Environmental Pollution, 257*, 113570. https://doi.org/10.1016/j.envpol.2019.113570.

Li, W., Luo, Y., & Pan, X. (2020b). Microplastics in agricultural soils. https://doi.org/10.1007/698_2020b_448.

Li, W., Wufuer, R., Duo, J., Wang, S., Luo, Y., Zhang, D., & Pan, X. (2020c). Microplastics in agricultural soils: Extraction and characterization after different periods of polythene film mulching in an arid region. *Science of the Total Environment, 749*, 141420. https://doi.org/10.1016/j.scitotenv.2020.141420.

Lightfoot, D. R. (1994). Morphology and ecology of lithic-mulch agriculture. *Geographical Review*, 172–185. https://doi.org/10.2307/215329.

Liu, H., Yang, X., Liu, G., Liang, C., Xue, S., Chen, H., Ritsema, C. J., & Geissen, V. (2017). Response of soil dissolved organic matter to microplastic addition in Chinese loess soil. *Chemosphere, 185*, 907–917. https://doi.org/10.1016/j.chemosphere.2017.07.064.

Maaß, S., Daphi, D., Lehmann, A., & Rillig, M. C. (2017). Transport of microplastics by two collembolan species. *Environmental Pollution, 225*, 456–459. https://doi.org/10.1016/j.envpol.2017.03.009.

Matzek, L. W., & Carter, K. E. (2016). Activated persulfate for organic chemical degradation: A review. *Chemosphere, 151*, 178–188. https://doi.org/10.1016/j.chemosphere.2016.02.055.

Miles, M. (2005). Identification, pest status, ecology and management of the green mirid, a pest of cotton in Australia, Ph.D. thesis, University of Queensland.
Murugan, R., Beggi, F., & Kumar, S. (2014). Belowground carbon allocation by trees, understory vegetation and soil type alter microbial community composition and nutrient cycling in tropical Eucalyptus plantations. *Soil Biology & Biochemistry, 76*, 257–267. https://doi.org/10.1016/j.soilbio.2014.05.022.
Nelson, E. D., Do, H., Lewis, R. S., & Carr, S. A. (2011). Diurnal variability of pharmaceutical, personal care product, estrogen and alkylphenol concentrations in effluent from a tertiary wastewater treatment facility. *Environmental Science and Technology, 45*, 1228–1234. https://doi.org/10.1021/es102452f.
Nizzetto, L., Bussi, G., Futter, M. N., Butterfield, D., & Whitehead, P. G. (2016a). A theoretical assessment of microplastic transport in river catchments and their retention by soils and river sediments. *Environmental Science: Processes & Impacts, 18*, 1050–1059. https://doi.org/10.1039/C6EM00206D.
Nizzetto, L., Futter, M., & Langaas, S. (2016b). Are agricultural soils dumps for microplastics of urban origin? ACS Publications. https://doi.org/10.1021/acs.est.6b04140.
O'Connor, D., Pan, S., Shen, Z., Song, Y., Jin, Y., Wu, W.-M., & Hou, D. (2019). Microplastics undergo accelerated vertical migration in sand soil due to small size and wet-dry cycles. *Environmental Pollution, 249*, 527–534. https://doi.org/10.1016/j.envpol.2019.03.092.
O'Kelly, B. C., El-Zein, A., Liu, X., Patel, A., Fei, X., Sharma, S., et al. (2021). Microplastics in soils: an environmental geotechnics perspective. *Environmental Geotechnics, 8*, 586-618. https://doi.org/10.1680/jenge.20.00179.
Prata, J. C., Silva, A. L. P., Da Costa, J. P., Mouneyrac, C., Walker, T. R., Duarte, A. C., & Rocha-Santos, T. (2019). Solutions and integrated strategies for the control and mitigation of plastic and microplastic pollution. *International Journal of Environmental Research and Public Health, 16*, 2411. https://doi.org/10.3390/ijerph16132411.
Primpke, S., Lorenz, C., Rascher-Friesenhausen, R., & Gerdts, G. (2017). An automated approach for microplastics analysis using focal plane array (FPA) FTIR microscopy and image analysis. *Analytical Methods, 9*, 1499–1511. https://doi.org/10.1039/C6AY02476A.
Qi, R., Jones, D. L., Li, Z., Liu, Q., & Yan, C. (2020). Behavior of microplastics and plastic film residues in the soil environment: A critical review. *Science of the Total Environment, 703*, 134722. https://doi.org/10.1016/j.scitotenv.2019.134722.
Qian, H., Zhang, M., Liu, G., Lu, T., Qu, Q., Du, B., & Pan, X. (2018). Effects of soil residual plastic film on soil microbial community structure and fertility. *Water, Air, & Soil Pollution, 229*, 261. https://doi.org/10.1007/s11270-018-3916-9.
Rahim, H. U., Mian, I. A., Arif, M., Rahim, Z. U., Ahmad, S., Khan, Z., Zada, L., Khan, M. A., & Haris, M. (2019). Residual effect of biochar and summer legumes on soil physical properties and wheat growth. *Pure and Applied Biology, 8*, 16–26. https://doi.org/10.19045/bspab.2018.700159.
Rahim.H.U., Mian, I., Muhammad, A., Sajjad, A., & Zaid, K. (2020a). Soil fertility status as influenced by the carryover effect of biochar and summer legumes. *Asian Journal of Agriculture and Biology, 8*, 11–16. https://doi.org/10.35495/ajab.2019.05.198.
Rahim, H. U., Ahmad, S., Khan, Z., & Khan, M. A. (2020b). Field-based investigation of aged biochar coupled with summer legumes effect on wheat yield in Pakistan. *Buletin Agroteknologi, 1*, 1–6. https://doi.org/10.32663/ba.v1i1.1152.
Rathore, A., Pal, A., & Sahu, K. (1998). Tillage and mulching effects on water use, root growth and yield of rainfed mustard and chickpea grown after lowland rice. *Journal of the Science of Food and Agriculture, 78*, 149–161. https://doi.org/10.1002/(SICI)1097-0010(199810)78:2%3c149::AID-JSFA94%3e3.0.CO;2-U.
Ren, X., Tang, J., Yu, C., & He, J. (2018). Advances in research on the ecological effects of microplastic pollution on soil ecosystems. *Journal of Agro-Environmental Science, 37*, 1045–1058.
Rillig, M. C. (2012). Microplastic in terrestrial ecosystems and the soil? ACS Publications. https://doi.org/10.1021/es302011r.

Rillig, M. C. (2018). Microplastic disguising as soil carbon storage. ACS Publications. https://doi.org/10.1021/acs.est.8b02338.
Rillig, M. C., Ingraffia, R., & de Souza Machado, A. A. (2017a). Microplastic incorporation into soil in agroecosystems. *Frontiers in Plant Science, 8*, 1805. https://doi.org/10.3389/fpls.2017.01805.
Rillig, M. C., Ziersch, L., & Hempel, S. (2017b). Microplastic transport in soil by earthworms. *Science and Reports, 7*, 1–6. https://doi.org/10.1038/s41598-017-01594-7.
Sharma, S., & Chatterjee, S. (2017). Microplastic pollution, a threat to marine ecosystem and human health: A short review. *Environmental Science and Pollution Research, 24*, 21530–21547. https://doi.org/10.1007/s11356-017-9910-8.
Shogren, R. (2000). Biodegradable mulches from renewable resources. *Journal of Sustainable Agriculture, 16*, 33–47. https://doi.org/10.1300/J064v16n04_05.
Steinmetz, Z., Wollmann, C., Schaefer, M., Buchmann, C., David, J., Tröger, J., Muñoz, K., Frör, O., & Schaumann, G. E. (2016). Plastic mulching in agriculture. Trading short-term agronomic benefits for long-term soil degradation? *Science of the Total Environment, 550*, 690–705. https://doi.org/10.1016/j.scitotenv.2016.01.153.
Sun, J., Wu, X., & Gan, J. (2015). Uptake and metabolism of phthalate esters by edible plants. *Environmental Science and Technology, 49*, 8471–8478. https://doi.org/10.1021/acs.est.5b01233.
Thompson, R. C., Olsen, Y., Mitchell, R. P., Davis, A., Rowland, S. J., John, A. W., McGonigle, D., & Russell, A. E. (2004). Lost at sea: Where is all the plastic? *Science, 304*, 838 (Washington).
UN-Environment. (2019). Our planet is drowning in plastic pollution.
Von Moos, N., Burkhardt-Holm, P., & Köhler, A. (2012). Uptake and effects of microplastics on cells and tissue of the blue mussel Mytilus edulis L. after an experimental exposure. *Environmental Science and Technology, 46*, 11327–11335. https://doi.org/10.1021/es302332w.
Wang, J., Lv, S., Zhang, M., Chen, G., Zhu, T., Zhang, S., Teng, Y., Christie, P., & Luo, Y. (2016). Effects of plastic film residues on occurrence of phthalates and microbial activity in soils. *Chemosphere, 151*, 171–177. https://doi.org/10.1016/j.chemosphere.2016.02.076.
Wang, J., Liu, X., Li, Y., Powell, T., Wang, X., Wang, G., & Zhang, P. (2019). Microplastics as contaminants in the soil environment: A mini-review. *Science of the Total Environment, 691*, 848–857. https://doi.org/10.1016/j.scitotenv.2019.07.209.
Xing, N., Zhang, Y., & Wang, L. (2001). *The study on dryland agriculture in North China*. Chinese Agriculture Press.
Xu, B., Liu, F., Cryder, Z., Huang, D., Lu, Z., He, Y., Wang, H., Lu, Z., Brookes, P. C., & Tang, C. (2020). Microplastics in the soil environment: Occurrence, risks, interactions and fate—A review. *Critical Reviews in Environment Science and Technology, 50*, 2175–2222. https://doi.org/10.1080/10643389.2019.1694822.
Yonkos, L. T., Friedel, E. A., Perez-Reyes, A. C., Ghosal, S., & Arthur, C. D. (2014). Microplastics in four estuarine rivers in the Chesapeake Bay, USA. *Environmental Science & Technology, 48*, 14195–14202. https://doi.org/10.1021/es5036317.
Zhang, D., Liu, H., Ma, Z., Tang, W., Wei, T., Yang, H., Li, J., & Wang, H. (2017). Effect of residual plastic film on soil nutrient contents and microbial characteristics in the farmland. *Scientia Agricultura Sinica, 50*, 310–319.
Zhang, G., Zhang, F., & Li, X. (2019). Effects of polyester microfibers on soil physical properties: Perception from a field and a pot experiment. *Science of the Total Environment, 670*, 1–7. https://doi.org/10.1016/j.scitotenv.2019.03.149.
Zhichao, W., Xianyue, L., & Haibin, S. (2015). Effects of residual plastic film on soil hydrodynamic parameters and soil structure. *Transactions of the Chinese Society for Agricultural Machinery, 46*, 101–106.
Zhu, D., Bi, Q.-F., Xiang, Q., Chen, Q.-L., Christie, P., Ke, X., Wu, L.-H., & Zhu, Y.-G. (2018a). Trophic predator-prey relationships promote transport of microplastics compared with the single Hypoaspis aculeifer and Folsomia candida. *Environmental Pollution, 235*, 150–154. https://doi.org/10.1016/j.envpol.2017.12.058.

Zhu, D., Chen, Q.-L., An, X.-L., Yang, X.-R., Christie, P., Ke, X., Wu, L.-H., & Zhu, Y.-G. (2018b). Exposure of soil collembolans to microplastics perturbs their gut microbiota and alters their isotopic composition. *Soil Biology & Biochemistry, 116*, 302–310. https://doi.org/10.1016/j.soilbio.2017.10.027.

Zubris, K. A. V., & Richards, B. K. (2005). Synthetic fibers as an indicator of land application of sludge. *Environmental Pollution, 138*, 201–211. https://doi.org/10.1016/j.envpol.2005.04.013.

Organic and Synthetic Mulching: Effects on Soil-Plant Productivity and Environment

Sharjeel Ahmad, Hamza Tariq, Saria Abbas, Muhammad Arshad, Amer Mumtaz, and Iftikhar Ahmed

Abstract Soil and water conservation are the important aspects in modern world, because of the scarcity of water, agricultural land degradation and soil loss mainly due to erosion. Mulching, a way to conserve both soil and water by covering it with different kinds of materials like organic (crop plant, compost, manures) or synthetic (paper, plastics, Aluminum foils). It controls evaporation rate and aids in managing soil and air microclimate. Through favorable microclimate, it improves soil physicochemical and biological properties. Mulches act as a soil cover to resist against erosion and provide congenial condition for plant growth. Mulching encourages soil and crop productivity, reduces the emergence of greenhouse gases and suppression of weeds. Plastic mulches are also becoming popular among farmers due to their low cost and easy handling. These materials have a greater importance than the organic ones as they are highly employable in controlled soil environment and could enhance soil-crop productivity. Mulching helps to balance hydro-thermal regimes by maintaining radiation flux, heat and water vapor transfer rate and soil heat capacity. Nowadays, biodegradable plastic mulches are employed which are relatively more sustainable as compared to conventional plastic mulches. The degradable nature of plastic mulches favors the microbial activities in soil, subsequently enhancing the productivity. The mulching could be effective in plant roots protection from hot, cold

---

S. Ahmad · H. Tariq · I. Ahmed (✉)
National Microbial Culture Collection of Pakistan (NCCP), Land Resources Research Institute (LRRI), National Agriculture Research Center (NARC), Islamabad 45500, Pakistan
e-mail: iftikhar.ahmed@parc.gov.pk

S. Ahmad
School of Science and the Environment, Memorial University of Newfoundland and Labrador, Corner Brook, Canada

S. Abbas
Department of Zoology, University of Science and Technology, Bannu, Pakistan

M. Arshad
Institute of Environmental Sciences and Engineering, School of Civil and Environmental Engineering, National University of Sciences and Technology, Islamabad 44000, Pakistan

A. Mumtaz
Food Sciences Research Institute (FSRI), National Agriculture Research Center (NARC), Islamabad 45500, Pakistan

K. Akhtar et al. (eds.), *Mulching in Agroecosystems*,
https://doi.org/10.1007/978-981-19-6410-7_19

or drought conditions. This part covers the broader aspects related to application of mulches in maintaining microclimate and soil-plant productivity.

Keywords Soil and water conservation · Organic mulches · Biodegradable plastics · Microclimate · Environment

1 Introduction

The term 'Mulch' is derived from the German term 'Molch' means easy to decompose and from Middle English 'Molsh' means strawy dung (Bhagat et al., 2016; Hodgson, 2020). As evident from its history, mulching has been known for a long time ago but academic record reveals that it was practiced in Persian marketers at the end of 18th century (Sebastian, 2012). In ancient times, this technique is employed to control weeds which was deeply concerned at that period of time (Bhardwaj, 2013). Mulching is defined as the soil layering with materials (Patricia, 1957) or the conservation of moisture in the field of watermelons by covering soil crust (Saha et al., 1974). Any material that assists in soil and water conservation when spreading at the soil surface is known as mulch (Kumar et al., 2016). Researchers elaborated the mechanisms very well that how mulching affects soil and crop productivity (Dilipkumar et al., 1990).

There are different kinds of mulches i.e., organic and inorganic (Telkar et al., 2006). The mostly used mulching materials are the plant residues like straw, peanut hulls, lead mold, compost and even some wood products like sawdust, wood chips and all types of manures including animals manures, green manures and poultry manures (Vos & Sumarni, 1997). Organic mulches include plant residues and all decomposable materials which when added up in soil aid in enhancing soil nutrition (Jodaugienė et al., 2006). Natural materials i.e., straws (cereal and flax), nonwoven wool or pine needles are being employed under different environmental conditions depending upon the situation of areas (Granatstein & Mullinix, 2008). However, Inorganic mulches include plastic mulches which only help in soil and water conservation (Iqbal et al., 2016).

The primary objectives of mulching are the prevention of water loss (evaporation), soil erosion, hindering weed infestation, limit fertilizer loss, promote soil and crop productivity (Bhagat et al., 2016). Hence, mulching is helpful in saving underground water resources, soil and environment for sustaining crop productivity (Ranjan et al., 2017). The literature shows that the mulches could highly impact on soil health by enhancing soil productivity, improving soil properties, maintaining soil microclimate and have antagonistic effect on weeds (Ngosong et al., 2019; Stigter, 1984) and crop productivity by sustaining its growth and development, profitability, nutritional amendment and quality maintenance (Khaledian et al., 2010).

Mulching practices have been employed to adjust the soil micro-climate and humidity conditions and thus potentially improves plant productivity in dry land agriculture, but few studies have focused on the impact of mulching on soil gaseous emissions (Chen et al., 2017). Plastic film mulching (PFM) usually diminish organic carbon stock of soil but enhances the greenhouse gas (GHG) emissions, that's why its employment is still being debated (Lee et al., 2019).

Recently farmers adopting conservation tillage practices to minimize soil degradation, water loss and environmental protections (Unger & McCalla, 1980). Conservation agriculture is one of the reliable methods, dominant due to continuous management, permanent or semi-permanent soil cover (growing crops or a dead mulch) (FAO, 2001). Conservation agriculture (CA) aims on enhancing soil and crop productivity, conserve natural resources due to the integrated systems for maintaining soil, water and biological resources aggregating with exterior inputs (Derpsch, 2003). The most common forms of CA are no-till (planting unprepared soil with minimum disturbance), ridge-till (planting in ridges, intercropped without disturbing the previous crops rows) and mulch-till (tillage with 30% crop residues on the soil surface) (CTIC, 2001). In this chapter, an effort has been made to cover the application of mulching in enhancing soil-plant productivity, its pros and cons, controlling pollution and its impact on the environment.

2 Mulching and Its Classification

There are different types of mulches, but the main categories are organic and inorganic mulches. The flowchart of mulching types are as shown in Fig. 1.

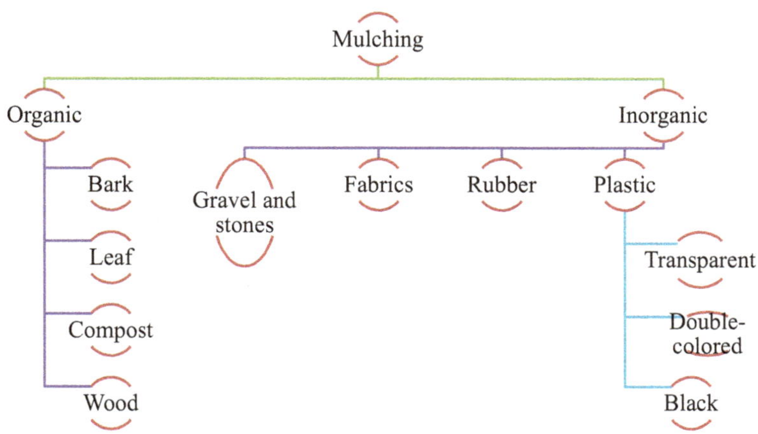

Fig. 1 Types of mulching

2.1 Organic Mulches

Organic mulches are defined as the plant residues employed as a soil cover and provide nutrients to plant upon decomposition. These mulches are not only inhibiting weeds and conserve water but also encourage the growth of worms and microorganisms. These help in maintaining soil environment, its properties and enhance crop yield by providing nutrients (Sinkevičienė et al., 2009). There are different kinds of materials that are used as organic mulches.

Bark mulches are those types of mulch materials are derived from two classed barks i.e., hardwood barks (by-product of paper industries and varies in size; usually trees or perennial plants bark) and softwood bark (also a by-product of paper industries but pine trees bark; slightly acidic in nature) (Harkin, 1971).

Leaf mulches are the shredded leaves applied on the upper layer of soil, which could not only improve the soil moisture and suppression of weeds but also enriched garden soil upon decomposition (Budelman, 1988).

Wood mulch contains the wood chips and bark, varies in size, help in reducing compaction by allowing infiltration of water. It is utilized as a land filling material by covering 2 inches (5 cm) soil upper crust, block sunlight to maintain soil microclimate and weed infestation by covering soil (Campbell, 2012; Carroll, 2020).

Compost, known as a Gardener's Gold, is formed by the biodegradation of organic waste. Compost is humus like substance having dark brown to black appearance, aids in enhancing nutrition content of soil and utilized as land filling material. It not only improves soil fertility but also ameliorates soil physical, chemical and biological properties (Ahmad et al., 2021; Ahmed et al., 2019).

2.2 Inorganic/Synthetic Mulches

Inorganic mulches are the decorative stones, pulverized tyres, lava rock, and geotextile fabrics, useful in xeriscaping and for soil protection in high traffic areas. Heat reflected from inorganic mulches kills the thin-barked tree, that's why these are not recommended for mulching around trees (Hashim et al., 2013; Wang et al., 2020b).

Gravel and stones do not retain moisture and can lead to heat stress on plants by the reflection and ground heating that burns roots and thus infesting weeds. They are best employed for trees, shrubs, and other plants and also include volcanic rock, crushed gravel, and marble chips (Pavlů et al., 2020; Wang et al., 2020a). Landscape fabric employed for long period of time to control weeds, as it allows very little amount of air and water to pass through. It can be utilized in association with organic mulches and degrade more rapidly as compared to inorganic mulches (Iqbal et al., 2020; Rendon et al., 2020; Vaddevolu et al., 2020).

Rubber mulch is made up of recyclable tyres. These are still under researched; however, previous studies specify it as toxic product as well as the flammable substance and not being recommended for use in the home landscape (Chopra & Koul, 2020; Jia et al., 2020; Rao et al., 2020).

Plastic mulches, also a type of synthetic mulches, employ polyethylene film to cover soil. It is easily employed as it will be easily disposable and decrease the environmental impact. It is usually a layer of plastic substance and works same way as the organic mulch like insulation of soil, prevention of soil erosion, and reduction in the moisture evaporation (Bandopadhyay et al., 2020; Qi et al., 2020a, 2020b).

Transparent plastic film works well for soil warming and encouraging rapidly in the growing season especially during early growth. Clear plastic film isn't compatible when being employed in the suppression of weed growth (Cheng et al., 2020; Kader et al., 2020; Zhang et al., 2020b).

Black plastic film is ideal for retaining the soil's moisture and employed in agriculture especially in arboriculture for removing weeds, maintaining soil microclimate, protection from erosion, and avoiding fruiting bodies from being in direct contact with the soil (Lalk et al., 2020; Ning et al., 2020).

These colored mulches are meant for absorption of certain kind of wavelength from the sun's radiation therefore known as wavelength selective mulch. This absorption of certain wavelength leads to encouraging plant growth and development and aids in maintaining numerous plant characteristics such as fruit size, color, root development, height, etc. as they reflect minimum heat therefore maintains lower leaf temperatures (Kasirajan & Ngouajio, 2012). The different kinds of double colored plastic mulches are as given in Table 1.

Table 1 Types of mulches and their characteristics

Sr. No.	Types of colored plastic mulches	Characteristics
I.	**Yellow-black/brown**	Brown or black side touches the soil while the yellow one faces upwards which attracts whitefly, therefore acts as a trap, and that's why prevents diseases.
II.	**White-black**	It aids in transmission more than 60% of the photoactive radiation into the plant which resulting in enhanced growth, thicker and stronger leaves. This film cools the soil.
III.	**Silver-black**	It's the most popular type of Mulch used in farming and is suitable for all crops. It reflects almost 25-30% light back into the plant or fruit.
IV.	**Red-black**	It is a partially translucent material that allows radiation to pass through it, thus makes the soil warm. The film also reflects light and helps in increased yield, early fruiting, good flower development etc.

3 Mulching and Soil

3.1 Soil Moisture

Soil moisture conservation through mulching is one of the important practices. It greatly affects the soil environmental conditions (Saikia et al., 2014). Through evaporation, water stress appears in bare soil when exposed to heat, wind and compacting forces and is almost unable to absorb rainfall or irrigation, which leads to soil compaction increasingly (Lalitha et al., 2010). For different soils, different mulching techniques which are preferable are as given in Table 2.

In summer season, weeds can also enhance the rate of evapo-transpiration in soil which lowers moisture content by 25%. When the soil crust is covered with mulch, aids in controlling weeds, reducing evaporation and enhances rainwater infiltration during different growing seasons. It provides many pros to crop productivity by soil-water conservation, enhancement in soil biological activity and improved chemical and physical properties of the soil (Kannan, 2020).

It was noted Kader et al. (2017a) that the high moisture content in grass-mulched soil up to a depth of 60 cm was associated with good infiltration and reduced evaporation. The increase in mulched plants dry weight is because of its efficiency to sustain soil moisture by enhancing water uptake capability of plants (Kader et al., 2017b). During initial growth phases, more water contents are being reserved in the soil profile with straw mulch than without it (Tang et al., 2021).

It was observed that as a result of soil water conservation, the straw mulch receiving treatment had significantly higher net returns as compared to control one

Table 2 Preferable mulching types at specific soil/ area conditions

Sr. No.	Type of soil/area	Mulch preferable
1.	Rainy season	Perforated mulch
2.	Plantation and orchard mulch	Thicker mulch
3.	Soil solarisation	Thin and transparent film
4.	Weed control in cropped land	Black plastic film mulch
5.	Saline water area	Black plastic film mulch
6.	Summer cropped land	White film
7.	Insect repellent	Silver color film
8.	Early germination	Thinner film
9.	Sandy soil	Black film
10.	Weed control through solarisation	Transparent film
11.	Nutrient deficient	Stubble
12.	Water deficient area	Sea weeds
13.	Field prone to Soil born diseases	Clear plastic mulch

(Park et al., 2021). In contrast, there are the large number of mulching materials which do not control soil water infiltration and retention, but they have one similarity that they are all permeable substance. Organic mulches are the ones that conserve water more effectively than the inorganic (Iiles, 1999; Singh et al., 1988), while organic and inorganic are better conservers than the synthetic ones and they all are good than the bare soil (Chalker-Scott, 2007).

Earlier it is well established that the decrease in soil water evaporation by 34–50% by the application of crop residue mulching (Mubvumba et al., 2021). Mulches decreases the evaporation and also limited the irrigation requirement (Harvold & Falleth, 2005; Srivastava, 2013). Many researchers (Khurshid et al., 2006; Pervaiz et al., 2009), Liu et al. (2002), stated the similar results that the mulching maintaining the soil ecological environment and also enhances the soil humidity contents. The utilization of polyethylene mulch in the field; improves the soil micro-climate especially during spring, increases moisture conservation and having antagonistic effect on weeds and certain insect pest (Saglam et al., 2017).

3.2 Soil Micro-climate

Usually, plastic mulches are preferred for maintaining soil microclimatic conditions. Under plastic films, the soil temperature is usually elevated and is dependent on the its color (Sintim et al., 2021). The black plastic-film mulched fields had significantly cooler soil climatic condition (1–2.80 °C) than that of clear plastic-film ones, as the solarization by black plastic-film mulch is lost through the reflecting back radiation (Singh & Kamal, 2012).

The un-mulched fields had the cooler soil temperature(about 2–3.80 °C lower) as compared to plastic film mulched one, at different times since planting (Anikwe et al., 2007). Plastic film mulching enhances soil surface temperature by maintaining the soil heat balance and therefore positively influencing the crop emergence (Aniekwe et al., 2004).

Soil crust temperatures can be very high in summer season, which may influence on the plant roots activity (Kassaye et al., 2021). A compost mulch can standardize the soil micro-climate by minimizing the soil heat regulation and making the temperature constant, suitable for root activity (Gheshm & Brown, 2020). The organic mulches, having the ability to adjust the soil temperature, are closely related with its capability to decrease water evaporation losses (Ranjan et al., 2017).

The correlation effects of water availability and temperature regulation enhance the effective use of soil layers for nutrients uptake (Ranjan et al., 2017; Xiukang et al., 2015). Chemically synthesized mulches also controlled the water evaporation losses as effective as the organic ones, but lack in some other benefits like addition of soil organic material (Bucki & Siwek, 2019).

3.3 Minerals Availability

The biodegradation of waste residues under mulches, releases organic acids which add up into the soil leading to low soil pH that aids in enhancing the bioavailability of most of the micronutrients (Mn, Zn, Cu, and Fe) (Ahmad et al., 2021). Organic wastes degradation produce water soluble mineral ions such as NO_3, NH_4^+, Ca^{2+}, Mg^{2+}, K^+ and fulvic acid into the soil which in return enhance the soil available nutrient content under mulching condition (Hannam et al., 2016).

Researchers (Qin et al., 2015) observed that the soil volume expanded because of the plant root system, thus, can be influenced by the nitrogen (N) source applied, which may have a collateral effect on the other mineral absorption. Mulches can be comprised of different organic waste materials with numerous properties that can make its variable impacts on the soil food web and different elements mineralization such as nitrogen and phosphorus (Fang et al., 2011). It was noticed (Murungu et al., 2011) that almost all the nitrogen and carbon transformations from organic material are caused by soil microbial activities. Use of manure mulch on the soil surface enhances the population of soil microbes (Masvaya et al., 2017).

Sawdust used as a mulching material, enhanced the concentrations of calcium, potassium and magnesium, mutually with the root development in the soil crust layer (Lima et al., 2016). The organic material addition in soil appears to be the only practical method which aids in increasing the soil aggregate stability and its structure (Ahmad et al., 2021). Organic mulches affect the soil physical properties by modifying different soil properties like soil organic matter (SOM), porosity and cation exchange capacity (CEC), but decrease the soil bulk density (Sas-Paszt et al., 2014). Mulching comprised of organic wastes improves the soil aggregate stability and its structure (Smets et al., 2008) and reduces erosion (Fernández & Vega, 2021). Earlier it was demonstrated (Rawat et al., 2021) that the enhancement in soil available mineral content was directly related with the organic mulches, correlated with the increased microbial activity and root development, ultimately responsible for the improved plant productivity (Liao et al., 2021).

4 Mulching and Plant Productivity

4.1 Mulches and Crops Production

Different scientists evaluated that the mulched trees grow 67% times better than the ones that were grown without mulch (Nyawira, 2016; Sindhu et al., 2017). Some of which have shown the similar results in the growth of trees, herbs, shrubs, and other plants. Application of mulches results in the enhancement in plant height, its diameter, leaf size and shape, and flower, fruit and/or seed production (Iriany et al., 2018).

Organic mulches are considered as the best type for overall plant growth and development. Tested mulches include three kinds of materials: (i) easily decomposable materials such as different herbs, shrubs, and compost, (ii) moderately decomposable materials including paper, hay and straw, and other crop residues and (iii) slowly decomposable materials, especially bark and woody chips (Shojaei et al., 2019). Gravel and stone are not effective mulch as compared to organic mulches in maintaining plant growth and development. The thinnest (5 cm) layer of organic mulch results in rapid carbon degradation (Sun et al., 2021).

Liu et al. (2021) noticed that the fine structured roots of *Camellia oleifera* tended to distribute in the shallow soil due to mulching and also demonstrated the comprehensive impact of different mulching treatments on the yield and economic traits of fruits of *C. oleifera.* The treatments were peanut stalk and straw treatment (PSS) > black mulch (BM) > *C. oleifera* shell (COS) > non-*mulched* or control (CK) > eco-film (EF). It was reported Bokszczanin et al. (2021) that the fruits grown from trees Miscanthus sp. in organic mulching comprised of different plots were significantly greater in number as compared to the fruits from fallow and plastic mulch plots.

Plastic mulching is an effective practice which is applicable on ridges as a cover with furrow irrigation method. This mulching type enhances the plant growth, improves water harvesting and crop productivity in semiarid areas through managing soil micro-climate, advanced flowering, weed infestation, reduced soil water loss and conservation of moisture in the field as compared to non-mulched soil (Mehta et al., 2010; Shirish et al., 2013). Different additives/substances are integrated into the plastic to modify some specific characteristic of the final product, which includes different color pigments, anti-block agents, antioxidants, ultraviolet (UV) inhibitors/stabilizers, flame-retardants, and photodegradable additives (Steinmetz et al., 2016). The potential benefits of mulching in improving soil-plant health are demonstrated in Fig. 2.

The production and yield of potato under paddy straw mulch plots were observed to be higher 27.9% and 18.18% respectively, as compared to un-mulched ones (Goel et al., 2019). The productivity of tomato and okra with straw mulching enhanced by 100 and 200% as compared to control one (Biswas et al., 2015; Mendonça et al., 2021).

The productivity of Okra under straw mulching was demonstrated higher as compared to dust mulching plots (Dalorima et al., 2014). Chavan (2009) demonstrated that the increase of grain yield of about 12.64, 9.06, 7.46 and 3.74% respectively, in different mulching fields as compared to operational practices (Ranjan et al., 2017).

5 Mulching and Agro-Forestry

The collective utilization of straw mulch and erosion decreases 95% rate of the soil erosion as compared to the forest barren soil (Lucas-Borja et al., 2018). The pines needle-like leave said in reducing erosion rate, and its sticks debris was helpful

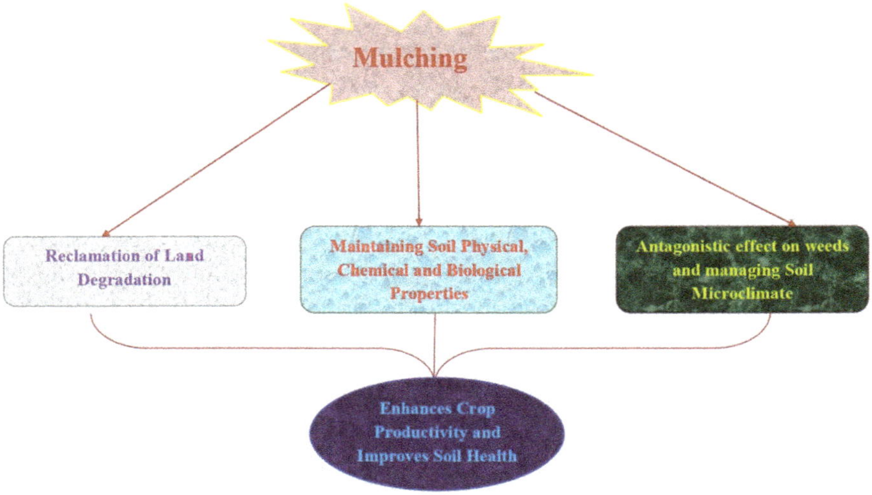

Fig. 2 Potential benefits of soil application of mulching

in successfully controlling runoff and erosion losses (Rudawska & Leski, 2021). Mulching materials break the water flow rate specially in mountainous areas, enhance the soil infiltration rate, and maintain the slope stabilization instead of employing mulching techniques (Tang et al., 2021). Low-nutrient-rich organic mulches lessen the weed growth. When applied in thick layer, they are more effective than the herbicides (Puka-Beals & Gamig, 2021).

Numerous crop residues and forest-produced materials employed as un-composted forms and have an antagonistic effect on weed species especially in agricultural lands. Mulches are also a best choice for the elimination of soil heavy metals such as arborvitae, eucalyptus, pine, and poplar leaves (Hathi et al., 2021). By the application of woodchips and compost in forest areas converts the copper toxic form into non-toxic form by making composite with the copper metal (Cu) for proper growth and development of crop plants (Pandiyan et al., 2021).

FMAFS (The Farmer Managed Agro-forestry System) introduces the wide variety of high yielding seeds and wood producing Australian acacias species especially the *A. colei, A. torulosa, A. tumida, A. elachantha.* These species are usually grown along the farm borders. It aids in providing human and animal food, environmental restoration services, crop protection and mulch (Rinaudo & Cunningham, 2008).

There are two approaches are made to reduce run off and/or erosion. (i) The primal approach to control erosion is to determine runoff rate and soil removing rate by employing barriers which shows by the means of land earth structures (ditch-and-bank structures, terraces), and production pattern. (ii) The second one is cover approach employed for the controlling of erosion which is the management of a soil cover by different plant species like herbs, shrubs, crop residues and tree litter and pruning etc. Techniques usually involves the intercropping with cover crops, mulching, minimum or zero tillage and agro-forestry (Prats et al., 2014).

Agro-forestry usually contributes as a barrier approach by the use of hedgerows as partially permeable barriers (direct approach) and by means of the trees stabilizes the earth structures and making productive land by adding its residues (an indirect approach). Agro-forestry usually involved in cover approaches by employing the tree litter and pruning in composite of the different crop cover and its residues (Kearney et al., 2019).

Conservation farming system includes three features of agro-forestry (hedgerow intercropping, fuel wood trees and fodder trees) combined with the pests management, mulching and minimum or zero tillage (Schroth et al., 2013). Integrated land utilization was applied in different areas prominence on planting trees along with the contour barrier strips and ridges (Salmoral et al., 2017). Improved-tree-fallow is found to be reproducible in shifting the cultivation effect (Lojka et al., 2010).

Poro (*Erythrina poeppigiana*) is a nitrogen-fixing species, employed as intercropped with the coffee (*Coffea arabica*) and cacao (*Theobroma cacao*) in America. By pruning, it can be employed as mulch and applied on field with fertilizer. This can include nutrients' addition in fertilizer, resulting in more productivity of crop plant (Payán et al., 2009).

Gliricidia (*Gliricidiasepium*), among those species are identified so far, have a equal compatibility to Leucaena (*Leucaena leucocephala*) which is employed as a pruned shrub in hedge row intercropping. It gives low result when being pruned at above ground. By utilizing differing scope of Gliricidia pruning, maize crop productivity gave a direct relation with the amount it retained as mulch (Fialho et al., 2021).

6 Mulching and Environment

6.1 Mulches and Greenhouse Gases

The main important source of greenhouse gas (GHGs) emission is agriculture soil, and agricultural management strategies would have a significant influence on GHGs (Hamrani et al., 2020). Due to recent concerns on climatic changes, no-till and straw mulch are getting an increasingly higher attention in agricultural activities as the two important conservation agricultural strategies (Riahi et al., 2011).

No-till was beneficial for CH_4 oxidation (enhances the methane-oxidizing bacteria activity) and decreased CH_4 emission (Jacinthe et al., 2014). The impact of no-till treatment on N_2O emission was highly associated with climate conditions and soil properties. In arid and semi-arid climate conditions, no-till enhanced the rate of N_2O emission in poorly aerated soil but did not effect on good-aerated soil. In humid climate condition, the effect of no-till on N_2O emission shows variation among different soil conditions (Ma et al., 2013).

Straw mulch generally reduced the soil CO_2 emission rate (Akhtar et al., 2020). Most researches showed that the straw mulch increases the rate of N_2O emission, while there are also some research results which indicate no effect or reducing N_2O emission rate (Dossou-Yovo et al., 2016). The mechanism of the impact of straw mulch on N_2O emission is complex and needs a further research.

As the promotion of conservation agriculture (CA), many studies were focused on crop productivity, soil water use efficiency, and soil properties, etc., while the impact of CA on GHGs was relative very less researched (especially for the combined effect of CA on the three major GHGs). Hence, research in the future should focus on the mutual effect of no-till and straw mulch on GHGs and forecasting the global warming potential of GHGs, which could provide basic data and theory supports for finding appropriate tillage treatments and straw mulch strategies (Kodzwa et al., 2020; Lal, 2015).

Some new nuclear technologies, such as isotope tracer technique (IST), can be utilized to demonstrate the direct and indirect role of straw mulch in the mitigation of GHGs (Fernández-Fernández et al., 2017). In future, we need the appropriate tillage practices and mulching techniques that can mitigate the GHGs and maintains reasonable crop yield should be searched in numerous soil types and weather conditions in different research regions (Bhattacharyya & Barman, 2018).

7 Mulches and Pollution

7.1 Air Pollution

When wind act upon soil crust (a source of particulate matter (PM) emissions) which causes erosion and deposition of soil particulates. Both wind erosion and mechanical actions lead to the increased soil quality issues, by the topsoil losses and soil deposition on different areas where it is not wanted including in lakes and rivers, becoming a water quality issue (Chalker-Scott, 2007).

The degree on which soils erode or become airborne by wind or other sources is usually based on the following factors, including (i) surface cover maintenance (residue, vegetation or other cover), (ii) soil properties and kind, (iii) compaction rate vers top soil looseness and (iv) the soil roughness (micro-topography) (Iqbal et al., 2020).

Conservation practices that help in maintaining soil cover significantly reduce the rate of wind erosion and mechanically driven forces from the soil crust. Employing those tillage systems that maintain crop residues on the soil crust reduces the tillage intensity rate and decreases the mechanical enforcement on fields would eventually reduce particulate matter (PM) generation potential. Adaptation of conservation tillage and residue management aid in retaining topsoil structure and soil adhesiveness, not to generate PM (Prosdocimi et al., 2016).

Mulching is one of the sources of soil surface cover by employing those materials non-existent on the field. In such conditions, bare soils usually in drought time periods, mulching gives very efficient results. If mulching is not viable due to soil and fertility issues, in those cases, it becomes necessary to invoke other conservation actions (such as surface roughening, wind barriers, etc.) to prevent PM emissions (Livesley et al., 2010).

Several National Resource Conservation Services by United States Department of Agriculture (NRCS-USDA) conservation practices especially designed for the managing of tillage practices and crop residues. The distribution, orientation and crop management is employed through the residue and tillage management while other plant residues remain in the soil to promote nutrient settlement and reduces the air pollution (Vieira et al., 2018).

Mulching is the method of employing different plant residues on the land surface. Other materials except plant residues used as mulching are the wood chips, rice hulls and other synthetic materials like plastics or fabrics etc. Mulching is an easy cheap method aids in reducing airborne erosion and diseases, conserving soil moisture, maintaining soil micro-climate and improving soil microbial activity (Lucas-Borja et al., 2018, 2019; Robichaud et al., 2013).

7.1.1 Plastic Pollution

The continuous utilization of plastics in agriculture is threatening overall ecosystem sustainability because of its residual persistency in different climatic conditions. By which plants, human beings and different living things are extremely vulnerable to this menace of plastic pollution (Zhang et al., 2020a). To which continuous utilization of plastic as mulching material plays a prime role in enhancing this condition. Lack of alternatives to this conventional product, makes the scenario even more worst (Chang-Rong et al., 2014; Gao et al., 2019).

Different kinds of biodegradable mulches (BDM) have the potential to revolutionize the problem of accumulation of plastic mulches residues. The main limiting factor in adopting BDMs are the price and lack of aesthetic value (Sintim and Flury, 2017). Moreover, there are still researches have done to know the key mechanics involved in its biological decomposition, active organisms employed in plastic degradation and the actual fate of soil micro-plastics (Sintim et al., 2019).

The utilization of plastic mulches will enhance the production cost of crop plant. This is because of investment in equipment like mulch transplanters, plug-mix seeders etc. These costs increases the productivity of crop (much greater than the input) (Briassoulis & Giannoulis, 2018).

Identification of plastic degrading microbes that can easily biodegrade the plastic mulch residues and also involved in bioremediation of plastic pollution. Recycling of these synthetic products also play a crucial role and ameliorate different hazards but need huge investment to get the intended outcomes. Recycling is limited due to the chemical employed in the production of PFM which will enhance the soil heavy metal concentration (Steinmetz et al., 2016; Waggoner et al., 1960).

8 Pros and Cons of Mulching

8.1 Advantages of Mulching

- It reflects back the sunlight significantly and lowers the soil heat absorbance rate (Sun et al., 2008).
- It is helpful in maintenance of soil micro-climate (Wang et al., 2003).
- Mulching aids in restricting weed growth as it provides soil cover and doesn't allow light to reach at soil surface (Radics & Szné Bognár, 2004).
- It provides protection to the soil from erosion, both by wind and water (Prosdocimi et al., 2016).
- It also helps in restricting rainwater flow rate and thus help in maintaining soil properties (Fernández & Vega, 2016).
- Rainwater runoff slows down and enhances the infiltration rate of water and improves the soil humidity level (Lal, 1997).
- Organic mulching improves the soil properties and crop-plant productivity (Montenegro et al., 2013).
- It also aids in enhancing soil fertility by adding nutrients in their available form and make the soil loose by adding organic matters (Wu et al., 2006).
- Mulching substances are also employed as a food of earthworms and other micro-flora of the soil (Jabran, 2019).
- These kind of mulches also improve the soil organic carbon contents (Ma et al., 2018).
- It helps in better root penetration, growth and development, and makes the soil more fragile (Niu et al., 2004).
- It favors water retention capacity of the soil and have antagonistic effects on weed growth (Jabran et al., 2015).
- Different kinds of mulches are easily degradable and make soil productive (Stirling & Eden, 2008).

8.2 Disadvantages of Mulching

- It makes the soil too moist under poorly drained soils and makes the anoxia condition which reduces the crop growth and productivity (Olson, 2006).
- It makes the environment conducive for pest and diseases' growth due to trapped moisture as it covers the soil (Yu et al., 2018).
- It eventually becomes the breeding spot for different pathogens, micro-flora, insects and pests (Mupangwa et al., 2012).
- Sometimes mulches become the weed which are not easy to be removed from the field (Coolong, 2012).
- Organic mulches are easily biodegradable and serve for only short duration (Martín-Closas et al., 2008).

- Plastic mulches residues become toxic materials which are harmful for humans, plants and micro-flora (Zhang et al., 2020a).
- Plastic mulches contribute to microplastic pollution that has emerging eco-toxicological impacts on the soil and water bodies.

9 Conclusion

Owing to health consciousness, population pressure and demand for 4Fs (Food, Fuel, Fiber and Feed), effective and environmentally friendly interventions are highly desired in order to ensure sustainability and development. Appropriate mulching offers the potential solutions for proper management of soils, high productivity of crops and promotion of agriculture and agro-forestry. Soil properties like physical, chemical and biological properties show significant improvements under the mulching condition. Plant growth and productivity are influenced by mulching as it maintains soil micro-climate, controls proliferation of weeds and erosion, and favors nutrient availability. Recent trends have been shifted towards the amendment of mulches with organic fertilizers to facilitate nutrient availability, hinder weed infestation and improving soil moisture conservation. Plastic mulches with multi-shades are employed to enhance the sunlight reflectivity and cool down the soil temperature especially in arid and semi-arid zones. The reflective nature of plastic mulches helps in controlling the movement of insects and pests. Mulching can help in controlling some of GHGs emissions. It can serve as one of the best approaches for reclamation of degraded soils. Although, there are certain limitations or disadvantages of mulching but overall impact is overwhelmingly beneficial economically and environmentally.

References

Ahmad, S., Khalid, R., Abbas, S., Hayat, R., & Ahmed, I. (2021). Potential of compost for sustainable crop production and soil health. In R. Prasad, V. Kumar, J. Singh, & C. P. Upadhyaya (Eds.).

Ahmed, M., Ahmad, S., Qadir, G., Hayat, R., Shaheen, F. A., & Raza, M. A. (2019). Innovative processes and technologies for nutrient recovery from wastes: A comprehensive review. *Sustainability, 11*, 4938.

Akhtar, K., Wang, W., Ren, G., Khan, A., Enguang, N., Khan, A., Feng, Y., Yang, G., & Wang, H. (2020). Straw mulching with inorganic nitrogen fertilizer reduces soil CO_2 and N_2O emissions and improves wheat yield. *Science of the Total Environment, 741*, 140488.

Aniekwe, N. L., Okereke, O. U., & Anikwe, M. A. N. (2004). Modulating effect of black plastic mulch on the environment, growth and yield of cassava in a derived savannah belt of Nigeria. *Tropicultura, 22*, 185–190.

Anikwe, M. A. N., Mbah, C. N., Ezeaku, P. I., & Onyia, V. N. (2007). Tillage and plastic mulch effects on soil properties and growth and yield of cocoyam (Colocasia esculenta) on an ultisol in southeastern Nigeria. *Soil and Tillage Research, 93*, 264–272.

Bandopadhyay, S., Liquet y Gonzalez, J. E., Henderson, K. B., Anunciado, M. B., Hayes, D. G., & DeBruyn, J. M. (2020). Soil microbial communities associated with biodegradable plastic mulch films. *Frontiers in Microbiology, 11*, 2840.

Bhagat, P., Gosal, S. K., & Singh, C. B. (2016). Effect of mulching on soil environment, microbial flora and growth of potato under field conditions. *Indian Journal of Agricultural Research, 50*, 542–548.

Bhardwaj, R. L. (2013). Effect of mulching on crop production under rainfed condition—A review. *Agricultural Reviews, 34*, 188–197.

Bhattacharyya, P., & Barman, D. (2018). Crop residue management and greenhouse gases emissions in tropical rice lands. In *Soil management and climate change* (pp. 323–335). Elsevier.

Biswas, S. K., Akanda, A. R., Rahman, M. S., & Hossain, M. A. (2015). Effect of drip irrigation and mulching on yield, water-use efficiency and economics of tomato. *Plant, Soil and Environment, 61*, 97–102.

Bokszczanin, K. Ł, Wrona, D., & Przybyłko, S. (2021). Influence of an alternative soil management system to herbicide use on tree vigor, yield, and quality of apple fruit. *Agronomy, 11*, 58.

Briassoulis, D., & Giannoulis, A. (2018). Evaluation of the functionality of bio-based plastic mulching films. *Polymer Testing, 67*, 99–109.

Bucki, P., & Siwek, P. (2019). Organic and non-organic mulches—impact on environmental conditions, yield, and quality of Cucurbitaceae. *Folia Horticulturae, 31*, 129–145.

Budelman, A. (1988) The decomposition of the leaf mulches of *Leucaena leucocephala Gliricidia sepium* and *Flemingia macrophylla* under humid tropical conditions. *Agroforestry Systems, 7*, 33–45.

Carroll, J. (2020). Types of bark mulch: Tips for using wood mulch in gardens. In *Gardening know how*.

Chalker-Scott, L. (2007). Impact of mulches on landscape plants and the environment—A review. *Journal of Environmental Horticulture, 25*, 239–249.

Chang-Rong, Y., En-Ke, L., Fan, S., Liu, Q., Liu, S., & Wen-Qing, H. (2014). Review of agricultural plastic mulching and its residual pollution and prevention measures in China. *Journal of Agriculture Resources and Environment, 31*, 95.

Chavan, M. L. (2009). Effect of organic mulches on soil moisture conservation and yield ofrabisorghum (M-35-1). *International Journal of Agricultural Engineering, 2*.

Chen, H., Liu, J., Zhang, A., Chen, J., Cheng, G., Sun, B., Pi, X., Dyck, M., Si, B., & Zhao, Y. (2017). Effects of straw and plastic film mulching on greenhouse gas emissions in Loess Plateau, China: A field study of 2 consecutive wheat-maize rotation cycles. *Science of the Total Environment, 579*, 814–824.

Cheng, Y., Zhu, L., Song, W., Jiang, C., Li, B., Du, Z., Wang, J., Wang, J., Li, D., & Zhang, K. (2020). Combined effects of mulch film-derived microplastics and atrazine on oxidative stress and gene expression in earthworm (Eisenia fetida). *Science of the Total Environment, 746*, 141280.

Chopra, M., & Koul, B. (2020). Comparative assessment of different types of mulching in various crops: A review. *Plant Archives, 20*, 1620–1626.

Coolong, T. (2012). Mulches for weed management in vegetable production. *Weed Control. Croatia, European Union: In Tech Open*, 57–74.

CTIC. (2001). Conservation Technology Information Center. CTIC.

Dalorima, L. T., Bunu, A., Kyari, Z., & Mohammed, T. (2014). Effects of different mulching materials on the growth performance of okra in Maiduguri. *International Research Journal of Agricultural Science and Soil Science, 4*, 145–149.

Derpsch, R. (2003). Conservation tillage, no-tillage and related technologies. In *Conservation agriculture* (pp. 181–190). Springer.

Dilipkumar, G., Sachin, S. S., & Rajesh, K. (1990). Importance of mulch in crop production. *Indian Journal of Soil Conservation, 18*, 20–26.

Dossou-Yovo, E. R., Brueggemann, N., Jesse, N., Huat, J., Ago, E. E., & Agbossou, E. K. (2016). Reducing soil CO_2 emission and improving upland rice yield with no-tillage, straw mulch and nitrogen fertilization in northern Benin. *Soil and Tillage Research, 156*, 44–53.

Campbell, S. (2012). *Mulch it!: A practical guide to using mulch in the garden and landscape*. Storey Publishing.

Fang, S., Xie, B., Liu, D., & Liu, J. (2011). Effects of mulching materials on nitrogen mineralization, nitrogen availability and poplar growth on degraded agricultural soil. *New Forests, 41*, 147–162.

FAO. (2001). Conservation agriculture, matching production with sustainability. In *What is the goal of conservation agriculture? FAO homepage.*

Fernández, C., & Vega, J. A. (2016). Are erosion barriers and straw mulching effective for controlling soil erosion after a high severity wildfire in NW Spain? *Ecological Engineering, 87*, 132–138.

Fernández, C., & Vega, J. A. (2021). Is wood strand mulching a good alternative to helimulching to mitigate the risk of soil erosion and favour the recovery of vegetation in NW Spain? *Landscape and Ecological Engineering*, 1–10.

Fernández-Fernández, M., Rütting, T., & González-Prieto, S. (2017). Effects of a high-severity wildfire and post-fire straw mulching on gross nitrogen dynamics in Mediterranean shrubland soil. *Geoderma, 305*, 328–335.

Fialho, J. S., Primo, A. A., de Aguiar, M. I., Magalhaes, R. B., dos Santos Maia, L., Correia, M. E. F., Campanha, M. M., & de Oliveira, T. S. (2021). Pedofauna diversity in traditional and agroforestry systems of the Brazilian semi-arid region. *Journal of Arid Environments, 184*, 104315.

Gao, H., Yan, C., Liu, Q., Ding, W., Chen, B., & Li, Z. (2019). Effects of plastic mulching and plastic residue on agricultural production: A meta-analysis. *Science of the Total Environment, 651*, 484–492.

Gheshm, R., & Brown, R. N. (2020). Compost and black polyethylene mulches improve spring production of romaine lettuce in southern New England. *HortTechnology, 1*, 1–9.

Goel, L., Shankar, V., & Sharma, R. K. (2019). Investigations on effectiveness of wheat and rice straw mulches on moisture retention in potato crop (*Solanum tuberosum L.*). *International Journal of Recycling of Organic Waste in Agriculture, 8*, 345–356.

Granatstein, D., & Mullinix, K. (2008). Mulching options for northwest organic and conventional orchards. *HortScience, 43*, 45–50.

Hamrani, A., Akbarzadeh, A., & Madramootoo, C. A. (2020). Machine learning for predicting greenhouse gas emissions from agricultural soils. *Science of the Total Environment, 741*, 140338.

Hannam, K. D., Neilsen, G. H., Forge, T. A., Neilsen, D., Losso, I., Jones, M. D., Nichol, C., & Fentabil, M. M. (2016). Irrigation practices, nutrient applications, and mulches affect soil nutrient dynamics in a young Merlot (Vitis vinifera L.) vineyard. *Canadian Journal of Soil Science, 96*, 23–36.

Harkin, J. M. (1971). Bark and its possible uses. Forest Products Laboratory, US Forest Service.

Harvold, K., & Falleth, E. (2005). Marigold beds and villa horses. *European Journal of Spatial Development.*

Hashim, S., Marwat, K. B., Saeed, M., Haroon, M., Waqas, M., & Shah, F. (2013). Developing a sustainable and eco-friendly weed management system using organic and inorganic mulching techniques. *Pakistan Journal of Botany, 45*, 483–486.

Hathi, H. S., Parmar, D. L., Bedva, S. M., Purohit, S. A., & Patel, D. B. (2021). Mulching: An expeditious overview. *2*, 69–72.

Hodgson, L. (2020). The story of mulch. In *Laidbackgardener.*

Iiles, J. K. (1999). Effect of organic and mineral mulches in soil properties and growth of fairview flame red maple trees. *Journal of Arboriculture, 25*, 163–167.

Iqbal, M., Bakshi, P., Wali, V. K., Kumar, R., Bhat, D., & Jasrotia, A. (2016). Efficacy of organic and inorganic mulching materials on weed count, growth, and yield of aonla (Emblica officinalis) cv. NA 7. *Indian Journal of Agricultural Sciences, 86*, 545–549.

Iqbal, R., Raza, M. A. S., Valipour, M., Saleem, M. F., Zaheer, M. S., Ahmad, S., Toleikiene, M., Haider, I., Aslam, M. U., & Nazar, M. A. (2020). Potential agricultural and environmental benefits of mulches—A review. *Bulletin of the National Research Centre, 44*, 1–16.

Iriany, A., Chanan, M., & Djoyowasito, G. (2018). Organic mulch sheet formulation as an effort to help plants adapt to climate change. *International Journal of Recycling of Organic Waste in Agriculture, 7*, 41–47.

Jabran, K., Ullah, E., Hussain, M., Farooq, M., Zaman, U., Yaseen, M., & Chauhan, B. S. (2015). Mulching improves water productivity, yield and quality of fine rice under water-saving rice production systems. *Journal of Agronomy and Crop Science, 201*, 389–400.

Jabran, K. (2019). Mulches for enhancing biological activities in soil. In *Role of mulching in pest management and agricultural sustainability* (pp. 41–46). Springer.

Jacinthe, P.-A., Dick, W. A., Lal, R., Shrestha, R. K., & Bilen, S. (2014). Effects of no-till duration on the methane oxidation capacity of Alfisols. *Biology and Fertility of Soils, 50*, 477–486.

Jia, Q., Zhang, H., Wang, J., Xiao, X., Chang, S., Zhang, C., Liu, Y., & Hou, F. (2020). Planting practices and mulching materials improve maize net ecosystem C budget, global warming potential and production in semi-arid regions. *Soil and Tillage Research*, 104850.

Jodaugienė, D., Pupalienė, R., Urbonienė, M., Pranckietis, V., & Pranckietienė, I. (2006). The impact of different types of organic mulches on weed emergence. *Agronomy Research, 4*, 197–201.

Kader, M. A., Senge, M., Mojid, M. A., & Ito, K. (2017a). Recent advances in mulching materials and methods for modifying soil environment. *Soil and Tillage Research, 168*, 155–166.

Kader, M. A., Senge, M., Mojid, M. A., & Nakamura, K. (2017b). Mulching type-induced soil moisture and temperature regimes and water use efficiency of soybean under rain-fed condition in central Japan. *International Soil and Water Conservation Research, 5*, 302–308.

Kader, M. A., Nakamura, K., Senge, M., Mojid, M. A., & Kawashima, S. (2020). Effects of coloured plastic mulch on soil hydrothermal characteristics, growth and water productivity of rainfed soybean. *Irrigation and Drainage, 69*, 483–494.

Kannan, R. (2020). Chapter-1 uses of mulching in agriculture: A review. *Current Research in Soil Fertility, 1*.

Kasirajan, S., & Ngouajio, M. (2012). Polyethylene and biodegradable mulches for agricultural applications: A review. *Agronomy for Sustainable Development, 32*, 501–529.

Kassaye, K. T., Boulange, J., Saito, H., & Watanabe, H. (2021). Soil water content and soil temperature modeling in a vadose zone of Andosol under temperate monsoon climate. *Geoderma, 384*, 114797.

Kearney, S. P., Fonte, S. J., García, E., Siles, P., Chan, K. M. A., & Smukler, S. M. (2019). Evaluating ecosystem service trade-offs and synergies from slash-and-mulch agroforestry systems in El Salvador. *Ecological Indicators, 105*, 264–278.

Khaledian, M. R., Mailhol, J.-C., Ruelle, P., Mubarak, I., & Perret, S. (2010). The impacts of direct seeding into mulch on the energy balance of crop production system in the SE of France. *Soil and Tillage Research, 106*, 218–226.

Khurshid, K., Iqbal, M., Arif, M. S., & Nawaz, A. (2006). Effect of tillage and mulch on soil physical properties and growth of maize. *International Journal of Agriculture and Biology, 8*, 593–596.

Kodzwa, J. J., Gotosa, J., & Nyamangara, J. (2020). Mulching is the most important of the three conservation agriculture principles in increasing crop yield in the short term, under sub humid tropical conditions in Zimbabwe. *Soil and Tillage Research, 197*, 104515.

Kumar, V., Kumar, M., & Rai, A. P. (2016). Mulching in fruit and vegetable crop production under rainfed conditions: A review. *Journal of Soil and Water Conservation, 15*, 313–317.

Lal, R. (1997). Mulching effects on runoff, soil erosion, and crop response on alfisols in Western Nigeria. *Journal of Sustainable Agriculture, 11*, 135–154.

Lal, R. (2015). Sequestering carbon and increasing productivity by conservation agriculture. *Journal of Soil and Water Conservation, 70*, 55A-62A.

Lalitha, M., Thilagam, V. K., Balakrishnan, N., & Mansour, M. (2010). Effect of plastic mulch on soil properties and crop growth-A review. *Agricultural Reviews, 31*, 145–149.

Lalk, G. T., Bi, G., Zhang, Q., Harkess, R. L., & Li, T. (2020). High-tunnel production of strawberries using black and red plastic mulches. *Horticulturae, 6*, 73.

Lee, J. G., Cho, S. R., Jeong, S. T., Hwang, H. Y., & Kim, P. J. (2019). Different response of plastic film mulching on greenhouse gas intensity (GHGI) between chemical and organic fertilization in maize upland soil. *Science of the Total Environment, 696*, 133827.

Liao, Y., Cao, H.-X., Xue, W.-K., & Liu, X. (2021). Effects of the combination of mulching and deficit irrigation on the soil water and heat, growth and productivity of apples. *Agricultural Water Management, 243*, 106482.
Lima, J. D., Zanetti, S., Nomura, E. S., Fuzitani, E. J., Rozane, D. E., & Iori, P. (2016). Growth and yield of anthurium in response to sawdust mulching. *Ciência Rural, 46*, 440–446.
Liu, J., Xu, S. A., Zhou, G. Y., & Lu, H. H. (2002). Effects of transplanting multi-cropping spring maize with plastic film mulching on the ecological effect, plant growth and grain yield. *J. Hubei Agri. College* **2**, 0.
Liu, J., Liu, J., Wang, J., Wang, H., Zuo, J., & Hu, D. (2021). Fruit yield and properties of Camellia oleifera Abel can be enhanced via fine roots promotion under mulch. *Archives of Agronomy and Soil Science.*
Livesley, S. J., Dougherty, B. J., Smith, A. J., Navaud, D., Wylie, L. J., & Arndt, S. K. (2010). Soil-atmosphere exchange of carbon dioxide, methane and nitrous oxide in urban garden systems: Impact of irrigation, fertiliser and mulch. *Urban Ecosystems, 13*, 273–293.
Lojka, B., Dumas, L., Preininger, D., Polesny, Z., & Banout, J. (2010). The use and integration of Inga edulis in agroforestry systems in the amazon—Review article. *Agricultura Tropica et Subtropica, 43*, 352–359.
Lucas-Borja, M. E., Zema, D. A., Carrà, B. G., Cerdà, A., Plaza-Alvarez, P. A., Cózar, J. S., Gonzalez-Romero, J., Moya, D., & de las Heras, J. (2018). Short-term changes in infiltration between straw mulched and non-mulched soils after wildfire in Mediterranean forest ecosystems. *Ecological Engineering, 122*, 27–31.
Lucas-Borja, M. E., González-Romero, J., Plaza-Álvarez, P. A., Sagra, J., Gómez, M. E., Moya, D., Cerdà, A., & de Las Heras, J. (2019). The impact of straw mulching and salvage logging on post-fire runoff and soil erosion generation under Mediterranean climate conditions. *Science of the Total Environment, 654*, 441–451.
Ma, Y., Sun, L., Zhang, X., Yang, B., Wang, J., Yin, B., Yan, X., & Xiong, Z. (2013). Mitigation of nitrous oxide emissions from paddy soil under conventional and no-till practices using nitrification inhibitors during the winter wheat-growing season. *Biology and Fertility of Soils, 49*, 627–635.
Ma, D., Chen, L., Qu, H., Wang, Y., Misselbrook, T., & Jiang, R. (2018). Impacts of plastic film mulching on crop yields, soil water, nitrate, and organic carbon in Northwestern China: A meta-analysis. *Agricultural Water Management, 202*, 166–173.
Martín-Closas, L., Bach, M. A., & Pelacho, A. M. (2008). Biodegradable mulching in an organic tomato production system (pp. 267–274).
Masvaya, E. N., Nyamangara, J., Descheemaeker, K., & Giller, K. E. (2017). Tillage, mulch and fertiliser impacts on soil nitrogen availability and maize production in semi-arid Zimbabwe. *Soil and Tillage Research, 168*, 125–132.
Mehta, D. K., Kaith, N. S., & Kanwar, H. S. (2010). Effect of training methods and mulching on growth, yield and fruit rot incidence in tomato (Solanum lycopersicum). *Indian Journal of Agricultural Sciences, 80*, 829–831.
Mendonça, S. R., Ávila, M. C. R., Vital, R. G., Evangelista, Z. R., de Carvalho Pontes, N., & dos Reis Nascimento, A. (2021). The effect of different mulching on tomato development and yield. *Scientia Horticulturae, 275*, 109657.
Montenegro, A. A. D. A., Abrantes, J., De Lima, J., Singh, V., & Santos, T. E. M. (2013). Impact of mulching on soil and water dynamics under intermittent simulated rainfall. *Catena, 109*, 139–149.
Mubvumba, P., DeLaune, P. B., & Hons, F. M. (2021). Soil water dynamics under a warm-season cover crop mixture in continuous wheat. *Soil and Tillage Research, 206*, 104823.
Mupangwa, W., Twomlow, S., & Walker, S. (2012). Reduced tillage, mulching and rotational effects on maize (Zea mays L.), cowpea (Vigna unguiculata (Walp) L.) and sorghum (Sorghum bicolor L. (Moench)) yields under semi-arid conditions. *Field Crops Research, 132*, 139–148.
Murungu, F. S., Chiduza, C., Muchaonyerwa, P., & Mnkeni, P. N. S. (2011). Mulch effects on soil moisture and nitrogen, weed growth and irrigated maize productivity in a warm-temperate climate of South Africa. *Soil and Tillage Research, 112*, 58–65.

Ngosong, C., Okolle, J. N., & Tening, A. S. (2019). Mulching: A sustainable option to improve soil health. In *Soil fertility management for sustainable development* (pp. 231–249). Springer.
Ning, R., Liang, J., Sun, Z., Liu, X., & Sun, W. (2020). Preparation and characterization of black biodegradable mulch films from multiple biomass materials. *Polymer Degradation and Stability*, 109411.
Niu, J. Y., Gan, Y. T., & Huang, G. B. (2004). Dynamics of root growth in spring wheat mulched with plastic film. *Crop Science, 44*, 1682–1688.
Nyawira, M. C. (2016). Analysis of heavy metal content in water hyacinth (*Eichonia crassipes*) from lake victoria and assessment of its potential as a feedstock for biogas production. *University of Nairobi.*
Olson, S. M. (2006). Mulching. *EDIS, 2006*.
Pandiyan, B., Mangottiri, V., & Narayanan, N. (2021). Carbon transformations of biochar based co-composting-A review. *Mini-Reviews in Organic Chemistry, 18*, 1–14.
Park, S.-I., Yang, H. I., Park, H.-J., Seo, B.-S., Jeong, Y.-J., Lim, S.-S., Kwak, J.-H., Kim, H.-Y., Yoon, K.-S., & Lee, S.-M. (2021). Rice straw cover decreases soil erosion and sediment-bound C, N, and P losses but increases dissolved organic C export from upland maize fields as evidenced by δ13C. *Science of the Total Environment, 753*, 142053.
Patricia, R. D. (1957). The mulching of vegetable. Commonwealth Agriculture Bureau.
Pavlů, L., Kodešová, R., Fér, M., Nikodem, A., Němec, F., & Prokeš, R. (2020). The impact of various mulch types on soil properties controlling water regime of the Haplic Fluvisol. *Soil and Tillage Research, 205*, 104748.
Payán, F., Jones, D. L., Beer, J., & Harmand, J.-M. (2009). Soil characteristics below Erythrina poeppigiana in organic and conventional Costa Rican coffee plantations. *Agroforestry Systems, 76*, 81–93.
Pervaiz, M. A., Iqbal, M., Shahzad, K., & Hassan, A. U. (2009). Effect of mulch on soil physical properties and N, P, K concentration in maize (Zea mays L.) shoots under two tillage systems. *International Journal of Agriculture and Biology, 11*, 119–124.
Prats, S. A., dos Santos Martins, M. A., Malvar, M. C., Ben-Hur, M., & Keizer, J. J. (2014). Polyacrylamide application versus forest residue mulching for reducing post-fire runoff and soil erosion. *Science of the Total Environment, 468*, 464–474.
Prosdocimi, M., Tarolli, P., & Cerdà, A. (2016). Mulching practices for reducing soil water erosion: A review. *Earth-Science Reviews, 161*, 191–203.
Puka-Beals, J., & Gamig, G. (2021). Weed suppression potential of living mulches, newspaper hydromulches, and compost blankets in organically managed carrot production. *HortTechnology, 1*, 1–8.
Qin, W., Hu, C., & Oenema, O. (2015). Soil mulching significantly enhances yields and water and nitrogen use efficiencies of maize and wheat: A meta-analysis. *Scientific Reports, 5*, 16210.
Qi, Y., Beriot, N., Gort, G., Lwanga, E. H., Gooren, H., Yang, X., & Geissen, V. (2020a). Impact of plastic mulch film debris on soil physicochemical and hydrological properties. *Environmental Pollution, 266*, 115097.
Qi, Y., Ossowicki, A., Yang, X., Lwanga, E. H., Dini-Andreote, F., Geissen, V., & Garbeva, P. (2020b). Effects of plastic mulch film residues on wheat rhizosphere and soil properties. *Journal of Hazardous Materials, 387*, 121711.
Radics, L., & Szné Bognár, E. (2004). Comparison of different mulching methods for weed control in organic green bean and tomato (pp. 189–196).
Ranjan, P., Patle, G. T., Prem, M., & Solanke, K. R. (2017). Organic mulching-A water saving technique to increase the production of fruits and vegetables. *Current Agriculture Research Journal, 5*, 371–380.
Rao, X., Liu, C.-A., Tang, J.-W., Nie, Y., Liang, M.-Y., Shen, W.-J., & Siddique, K. H. M. (2020). Rubber-leguminous shrub systems stimulate soil N_2O but reduce CO_2 and CH_4 emissions. *Forest Ecology and Management, 480*, 118665.
Rawat, L., Bisht, T. S., & Naithani, D. C. (2021). Plant disease management in organic farming system: Strategies and challenges. In *Emerging trends in plant pathology* (pp. 611–642). Springer.

Rendon, D., Hamby, K. A., Arsenault-Benoit, A. L., Taylor, C. M., Evans, R. K., Roubos, C. R., Sial, A. A., Rogers, M., Petran, A., & Van Timmeren, S. (2020). Mulching as a cultural control strategy for Drosophila suzukii in blueberry. *Pest Management Science, 76*, 55–66.

Riahi, K., Rao, S., Krey, V., Cho, C., Chirkov, V., Fischer, G., Kindermann, G., Nakicenovic, N., & Rafaj, P. (2011). RCP 8.5—A scenario of comparatively high greenhouse gas emissions. *Climatic Change, 109*, 33–57.

Rinaudo, A., & Cunningham, P. (2008). Australian acacias as multi-purpose agro-forestry species for semi-arid regions of Africa. *Muelleria, 26*, 79–85.

Robichaud, P. R., Lewis, S. A., Wagenbrenner, J. W., Ashmun, L. E., & Brown, R. E. (2013). Post-fire mulching for runoff and erosion mitigation: Part I: Effectiveness at reducing hillslope erosion rates. *CATENA, 105*, 75–92.

Rudawska, M., & Leski, T. (2021). Ectomycorrhizal fungal assemblages of nursery-grown scots pine are influenced by age of the seedlings. *Forests, 12*, 134.

Saglam, M., Sintim, H. Y., Bary, A. I., Miles, C. A., Ghimire, S., Inglis, D. A., & Flury, M. (2017). Modeling the effect of biodegradable paper and plastic mulch on soil moisture dynamics. *Agricultural Water Management, 193*, 240–250.

Saha, A. K., Bandopadhaya, A. K., & Sarkar, B. (1974). Effect of different soil management practices on conservation of soil moisture and growth and fruiting of lemon (Citrus Limon Burm). *Indian Journal of Horticulture, 31*, 337–341.

Saikia, U. S., Kumar, A., Das, S., Pradhan, R., Goswami, B., Wungleng, V. C., Rajkhowa, D. J., & Ngachan, S. V. (2014). Effect of mulching on microclimate, growth and yield of mustard (Brassica juncea) under mid-hill condition of Meghalaya. *Journal of Agrometeorology, 16*, 144.

Salmoral, G., Willaarts, B. A., Garrido, A., & Guse, B. (2017). Fostering integrated land and water management approaches: Evaluating the water footprint of a Mediterranean basin under different agricultural land use scenarios. *Land Use Policy, 61*, 24–39.

Sas-Paszt, L., Pruski, K., Żurawicz, E., Sumorok, B., Derkowska, E., & Głuszek, S. (2014). The effect of organic mulches and mycorrhizal substrate on growth, yield and quality of Gold Milenium apples on M. 9 rootstock. *Canadian Journal of Plant Science, 94*, 281–291.

Schroth, G., da Fonseca, G. A. B., Harvey, C. A., Gascon, C., Vasconcelos, H. L., & Izac, A.-M.N. (2013). *Agroforestry and biodiversity conservation in tropical landscapes*. Island Press.

Sebastian, M. (2012). The history of mulch. In *IMC outdoor living, a division of liberty*.

Shirish, P. S., Tushar, K. S., & Satish, B. A. (2013). Mulching: A soil and water conservation practice. *Research Journal of Agriculture and Forestry Sciences*, 60–63.

Shojaei, S., Ardakani, M. A. H., & Sodaiezadeh, H. (2019). Optimization of parameters affecting organic mulch test to control erosion. *Journal of Environmental Management, 249*, 109414.

Sindhu, R., Binod, P., Pandey, A., Madhavan, A., Alphonsa, J. A., Vivek, N., Gnansounou, E., Castro, E., & Faraco, V. (2017). Water hyacinth a potential source for value addition: An overview. *Bioresource Technology, 230*, 152–162.

Singh, A. K., & Kamal, S. (2012). Effect of black plastic mulch on soil temperature and tomato yield in mid hills of Garhwal Himalayas. *Journal of Horticulture and Forestry, 4*, 77–79.

Singh, B., Gupta, G. N., Prasad, K. G., & Mohan, S. (1988). Use of mulches in establishment and growth of tree species on dry lands. *Indian Forester, 114*, 307–316.

Sinkevičienė, A., Jodaugienė, D., Pupalienė, R., & Urbonienė, M. (2009). The influence of organic mulches on soil properties and crop yield. *Agronomy Research, 7*, 485–491.

Sintim, H. Y., Bandopadhyay, S., English, M. E., Bary, A. I., DeBruyn, J. M., Schaeffer, S. M., Miles, C. A., Reganold, J. P., & Flury, M. (2019). Impacts of biodegradable plastic mulches on soil health. *Agriculture, Ecosystems & Environment, 273*, 36–49.

Sintim, H. Y., & Flury, M. (2017). Is biodegradable plastic mulch the solution to agriculture's plastic problem? ACS Publications.

Sintim, H. Y., Bandopadhyay, S., English, M. E., Bary, A., y González, J. E. L., DeBruyn, J. M., Schaeffer, S. M., Miles, C. A., & Flury, M. (2021). Four years of continuous use of soil-biodegradable plastic mulch: Impact on soil and groundwater quality. *Geoderma, 381*, 114665.

Smets, T., Poesen, J., & Knapen, A. (2008). Spatial scale effects on the effectiveness of organic mulches in reducing soil erosion by water. *Earth-Science Reviews, 89*, 1–12.

Srivastava, Y. (2013). Advances in food science and nutrition. *Oba Ile: Science and Education Development Institute.*

Steinmetz, Z., Wollmann, C., Schaefer, M., Buchmann, C., David, J., Tröger, J., Muñoz, K., Frör, O., & Schaumann, G. E. (2016). Plastic mulching in agriculture. Trading short-term agronomic benefits for long-term soil degradation? *Science of the Total Environment, 550*, 690–705.

Stigter, C. J. (1984). Mulching as a traditional method of microclimate management. *Archives for Meteorology, Geophysics, and Bioclimatology, Series B, 35*, 147–154.

Stirling, G. R., & Eden, L. M. (2008). The impact of organic amendments, mulching and tillage on plant nutrition, Pythium root rot, root-knot nematode and other pests and diseases of capsicum in a subtropical environment, and implications for the development of more sustainable vegetable farming systems. *Australasian Plant Pathology, 37*, 123–131.

Sun, X., Wang, G., Ma, Q., Liao, J., Wang, D., Guan, Q., & Jones, D. L. (2021). Organic mulching promotes soil organic carbon accumulation to deep soil layer in an urban plantation forest. *Forest Ecosystems, 8*, 1–11.

Sun, Z.-J., Yi, J.-H., Liu, J.-F., Lian, S.-X., & Yin, D.-L. (2008). Effect of dual-sunlight conversion film mulching on soil temperature, growth and physiological characteristics of tobacco. *Chinese Journal of Eco-Agriculture, 1.*

Tang, M., Li, H., Zhang, C., Zhao, X., Gao, X., & Wu, P. (2021). Mulching measures improve soil moisture in rain-fed jujube (Ziziphus jujuba Mill.) orchards in the loess hilly region of China. *Sustainability, 13*, 610.

Telkar, S. G., Singh, A. K., Kant, K., Solanki, S. P. S., & Kumar, D. (2006). Types of Mulching and their uses for dryland condition. Rep. No. 2456–8759.

Unger, P. W., & McCalla, T. M. (1980). Conservation tillage systems. In *Advances in agronomy* (Vol. 33, pp. 1–58). Elsevier.

Vaddevolu, U. B. P., Lester, J., Scherer, T. F., Lee, C., & Jia, X. (2020). Automatic sensor controlled drip irrigation with four different mulches for tomato and watermelon production (p. 1). American Society of Agricultural and Biological Engineers.

Vieira, D. C. S., Serpa, D., Nunes, J. P. C., Prats, S. A., Neves, R., & Keizer, J. J. (2018). Predicting the effectiveness of different mulching techniques in reducing post-fire runoff and erosion at plot scale with the RUSLE, MMF and PESERA models. *Environmental Research, 165*, 365–378.

Vos, J. G. M., & Sumarni, N. (1997). Integrated crop management of hot pepper (Capsicum spp.) under tropical lowland conditions: Effects of mulch on crop performance and production. *Journal of Horticultural Science, 72*, 415–424.

Waggoner, P. E., Miller, P. M., & De Roo, H. (1960). C. Plastic mulching: principles and benefits. *Bulletin Connecticut Agricultural Experiment Station, 634.*

Wang, J., Li, F., Song, Q., & Li, S. (2003). Effects of plastic film mulching on soil temperature and moisture and on yield formation of spring wheat. *Ying Yong Sheng Tai Xue Bao = The Journal of Applied Ecology, 14*, 205–210.

Wang, H., Lu, D., Wang, Q., & Shan, C. (2020a). Effects of embedded gravel or gravel mulching in southern red soil on slope sediment yield and runoff. *Polish Journal of Environmental Studies, 30.*

Wang, Y., Huang, Q., Liu, C., Ding, Y., Liu, L., Tian, Y., Wu, X., Li, H., Awasthi, M. K., & Zhao, Z. (2020b). Mulching practices alter soil microbial functional diversity and benefit to soil quality in orchards on the Loess Plateau. *Journal of Environmental Management, 271*, 110985.

Wu, J., Zhu, Z., Zheng, J., & Jiang, X. (2006). Influences of straw mulching treatment on soil physical and chemical properties and crop yields. *Southwest China Journal of Agricultural Sciences, 19*, 192–195.

Xiukang, W., Zhanbin, L., & Yingying, X. (2015). Effects of mulching and nitrogen on soil temperature, water content, nitrate-N content and maize yield in the Loess Plateau of China. *Agricultural Water Management, 161*, 53–64.

Yu, Y.-Y., Turner, N. C., Gong, Y.-H., Li, F.-M., Fang, C., Ge, L.-J., & Ye, J.-S. (2018). Benefits and limitations to straw-and plastic-film mulch on maize yield and water use efficiency: A meta-analysis across hydrothermal gradients. *European Journal of Agronomy, 99*, 138–147.
Zhang, D., Ng, E. L., Hu, W., Wang, H., Galaviz, P., Yang, H., Sun, W., Li, C., Ma, X., & Fu, B. (2020a). Plastic pollution in croplands threatens long-term food security. *Global Change Biology, 26*, 3356–3367.
Zhang, Y., Feng, R., Nie, W., Wang, F., & Feng, S. (2020b). Plastic Film Mulch Performed Better in Improving Heat Conditions and Drip Irrigated Potato Growth in Northwest China than in Eastern China. *Water, 12*, 2906.

Mulching is a Mechanism to Reduce Environmental Stresses in Plants

Syed Tanveer Shah, Izhar Ullah, Abdul Basit, Muhammad Sajid, Muhammad Arif, and Heba I. Mohamad

Abstract Mulching is considered as one of the most important elements of sustainable agriculture that promote growth and productivity of crop plants, reduce weed growth, maintain the optimal soil temperature and retain soil moisture and increase the aesthetic value of land and mitigate the adverse effect of biotic and abiotic stresses. Mulches (earlier agricultural bioproducts were used as mulch) gained popularity in the early-nineteenth century as a consequence of their ability to optimize the conditions for agricultural lands. The sources of organic mulches include plants and animals residues. Organic mulches that are widely used around the globe involves straws, husks, sawdust, grasses, manures and composts. Mulch made of polythene plastic is the most widely used inorganic mulch around the globe. It has been observed that mulch has a positive effect of environmental conditions such as light, heat, soil as well as crop growth, yield and quality. Mulch also helps to mitigate environmental stresses making favorable environment to the plant. This chapter emphasized on the importance of mulches in overcoming environmental stress in plants and gathers detailed information about the positive effect of mulch not only on plants but also on soils as well. Furthermore, information has also been given on the mitigation of different environmental stresses which harm crop growth.

Keywords Allelopathy · Abiotic and biotic stresses · Mulching · Sustainable agriculture · Thermal conditions

---

S. T. Shah (✉) · A. Basit · M. Sajid
Department of Horticulture, Faculty of Crop Production Sciences, The University of Agriculture Peshawar, Peshawar 3125, Pakistan
e-mail: syedtanveer07@yahoo.com; syedtanveer07@aup.edu.pk

I. Ullah
Department of Horticulture, Faculty of Agriculture, Ondokuz Mayis University, Samsun, Turkey

M. Arif
Department of Agronomy, Faculty of Crop Production Sciences, The University of Agriculture Peshawar, Peshawar 3125, Pakistan

H. I. Mohamad
Department of Biological and Geological Sciences, Faculty of Education, Ain Shams University, Cairo, Egypt

K. Akhtar et al. (eds.), *Mulching in Agroecosystems*,
https://doi.org/10.1007/978-981-19-6410-7_20

1 Introduction

The term mulch came from German word molsch, meaning "quick to decay". Mulches are used from a long time to increase the production of vegetables. Mulch is any material that is applied as a coating over the soil surface. It may be organic or inorganic in nature. Mulching is defined as "the application of various covering materials to the soil's surface in order to reduce the loss of moisture, to control weed population, and to increase yield of crops (Kader et al., 2019; Nalayini, 2007). Mulches enhance the performance of crops due to its ability of controlling the growth of weeds through shading, it also advances the movement of water into or through the soil profile and it also barricades the evapotranspiration (Rathore et al., 1998). In addition, mulches also regulates the temperature of the soil and roots of the plant, it also helps in reducing the compactness and erosion of the soil (Ngouajio & McGiffen, 2004). Mulches gained popularity in the early-nineteenth century as a consequence of their ability to optimize the conditions for agricultural lands. Due to the shielding potential of deep type mulches, trees and shrubs were also planted using deep mulches to protect them from drought and frost injury in severe environmental situations. In the mid-nineteenth century, agricultural by-products were used as mulches (Clifford & Massello, 1965). Mulch can be of natural and synthetic sources. The sources of organic mulches includes plants and animals residues. Organic mulches that are widely used around the globe involves straws, husks, sawdust, grasses, manures and composts (Rathore et al., 1998). Mulch made of polythene plastic is the most widely used inorganic mulch around the globe. The use of plastic mulch in agriculture is increasing day by day. According to a report, the plastic mulch was used on almost 22,000,000 ha of agricultural land around the world and 15,000,000 ha of agricultural land in china in 1999 and 2000 respectively (Miles, 2005; Xing et al., 2003). Annually, 700,000 tonnes of plastic mulch are used worldwide while, 140,000 tonnes is used in the United States alone (Espi et al., 2006; Shogren, 2001). Due to the low cost of black plastic mulch relative to other mulches, it is widely used in cultivated lands worldwide (Ngouajio & McGiffen, 2004). Mulching protects the plant from various fungal infections and other pest diseases caused due to the crop and plant residues. In addition, mulches also play a vital role in keeping the plants hydrated in water scarce areas by increasing soil's porosity and drainage. In addition, it gives protection to plants against winter frosts (Gill et al., 2011). Mulches are used in landscaping and agriculture because of their aesthetic, economic, and environmental benefits. Mulching is important for establishing the plants for conservation purposes, and it requires very little maintenance. In gardening and landscaping, minimal care is needed for a variety of reasons, including variation in garden (Chalker-Scott, 2007). Mulch helps the soil to preserve its composition and prevents crust formation (Kasirajan & Ngouajio, 2012). Mulch reduces the need of frequent irrigation as it conserves the soil moisture, mulch also prevent the soil particles from loosening its structure due to heavy irrigation and also inhibits the growth of weeds thus preventing the competition of plants with weeds for resources and ultimately promoting growth and yield of a plant (Sultana et al., 2015). This review emphasize the potential impact of

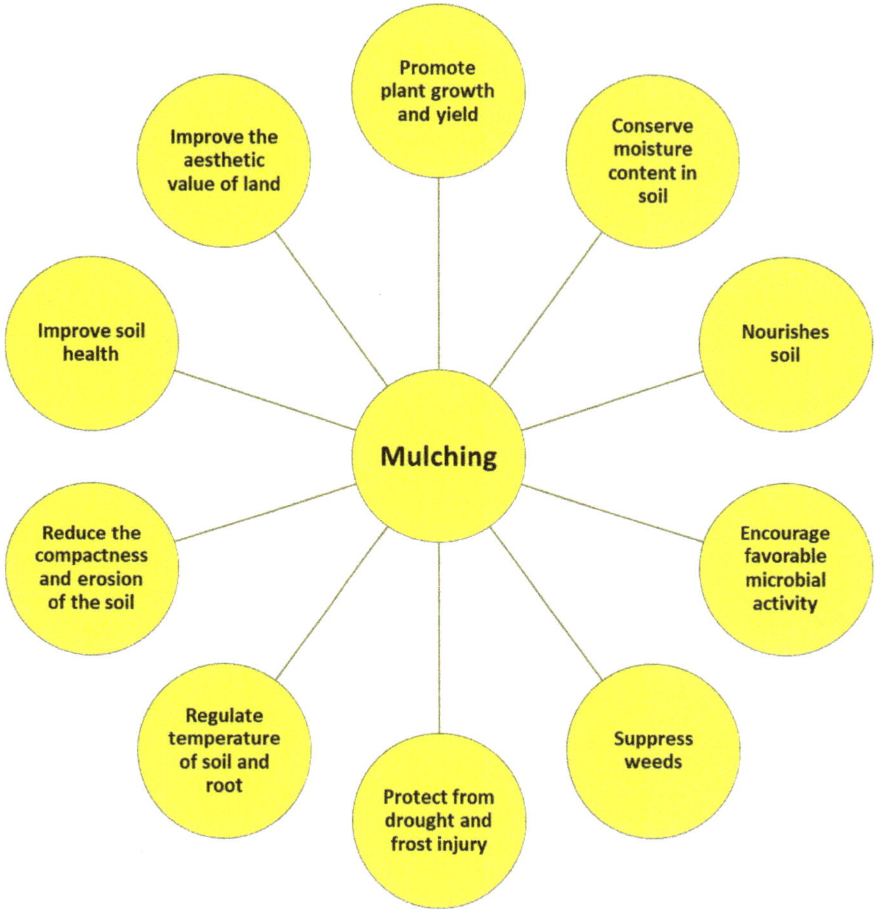

Fig. 1 Potential impact of mulching on agricultural and environmental conditions

mulches on agriculture and environment aspects as observed by previous researchers (Fig. 1).

2 Impact of Mulching on Environmental Conditions

2.1 Light Conditions

Mulching is considered as one of the most important elements of sustainable agriculture that promote growth and productivity of crop plants, reduce weed growth, maintain the optimal soil temperature and retain soil moisture and increase the aesthetic value of land and mitigate the adverse effect of biotic and abiotic stresses. Different inorganic mulches of different color are used widely, some of them like black plastic mulch disturb the photosynthetic activity of plants by blocking the rays of sunlight penetration into the soil and increase the temperature of soil but the performance of other like transparent mulch are very effective as a solarization of soil for better germination of nursery seedlings during winter season. According to Siwek (2002) and Espi et al. (2006) checked the effect of black, black polyethylene film thickness on transmittance of sunlight radiation and observed that radiation between 400 and 1100 nm did not show any transmittance from 0.05 mm thick black polyethylene film. While white material of polyethylene film recorded (PAR) Photosynthetically active radiation (66.7%) followed by green and blue that received PAR (37.5% and 28.9%).Though, mostly photosynthesis favors blue light (Hogewoning et al., 2010). Similarly Piszczek and Głowacka (2008) recorded better growth of cucumber seedlings illuminated with fluorescent lamps emitting blue light of various wavelength (50–60 μmol m^{-2} s^{-1}). Fatemi et al. (2013) also observed the performance of different mulches (blue and red mulch) on vegetative and reproductive growth of squash crop. They reported that squash crop applied with blue mulch gained higher vegetative growth due to receiving of maximum rays of sunlight from the reflected surface. In contrast, squash plants covered with red mulch gained higher quality of light which results more reproductive growth and ultimately higher squash yield.

2.2 *Thermal Conditions*

One of the effects of mulched soil is rise in temperature around the cultivated plants. This is especially significant in countries with cold temperature in winter and the cultivation of cucurbits is sensitive to low temperature (Kalbarczyk, 2009).Higher temperatures hastens the onset of the fruiting season by adversely affecting growth and plants developmental processes (Pramanik et al., 2015). The decrease in photosystem II efficiency is due to decrease in assimilation of CO_2 triggered due to low soil temperature. Under stress conditions, sustaining the favorable temperature for the roots of cucurbit prevents the transmission of high levels of abscisic acid to those sections, which inhibits photosynthesis and induces stomatal closure and thus reduce loss of water through transpiration (Zhang et al., 2008). Presently, black film or other dark-colored materials are commonly used as a covering material in many temperate countries like China for growing vegetables like cucumber, tomato and brinjal where in summer days the mean temperature of soil supplemented with mulching can be rise by 2–5 °C (Stobdan, 2015). These covering helps the soil to absorb a lot of light and causes the soil to excessive heat up (Lamont, 1993; López et al., 2009).The findings of Haapala et al. (2015) and López-Tolentino et al. (2016) also reported that use of black paper mulch significantly increase the temperature of the soil. Attallah (2016) and Martín-Closas et al. (2017) also reported that the soil warms up the most when exposed to a colorless film.

Black polyethylene film having 0.038 mm thickness used as a soil covering material in Mexico, recorded higher temperature than the soil mulched with blue, red and green Oxo-degradable mulches having same thickness. (López-Tolentino et al., 2016). Similarly the use of green biodegradable mulch maximized the temperature of the soil as compared to black film, and for exposed soil. The temperature rise was most noticeable in the first few days after the mulches were applied (Filippi et al., 2011). El-Shaikh and Fouda (2008) carried out an experiment an experiment to find out the effect of polyethylene black, yellow and transparent mulches and organic wheat straw mulch on soil temperature in cucumber cultivation, they found that during daytime all the inorganic mulches observed rise in soil temperature. The colorless film recorded the highest temperature i.e., 7 °C as compared to control. however, in case of organic mulch, the temperature of the soil increased by 1 °C at night (2 a.m) as compared to control. Van Donk et al. (2011) also observed decrease in temperature of soil supplemented with organic mulch (woodchips) as covering material. In Ontario, it was reported over the duration of about 2 weeks, that the soil covered with black mulch recorded around 1.5 °C temperature higher during the day and more than 2–3 °C higher at night than the temperature of the exposed soil (Snyder et al., 2015). A greenhouse study was carried out by Homez and Arouiee (2016), to examine the heat absorption in soil by using mulches of different types and colors. He noted that black polyethylene black film, followed by organic mulch (rice husks), however the exposed soil absorbed the lowest amount of heat. The findings indicate that synthetic polymer mulches are capable of much more widespread use for heat loving vegetables in colder climates, while natural polymer mulches are best adapted in hotter climates to navigate or reduce the adverse impact of temperature fluctuation during the day (Fig. 2).

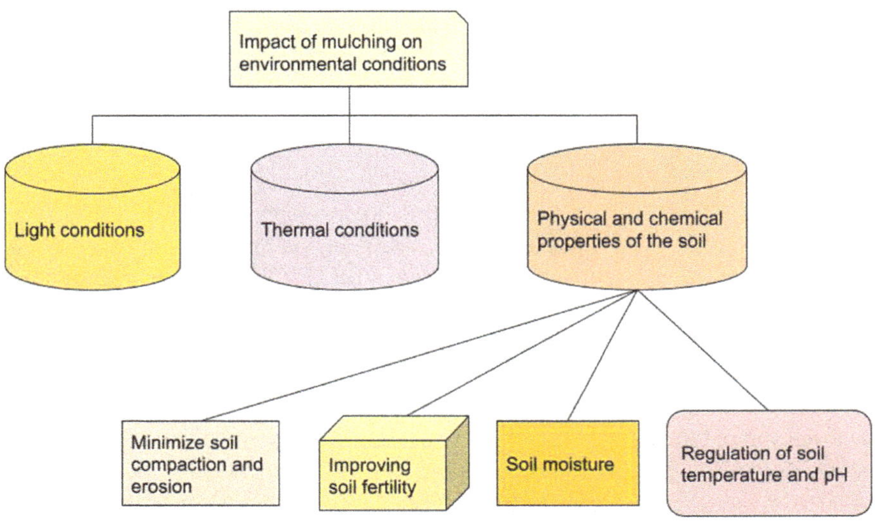

Fig. 2 Impact of mulching on environmental conditions of crop plants

2.3 Physical and Chemical Properties of the Soil

The use of organic mulches tends to be known by its characteristics of promoting the activity of different enzymes for breakdown the agriculture waste which leads to the promotion of more worms' occurrence and their growth (Jodaugienė et al., 2010). Because of the activity of microorganisms, organic mulches such as saw dust which have high C:N ratio may promote the deficiency of nitrogen for short-term in the soil. This difference also affects the amounts of essential nutrients and acidity of the mulched-soil (Sas-Paszt et al., 2014). With the passage of time as the vegetation starts, and the process of decaying advances the mulch gradually improves the fertility level by providing the digestible form of nutrients in the form of humus to the soil (Tittarelli et al., 2014). As opposed to synthetic mulches, flaked paper mulching enhanced the amount of microbial population involved in the nitrogen and phosphorus cycle on locations where urban compost and sludge were used (Forge et al., 2003). Cabilovski et al. (2014), in strawberry, found higher amount of micronutrients in soil covered with polyethylene mulches as compared to straw (organic mulch). According to the findings reported, the soil physical properties and the amount of organic carbon enhanced in organic mulched soils mulched in NT (No tillage) system. Compared to the control, ksat was significantly greater on the mulched site, while moisture content was 40–60% greater. After decades of research, Kahlon et al. (2013) found that using organic mulch (wheat straw) in NT (No tillage) and CT (Conservation tillage) improved the Organic carbon, the composition of soil particles, as well as other chemical and physical properties of the soil. In another study, it was reported that thick coating of top 10 cm soil with maize residues enhanced all characteristics of soil except soil porosity than exposed soil and 5–15 cm thick mulched soil (Kakaire et al., 2015).

2.3.1 Minimizing Soil Compaction and Erosion

Mulching is considered as one of the most effective techniques of sustainable agriculture that tends to prevents the erosion problems of soil, minimize soil compaction and have a significant influence on plant growth and development. On slopes, growing of living mulch like grass tends to decreases soil erosion by combining top soil and linking them together in a complex unit (Tanavud et al., 2001). The barley crop was used as the living mulch material by Sartz (1963). Similarly Borst and Woodburn (1942) reported that thin coating of 0.6 inches mulch could diminished the 86% erosion problem of soil. The most common mulches used to mitigate soil erosion are straws and other cereal crop residues (Samarappuli & Yogaratnam, 1984).In contrast to barren soil in forests, the soil erosion was reduced by 95% by the integrated application of straw mulch and erosion net (Megahan, 1974). Pinus leaves have also been found to be very effective in minimizing soil erosion (McCambridge et al., 1982), and Pinus sticks residues were also successfully used to prevent soil runoff and erosion losses (Rothwell, 1978). Mulch materials break the speed of water in mountainous areas and increase the penetration rate of soil, but engineering methods must be used to preserve slope stability rather than relying solely on mulching (Chalker-Scott, 2007). Compaction caused by heavy equipment or machinery is becoming a critical concern in many arable lands (Chalker-Scott, 2007). Through application of organic mulch materials such as bark, will help in the reduction of compaction (Oliveira & Merwin, 2001). The mulch material can help ease the massive loads of heavy implements' feet and tyres by reducing the beating action of raindrops. Mulching should be done prior to the formation of soil compaction, as mulching will not significantly increase soil aggregation once compaction has occurred (Donnelly & Shane, 1986).

2.3.2 Improvement of Soil Fertility

Organic mulches have a wide range of beneficial effects on soil quality, including increased nutrient levels. On the other hand, the type of material, soil properties, and climatic conditions, decide whether soil nutrients increase, decrease, or remain unchanged. Organic mulches are more beneficial because they can decompose in a suitable environment, providing the necessary nutrients. In comparison to inorganic mulches, studies have shown that organic mulches i.e., wood chips, grass, green manures, and sawdust enriches soil with digestible nutrients in the form of humus. (Ansari et al., 2001; Downer & Hodel, 2001; Pickering & Shepherd, 2000; Singh et al., 1991). Niggli et al. (1988) revealed that the application of low nutrient mulches, such as uncomposted straw or bark, reduced soil nitrogen levels while having very less effect on plant nutrition. Composted mulches, on the other hand, help crops grow and yield more because they contain high amount of nitrogen. (Tilander & Bonzi, 1997). Mulches with low nitrogen content, such as straw mulch, can also boost plant nutrition and soil nutritional level, sawdust mulch (Arthur & Wang, 1999), and bark mulch (Pfammatter & Dessimoz, 1997), are able to raise levels of nutrient in both leaves and in soil. In contrast to leaf or grass mulch, husk mulch contain higher amount of nitrogen and has been found to be a more efficient in enhancing nutrients level in soil (Singh & Singh, 1999). Water hyacinth, as a possible organic substrate, has the potential to improve microbial diversity in soils. In general, mulched plots had substantially higher soil respiration and microbial activity than control plots. Vermicompost mulched plots had higher bacterial and fungal counts at both the surface and sub-surface soil layers, compared to the other treatments (Balasubramanian et al., 2013).

2.3.3 Soil Moisture

A number of abiotic factors such as winds, high temperature levels, extreme climatic conditions, and excessive weeding play a critical role in transforming fertile lands to barren, as they are prime responsible for continuous moisture loss from fertile fields. Mulches has known for its characteristics of minimizing the weed population and 25% loss of moisture through evaporation (Harris et al., 2004).While also improving soil absorption and retention. It has been documented that straw mulch can reduce loss of moisture through evaporation by 35% (Russell, 1939). To preserve soil moisture content, livestock wastes, crop remaining, and various stone gravels types are commonly used as mulches (Buban et al., 1996; Siipilehto, 2001). Furthermore, Mulching outperform cover crops because cover crops competing for water supplies with the main crop (Downer & Hodel, 2001).Various researchers (Ahmad et al., 2015, 2020; Iqbal et al., 2019; Kader et al., 2019) concluded that mulches can help crop plants save water by reducing their irrigation needs, and in some cases, they can completely eliminate the need for irrigation and helps to survive the plants in arid climates. Some organic mulches serve as sponges, retaining rainfall and irrigation water and preventing runoff while also providing water when crops are in need. By mulching the straws, the runoff was minimized by 43% (Borst & Woodburn, 1942). Mulches reduce the need for supplemental irrigation by retaining water and decreasing soil profile drainage at the same time. Reduction in loss of moisture from soil has been found to be very effective, particularly in desert and semi-arid areas. Organic and mineral mulches are the most effective ways of reducing evaporation. According to some experiments, organic mulching raises soil water content only in deeper layers (80–220 cm), although soil water content may be lesser at a depth of 3080 cm. White (2004) used mulches of various colors and found that mulch color had no impact on soil moisture content. The lower value of water field capacity was recorded in exposed soil. It was 37%, compared to an average of 40–68% under the mulches. Mulching with colorless film reduced soil water loss by 6–11% in an experiment done in conditions of severe aridity i.e., Central Sudan, annual rainfall of 400 mm (Abdelrahman et al., 2016). Other researchers, who examined the performance of color substances of inorganic mulches came up with similar findings (Mahadeen, 2014). Mulched soil significantly reduced the irrigation requirement of cucumber crop (El-Shaikh & Fouda, 2008; Spiżewski et al., 2010) and melon (Alenazi et al., 2015).

2.3.4 Regulation of Soil Temperature and pH

Mulching is considered as one of the most effective practices of sustainable agriculture that provide favorable environmental temperature for better growth and development. Mulch application has been shown in several studies to keep the soil cool under extreme temperature (Kader et al., 2019; Long et al., 2001), however, in cool climatic conditions, it keeps the soil warm (Kader et al., 2019). High temperature have a negative impact on newly emerging plant roots, limiting nutrient and water uptake. Extreme temperatures during the early stages of plant growth can stress the plants because newly grown roots are unable to take water and essential plant nutrients properly (Chalker-Scott, 2007). As a result, careful soil temperature conservation and control is a life-threatening factor for optimal plant development. Mulches, on the other hand, can reduce the temperature by 10 degrees Celsius in hot and dry environments such as in deserts (barren soil) (Martin & Poultney, 1992). The temperature of the soil is affected differently by different types of mulches. Due to solar radiation, some mulches keep soil temperature higher than un-mulched soil or living mulches (Montague & Kjelgren, 2004). Inorganic mulches have been found to be ineffective at regulating soil temperature because they can raise it instead of keeping it at a comfortable level (Kader et al., 2019; Montague & Kjelgren, 2004). Aside from synthetic mulches, several other forms of mulches have also been found to be ineffective at regulating soil temperature. Pine bark mulching, for example, raised the soil surface temperature, causing nearby plants to raise transpiration level through their leaves (Zajicek & Heilman, 1994). Organic mulches are more valuable because they can decompose and provide nutrients (Ansari et al., 2001; Downer & Hodel, 2001). Niggli et al. (1988) According to the findings, un-composted or low-nutrient mulches decreased the amount of soil nitrogen without influencing plant nutrition, thus resulting in less watershed contamination.

3 Weed, Pest and Disease Management

Deepness is a feature of good mulch that keeps weeds at bay while promoting healthy plants and soil. Studies have shown that depth of mulch plays a key role in suppression of weeds. Opposed to mulches applied at shallower depths, organic mulches supplemented at a deeper depth tends to minimize growth of weed species (Zaragoza et al., 1995). Black plastic mulch is one of the most convenient and easily manageable mulch as compared to other mulches in the field. Following the sowing of crops, unwanted plants should be removed (Brainard et al., 2013). Herbicides are used to control weed infestations on a small scale due to the restricted supply of herbicides that have been approved for use (Matyjaszczyk & Dobrzański, 2017). The use of multiple mechanical weed remover machines are very expensive in term of energy and is time consumable. It also leads to the soils excessive drying and disruption of soil structure. Weeding by manual method is difficult to use due to the excessive growth of weed roots in the shallow root system of cucurbits, resulting in inevitable harm. However, mulching the soil is a reliable and safe process (Abouziena & Haggag, 2016). According to Schonbeck (2015), the critical period of competition for cucurbits is the first 4–6 weeks after planting. Plants grow quickly throughout this period and cover the soil in the gaps between the rows. Mulches have a declining impact on weed populations because they provide a physical barrier and block photoactive radiation (PAR). Mulches also diffuse out heat waves in the far-infrared range, which suppresses weeds growth at primary phase of production. Similar results were in accordance with Ngouajio and Ernest (2004) evaluated the performance of polyethylene mulches and varying light transmittance in the range of 400–1100 nm on weed infestation. White film had the most weeds and the most biomass, followed by grey film. The remaining black, brown, and green mulches effectively inhibited the growth of weeds, which were limited to no more than 25 plants/m^2. According to Broschat (2007), wood mulches such as bark and pine mulches greatly reduced the growth of dicotyledonous weeds. The thickness of organic material was found to have a positive association with the decrease in total weed infestation (van Donk et al.,

2011). In addition, allelopathins like phenols, or benzoxazine can diminish herbicides amount used by lowering the germination rate of weeds (Tabaglio et al., 2008), in contrast it also has an adverse influence on plant growth. As a result, when compared to the other organic mulches, the top area of soil covered with the leaves of Vigna unguiculata plants, as well as the unmulched one, observed maximum infestation of weeds with dicotyledonous organisms. Newspaper pages, flaked newspapers, and wheat straw deteriorated the most under the specific conditions of the atmosphere within the high plastic tube, followed by flaked newspapers and wheat straw. Sánchez et al. (2008) reported that degradation ratio of these mulches was 4.9:3:1 when checked at the end of growing condition of crops. However, in the summer cucumber production, both of these mulches greatly decreased the amount of weed infestation. The use of organic mulches often creates a paradox. Findings of various researchers concluded that organic mulches attract pests, and also acts as an insect repellents (Anderson et al., 2002). Thuja species are used as a pests repellents, including clothing moths, cockroaches, ants, carpet beetles, and termites (Chalker-Scott, 2007). Termite attack is generally thought to be attracted to woody mulches. This fact has been the subject of several studies, all of which have yielded interesting results. A study conducted by Long et al. (2001) compared termite activity beneath organic (wood) and inorganic (gravel) mulches and reported that termite incidence was found lower under wood mulch compared to gravel mulch. When organic mulches were used in a laboratory experiments against termites control, the death rate was extremely high. Growth of termites favor higher levels of nitrogen and phosphorus contained mulches. As a result, in areas where termites are a major pest, organic mulches with low nutrient content should be used (Martin & Poultney, 1992). Mulch products are often subjected to a specific treatment of temperature in order to kill all of the detrimental organisms present in the mulch. Commercially available mulches of organic nature are sterile. To prevent pathogens, concentrate on commercially available mulches (Chalker-Scott, 2007). According to a six-year analysis when mulches contained canker diseased were used for covering material of healthy plants, no pathogen transfer was detected. In contrast the only proof of tip blight transmission was detected when same species were covered as mulch with Austrian pine foliage (Jacobs, 2005). The Austrian pine tip blight pathogen had no effect on other crop plants. This proposed that the pathogen attack was more visible on Austrian pine due its susceptible nature. Epidemiology, rather than the pathogenic source, may be the cause of disease (Chalker-Scott, 2007). Pathogens can also be transmitted by soil. In healthy soils, pathogens or microorganisms are still present. These pathogens turn infectious when soil conditions are low or anaerobic, adversely affecting the plants growth and development (Foreman et al., 2002). Koski and Jacobi (2004) reported that products contained in un-composted mulch would certainly not be used; if they are used, they will be the source of pathogen transmission and hence resultantly can cause great losses to the crop. Pathogens are present in the mulch materials of diseased plants. If we use these kinds of mulches, diseases can spread to healthy plants as well. As a result, mulch products should be thoroughly composted before being used (Hoitink & Krause, 1999). Mulch like Honey locust canker carries pathogens which adversely affecting growth and developmental processes of plants (Koski & Jacobi, 2004).

4 Allelopathy in Mulching

Allelopathy is the production of biochemical by crop plants, or often by mulches of organic nature, which inhibits seed germination and plant growth. Allelochemicals are used to keep weeds at bay in crop plants. The trees bark of black walnut produce a biochemical compound "Juglonic acid" which has an effective role in controlling the growth of weeds and undesirable plants. But this biochemical has an adverse impact on plants seedlings and shallow rooted crops (Harris et al., 2004). Allelopathy suppressed seed germination in a laboratory experiment using an allopathic compound (Duryea et al., 1999), similarly, some of the researchers believe that existing crop plants should be treated in the same way. The mulch materials are more vulnerable to seedlings or small seeds than mature plants. Previous researches have proven the allopathic effect of different mulches and reported that application of different mulch extracts (eucalyptus, acacia, and pine) were found effective in completely or partially suppressing the growth of different weed species (Schumann et al., 1995). Broad-leaved plants or dicot species were more adversely affected as compared to Narrow-leaved plants like grasses (Schumann et al., 1995). The contrary results were also reported in a rice crop research report (Lillaram & Rao, 1980). During the initial stages of landscaping, insufficient use of mulches have adversely affected the growth of plants; thus, sufficient composting and woody mulch can ensure that landscapes are not harmed (Chalker-Scott, 2007).

5 Role of Mulching in Improving Growth Yield and Quality of Crops

Mulching is commonly used to flourish natural ecosystem consisting of herbaceous and other tree species. Many studies have shown that mulches have a beneficial effect on germination, survival percentage of seedlings and overall growth and production of plants as compared to unmulched treatments. This is advantageous for optimum yield with minimal input resources (Kwambe et al., 2015; Kader et al., 2019). Iqbal et al. (2019) and Ahmad et al. (2020) reported that reduction in grain yield and grain quality of winter wheat can be compensated by straw mulching together with wide precision planting. The combined use of planting basin tillage and sorghum stover mulching can significantly increase grain weight/1000 kernels and grain yield of dry-land sorghum (Masaka et al., 2019). Under mulching, plants respond to changing soil properties and environmental factors by increasing yield and exhibiting improvements in plant growth and development. Habimana et al. (2014)

used photometric observations of watermelon covered with mulch at various spacing. They reported that mulching applied after 60 days of sowing has observed positive impact on vegetative attributes of watermelon. Similarly the values of the studied parameters tended to increase as the spacing increased. In addition, in contrast to exposed soil, the plants on polyethylene film mulch grew and yielded the fastest, followed by wheat straw. On black film (PE 0.125 mm), pumpkin plants covered with black film of thickness (0.125 mm) had observed higher percentage of seedling emergence and produced earlier flower, as well as recorded higher fresh weight (76%) of uppermost part of the plant from the soil (Mahadeen, 2014). Similarly pot grown watermelon covered with red polyethylene film (0.05 mm thick) had observed lower value of leaves relative water content. Watermelon plants produced more biomass of the aboveground sections on the mulched field, and the photosynthetic area of the leaves and transpiration rate of leaves was observed higher but the evaporation rate was recorded lower (Ferus et al., 2011). Using straw as a mulch, the increase in transpiration was reported in the absence of major changes in cucumber photosynthesis rate (Hnilička et al., 2012). Černiauskienė et al. (2015) reported that crop residues mulched with peat had a positive impact on the production of dry mass and quality attributes (crude fibre, protein, and ash) of oil content in pumpkin seed. According to Minuto et al. (2008), zucchini crop covered with polyethylene and biodegradable mulches had observed 2 times higher total yield as compared to un-mulched soil. Similarly Fatemi et al. (2013) noted that covering of summer squash with blue and red plastic films enhanced total yield by (100 and 31.5% respectively), as compared to un-covered area in an arid environment (Iran).

6 Significance of Mulching in Constructing a Healthy Environment

6.1 Mitigation of Drought Impacts

Cotton plants experience a wide range of developmental and functional changes as a result of drought for example, Drought stymies a variety of physiological factors that regulate lint development and fiber quality (Dong et al., 1996). There are four types of drought resistance mechanisms in plants: restoration, avoidance, adaptation, and drought escape (Donnelly & Shane, 1986). When subjected to moderate drought, water stress avoidance is the preservation of essential biochemical functions such as stomatal regulation. Drought tolerance refers to a flora's ability to withstand extreme dehydration by osmotic modification and osmo-protectant (Downer & Hodel, 2001). Drought escape is the ability of plants to control their growth in order to avoid water loss (Downer et al., 2002). Climate change create a drastic condition for growing of crop like potato, sensitive to drought and high temperature condition which results lower yield of potato crop. Drought and potato yield were reported to be significantly significant (89%) by Hlavinka et al. (2009). When drought is combined with high daytime temperatures, the soil cannot support the growth of potato tubers. For the entire duration of vegetation, ample and usable soil water is a critical factor for the better stem growth and tuber production. According to Welbaum (2015) and Adamchuk et al. (2016), the higher development of potato tubers need consistent supply of water and temperature of soil between 16 and 20 °C. Development of tuber is affected by drought and high temperatures of soil other than optimum. These unfavorable vegetation conditions may cause tuber malformation and production of small sized tuber and also adversely affecting the quality of tubers especially its starch content. Mulching is one of the solutions to these issues. Mulching of organic materials such as straw, compost, and other organic waste is a common practice all over the world. Cereal straw is a useful mulching material that is readily available. The main advantages of straw mulch treatment include ease of application, a reduction in soil temperature, reduced daytime temperature fluctuations, and an increase in the moisture content of the soil (Adamchuk et al., 2016; Dudás et al., 2016; Elbl et al., 2014).

6.2 Management of Soil Salinity and Sodicity Dynamics

Healthy soils will help to ensure food security by ensuring long-term crop production. However, as a result of anthropogenic perturbations and a variety of other causes, the issue of soil salinity is spreading through about a third of the land surface every day. Soil erosion, reduced crop production, lower water holding capacity of soil, a decrease in biodiversity of soil, and, eventually, erosion problem and carbon depletion result from the use of soil in unsustainable way. Irrigating plants with untreated urban waste water containing high levels of salts can have negative consequences for growth and production of the plant. Furthermore, the widespread use of inorganic fertilizers, pesticides, detergents, and other chemicals can significantly increase salt levels in soils (Chalker-Scott, 2007). Mulching may help to eradicate the salt stress by increasing water holding capacity of the soil and lowering evaporation. The use of mulches of organic nature have been observed effective for restoration of soil structure and desalination than synthetic mulches (Dong et al., 1996).Application of organic mulches encourage the growth of microbes which results to deterioration of several hazardous residues and also reduce the adverse impact of salt stress (Gan et al., 2003).

6.3 Alleviation of Heavy Metals

Heavy metals are harmful to health of both animals and humans. Mulches are an excellent source for navigating the adverse effect of heavy metal (Chalker-Scott, 2007). Some organic mulches like Poplar, arborvitae, eucalyptus and fir leaves are often used to alleviate the deleterious effect of heavy metals (Salim & El-Halawa,

2002). In forest areas, using mulches like woodchips and compost will create a complex of copper metal and turn it into a shape that is not harmful to crop plants (Kiikkilä et al., 2002). Heavy metals can be removed from lawn and garden soils using organic and living mulches. Mulching with leaves of eucalyptus, oak, and poplar can extract common urban contaminants including cadmium and lead from the soil (Salim & El-Halawa, 2002). Similarly, complex mixture of copper contained compost-woodchip was discovered to disinfect forest soils (Kiikkilä et al., 2002).

6.4 Reduction of Pesticide Uses

Mulches minimize the usage of herbicides, insecticides, and fungicides by reducing weeds, plant stress, and vulnerability to pests and pathogens. Mulch materials can help plants cope with a variety of pathogens and reduce stress. Plants develop resistance to weeds and other harmful pests, eliminating the need for fungicides, insecticides, and herbicides. The reduction in the usage of such chemicals would benefit farmers in the sense that no money will be spent on them, and it will also benefit the population of beneficial soil species and the atmosphere (Chalker-Scott, 2007). Apart from those near seashores, many ecosystems are affected by salinity stress. Evaporating water leaves salt crusts in arid landscapes, making them particularly salty. The higher amount of salts are released from different sources of chemicals, fertilizers and detergents in irrigation water in arid areas, as well as poorly treated domestic or non-sewage waste water which not only affect the quality of water but also has an adverse impact on other components of soil biodiversity and environment. Salt levels in container plants that have been over-fertilized can rise as well. Mulches limit evaporation, leaving more water in the soil and diluting salts. In addition, they can also promote plant growth by reducing the effects of salt toxicity (Ansari et al., 2001). In this case, plastic mulches are ineffective, they are unable to bind ions in the same way that organic materials do. Pesticides and other toxins may be degraded using organic mulches (Smith & Skroch, 1995), probably by increasing the number of pesticide-degrading microbial populations.

6.5 Visual Enhancement

Mulches may be both decorative and functional; though this is not a scientifically observable factor, aesthetics can affect mulch selection and application. Different mulches like ground covers or tumbled glass, will enhance the design elements of a landscape, while also preserving the soil. Visually distinguishable mulches can be used to guide pedestrians through a landscape, protecting vulnerable root zones while also adding a design feature. Other sensory components are added to certain mulches in addition to visual interest: The landscape is scented with aromatic ground covers and organic mulch. Mulches enhance the aesthetic value of garden crops which is

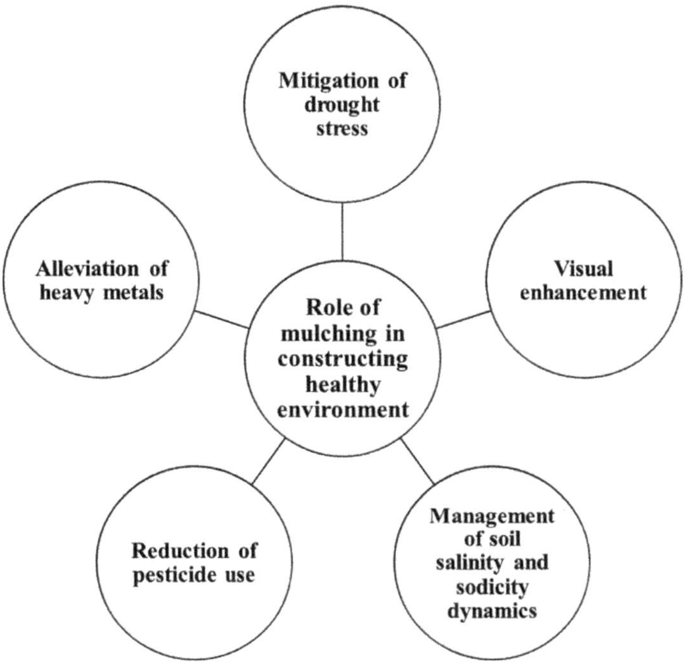

Fig. 3 Schematic representation of significant indicators of mulches responsible forconstructing a healthy environment

considered an important factor for attractiveness of buyers, who might otherwise view them as "disordered" and give more attractive look to un-cultivated area. The tumbled grass mulches and ground covers are attractive and provide protection to soil from unfavorable environmental conditions and other pest attacks. Mulching play a significant role in increasing the visible features of the land area by adopting selection criteria during landscape designing of an area of the land (Kader et al., 2019). People enjoy the aroma of certain fragrant ground covers and new mulch. People are often attracted to smooth rock mulch and soft ground covers because of their visual beauty (Chalker-Scott, 2007; Kader et al., 2019) (Fig. 3).

7 Economic Comparisons of Crops Under Mulching and Without Mulching

People calculate the cost and profit of using synthetic chemicals, fertilizers, and mulches if they make an investment. Mulch materials are less expensive than other inorganic materials because of its effect influence on soil health and efficiency of crop plant. There will be no need to purchase pesticides if we use mulches or on several other weed-control methods (Gardiner & Yeiser, 1998). The use of locally generated wood debris in the restoration of deteriorated lands enhances farmer revenue and improves performance of plants (Munir et al., 1998). Brush mulch has been found to be the most efficient and cost-effective method for re-vegetating roadsides in urban areas (Rothwell, 1978). Locally accessible products, such as peat and timber remaining, are cost-effective and improve crop growth and production (Kader et al., 2019). Raman et al. (2004) evaluated the performance of different mulches (Sawdust, coir pith, water hyacinth, sugarcane trash mulch, and wheat straw mulch) on cotton growth, yield, and weed control. Sugarcane trash mulch had the lowest weed density (20 m^2) as compared to other mulch types, with the highest density (225 m^2) in the control treatment. The control treatment had the higher density (225 m^2) and weed biomass (157 gm^2), followed by other mulch types (saw dust, rice straw, water hyacinth, and coir pith), with sugarcane trash having the least. Sugarcane trash (91% weed control effectiveness), coir pith (84% weed control effectiveness), water hyacinth (77% weed control effectiveness), rice straw (65% weed control effectiveness), and saw dust (52% weed control effectiveness) were all higher than the control treatment. Ather et al. (2013) conducted a field experiment to examine the effect of different weed management practices, such as wheat straw mulching, manual hoeing, acetochlor, and pendimethalin + prometryne, on weed growth, biomass production, and cotton productivity. At maturity, the plots applied with wheat straw had recorded maximum value of weed control productivity as compared to other treatments. In comparison to the other plots of the experiment, the performance of growth and yield related characteristics of mulched plots were more effective as compared to other treatments (Fig. 4).

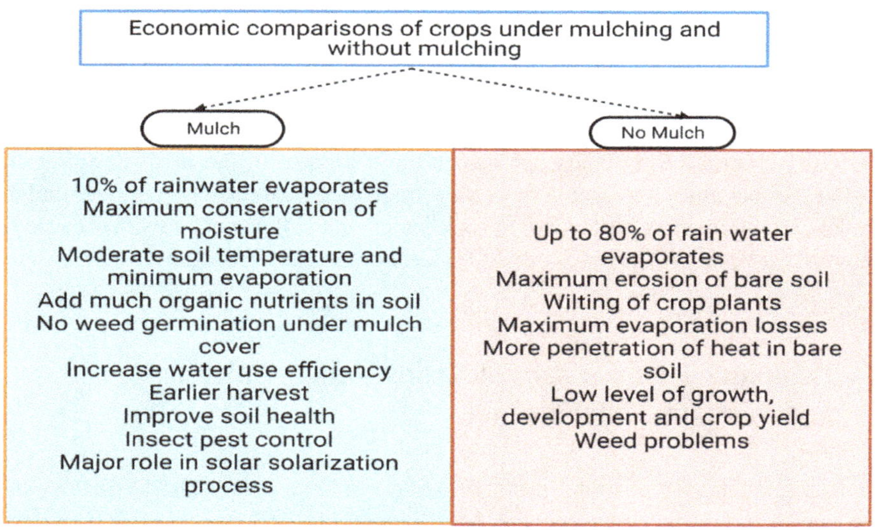

Fig. 4 Schematic representation of economic comparisons of crops under mulching and without mulching

Mohtisham et al. (2013) reported that treatments with sesbania and plastic mulch had lower weed densities than the control treatment. When compared to sesbania, plastic mulch was more effective at controlling weeds. The weed density of plastic mulched crops was recorded lower which is attributed to the plastic sheet's greater weed control effectiveness. Weeds were often found in the spaces between plants. Vasilakoglou et al. (2006) evaluated the various approaches of weed control (Herbicides, cereals as mulch, and inter-row cultivation) on overall phonological phases of cotton. The growth of cotton crop was significantly higher in different methods of weed control than control treatments According to the findings of this report, using cereals as mulch is an efficient technique for reducing weed intensity while also increasing cotton seed yield. Cotton yield losses of up to 85% have been recorded by Bryson et al. (2003), and Latha (2005). The crucial time of weed growth in cotton crop was discovered to be first two months after planting. Keeping plots weed-free for the first 80 days resulted in a significant increase in seed cotton production. Dil Baugh et al. (2011) checked the performance of different strategies of weed control (plastic mulching, hand weeding, and mechanical weed control) on the intensity of weeds. They found that after 30 days of sowing, plots applied with different mulches significantly decreased weed intensity of wide leaves weed completely and 95.8% of narrow leaves weed compared to un-mulched treatment. Zongbin et al. (2004) concluded that application of mulched material had significantly increased the water holding capacity of soil and maintained optimal soil temperature for better growth of cotton crop as compared to un-mulched treatment.

8 Conclusion

Any material either organic or inorganic to reduce moisture loss, weed control, increase crop performance and soil fertility is known as mulch. Different inorganic mulches of different color are used widely which has many advantages like better germination, affective photosynthetic activity. Mulch covers can increase soil temperature around the crop and make it favorable for many of the vegetable crops especially cucumber, tomato, brinjal etc. Various colored plastics are used such as black, yellow and transparent mulches which can help to increase soil temperature. Apart from inorganic mulches, organic mulches like wood chips, grass, green manures, and sawdust are also used as covering material which can increase soil temperature. Apart from increase in soil temperature, mulch material can also soil physical and chemical properties by increasing enzyme activity to breakdown different agriculture waste materials, promote C:N ratio to overcome nitrogen deficiency in the soil for a short period of time. It can also help to minimize the soil to become compact which ultimately prevent erosion but is not helpful to increase soil aggregation once compaction is caused. Soil fertility can also be increased with the help of organic mulch by providing necessary nutrients to the plant as compared to inorganic mulch. One of the advantages of mulch is to improve soil absorption and rentention that can reduce the irrigation need for the crop. Soil pH plays a very important in crop growth which can be effectively maintained to an optimum level with the help of mulching. Weeds, insects/pests and diseases can be managed by covering the soil with mulch material. Apart from these advantages to parts, mulch can mitigate drought, manage salinity and sodicity dynamics, heavy metals alleviation and reduce pesticide use. Recent studies showed that mulching is economically effective as compared to non-mulching.

References

Abdelrahman, N. A., Abdalla, E. A., Ibrahim, E. A., & El Naim, A. M. (2016). The effect of plastic mulch on growth and yield of rain-fed cowpea and watermelon in north Kordofan state of Sudan. *World Journal of Agricultural Research, 4*(5), 139–142.

Abouziena, H. F., & Haggag, W. M. (2016). Weed control in clean agriculture: A review. *Planta Daninha, 34*(2), 377–392.

Adamchuk, V., Prysyazhnyi, V., Ivanovs, S., & Bulgakov, V. (2016). Investigations in technological method of growing potatoes under mulch of straw and its effect on the yield. In *Proceedings of the 15th International Scientific Conference Engineering for Rural Development* (pp. 25–27).

Ahmad, S., Raza, M. A. S., Saleem, M. F., Zahra, S. S., Khan, I. H., Ali, M., Shahid, A. M., Iqbal, R., & Zaheer, M. S. (2015). Mulching strategies for weeds control and water conservation in cotton. *Journal of Agricultural and Biological Sciences, 8*, 299–306.

Ahmad, S., Raza, M. A. S., Saleem, M. F., Zaheer, M. S., Iqbal, R., Haider, I., Aslam, M. U., Ali, M., & Khan, I. H. (2020). Significance of partial root zone drying and mulches for water saving and weed suppression in wheat. *Journal of Animal and Plant Sciences, 30*, 154–162.

Alenazi, M., Abdel-Razzak, H., Ibrahim, A., WahbAllah, M., & Alsadon, A. (2015). Response of muskmelon cultivars to plastic mulch and irrigation regimes under greenhouse conditions. *Journal of Animal and Plant Sciences, 25*(5), 1398–1410.

Anderson, J. T., Thorvilson, H. G., & Russell, S. A. (2002). Landscape materials as repellents of red imported fire ants. *Southwestern Entomologist, 27*, 155–163.

Ansari, R., Marcar, N. E., Khanzada, A. N., Shirazi, M. U., & Crawford, D. F. (2001). Mulch application improves survival but not growth of Acacia ampliceps Maslin, *Acacia nilotica* L. and *Conocarpus lancifolius* L. on a saline site in southern Pakistan. *International Journal of Review, 3*, 158–163.

Arthur, M. A., & Wang, Y. (1999). Soil nutrients and microbial biomass following weed control treatments in a Christmas tree plantation. *Journal of America Society Soil Science, 63*, 629–637.

Ather, M. N., Idrees, N. M., Ayub, M., Tanveer, A., & Mubeen, K. (2013). Effect of different weed control practices and sowing methods on weeds and yield of cotton. *Pakistan Journal of Botany, 45*, 1321–1328.

Attallah, S. Y. (2016). Effect of plastic mulch color on growth and productivity of different summer squash varieties grown off-season. *Assiut Journal of Agricultural Sciences, 47*(4), 167–177.

Balasubramanian, D., Arunachalam, K., Arunachalam, A., & Das, A. K. (2013). Effect of water hyacinth (*Eichhornia crassipes*) mulch on soil microbial properties in lowland rainfed rice-based agricultural system in Northeast India. *Agricultural Research, 2*, 246–257.

Borst, H. L., & Woodburn, R. (1942). The effect of mulching and methods of cultivation on runoff and erosion from Muskingham silt loam. *Journal of Agricultural Engineering, 23*, 19–22.

Brainard, D. C., Peachey, R. E., Haramoto, E. R., Luna, J. M., & Rangarajan, A. (2013). Weed ecology and nonchemical management under strip-tillage: Implications for northern U.S. vegetable cropping systems. *Weed Technology, 27*, 218–230.

Broschat, T. K. (2007). Effects of mulch type and fertilizer placement on weed growth and soil pH and nutrient content. *Hort Technology, 17*(2), 174–177.

Bryson, C. T., Reddy, K. N., & Molin, W. T. (2003). Purple nutsedge (Cyperus rotundus) population row in narrow transgenic cotton (Gossypium hirsutum L.) and soybean (Glycine max) rotation. *Weed Technology, 17*, 805–810.

Buban, T., Helmeczi, B., Papp, J., Dorgo, E., Jakab, L., Kajati, I., Merwin, I., Polesny, F., & Muller, W. (1996). Olszak RW. IFP-compatible ground-cover management systems in a new-planted apple orchard, *19*, 263–267.

Cabilovski, R., Manojlovic, M., Bogdanovic, D., Magazin, N., Keserovic, Z., & Sitaula, B. (2014). Mulch type and application of manure and composts in strawberry (Fragaria × ananassa Duch.). *Zemdirbyste-Agriculture, 101*(1), 67–74.

Černiauskienė, J., Kulaitienė, J., Danilčenko, H., & Jarienė, E. (2015). *Proceedings of the 7th International Scientific Conference on Rural Development.*

Chalker-Scott, L. (2007). Impact of mulches on landscape plants and the environment—A review. *Journal of Environment Horticulture, 25*, 239–249.

Clifford, E. D., & Massello, J. W. (1965). Mulching materials for nursery seedbeds. *Tree Planters' Notes, 72*, 18–22.

Dil Baugh, M., Afzal, M. N., Raza, I., & Dupont, P. L. (2011). Comparative study of different weeding methods on cotton crop under drip irrigation system. In *World Cotton Research Conference-5* (Vol. 7–11, pp. 392–395). Excel India Publishers.

Dong, B. B., Zhu, H. T., Zhong, Z. K., & Ye, G. F. (1996). Study on ecological effect of the forest land under crop sowing and mulching of coastland soil by newly planted. *Acta Agriculture Zhejiangensis, 8*(154), 157.

Donnelly, J. R., & Shane, J. B. (1986). Forest ecosystem responses to artificially induced soil compaction. I. Soil physical properties and tree diameter growth. *Canadian Journal for Research, 16*, 750–754.

Downer, J., & Hodel, D. (2001). The effects of mulching on establishment of Syagrus romanzoffiana (Cham.) Becc., Washingtonia robusta H. Wendl. and Archonto phoenix cunning hamiana (H. Wendl.) H.Wendl. and Drude in the landscape. *Scientia Horticulturae, 87*, 85–92.

Downer, J., Faber, B., & Menge, J. (2002). Factors affecting root rot control in mulched avocado orchards. *Hort Technology, 12*, 601–605.

Dudás, P., Gedeon, C., Menyhárt, L., Ambrus, G., & Tóth, F. (2016). The effect of mulching on the abundance and diversity of ground beetle assemblages in two hungarian potato fields. *Columella: Journal of Agricultural and Environment Sciences, 3*(1), 45–53.

Duryea, M. L., Huffman, J. B., English, R. J., & Osbrink, W. (1999). Will subterranean termites consume landscape mulches. *Journal of Arboriculture, 25*, 143–149.

Elbl, J., Plošek, L., Kintl, A., Přichystalová, J., Záhora, J., & Friedel, J. K. (2014). The effect of increased doses of compost on leaching of mineral nitrogen from arable land. *Polish Journal of Environmental Studies, 23*(3).

El-Shaikh, A., & Fouda, T. (2008). Effect of different mulching types on soil temperature and cucumber production under Libyan conditions. *Misr Journal of Agricultural Engineering, 25*(1), 160–175.

Espi, E., Salmeron, A., Fontecha, A., Garcia, Y., & Real, A. I. (2006). Plastic films for agricultural applications. *Journal of Plastic Film & Sheeting, 22*, 85–102.

Fatemi, H., Aroiuee, H., Azizi, M., & Nemati, H. (2013). Influenced of quality of light reflected of coloredmulch on Cucurbita pepo var Rada under field condition. *International Journal of Agriculture Research and Review, 3*(2), 374–380.

Ferus, P., Ferusová, S., & Kóňa, J. (2011). Water dynamics and productivity in dehydrated watermelon plants as modified by red polyethylene mulch. *Turkish Journal of Agriculture and forestry, 35*, 391–402.

Filippi, F., Magnani, G., Guerrini, S., & Ranghino, F. (2011). Agronomic evaluation of green biodegradable mulch on melon crop. *Italian Journal of Agronomy, 6*(e18), 111–116.

Foreman, G. L., Rouse, D. I., & Hudelson, B. D. (2002) Wood chip mulch as a source of Verticillium dahliae. *Phytopathology, 92*, S26 (abstract).

Forge, T. A., Hogue, E., Neilsen, G., & Neilsen, D. (2003). Effects of organic mulches on soil microfauna in the root zone of apple: Implications for nutrient fluxes and functional diversity of the soil food web. *Applied Soil Ecology, 22*, 39–54.

Gan, J., Zhu, Y., Wilen, C., Pittenger, D., & Crowley, D. (2003). Effect of planting covers on herbicide persistence in landscape soils. *Journal of Environmental Science and Technology, 37*, 2775–2779.

Gardiner, E. S., & Yeiser, J. L. (1998). Converting stands of low-grade hardwoods to loblolly pine: Stimulating growth and reducing costs through litter retention. *Southern Journal of Applied Research, 22*, 148–155.

Gill, H. K., McSorley, R., & Branham, M. (2011). Effect of organic mulches on soil surface insects and other arthropods. *Florida Entomological*, 226–232.

Haapala, T., Palonen, P., Tamminen, A., & Ahokas, J. (2015). Effects of different paper mulches on soil temperature and yield of cucumber (Cucumis sativus L.) in the temperate regime. *Agricultural and Food Science, 24*, 52–58.

Habimana, S., Ngezahimana, J. B., Nyabyenda, E., & Umulisa, C. (2014). Growth and yield of watermelonas affected by different spacing and mulching types under Rubona conditions in Rwanda. *Scholarly Journal of Agricultural Science, 4*(10), 517–520.

Harris, R. W., Clark, J. R., & Matheny, N. P. (2004). *Arboriculture: Integrated management of landscape trees, shrubs, and vines* (4th edn, p. 578). Prentice Hall, Inc.

Hlavinka, P., Trnka, M., Semerádová, D., Dubrovský, M., Žalud, Z., & Možný, M. (2009). Effect of drought on yield variability of key crops in Czech Republic. *Agricultural and Forest Meteorology, 149*(3–4), 431–442.

Hnilička, F., Koudela, M., Martinková, J., & Svozilová, L. (2012). Effects of deficit irrigation and straw mulching on gas exchange of cucumber plants (Cucumis sativus L.). *Acta Universitatis Agriculturae et Silviculturae Mendelianae Brunensis, 60*(3), 43–50.

Hogewoning, S. W., Trouwborst, G., Maljaars, H., Poorter, H., Van Leperen, W., & Harbinson, J. (2010). Bluelight dose-responses of leaf photosynthesis, morphology, and chemical composition of Cucumis sativus grown under different combinations of red and blue light. *Journal of Experimental Botany, 61*(11), 3107–3117.

Hoitink, H. A. J., & Krause, M. S. (1999). Control of nuisance and detrimental molds (Fungi) in mulches and composts. *Special Circular Ohio Agricultural Research and Development Center, 165*, 66–69.

Homez, T. J., & Arouiee, H. (2016). Evaluation of soil temperature under mulches and garlic extract on yield of cucumber (*Cucumis sativus* L.) in greenhouse conditions. *Journal of Horticulture, 3*(1), 175.

Iqbal, R., Muhammad, A. S. R., Muhammad, F. S., Imran, H. K., Salman, A., Muhammad, S. Z., Muhammad, U., & Imran, H. (2019). Physiological and biochemical appraisal for mulching and partial rhizosphere drying of cotton. *Journal of Arid Land, 11*, 785–794.

Jacobs, K. A. (2005). The potential of mulch to transmit three tree pathogens. *Journal of Arboriculture, 31*, 235–241.

Jodaugienė, D., Pupalienė, R., Sinkevičienė, A., Marcinkevičienė, A., Žebrauskaitė, K., Baltaduonytė, M., et al. (2010). The influence of organic mulches on soil biological properties. *Zemdirbyste Agriculture, 97*(2), 33–40.

Kader, M. A., Singha, A., Begum, M. A., Jewel, A., Khan, F. H., & Khan, N. I. (2019). Mulching as water-savingtechnique in dry land agriculture. *Bulletin of the National Research Centre, 43*, 1–6.

Kahlon, M. S., Lal, R., & Varughese, M. A. (2013). Twentytwo years of tillage and mulch impact o physical characteristics and carbon sequestration. *Soil and Tillage Research, 125*, 151–158.

Kakaire, J., Makokha, G. L., Mwanjalolo, M., Mensah, A. K., & Emmanuel, M. (2015). Effects of mulching on soil hydro-physical properties in Kibaale subcatchment, south central Uganda. *Applied Ecology and Environmental Sciences, 3*(5), 127–135.

Kalbarczyk, R. (2009). Potential reduction in cucumber yield (*Cucumis sativus* L.) in Poland caused by unfavourable thermal conditions of soil. *Acta Scientiarum Polonorum Hortorum Cultus, 8*(4), 45–58.

Kasirajan, S., & Ngouajio, M. (2012). Polyethylene and biodegradable mulches for agricultural applications: A review. *Agronomy for Sustainable Development, 32*(2), 501–529.

Kiikkilä, O., Pennanen, T., Perkiömäki, J., Derome, J., & Fritze, H. (2002). Organic material as a copper immobilising agent: A microcosm study on remediation. *Basic and Applied Ecology, 3*(3), 245–253.

Koski, R., & Jacobi, W. R. (2004). Tree pathogen survival in chipped wood mulch. *Journal of Arboriculture, 30*, 165–171.

Kwambe, X. M., Masarirambi, M. T., Wahome, P. K., & Oseni, T. O. (2015). The effects of organic and inorganic mulches on growth and yield of green bean (*Phaseolus vulgaris* L.) in a semi arid environment. *The Agriculture and Biology Journal of North America, 6*, 81–89.

Lamont, W. J. (1993). Plastic mulches for the production of vegetable crops. *Hort Technology, 3*(1), 35–39.

Latha. (2005). Evaluation of chemical weed control methods in ricefallow cotton (*Gossypium hirsutum* L.) in the coastal region of Karaikal. M.Sc (Ag.) Thesis submitted to Tamil Nadu Agriculture University.

Lillaram, N. T., & Rao, B. V. V. (1980). Leachings from eucalyptus leaves have no adverse effect on germination and growth of paddy. *Current Research, 9*, 202–203.

Long, C. E., Thorne, B. L., Breisch, N. L., & Douglass, L. W. (2001). Effect of organic and inorganic landscape mulches on subterranean termite (Isoptera: Rhinotermitidae) foraging activity. *Journal of Environmental Entomology, 30*, 832–836.

López, J. C., Pérez Parra, J., & Morales, M. A. (2009). Plastics in agriculture. Cajamar Rural Sociedad Cooperativa de Crédito.

López-Tolentino, G., Cárdenas-Flores, A., IbarraJiménez, L., & Guerrero-Santos, R. (2016). Field performance of a foto-biodegradable film for soil mulching in zucchini crop. *Revista Internacional De Investigación e Innovación Tecnológica, 3*(19), 11–19.

Mahadeen, A. Y. (2014). Effect of polyethylene black plastic mulch on growth and yield of two summer vegetable crops under rain-fed conditions under semi-arid region conditions. *American Journal of Agricultural and Biological Sciences, 9*(2), 202–207.

Martin, P. J., & Poultney, R. (1992). Survival and growth of clove seedlings in Zanzibar. Effects of mulching and shade crops. *Journal of Tropical Agriculture, 69*, 365–373.

Martín-Closas, L., Costa, J., & Pelacho, A. M. (2017). Agronomic effects of biodegradable films on crop and field environment. In M. Malinconico (Ed.), *Soil degradable bioplastics for a sustainable modern agriculture, green chemistry and sustainable technology* (pp. 67–104). Springer-Verlag GmbH.

Masaka, J., Dera, J., & Muringaniza, K. (2019). Dry land grain Sorghum (Sorghum bicolor) yield and yield component responses to tillage and mulch practices under subtropical African conditions. *Agricultural Research*, 1–9.

Matyjaszczyk, E., & Dobrzański, A. (2017). Analiza możliwości regulacji zachwaszczenia w ochronie cukinii w Polsce. In Zagadnienia Doradztwa Rolniczego. A. P. Wiatrak (Ed.), *CDR w Brwinowie Oddział w Poznaniu & Stowarzyszenie Ekonomistów Rolnictwa I Agrobiznesu, 2*(88), 104–115.

McCambridge, W. F., Morris, M. J., & Edminster, C. B. (1982). *Herbage production under ponderosa pine killed by the mountain pine beetle in Colorado* (Vol. 416). USDA Forest Service, Rocky Mountain Forest and Range Experiment Station.

Megahan, W. F. (1974). *Deep-rooted plants for erosion control on granitic road fills in the Idaho Batholith* (Vol. 161). US Department of Agriculture, Forest Service, Intermountain Forest and Range Experiment Station.

Miles, M. (2005). Identification, pest status, ecology and management of the green mirid, a pest of cotton in Australia. Ph.D. thesis, University of Queensland.

Minuto, G., Guerrini, S., Versari, M., Pisi, L., Tinivella, F., Bruzzone, C., et al. (2008). Use of biodegradable mulching in vegetable production. In *Proceedings of the 16th IFOAM Organic World Congress*.
Mohtisham, A., Ahmad, R., Ahmad, Z., & Aslam, M. R. (2013). Effect of different mulches techniques on weed infestation in aerobic rice (*Oryza sativa* L.). *American-Eurasian Journal of Agricultural & Environmental Science, 13*, 153–157.
Montague, T., & Kjelgren, R. (2004). Energy balance of six common landscape súrfaces and the influence of surface properties on gas exchange of four containerized tree species. *Scientia Horticulturae, 100*, 229–249.
Munir, A. D., Majid, N. M., Abdoland, I., & Khan, G. S. (1998). Effects of mulching on the growth of interplanted Acacia mangiumon sandy tin-tailings in Peninsular Malaysia. *Lyallpur Akhbar, 65*, 3.
Nalayini, P. (2007) Poly-mulching a case study to increase cotton productivity. Senior scientist, Central Institute for Cotton Research, Regional Station, Coimbatore.
Ngouajio, M., & Ernest, J. (2004). Light transmission trough colored polyethylene mulches affects weed populations. *Hortscience, 39*(6), 1302–1304.
Ngouajio, M., & McGiffen, M. E. (2004). Sustainable vegetable production: Effects of cropping systems on weed and insect population dynamics. *Acta Horticulturae, 638*, 77–83.
Niggli, U., Weibel, F. P., & Potter, C. A. (1988). Weed control in aperennial crop using an organic mulch). *Zeitschrift Fur Pflanzenkrankheiten Und Pflanzenschutz, 11*, 357–365.
Oliveira, M. T., & Merwin, I. A. (2001). Soil physical conditions in a New York orchard after eight years under different groundcover management systems. *Journal of Plant & Soil, 234*, 233–237.
Pfammatter, W., & Dessimoz, A. (1997). Influence de l'irrigationetde la couverture du sol sur le developpementet le rendement de jeunespommiers (Influence of irrigation and ground cover on development and yields of young apple trees). *Revue Suisse De Viticulture, D'arboricultureetd' Hort, 29*, 301–304.
Pickering, J. S., & Shepherd, A. (2000). Evaluation of organic landscape mulches: Composition and nutrient release characteristics. *Arboricultural Journal, 23*, 175–187.
Piszczek, P., & Glowacka, B. (2008). Effect of the colour of light on cucumber (Cucumis sativus L.) seedlings. *Vegetable Crops Research Bulletin, 68*, 71.
Pramanik, P., Bandyopadhyay, K. K., Bhaduri, D., Bhattacharyya, R., & Aggarwal, P. (2015). Effect of mulch on soil thermal regimes—A review. *International Journal of Agriculture Environment and Biotechnology, 8*(3), 645–658.
Raman, R., Kuppuswamy, G., & Krishnamoorthy, R. (2004). Effect of mulching on the growth and yield of cotton. *Journal of Ecobiology, 16*, 275–278.
Rathore, A. L., Pal, A. R., & Sahu, K. K. (1998). Tillage and mulching effects on water use, root growth, and yield of rain-fed mustard and chickpea grown after lowland rice. *Journal of the Science of Food and Agriculture, 78*, 149–161.
Rothwell, R. L. (1978) Erosion control on forest roads. *Agriculture and Forestry Bulletin, 1*, 29–32 (University of Alberta).
Russell, J. C. (1939). The effect of surface cover on soil moisture losses by evaporation. *America Soil Science Society, 4*, 65–70.
Salim, R., & El-Halawa, R. A. (2002). Efficiency of dry plant leaves (mulch) for removal of lead, cadmium and copper from aqueous solutions. *Process Safety and Environmental Protection, 80*, 270–276.
Samarappuli, L., & Yogaratnam, N. (1984). Some aspects of moisture and soil conservation in rubber plantations. In *Proceedings of the International Rubber Conference. 75 Years of rubber research in Sri Lanka* (Vol. 1, (pp. 529–543, Part 2). Rubber Research Institute of Sri Lanka.
Sánchez, E., Lamont, W. J., & Orzolek, M. D. (2008). Newspaper mulches for suppressing weeds for organic high-tunnel cucumber production. *Hort Technology, 18*(1), 154–157.
Sartz, R. S. (1963). *Water yield and soil loss from soil-block lysimeters planted to small trees and other crops* (23 pp). USFS Research Paper LS-6.

Sas-Paszt, L., Pruski, K., Żurawicz, E., Sumorok, B., Derkowska, E., & Gluszek, S. (2014). The effect of organic mulches and mycorrhizal substrate on growth, yield and quality of Gold Milenium apples on M.9 rootstock. *Canadian Journal of Plant Science, 94*, 281–291.
Schonbeck. M. (2015). Weed management strategies for organic cucurbit crops in the Southern United States. Extension.
Schumann, A. W., Little, K. M., & Eccles, N. S. (1995). Suppression of seed germination and early seedling growth by plantation harvest residues. *South African Journal of Plant and Soil, 12*, 170–172.
Shogren, R. L. (2001). Biodegradable mulches from renewable resources. *Journal of Sustainable Agricultural, 16*, 33–47.
Siipilehto, J. (2001). Effect of weed control with fiber mulches and herbicides on the initial development of spruce, birch and aspen seedlings on abandoned farmland. *Silva Fennica, 35*, 403–414.
Singh, A. K., & Singh, R. B. (1999). Effect of mulches on nutrient uptake of Albiziaproceraand subsequent nutrient enrichment of coal mine overburden. *Journal of Tropical Science, 11*, 345–355.
Singh, S. B., Pramod, K., Prasad, K. G., & Kumar, P. (1991). Response of Eucalyptus to organic manure mulch and fertilizer sources of nitrogen and phosphorus. *Van Vigyan, 29*, 200–207.
Siwek. P. (2002) Modification of environmental conditions by mulching and direct plant covering in the culture of cucumber and stalk celery. Zesz. Nauk. AR w Krakowie, ser. Rozprawy, 279.
Smith, L. J., & Skroch, W. A. (1995). Turf herbicide injury to landscape trees as influenced by mulch. *Journal of Environmental Horticulture, 13*, 60–63.
Snyder, K., Grant, A., Murray, C., & Wolff, B. (2015). The effects of plastic mulch systems on soil temperature and moisture in central Ontario. *Hort Technology, 25*(2), 162–170.
Spiżewski, T., Frąszczak, B., Kałużewicz, A., Krzesiński, W., & Lisiecka, J. (2010). The effect of black polyethylene mulch on yield of field-grown cucumber. *Acta Scientiarum Polonorum Hortorum Cultus, 9*(3), 221–229.
Stobdan, T. (2015). Plasticulture in cold arid horticulture. *Scientific Specialist*, 155–159.
Sultana, S., Kashem, M. A., & Mollah, A. K. (2015). Comparative assessment of cow manure vermicompost and NPK fertilizers and on the growth and production of Zinnia (*Zinnia elegans*) Flower. *Open Journal of Soil Science, 5*(09), 193.
Tabaglio, V., Gavazzi, C., Schulz, M., & Marocco, A. (2008). Alternative weed control using the allelopathic effect of natural benzoxazinoids from rye mulch. *Agronomy for Sustainable Development, 28*, 397–401.
Tanavud, C., Kheowvongsri, P., Yongchalermchai, C., Leowarin, W., Densrisereekul, O., Bennui, A., Muraseand, J., & Kimura, M. (2001). Effects of land use patterns on soil and water quality in Khlong U-Taphao Basin. *Thai Journal of Agricultural Science, 34*, 15–31.
Tilander, Y., & Bonzi, M. (1997). Water and nutrient conservation through the use of agro-forestry mulches and sorghum yield response. *Journal of Plant & Soil, 197*, 219–232.
Tittarelli, F., Campanelli, G., Farina, R., Napoli, R., Ciaccia, C., & Testani, E., et al. (2014). Effect of covercrop management and compost application on soil N fertility of organic melon. In G. Rahmann & U. Aksoy (Eds.), *Proceedings of the 4th ISOFAR Scientific Conference on "Building Organic Bridges"* (pp. 709–712).
van Donk, S. J., Lindgren, D. T., Schaaf, D. M., Petersen, J. L., & Tarkalson, D. D. (2011). Wood chip mulch thickness effects on soil water, soil temperature, weed growth and landscape plant growth. *Journal of Applied Horticulture, 13*(2), 91–95.
Vasilakoglou, K., Dhima, I., & Eleftherohorinos, L. A. (2006). Winter cereal cover crop mulches and interrow cultivation effects on cotton development and grass weed suppression. *Agronomy Journal, 98*, 1290–1297.
Welbaum, G. E. (2015). Family Solanaceae. In *Vegetable production and practices* (pp. 176–221).
White, J. M. (2004). Summer squash yield and fruit size when grown on eight mulch colors in central Florida. *Proceedings of the Flordia State Horticultural Society, 117*, 56–58.

Xing, N. Q., Zhang, Y. Q., & Wang, L. X. (2003). *The study on dry-land agriculture in North China*. Chinese Agriculture Press.

Zajicek, J. M., & Heilman, J. L. (1994). Transpiration by crape myrtle cultivars surrounded by mulch, soil and turfgrass surfaces. *Journal of Horticultural Sciences, 26*, 1207–1210.

Zaragoza, C., Moya, S., & Martinez, G. (1995). Effects of mulches based on pine bark and pruning residues in a fruit orchard). In *Proceedings of the 1995 Congress of the Spanish Weed Science Society Huesca* (pp. 283–290).

Zhang, Y. P., Qiao, Y. X., Zhang, Y. L., Zhou, Y. H., & Yu, J. Q. (2008). Effects of root temperature on leaf gas exchange and xylem sap abscisic acid concentrations in six Cucurbitaceae species. *Photosynthetica, 46*(3), 356–362.

Zongbin, M., Lingli, L., Weiping, F., Deyi, X., & Tiegang, Y. (2004). Effects of wheat straw mulching on soil temperature, moisture and growth development of summer cotton. *Acta Agriculturae Universitatis Henanensis, 38*(4), 379–383.

Role of Agricultural Soil Mulching on Net Global Warming Potential and Greenhouse Gas Intensity of Different Cropping Systems

Saadatullah Malghani and Xiaolin Liao

Abstract Net global warming potential and greenhouse gas index express the actual environmental cost of an agricultural system and assist in solving the global challenges of increasing food production and identifying the primary targets for mitigation in different cropping systems and regions. This chapter reviews the effects of different soil surface mulching approaches on GWP, NGWP, and GHGI of different agricultural cultivation systems. We broadly discussed the prospects of mulching techniques used in isolation or combination with other conservation management approaches. In addition, the chapter also sheds light on methods of quantifying climate-related indexes and highlights the pros and cons of different protocols. The literature suggested that residue-based soil mulching may increase yield-scaled GWP due to its triggering effect on GHG emissions with and without improving crop yield. In contrast plastic film as mulch could reduce GWP by enhancing nutrient use efficiency and unique interaction with rainfall events that correlate with large pulses of GHGs emissions. Moreover CO_2 equivalent C inputs in the soil were also reported to be higher in plastic film mulch systems than in crop residues suggesting relatively low NGWP and GHGI for the former mulching approach. However, changes in agronomic inputs and farm machinery requirements between plastic and residue mulching approaches were ignored entirely by the published reports and thus did not truly represent the potential of mulching to reduce or induce NGWP in different cropping systems. Residue mulching can decrease indirect GHGs emissions associated with agronomic inputs and farm operations.

Keywords Global warming potential · Nutrient use efficiency · Yield

---

S. Malghani (✉) · X. Liao (✉)
College of Biology and the Environment, Nanjing Forestry University, Nanjing, China
e-mail: sum@plen.ku.dk; saadat.malghani@njfu.edu.cn

X. Liao
e-mail: liaoxiaolin@njfu.edu.cn

S. Malghani
Department of Plant and Environmental Sciences, University of Copenhagen, Copenhagen, Denmark

K. Akhtar et al. (eds.), *Mulching in Agroecosystems*,
https://doi.org/10.1007/978-981-19-6410-7_21

1 Introduction

Agricultural soil surface covered with different materials, including crop residues, plastic etc., is a common approach to reduce moisture loss via evaporation and control large soil temperature fluctuations. In agricultural soils, surface evaporation is the primary route of loss of water, estimating up to 65% of water loss in hot climate conditions, thus resulting in failure of cropping or contributing to a large proportion of the economic cost of farming via irrigation (Lu, 2020; Ma et al., 2018). Mulching reduces soil temperature fluctuation by regulating solar radiations and soil moisture contents (Wang et al., 2019). In addition, agricultural soil mulching improves soil structure by enhancing soil aggregation and porosity and reducing soil erosion and compaction (Shah & Wu, 2020). However, mulching is not limited to achieving high crop yield via soil moisture and temperature conservation. However, several co-benefits, including soil carbon sequestration, weed control, enhanced nutrient supply, and relatively easy farming management, are significant reasons behind the global expansion of mulching technique, especially in plastic mulch (Berger et al., 2013; Shah & Wu, 2020). Although there is no global-scale data on the use of different mulch, regional estimates point out a significant increase in mulching use in the agricultural sector since the introduction of plastic mulch in the 1990s. Shah and Wu (2020) reported that approximately 19 Million hectares of farmlands were under plastic mulch in 2017 in China, showing a 3–4 times increase in mulching from the 1990s.

The agricultural sector faces enormous challenges of enhanced global food demand by the highest human population and climate change threats. However, agriculture could also be part of the solution, taking its potential to ensure food security and mitigate climate change. Estimates suggest that soils under crop farming could sequester up to 1.6 Pg C per year depending on agricultural management (Smith et al., 2008). Studies investigating the effect of surface mulching exhibited an increase in soil carbon via residue application or enhanced below-ground biomass (Gonzalez-Sanchez et al., 2019; Johnson et al., 2006). However, soil moisture conservation and increased available carbon promote the production and emission of atmospheric GHGs via enhanced microbial activities (An et al., 2015). The increasing demand for plastic film by agriculture and uncertainty regarding the effect of mulching on soil carbon sequestration and GHGs emissions demands further investigations.

According to FAO estimates, the contribution of agricultural crop farming and livestock activities to global GHG emissions was around 20% in the last decade (2010–2017), which showed a decreasing trend, as agricultural contributions were 25% and 29% in the 2000s (2000–2009) and 1990s (1990–1999), respectively. However, this decreasing trend was not solely due to the decline in GHG emissions from agricultural sectors but also due to a relative increase in the energy sector's contribution to global GHG emissions (IPCC, 2014). Moreover, the calculated GHG emission rates from the agricultural sector represent food production processes within the farm gate and do not include sources generated outside the farm gate. According to IPCC estimates, if all relevant food processes, including production, transport and retail, are included, the agricultural sector contribution may increase to 30–40% globally (Tubiello, 2019). In order to bring further decrease in GHGs emissions from the agricultural sector, it is crucial to record the capacity of different sources and investigate the drivers behind the observed emission trends at the farm level to the global scale.

Global warming potential (GWP) measures the cumulative radiative forcing of various GHG relative to some reference gas, usually CO_2. According to IPCC, GWP is an index to measure relative radiative forcing caused by the unit mass of a substance about CO_2 over a specific time horizon. As GWP accounts for direct emissions of GHGs from different ecosystems thus is used as a base for comparing the environmental cost of land uses and management. As direct GHG emissions from a unit area of different land use do not truly represent its environmental cost, net GWP (NGWP) was introduced. The NGWP considers GWP associated with GHG fluxes and includes GWP of different management and on-farm operations, including production and transport of agronomic inputs, as these operations are generally fossil fuel-based energy-dependent. Robertson et al. (2000) included changes in SOC as a base for estimating net CO_2 exchange between an ecosystem and atmosphere instead of using direct CO_2 emissions considering the carbon inputs from below-ground biomass and above ground litter and/or residue inputs (Robertson & Grace, 2004; Robertson et al., 2000). So far, both GWP and NGWP have been used extensively for comparing the environmental cost of different ecosystems. As in the agricultural sector, sustainable high crop yield is the central goal of all agronomic management practices; therefore, another term of greenhouse gas intensity (GHGI) was introduced to evaluate GWP and/or NGWP based on crop yield (Lee et al., 2019; Sainju, 2020). GHGI expresses the actual environmental cost of an agricultural system to assist in solving the global challenges of increasing food production and concomitantly identifying the primary targets for mitigation in different cropping systems and regions. It is essential when seeking ways to decrease total GHG emissions associated with agricultural production without compromising sustainable food production. GWP-based GHGI is mentioned as yield scale GWP throughout the text of this chapter to avoid confusion between GWP and NGWP based GHGI.

This chapter mainly explores the literature investigating the effects of different soil surface mulching approaches on GWP, NGWP and GHGI of different agricultural cultivation systems. It broadly discusses the prospects of mulching techniques used in isolation or combination with other conservation management approaches. Moreover, a brief discussion is included on different critical parameters required for estimating GWP, NGWP and GHGI of agriculture systems.

2 Methods to Estimate Global Warming Potential (GWP), Yield Scale GWP, Net Global Warming Potential (NGWP), and Greenhouse Gas Intensity (GHGI)

2.1 Global Warming Potential (GWP) of GHGs and Yield Scale GWP

The term global warming potential (GWP) was introduced by IPCC for measuring the cumulative radiative forcing of different GHG relative to CO_2 and is simply estimated by adding the products of N_2O and CH_4 emission rates with their respective radiative factors (α_{CH4}, α_{N2O}) on a given time scale, commonly 100-year time horizon (Eq. 1). The conversion coefficients have evolved with time, improving scientific understanding (Table 1). Although there is no specific rule in choosing a time scale GWP, a 100-year time scale is adopted.

$$GWP_{direct} = CO_{2equivalent} GHGs(CH_4, N_2O) = \alpha_{CH_4} CH_4 + \alpha_{N_2O} N_2O \quad (1)$$

$$YieldscaleGWP = GWP(kgCO_2\,equivalent/ha)/yield(kg/ha) \quad (2)$$

Table 1 Radiative forcing of CH_4 and N_2O with respect to CO_2 at different time horizon and their evolution

		20-year scale	100-year scale	500-year scale
IPCC (2014)	CH_4	84	28	–
	N_2O	264	265	–
IPCC (2007)	CH_4	72	25	7.6
	N_2O	289	298	153
IPCC (2001)	CH_4	62	23	7
	N_2O	275	296	156
IPCC (1995)	CH_4	56	21	6.5
	N_2O	280	310	170
Climate change Report (1994)	CH_4	62	24.5	7.5
	N_2O	290	320	180
FAR (1992)	CH_4	–	11	–
	N_2O	–	270	–

Source Climate assessment reports of IPCC

2.2 *Net Global Warming Potential (NGWP) and GHGI*

For estimating NGWP, Robertson et al. (2000) used direct soil GHG emissions and included two additional factors representing (1) indirect GHG emissions (Eq. 1) associated with farm operation, including fertilizer, tillage, irrigation etc. and (2) changes in soil organic carbon stocks after converting the all factors into the standard scale of CO_2 equivalents.

$$NGWP = GWP_{direct} + GWP_{indirect} - CO_{2equivalent}\Delta SOC \quad (3)$$

Mosier et al. (2006) suggested a modified approach for short-term studies by replacing ΔSOC with the difference between CO_2-equivalents of organic inputs from previous crop residues and soil respiration rates (Eq. 4). Mosier et al.'s (2006) approach is commonly known as the soil respiration method.

$$NGWP = {*}GWP_{direct} + GWPindirect - CO_{2equivalent}Input(residue) \quad (4)$$

where *GWPdirect represents CO_2 equivalents of all three GHGs including N_2O, CH_4 and CO_2.

NGWP values estimated via soil respiration method may differ if calculated using NECB or soil-based ΔSOC approaches; therefore an agreement is required among researchers to use one standard method, or efforts should be made to find conversion of values of NGWP estimated by one method into others.

$$GHGI = \frac{NGWP}{yield} \quad (5)$$

2.2.1 $GWP_{indirect}$ (CO_2 Equivalents Associated with Agronomic Inputs & Farm Operations)

Fossil fuel or fossil fuel-based power (electricity) is required for most agronomic inputs and farm-gate operations, including fertilizer, residue application, plastic mulch, tillage, irrigation, pesticide application etc., thus, indirectly contributing to GHG emissions (Lal, 2004). In addition, off-gate farm operations, the main transport of agricultural input associated with GHG, are also recommended to estimate $GWP_{indirect}$ of agricultural operations (Robertson et al., 2000).

The CO_2 equivalent of different agricultural inputs and farm operations are estimated by multiplying the fuel and energy consumption rates with emission coefficients (EF). We found significant differences in EF values used in literature to convert farm operations into CO_2 equivalent. A summary is provided in Table 2. Severe limitations exist in reporting the energy consumption rates of on-gate and off-gate farm operations and comparing different agricultural management practices regarding their environmental effects.

$$GWP_{indirect} = Total\ units/ha_{farm\ operation/input} * EF(kgCO_{2\ equivalents\ per\ unit}) \quad (6)$$

2.2.2 Methods to Estimate Change in the Soil Organic Carbon (ΔS.O.C.)

There are several methods for recording changes in the SOC of the soil profile. The first approach considers all types of organic matter, including partially or incomplete decomposed residue within a time unit as, SOC (usually one year is recommended as net GWP and GHGI represent the year scale) and referred as SOC method .

$$\Delta SOC\left(kg\,C\,ha^{-1}\,year^{-1}\right) = \frac{SOC_{end\ of\ exp}ha^{-1} - SOC_{before\ mulching}ha^{-1}}{Experiment\ duration\ (years)} \quad (7)$$

A simple factor of 3.67 (or 44/12) is used for converting the calculated change in ΔSOC to $CO_{2equivalent}$;

$$\Delta SOC\left(kg\,CO_2\,ha^{-1}\,year^{-1}\right) = 3.67 \times \Delta SOC\left(kg\,C\,ha^{-1}\,year^{-1}\right) \quad (8)$$

However, considering the difficulties in recording the changes in SOC due to inherent high spatial variability and lack of analytical precision in methods to observe small changes in SOC concerning the significant background and short-term nature of experiments, the SOC method is not considered feasible. Some studies pointed out that at least five years or more of experimental time may require observing a considerable shift in SOC contents in agricultural soils depending on the size of native SOC (Sainju, 2020; Srinivasarao et al., 2015).

In the modified approach, researchers used estimates of net ecosystem carbon budget (NECB) as a replacement for ΔSOC (Eq. 7) (Huang et al., 2013a; Jia et al., 2021), which we referred to as the NECB method throughout the text. NECB represents the difference between C inputs and outputs (Eq. 9), and a detailed description of each parameter can be found elsewhere in the literature (Smith et al., 2010).

In brief, the individual parameters that are combined to calculate overall CO_2 entering into an agricultural ecosystem include net primary productivity (NPP), biomass entering from previous crop residues (i.e. mulching, crop stubbles, and litter), and organic fertilizers, including CO_2 associated with urea and via irrigation (dissolved organic carbon) (Eq. 10). The NPP represents all photo-assimilated CO_2 allocated to above- and below-ground biomass, including shoots, grains or fruits, roots, litter and rhizodeposits (Eq. 11). While the outputs not only include the soil respiration (Eq. 13) but also consider all the harvest products that are removed from

Table 2 Farm operations responsible for indirect GHG emission and EF adopted in literature to convert each unit into CO_2 equivalents

Farm input/operation	Unit	EF (Kg CO_2 equivalent per unit)	References
Power (Electricity)	KWH	0.28 (coal based)	West and Marland (2002)
		0.26 (petroleum based)	
		0.26 (coal based)	Zhang et al. (2013)
		0.072	Lal (2004), West and Marland (2002)
		1.30	Zhang et al. (2013)
		0.95	Wang et al. (2012)
Fuel (Diesel)	L	3.94	Huang et al. (2011, 2013b)
		0.72	Cheng et al. (2011)
		0.94	Lal (2004)
	Kg	3.75	Chen et al. (2014)
		0.89	Wang et al. (2017)
Pesticide	Kg	18	Huang et al. (2011)
		4.93	West and Marland (2002)
		5.1	Lal (2004)
		6.0	Dubey and Lal (2009)
		12.44	Wang et al. (2017)
Insecticide		16.61	Wang et al. (2016)
		4.93	West and Marland (2002)
Herbicide		10.15	Wang et al. (2016)
		4.70	West and Marland (2002)
Fungicide		5.18	
Seed		0.58	Wang et al. (2016)
N-fertilizer (N)	Kg	8.3	Zhang et al. (2013), Chen et al. (2014)
		1.74	Lu and Wang (2008)
		1.35	Dubey and Lal (2009), Lal (2004)
		1.53	Wang et al. (2017)
P-fertilizer (P_2O_5)	Kg	1.08	Huang et al. (2011)
		0.2	Dubey and Lal (2009), Lal (2004)
		0.79	Chen et al. (2014)
		1.63	Wang et al. (2017)
K-fertilizer	Kg	0.98	Huang et al. (2011)

(continued)

Table 2 (continued)

Farm input/operation	Unit	EF (Kg CO_2 equivalent per unit)	References
		0.15	Dubey and Lal (2009), Lal (2004)
		0.55	Chen et al. (2014)
		0.65	Wang et al. (2017)
Compound fertilizer	Kg	1.77	Wang et al. (2017)
Plastic film	Kg	5.18	Cheng et al. (2011)
		22.72	Wang et al. (2017)

the field (Eq. 12).

$$\Delta \text{SOC} = \text{NECB} = \sum C_{\text{inputs}} - \sum C_{\text{outputs}} \tag{9}$$

$$\sum C_{\text{input}} = \text{NPP} + \text{Additional residue} + \text{urea} \tag{10}$$

$$\text{NPP} = C_{\text{biomass}} + C_{\text{roots}} + C_{\text{rhizodeposition}} + C_{\text{litter}} \tag{11}$$

$$\sum C_{\text{output}} = C_{\text{in harvest removed from field}} + C_{\text{soil respired}} \tag{12}$$

$$\text{Soil respiration rates}\left(\text{KgCO}_2\text{ha}^{-1}\right) = 1 - f_{\text{root respiration}} * \text{CO}_{2\,(\text{emitted at soil surface})} \tag{13}$$

Both approaches to measuring changes in soil carbon (SOC) due to agriculture management have pros and cons, as both have the potential to affect net GWP values. The first approach (SOC method) is time-dependent as one may not observe bulk SOC changes within a few years of surface mulching practices. However, measuring changes in bulk SOC is probably the best as it accounts for all potential carbon inputs and outputs that are impossible in the NECB approach. The NECB approach is excellent for short-term experiments. It needs the amount of mulch residues and/or C added via vegetation, which is generally assumed based on aboveground biomass production (Jia et al., 2021). However, the NECB approach also requires soil respiration rates. The accurate accounting of CO_2 emitted via microbial respiration is challenging, as separating a fraction of CO_2 via root respiration is practically impossible without stable isotope tracing (Braig & Tupek, 2010). Some studies suggest including all respiration rates, while others use empirical factors to separate CO_2 emitted from contrasting sources (Huang et al., 2013a; Smith et al., 2010). The fraction of root respiration, for example, ranges from 0.15 to 0.95, depending on the crop and plant growth stages (Sainju, 2020).

3 Effects of Agricultural Surface Mulching on GWP and Yield Scale GWP

The exchange of GHGs between the atmosphere and soil is driven by soil microbial activities controlled by several key factors that can be influenced by soil surface mulching, among which soil moisture contents and soil temperature are the two most critical variables (Lavrent'ev et al., 2008). On the one hand, soil surface mulching may enhance the production of GHGs (e.g., CH_4 and N_2O) in soil via increasing soil moisture and temperature. On the other hand, surface-mulching materials, especially plastic, and high water contents may physically reduce the diffusion rates of GHGs (Ma et al., 2018). A detailed description of the effects of mulching on individual GHG emission processes and rates can be found elsewhere in this book.

Overall, the effects of agricultural soil surface mulching on GHG emissions based on GWP and yield-scaled GWP vary depending on the type of cropping system, type of mulching, changes in the intensity of agronomic inputs and co-application of surface mulching along with improved practices—e.g., replacing standing water condition in paddy with the saturated condition along with plastic covering etc.

3.1 Effects of Crop Residue Mulching

Between two major mulching types, residue mulching increase yield-scaled GWP due to its triggering effect on GHG emissions with and without improving crop yield (Zhang et al., 2016; Zheng et al., 2021). The magnitude of the positive effect of residue mulch on GWP associated with GHG emissions may depend on the composition of residue, application rates, cropping system, and environmental conditions.

In upland agricultural systems where CH_4 fluxes are generally negligible, N_2O contributes to total GWP. Studies investigating the effect of residue mulch on N_2O emissions showed multiple roles of residue mulch, including changes in physical conditions, enhanced supply of mineral N, and improved N-cycling microbial community activities (Chen et al., 2013; Garcia-Ruiz & Baggs, 2007). Garcia-Ruiz and Baggs (2007) pointed out that enhanced N_2O emissions from residue mulch are predominantly driven by the mulch driven changes in physical soil conditions. Placing residues on the soil surface promotes anaerobic microsites via moisture conservation, resulting in N_2O production by N-cycling microbial communities (Jiang et al., 2017). Garcia-Ruiz and Baggs (2007) pointed out that the residue mulch driven changes in physical soil conditions trigger N_2O emissions associated with synthetic fertilizer by 71–123%. While N_2O production promotion via N-supply directly from the mineralization of residue may also play a significant role, the emission factor of residue N is generally lower than synthetic fertilizer. Garcia-Ruiz and Baggs (2007) reported 0.9%, 2.5%, and 0.01% of N contents of rye, wheat and organic olive crop weed residues, released as N_2O, respectively, upon mulching, suggesting a significant role of the C:N ratio. The labile proportion of residue can also promote N_2O emissions via promoting activities of denitrifying microbial communities.

CH_4 is the major contributor to GWP in paddy cropping systems, constituting up to 98% of total GHG emissions driven by the continuous flooding conditions required for higher crop yield (Ma et al., 2008). As CH_4 producing biochemical processes are highly dependent on reduced conditions (Eh), various water management approaches have been evaluated for their potential to reduce CH_4 emission rates in paddy cultivation systems (Kreye et al., 2007; Zhang et al., 2020). Some widespread water-saving rice cultivation approaches include off-seasonal drainage or commonly known as upland rotation, mid-seasonal drainage (standing water is removed for a short period) and alternate wetting and drying (drainage during mid-season, post-grain filling and off-season) (Zou et al., 2005). Although there is inconsistency regarding the potential of different water management approaches in reducing the GWP of paddy cultivation, the following key observations are common among reports. Firstly, water drainage off or mid-season reduced CH_4 emission rates (Liao et al., 2020). Secondly, dry conditions caused by relatively better drainage significantly enhanced N_2O emission rates. Some studies even recorded higher GWP in drainage approaches as N_2O offset the decrease in CH_4 (Lagomarsino et al., 2016). Thirdly, pro-longed drainage could drastically reduce crop yield (Carrijo et al., 2017). Lastly, leaving land fallow during off-season reduced carbon contents sequestered during the paddy growing season, thus leading to net positive GWP.

In general, the use of residue mulch triggers the emission of GHGs by promoting the activities of CH_4 and N_2O, producing microbial communities via an enhanced supply of substrates. However, residue composition, application time, and method play a minor role. In this regard, (Hwang et al., 2017) pointed out that the choice of residue composition (CN ratio) and application time (at paddy transplant vs. during winter fallow season) significantly alter the residue driven changes in CH_4 emissions. Similarly, Lee et al. (2020) suggested that straw surface application may have less effect on CH_4 emission rates than surface incorporation. Among different water management approaches, residue mulch could increase GWP under paddy cultivation by 22–32% (Wu et al., 2019).

3.2 *Effects of Plastic Film Mulching*

Unlike residue mulch, plastic mulch showed decreased GWP and yield-scaled GWP associated with direct GHG emissions in upland agriculture systems (Fawibe et al., 2019; Li et al., 2014). Two factors mainly drive the decrease in GWP in the upland system through plastic mulching. Firstly, plastic mulch layer interaction with rainwater differs from residue mulch. It is important to note that rainfall events commonly correlate with high N_2O emissions, driven by enhanced denitrification rates (Malghani et al., 2020). Unlike residue mulch, plastic film mulches intercept the rainwater due to its impermeable nature and inhibit the formation of any anaerobic microenvironments (Berger et al., 2013). Secondly, plastic film mulching improves crop plants' nitrogen use efficiency, decreasing microbial available N contents for nitrifying and denitrifying microbial communities (Chen et al., 2014). Literature suggests that improved NUE decreases N_2O emissions from native soil N contents and mitigates fertilizer-induced N_2O emissions (Li et al., 2021; Zheng et al., 2021). Li et al. (2021) investigated the mitigating effect of plastic mulching on residue induced N_2O emissions in the maize cropping system and recorded approximately 52% less N_2O emissions rates from plastic mulch treatments. Plastic mulching significantly reduces yield-scaled GWP in upland soil systems, especially in semiarid regions with high N input. Studies recorded 18% and 27% increase in the yield of wheat and maize, respectively, the two most common crops in semiarid regions (He et al., 2018).

The use of plastic film mulch is also common in paddy systems of subtropical areas in China, where double cropping followed by winter fallow is the leading practice. Double cropping under paddy cultivation significantly increases the demand for irrigation water, and the ground cover rice production system (GCRP) with plastic film mulching is gaining popularity in these areas to save water. For instance, (Zhang et al., 2020) evaluated plastic film mulching in a paddy system combined with a reduction in irrigation water during the growing season against two traditional water regime management approaches, including continuous flooding (off and on seasons), off-seasonal drainage with flooding during paddy growth. The study found 18–60 and 34–60% decrease in GWP and yield-scaled GWP under the plastic mulching approach. In comparison with off-seasonal drainage treatment, which also reduced GWP concerning continuous flooding system but at the cost of reducing rice yield by 23%, the GCRP system has the potential to reduce GWP without compromising rice yield (Table 3).

4 Role of Mulching on NGWP and GHGI in Different Agricultural Systems

4.1 Agricultural Surface Mulching-driven Changes in Agronomic Inputs and Farm Operations Could Alter Indirect GHG Emissions

Agricultural soil surface mulching via either crop residue or plastic film can potentialy increase and/or decrease indirect GHGs emissions associated with agronomic operations and inputs. For instance, residue mulching supplies nutrients to the soil upon degradation, thus potentially reducing the need for nitrogenous and other fertilizers (Akhtar et al., 2020; Kong, 2014). Similarly, plastic mulching improves the N-use efficiency of crop plants, thus reducing fertilizers' indirect losses via leaching and volatilization (He et al., 2018). Moreover, enhanced water use efficiency and soil water conservation by surface mulches reduce the need for multiple irrigations.

Table 3 Literature review regarding the effect of different soil surface mulching approaches on GWP associated with direct GHG emissions and yield-scaled GWP. NT and RF represent non-tillage and ridge and furrow systems

Study	Cropping system	Exp. duration	Mulch type/rates (tons/ha)	Additional treatments	% change GWP	% change (Yield-scaled GWP)
Akhtar et al. (2020)	Rainfed maize-wheat	2 wheat growth seasons	Residue (maize)/4.5 & 9	N fertilizer rates	15 ~ 48	13 ~ 43
Chen et al. (2017)	Conventional (Wheat-maize)	2 years (full)	Residue/4		−29	−55
			Plastic (100%)		5	−25
			Plastic (50%)		−23	−37
Cuello et al. (2015)	Conventional (Maize)	2	plastic	NPK versus residue	16 ~ 79	−7 ~ 35
Jarecki and Lal (2006)	NT (No crop)		Residue		−2	
Kreye et al. (2007)	Paddy vs. water saving		Residue		70	270
			Plastic		107	146
Li et al. (2014)	Cotton	1 season	Plastic	Drip irrigation	−28	
Li et al. (2021)	Maize	2 seasons	Plastic	Straw incorporation	−26	−20
Liu et al. (2021)	Paddy (Water saving)	2 seasons	Residue (wheat)/8.3	Water saving	−33	−21
			Plastic		−48	−48
Okuda et al. (2007)	Orchard	3 months	Plastic		−42	
Peters et al. (2020)	NT (Maize)		residue	Cover crops	−24 ~ 87	
Varughese (2011)	NT (No crop)		Residue (maize)/8 & 16		43 ~ 104	
Wang et al. (2020)	Rice	2 Seasons	Residue (rapeseed)/6		21	19
Wu et al. (2019)	Rice	1 G season	Residue (rice)/5	Irrigation	28 ~ 50	18 ~ 47

(continued)

Table 3 (continued)

Study	Cropping system	Exp. duration	Mulch type/rates (tons/ha)	Additional treatments	% change GWP	% change (Yield-scaled GWP)
Xu et al. (2004)	Paddy (Water saving)	1 season	Residue (rice)/6	Field capacity	61	
			Plastic		88	
Xu et al. (2020)	RF (Wheat)	3 years (Seasonal)	plastic	Precipitation rates & irrigation amount	−111 ~ 15	−172 ~ 13
Yao et al. (2014)	Paddy (Water saving)	1 year	plastic	Fertilization types	−82 ~ 12	−84 ~ 15
Yao et al. (2017)			Plastic	saturated	−42	−47
			Bioplastic		−57	−59
			Plastic	Field capacity	−52	−56
Zhang et al. (2016)	Conventional (Paddy)	1 season	Residue (rapeseed)/3, 6 & 9		18 ~ 77	-4 ~ 53
Zhang et al. (2018)	Paddy (Water saving)	5 years (Seasonal)	Plastic		−54	−52
Zheng et al. (2021)	RF & Rainfed (Maize)	2 year seasons	Residue (wheat)/4.5	N fertilizer rates	−26 ~ −13	−32 ~ −18
			Plastic (RF)		50 ~ 98	31 ~ 74

Irrigation is the major component of farm operations with significantly high environmental costs (Lal, 2004). In addition, mulching mainly plastic mulch decreases the need for tillage via improved soil structure. Similarly, plastic mulching could efficiently control weed growth, reduce disease and insect invasion, and thus potentially reduce the use of pesticides (Shah & Wu, 2020). Unlike plastic mulch, residue mulch may increase weed growth and disease occurrence, thus increasing indirect GHG associated with pesticide or other form operations.

4.2 Agricultural Soil Surface-driven Changes in SOC (ΔSOC.)

Change in carbon stocks of agricultural soil is one of the essential parameters of net global warming potential and is crucial for enhancing soil resilience against climate change and degradation. According to IPCC guidelines, efforts must be adopted to increase the soil carbon stocks via a combination of adaptive measures, including no or reduced tillage, incorporation or surface application of residues and minimizing the loss of native soil carbon stocks via protection against erosion (IPCC, 2014). Globally soils are one of the largest terrestrial carbon reservoirs, but the proportion of cropland soils only accounts for 8–10% of this global carbon pool (Sainju, 2016; Smith et al., 2008). On the one hand, the relatively low proportion of cropland soil to global soil C stocks indicates the history of C loss driven by land-use change. On the other hand, it also points out the massive potential of cropland soils to mitigate current global climate change via atmospheric CO_2 sequestrations. Some estimates suggest cropland soil has the potential to sequester up to 1.6 Pg C equivalent each year (Smith et al., 2008).

In addition to the primary role of conserving soil moisture and reducing soil temperature fluctuations, the mulching technique also has the potential to alter soil carbon contents. The positive effects of surface mulching on soil carbon stocks include physical protection of organic matter via improved aggregation, reduction in soil carbon loss via soil erosion and enhanced incorporation of the organic matter directly via residue application or indirectly via ground crops biomass (He et al., 2018; Jia et al., 2021). While the adverse effects of mulching on soil carbon stock include enhancing microbial decomposition activities and a significant increase in GHG emissions (Chen et al., 2019; Ma et al., 2018). Some studies suggest that enhanced GHG emission by mulching sometimes offsets its SOC related mitigation capability (Steinmetz et al., 2016; Xia et al., 2018). Therefore, it is crucial to calculate net GWP to estimate the overall effect of surface mulching.

4.3 Agricultural Soil Surface Mulching and Crop Yield

Crop yield is one of the main parameters for all agricultural farming strategies with and without environmental conservation. The effect of surface mulching with residue and plastic film on crop yield depends on several factors, including mulch type, duration, crop type and climatic conditions (Ma et al., 2018; Yan et al., 2019).

The residue mulch improves soil structure, enhancing water availability and providing nutrients, thus potentially enhance crop growth. However, concerning conventional agriculture that depends heavily on fertilizers and tillage, residue mulch on crop yield depends on environmental conditions and rates of mulch (Yin et al., 2015). The residue mulch significantly increases grain yield in common crops of arid zones (e.g., corn, wheat, maize, cotton, and potato) by enhancing soil moisture conservation and water use efficiency (Pittelkow et al., 2015; Rusinamhodzi et al., 2011). In the presence of sufficient precipitations or irrigation, the difference in crop yield between conventional agricultural methods and surface mulching application may diminish (Yan et al., 2019). In addition, surface mulching may reduce crop yield if precipitation occurs at the initial seeding stages due to acidic conditions triggered by residue surface mulches (Shah & Wu, 2020).

Similarly, the effect of surface residue mulch also depends on residue application rates (Li et al., 2018). On the one hand, low application rates of residue may not conserve enough soil moisture. On the other hand, high application rates of residue mulch may trigger N immobilization, causing N deficiency in plants (Pi, 2017). The N-immobilization effect of residue mulch may also depend on the residue type, as different crop residues vary in their C:N ratio (Yan et al., 2019). Thus, appropriate rates of residue mulch depend on the residue type and climatic conditions of the area. The appropriate dose of fertilizer and improving time lapse between residue mulch application and planting are necessary to reduce N immobilization (Bhogal et al., 1997).

In comparison to residue mulch, plastic mulch's effect on crop yield is most often positive and linked with two key parameters, i.e., water and nitrogen use efficiency. Ma et al. (2018) showed an average of 43.1 (19.8–79.4%) increase in the yield of three major crops of Northwestern China, including potato, maize and wheat, under plastic mulching methods. The significant increase in crop yield was associated with cumulative precipitation rates and N inputs rates. Ma et al. (2018) recorded that the maximum increase in crop yield was achieved in regions where precipitation rates were between 100 and 200 mm and N-application rates were between 100 and 200 kg/ha. Among different crops, maize performed best under plastic mulching in the arid zone of China, as plastic mulch enhanced maize yield by 79.4% (Ma et al., 2018).

Li et al. (2018) compared the effects of two mulching strategies on potato yield based on field data across seven different geographic regions of China with varying annual precipitation rates and mean temperature. Their study suggests a mulch type-dependent interaction between regional climatic conditions and surface mulching exists as potato crops increase by residue mulch, and plastic film mulch differs among regions. Moreover, the effect of plastic mulching on potato yield was higher than residue mulch, 27.7 versus 21.5%, potentially due to plastic's high efficiency in conserving providing better insolation during relative cold periods.

5 Effects of Agricultural Soil Mulching on NGWP and GHGI Under Different Cultivation Systems

Agricultural soil surface mulching can reduce the environmental cost of crop cultivation depending on the intensity of changes in GHG emissions, agronomic inputs, farm operations, and SOC stocks. The relative contribution of each factor to NGWP varies between studies depending on the methods of NGWP estimations and mulching types. Studies that conducted direct measurements of net difference in SOC stock after continuous mulching of a cropping system found a decisive role of relative change in SOC under residue mulching, pointing potential of mulching to mitigate climate change (Pratibha et al., 2016). However, in general, SOC method based studies emphasized that direct GHG emissions and indirect GHG emissions via agronomic inputs and farm operations play a decisive role as net CO_2 exchange measured via ΔSOC is usually small. In contrast, the NECB method that accounts cumulative exchange of CO_2 between crop plants and cropping soil with the atmosphere favours the role of ΔSOC for NGWP estimates.

Among the significant factors that account for GHG emissions, the role of indirect exchange via agronomic practices, including fertilization inputs and farm operations, is often more significant than direct exchange rates as N_2O and CH_4 (Bayer et al., 2016; Pratibha et al., 2016; Wang et al., 2021). However, almost all known reports estimating the effect of mulching on NGWP and GHGI completely ignored this significant factor and thus do not truly represent the potential of mulching to reduce or induce NGWP in different cropping systems. Such limitations can be ignored for studies where mulching was the only difference among treatments, while the rest of agronomic inputs and farm operations were identical. However, under conditions when a mulching approach is accompanied by changes in the agronomic inputs such as residue mulch that could reduce fertilizer demand or water-saving in paddy, indirect GHG sources could play a decisive role in net estimates of NGWP. Thus, estimates of indirect GWP associated with mulching operation related activities such as plastic sheet production, transportation and application should not be ignored. Moreover, excluding indirect GHG emissions limits, the comparison among agronomic practices on a regional or global scale as the intensity of operation varies with the cultivation system.

5.1 *Conventional Agricultural Farming Systems*

Conventional cultivation systems rely heavily upon on-farm operations and agronomic inputs to maintain water and nutrient supply for high-yield production. Therefore, conventional agronomic practices are the primary cause of significant agricultural NGWP Farming operations, including different tillage and fertilizers, enhancing direct GHG emission rates and contributing to indirect GHG emissions. In addition, conventional cultivation approaches drastically affect soil SOC stocks, causing soil fertility loss. Mulching materials reduce the need for tillage and fertilizer rates by improving soil structure, rainwater infiltration, and water and fertilizer use efficiency, thus, ensuring sustainable crop yield and climate change mitigation.

Wang et al. (2021) investigated the effect of residue and plastic mulch in a long-term field experiment (15-years) in a semiarid climate without altering the agronomic operations and inputs. Although wheat crop yield was increased by 22% and 13% in plastic and residue mulch treatments with respect to no mulch, respectively, only residue mulch showed significant CO_2 mitigation effects, with NGWP and GHGI estimates decreasing by 124% and 121% in residue mulch, respectively. In contrast, plastic mulch increased NGWP and GHGI by 207 and 147%, suggesting plastic mulching may not be suitable without an additional conservation approach for improving wheat yield due to the large gap between the increase in crop yield and its environmental impact cost. While Lee et al. (2019) reported that enhanced estimates of NGWP under plastic mulch may not always translate into GHGI, especially if mulch improves yield significantly. The latter study recorded mitigation of the environmental cost of maize grain yield under plastic film mulching as GHGI was reduced by 25–30%. However, Lee et al. (2019) pointed out that using plastic mulch with straw incorporation may intensify the NGWP and GHGI due to relatively less increase in yield along triggering GHGs emissions. In contrast, Li et al. (2021) exhibited that 50% surface mulching using plastic film reduced GHG emissions from residue-incorporated soil under maize cultivation. The contrasting effect of plastic mulch with straw incorporation into the soil could be associated with local climatic conditions, as Lee et al. (2019) and Li et al. (2021) field experiments were situated under humid and semiarid climate conditions.

5.2 *Surface Mulching Along with a No-Tillage Approach*

Like agricultural surface mulching, the no-tillage approach was initially introduced to control soil erosion in dry regions where concentrated high rainfall events often cause water erosion under conventional tillage. However, significant yield loss, especially in the first few years of the no-tillage system, are the primary cause of relative less adaptation to this approach, especially in countries where farmers own small farms (Varughese, 2011). Moreover, a no-tillage approach often triggers GHG emissions, thus offsetting potential benefits (Srinivasarao et al., 2015). Using surface mulching in a no-tillage system could be a promising approach as crop yield is significantly improved under mulching via improved water and nutrient use potential. Pratibha et al. (2016) pointed out that a no-tillage system under residue mulch has a significantly high potential to sequester atmospheric CO_2 and thus could be adopted mitigate climate change.

Similarly, (Yagioka et al., 2015) recorded negative NGWP in the no-tillage system under weed mulch. Bayer et al. (2016) compared conventional agricultural systems with a no-tillage system under two mulch types and found the effect of mulching on NGWP was mulch and tillage system dependent. According to this study, a no-tillage system reduced NGWP and GHGI of maize crop when combined with the mulching approach. More importantly, relatively large residue mulch rates and no tillage were net sink for GHGs as NGWP estimates were negative.

5.3 *Ridge & Furrow Cultivation System*

Ridge and furrow (RF) is an innovative approach to improving the utilization and conservation of available water in arid and semiarid regions where precipitation events and rates are uneven and hard to manage for crop cultivation (Gan et al., 2013). Surface mulching is an integral part of the RF cultivation system to achieve high water use efficiency and reduce soil losses via surface run-off and evaporation. Depending on their availability, several types of material are used as surface mulch in RF systems, including plastic, crop residue and gravel. Plastic film mulch is among the most adopted mulching approach in areas of RF systems, while crop residues are partially used along with plastic mulch due to the scarcity of available crop residues. A detailed description of different RF practices can be found elsewhere in the literature (Gan et al., 2013). Briefly, the three most common approaches of mulching in RF system include; (1) both ridge and furrow covered with the same mulching material, i.e., plastic film, (2) both ridge and furrow covered with different mulching materials (plastic and residue mulch) and (3) only ridges covered with plastic mulch. Latter is among the most widely adopted approach and well documented regarding its effect on crop yield.

RF cultivation can potentially increase or decrease GWP associated with direct emissions of GHGs and net GWP of cropping in arid and semiarid regions, depending on annual rainfall rates and available irrigation to fulfill crop needs. Studies investigating the potential effect of RF systems on GHG emission found higher N_2O emission rates in RF with respect to conventional flat systems due to high moisture conservation (Berger et al., 2013; Jia et al., 2020). In this regard, Ali et al. (2021) compared N_2O emissions rates from RF and conventional flat systems at different simulated rainfall rates and suggested that RF systems only suppressed N_2O emissions at higher rainfall rates. While at simulated rainfall rates of 125 and 200 mm, N_2O emissions were higher in the RF system. However, regarding yield-scaled GWP, RF systems usually exhibit less GWP than conventional systems due to significant yield improvement (Xu et al., 2020).

A significant increase in crop yield could lead to enhanced soil carbon sequestration in RF cultivation systems. For instance, Yu et al. (2007) recorded that SOC contents increased from 0.69 to 1.73 g/kg after three years of converting the conventional system into an RF system. Among the three mulching approaches in RF, partial application of residue or residue mulch alone may even enhance the soil carbon sequestration potential of the RF system. Furthermore, ridges can be maintained without additional tillage, thus improving SOC stocks. In a long-term field experiment, Jiang and Xie (2009) recorded approximately a 48 g/kg increase in SOC in residue-mulched ridges after 5 years of preparation.

The overall effect of the RF system on NGWP and GHGI depends on the net balance among sources of GHGs, including increased rates of direct GHG emissions, indirect GHGs associated with additional agronomic operations, and soil carbon sequestration and crop yield. A few studies reported an increase in NGWP in RF systems compared to conventional systems, indicating RF systems may lead to an increase in the environmental cost of grain yield in semiarid or arid regions (Ali et al., 2021; Jia et al., 2020, 2021). Zheng et al. (2021) pointed out that residue mulch applied in conventional flat systems reduced NGWP and GHGI compared to RF systems. However, we found several limitations in studies estimating the changes in NGWP and GHGI after converting the conventional flat cultivation system to the RF system in arid and semiarid regions. Firstly, almost all studies measured seasonal GHG emission rates during crop growth, i.e., summer maize and winter wheat, instead of yearlong GHG emissions. Secondly, spatial variability in RF systems, ridges, furrows, and plant holes plastics were not considered while measuring the soil emitted GHGs (Berger et al., 2013). Thirdly, changes in agronomic inputs during the conversion of the conventional flat system to RF were also not considered. Two significant changes are apparent, including plastic mulch and the formation of ridges, and both are sources of indirect GHGs. Therefore, the CO_2 equivalent of these operations should be included in NGWP estimates. Fourthly, almost all-available literature estimating NGWP and GHGI of RF systems accounted for changes in SOC via the NECB approach, which generally overestimates the effects. We highly recommend that future studies consider the limitations mentioned above for a more comprehensive estimate.

5.4 Rice-Based Cropping System

The use of surface mulching in the paddy cultivation system is practised in two forms; (1). cultivation of upland crops, i.e. barley, rapeseed, vetch etc., immediately after rice harvesting, either harvested and spread as mulched during winter fallow season or kept as living mulch followed by incorporation (Hwang et al., 2017; Lee et al., 2020; Peters et al., 2020). (2) ground covering of the paddy field after preparation and fertilization with plastic film along with replacement of continuous standing water with maintaining the irrigation water levels at field capacity (70–90% WFPS) (Chen et al., 2019; Yao et al., 2014).

A few studies investigated the effect of residue mulch or residue application in paddy systems on NGWP and GHGI. We found only two published reports (Hwang et al., 2017; Lee et al., 2020) from temperate paddy system where single rice crop was cultivated and field were left fallow for the rest of the year. Interestingly, both reports used the NCBE approach to calculate the change in SOC and recorded the opposite effect of residue application on NGWP and GHGI. Hwang et al. (2017) reported that applying barley and hairy vetch individually or in combination promoted CH4 emissions rates with minimal soil carbon sequestration, leading to a significant increase in NGWP. In contrast, despite enhanced GHG emissions, Lee et al. (2020) recorded a 50–55% decrease in NGWP with rice residue application.

Yao et al. (2014) estimated NGWP and GHGI in paddy cropping system with and without ground covering approach of the plastic film under three different fertilization regimes, including no fertilizer, urea and chicken manure. This study found a significant decrease in NGWP and GHGI under plastic mulches primarily due to a reduction in the cumulative GHG emissions and SOC sequestration. However, their study also ignored indirect GWP associated with plastic mulch application and decreased irrigation water contents (Table 4).

Table 4 Brief review of the literature investigating the effects of surface mulching techniques on NGWP and GHGI in different cropping systems

Study	Cropping system	Exp. duration	Mulch type/rate (tons/ha)	Additional treatments	% change NGWP	% change GHGI
Ali et al. (2021) [c]	RF (Wheat)	2 years	Plastic on ridges	Irrigation amount (125–275 mm)	−4 ~ 96	−20 ~ 84
Bayer et al. (2016)[a]	NT (Maize)	1 year	Cover crops/4 ~ 5.5		−60 ~ −192	−52 ~ −196
Fan et al. (2018)[a]	NT vs. NT (Maize)	1 year	Residue (Maize)/9 ~ 12		−173.6	−168.4
Jia et al. (2020) [c]	RF (Maize)	2 years (Seasonal)	Plastic	N fertilizer rates	36 ~ 84	10 ~ 111
Jia et al. (2021) [c]	Conventional (Maize)	2 years (Seasonal)	Residue (Maize)/ 9		49	6
			Plastic		−22	−63
	RF (Maize)	2 years	Plastic		315	−34
			Plastic & residue/9	Residue in furrows	357	−36
Lee et al. (2019)[c]	Conventional (Maize)	2 years (Seasonal)	Plastic		22	−31
				Straw incorporation	141	50
Lee et al. (2020)[c]	Conventional (Paddy)	2 years	Residue (Rice)/ 8.25	Offseason mulch	6	−12
Li et al. (2021)[c]	Conventional (Maize)	2 years (Seasonal)	plastic	Straw incorporation	−82	−84
Liu et al. (2016)[a]	Paddy (Water saving)		Residue (Rice) 8.65		454	429
Pratibha et al. (2016)[a*]	NT vs. NT (P.pea/castor bean)		Residue (Crops)		−620 ~ −180	−545 ~ −165
Wang et al. (2021)[a*]	NT (Wheat/pea)	15 years (GHG as 1% EF of N fertilizer)	Residue (wheat)		−20	−176
			Plastic		39	7
	Conventional (Wheat/pea)		Residue (wheat)		−124	−121
			Plastic		207	147
Yao et al. (2014)[c]	Paddy (Water saving)		Plastic mulch	N-fertilizer type & rate	−61 ~ −10	−58 ~ −8

(continued)

Table 4 (continued)

Study	Cropping system	Exp. duration	Mulch type/rate (tons/ha)	Additional treatments	% change NGWP	% change GHGI
Zhang et al. (2014)[a]	NT vs. NT (Upland rice)		Residue (rapeseed)/3,6,9		−33 ~ −71	−35 ~ −72

a, b, c Symbols in the study column represent different NGWP estimation methods, SOC, soil respiration, and NECB. * indicate studies that included indirect GHGs emissions associated with farm operations. RF, NT, and plastic cover with water-saving irrigation in paddy systems were compared with traditional flat, tillage and paddy systems, respectively, if not mentioned

6 Conclusion and Future Perspectives

The effects of agricultural soil surface mulching on GWP, NGWP, and GHGI differ among studies on mulch type and other conservation approaches such as NT, RF, fertilization, and irrigation. Residue mulch enhances GHG emissions between the two major mulch types; however, GHGI may differ among cultivation systems. In contrast, there is considerable uncertainty in the literature regarding the effect of plastic mulch showing a significant increase, decrease, or no effect on GHG emissions. However, most literature found a significant decrease in GHG emissions from conservative cultivation systems under plastic mulch (Table 3). These conservation cultivation systems mainly include water-saving strategies in paddy and ridge and furrow systems. However, the positive and negative effects of residue and plastic mulch on GHG emissions may not always translate into an increase in NGWP and GHGI due to significant changes in soil carbon and crop yield (Fig. 1). On the one hand, the residue mulching approach triggers GHG emissions, but considerable significant inputs could improve SOC stock leading to a reduction in NGWP and GHGI. On the other hand, despite decreased GHG emissions or increased crop yield, the plastic mulching approach could enhance NGWP and GHGI if not used along with additional conservation practices.

In this review, we found several limitations in approaches and methodologies among studies investigating the role of surface mulching on GWP, NGWP and GHGI, thus suggesting that future studies should consider the following points;

1. The NECB approach overestimates the NGWP of crop cultivation systems due to the use of CO_2 sequestered by plants into their biomass as an output or loss of SOC. Suppose the harvest is removed from calculations of the NECB method, using C inputs via NPP of below-ground biomass and crop litter and accounting for CO_2 emission rates. In that case, the NECB approach can be used as a soil respiration method suggested by (Mosier et al., 2006).
2. Direct soil carbon measurements are potentially the best approach to estimate agricultural surface mulch-driven changes in the net exchange of C between soil and atmosphere. The soil carbon approach also includes the potential loss of soil carbon via wind and water erosions that are generally ignored by the modified

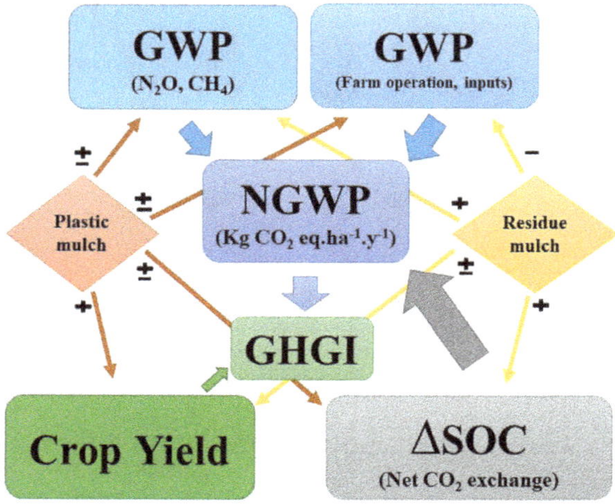

Fig. 1 Potential effects of plastic and residue mulch on different factors of NGWP and GHGI

approaches. Although the fate of eroded SOC is debatable, a decrease in erosion is one of the key services of surface mulching, especially in dry climates (Park et al., 2021). Thus, the CO_2 equivalent of this service should not be ignored.

3. Sampling intervals should be shortened for soil emitted gas measurements depending on rainfall events. Moreover, temporal and spatial variations should also be considered for estimating the absolute amount of GHGs. The majority of studies investigating the effect of mulching on GHG based GWP performed seasonal measurements. Seasonal measurements may be suitable for estimating the yield-scaled GWP of a single crop. However, it does not truly fit for estimating the mitigation potential of a cultivation system under mulching. The effect of residue and plastic mulch is not limited to one season or one year; thus, demand for yearly based GHGs measurements
4. N-fertilizer based N_2O emission factor (EF) may be considerably induced by residue mulch; thus, more research is needed. Garcia-Ruiz and Baggs (2007) pointed out that the residue mulch driven changes in physical soil conditions affected residue interactions with synthetic fertilizer, resulting in 71–123% increase in N_2O emission and thus have the potential to alter the N-fertilizer emission factor (EF).
5. Potential of residue mulch to supply nutrients to plants and its ability to increase N_2O emission factor of N-fertilizer, demand N fertilizer management. Synthetic fertilizer use can be reduced under residue mulching without negatively affecting yield (Akhtar et al., 2020).
6. Residue mulching can decrease indirect GHGs emissions associated with agronomic inputs and farm operations. Therefore, it is necessary to include CO_2 equivalents of different agronomic inputs and farm operations to represent the actual environmental cost of cropping under different cultivation systems

Acknowledgments This work was supported by the Natural Science Foundation of China (grant no. 42007090, 32050410301), Nanjing Forestry University start-up funding (GXL2020004), High-level Innovation and Entrepreneurship Talents Introduction Program of Jiangsu Province of China.

References

Akhtar, K., Wang, W., Ren, G., Khan, A., Enguang, N., Khan, A., Feng, Y., Yang, G., & Wang, H. (2020). Straw mulching with inorganic nitrogen fertilizer reduces soil CO_2 and N_2O emissions and improves wheat yield. *Science of the Total Environment, 741*, 140488.

Ali, S., Xu, Y., Ma, X., Jia, Q., & Jia, Z. (2021). Farming practices and deficit irrigation management improve winter wheat crop water productivity and biomass through mitigated greenhouse gas intensity under semiarid regions. Environmental Science and Pollution Research.

An, T., Schaeffer, S., Li, S., Fu, S., Pei, J., Li, H., Zhuang, J., Radosevich, M., & Wang, J. (2015). Carbon fluxes from plants to soil and dynamics of microbial immobilization under plastic film mulching and fertilizer application using ^{13}C pulse-labeling. *Soil Biology and Biochemistry, 80*, 53–61.

Bayer, C., Gomes, J., Zanatta, J. A., Vieira, F. C. B., & Dieckow, J. (2016). Mitigating greenhouse gas emissions from a subtropical Ultisol by using long-term no-tillage in combination with legume cover crops. *Soil and Tillage Research, 161*, 86–94.

Berger, S., Kim, Y., Kettering, J., & Gebauer, G. (2013). Plastic mulching in agriculture—Friend or foe of N_2O emissions? *Agriculture, Ecosystems & Environment, 167*, 43–51.

Bhogal, A., Young, S. D., & Sylvester-Bradley, R. (1997). Straw incorporation and immobilization of spring-applied nitrogen. *Soil Use and Management, 13*, 111–116.

Braig, E., & Tupek, B. (2010). Separating soil respiration components with stable isotopes: Natural abundance and labelling approaches. *Iforest, 3*, 92–94.

Carrijo, D. R., Lundy, M. E., & Linquist, B. A. (2017). Rice yields and water use under alternate wetting and drying irrigation: A meta-analysis. *Field Crops Research, 203*, 173–180.

Chen, H., Li, X., Hu, F., & Shi, W. (2013). Soil nitrous oxide emissions following crop residue addition: A meta-analysis. *Global Change Biology, 19*, 2956–2964.

Chen, X., Cui, Z., Fan, M., Vitousek, P., Zhao, M., Ma, W., Wang, Z., Zhang, W., Yan, X., Yang, J., Deng, X., Gao, Q., Zhang, Q., Guo, S., Ren, J., Li, S., Ye, Y., Wang, Z., Huang, J., et al. (2014). Producing more grain with lower environmental costs. *Nature, 514*, 486–489.

Chen, H., Liu, J., Zhang, A., Chen, J., Cheng, G., Sun, B., Pi, X., Dyck, M., Si, B., Zhao, Y., & Feng, H. (2017). Effects of straw and plastic film mulching on greenhouse gas emissions in Loess Plateau, China: A field study of 2 consecutive wheat-maize rotation cycles. *Science of the Total Environment, 579*, 814–824.

Chen, S.-J., Jiang, C.-S., Ni, X., Li, X.-X., & Hao, Q.-J. (2019). Effect of plastic film mulching on greenhouse gas rmissions from rice-rapeseed rotation in cropland. *Huan jing ke xue = Huanjing kexue, 40*, 4213–4220.

Cheng, K., Pan, G., Smith, P., Luo, T., Li, L., Zheng, J., Zhang, X., Han, X., & Yan, M. (2011). Carbon footprint of China's crop production—An estimation using agro-statistics data over 1993–2007. *Agriculture, Ecosystems & Environment, 142*, 231–237.

Cuello, J. P., Hwang, H. Y., Gutierrez, J., Kim, S. Y., & Kim, P. J. (2015). Impact of plastic film mulching on increasing greenhouse gas emissions in temperate upland soil during maize cultivation. *Applied Soil Ecology, 91*, 48–57.

Dubey, A., & Lal, R. (2009). Carbon Footprint and sustainability of agricultural production systems in Punjab, India, and Ohio, USA. *Journal of Crop Improvement, 23*, 332–350.

Fan, J., Luo, R., Liu, D., Chen, Z., Luo, J., Boland, N., Tang, J., Hao, M., McConkey, B., & Ding, W. (2018). Stover retention rather than no-till decreases the global warming potential of rainfed continuous maize cropland. *Field Crops Research, 219*, 14–23.

Fawibe, O. O., Honda, K., Taguchi, Y., Park, S., & Isoda, A. (2019). Greenhouse gas emissions from rice field cultivation with drip irrigation and plastic film mulch. *Nutrient Cycling in Agroecosystems, 113*, 51–62.

Gan, Y., Siddique, K. H. M., Turner, N. C., Li, X.-G., Niu, J.-Y., Yang, C., Liu, L., & Chai, Q. (2013). Ridge-furrow mulching systems—An innovative technique for boosting crop productivity in semiarid rain-fed environments. In D. L. Sparks (Ed.), *Advances in agronomy* (pp. 429–476). Academic Press.

Garcia-Ruiz, R., & Baggs, E. M. (2007). N2O emission from soil following combined application of fertiliser-N and ground weed residues. *Plant and Soil, 299*, 263–274.

Gonzalez-Sanchez, E. J., Veroz-Gonzalez, O., Conway, G., Moreno-Garcia, M., Kassam, A., Mkomwag, S., Ordonez-Fernandez, R., Trivino-Tarradas, P., & Carbonell-Bojollo, R. (2019). Meta-analysis on carbon sequestration through conservation agriculture in Africa. *Soil & Tillage Research, 190*, 22–30.

He, G., Wang, Z., Li, S., & Malhi, S. S. (2018). Plastic mulch: Tradeoffs between productivity and greenhouse gas emissions. *Journal of Cleaner Production, 172*, 1311–1318.

Huang, J., Chen, Y., Sui, P., & Gao, W. (2013a). Estimation of net greenhouse gas balance using crop- and soil-based approaches: Two case studies. *Science of the Total Environment, 456–457*, 299–306.

Huang, T., Gao, B., Christie, P., & Ju, X. (2013b). Net global warming potential and greenhouse gas intensity in a double-cropping cereal rotation as affected by nitrogen and straw management. *Biogeosciences, 10*, 7897–7911.

Hwang, H. Y., Kim, G. W., Kim, S. Y., Mozammel Haque, M., Khan, M. I., & Kim, P. J. (2017). Effect of cover cropping on the net global warming potential of rice paddy soil. *Geoderma, 292*, 49–58.

IPCC. (2014). Climate Change 2014: Synthesis report. contribution of working groups I, II and III to the fifth assessment report of the intergovernmental panel on climate change. In R. K. Pachauri & L. A. Meyer (Eds.).

Jarecki, M. K., & Lal, R. (2006). Compost and mulch effects on gaseous flux from an alfisol in Ohio. *Soil Science, 171*, 249–260.

Jia, Q.M., Wang, J., Ali, S., Chang, S. H., Zhang, C., & Hou, F. J. (2020). Nutrient management and cultivation techniques affect maize production through regulating greenhouse gas intensity and carbon budget under semiarid climate. *Journal of Cleaner Production, 276*.

Jia, Q. M., Zhang, H. X., Wang, J., Xiao, X. M., Chang, S. H., Zhang, C., Liu, Y. J., & Hou, F. J. (2021). Planting practices and mulching materials improve maize net ecosystem C budget, global warming potential and production in semiarid regions. *Soil & Tillage Research, 207*.

Jiang, X.-J., & Xie, D.-T. (2009). Combining ridge with no-tillage in lowland rice-based cropping system: Long-term effect on soil and rice yield. *Pedosphere, 19*, 515–522.

Jiang, C. M., Yu, W. T., Ma, Q., Xu, Y. G., & Zou, H. (2017). Alleviating global warming potential by soil carbon sequestration: A multi-level straw incorporation experiment from a maize cropping system in Northeast China. *Soil and Tillage Research, 170*, 77–84.

Johnson, J.M.-F., Allmaras, R. R., & Reicosky, D. C. (2006). Estimating source carbon from crop residues, roots and rhizodeposits using the national grain-yield database. *Agronomy Journal, 98*, 622–636.

Kong, L. (2014). Maize residues, soil quality, and wheat growth in China. *A Review. Agronomy for Sustainable Development, 34*, 405–416.

Kreye, C., Dittert, K., Zheng, X., Zhang, X., Lin, S., Tao, H., & Sattelmacher, B. (2007). Fluxes of methane and nitrous oxide in water-saving rice production in north China. *Nutrient Cycling in Agroecosystems, 77*, 293–304.

Lagomarsino, A., Agnelli, A. E., Linquist, B., Adviento-Borbe, M. A., Agnelli, A., Gavina, G., Ravaglia, S., & Ferrara, R. M. (2016). Alternate wetting and drying of rice reduced CH_4 emissions but triggered N_2O peaks in a clayey soil of Central Italy. *Pedosphere, 26*, 533–548.

Lal, R. (2004). Carbon emission from farm operations. *Environment International, 30*, 981–990.

Lavrent'ev, R. B., Zaitsev, S. A., Sudnitsyn, I. I., & Kurakov, A. V. (2008). Nitrous oxide production by fungi in soils under different moisture levels. *Moscow University Soil Science Bulletin, 63*, 178–183.

Lee, J. H., Lee, J. G., Jeong, S. T., Gwon, H. S., Kim, P. J., & Kim, G. W. (2020). Straw recycling in rice paddy: Trade-off between greenhouse gas emission and soil carbon stock increase. *Soil and Tillage Research, 199*, 104598.

Lee, J. G., Cho, S. R., Jeong, S. T., Hwang, H. Y., & Kim, P. J. (2019). Different response of plastic film mulching on greenhouse gas intensity (GHGI) between chemical and organic fertilization in maize upland soil. *Science of the Total Environment, 696*.

Li, Z., Zhang, R., Wang, X., Chen, F., Lai, D., & Tian, C. (2014). Effects of plastic film mulching with drip irrigation on N_2O and CH_4 emissions from cotton fields in arid land. *The Journal of Agricultural Science, 152*, 534–542.

Li, Q., Li, H., Zhang, L., Zhang, S., & Chen, Y. (2018). Mulching improves yield and water-use efficiency of potato cropping in China: A meta-analysis. *Field Crops Research, 221*, 50–60.

Li, N., Ma, X., Bai, J., Xu, H., Feng, Y., Ren, G., Yang, G., Han, X., Wang, X., Ren, C., & Kong, D. (2021). Plastic film mulching mitigates the straw-induced soil greenhouse gas emissions in summer maize field. *Applied Soil Ecology, 162*, 103876.

Liao, B., Wu, X., Yu, Y., Luo, S., Hu, R., & Lu, G. (2020). Effects of mild alternate wetting and drying irrigation and mid-season drainage on CH_4 and N_2O emissions in rice cultivation. *Science of the Total Environment, 698*, 134212.

Liu, W., Hussain, S., Wu, L., Qin, Z., Li, X., Lu, J., Khan, F., Cao, W., & Geng, M. (2016). Greenhouse gas emissions, soil quality, and crop productivity from a mono-rice cultivation system as influenced by fallow season straw management. *Environmental Science and Pollution Research, 23*, 315–328.

Liu, J., Huang, X., Jiang, H., & Chen, H. (2021). Sustaining yield and mitigating methane emissions from rice production with plastic film mulching technique. *Agricultural Water Management, 245*, 106667.

Lu, X. (2020). A meta-analysis of the effects of crop residue return on crop yields and water use efficiency. *PLoS ONE, 15*, e0231740.

Ma, J., Xu, H., Yagi, K., & Cai, Z. (2008). Methane emission from paddy soils as affected by wheat straw returning mode. *Plant and Soil, 313*, 167–174.

Ma, D., Chen, L., Qu, H., Wang, Y., Misselbrook, T., & Jiang, R. (2018). Impacts of plastic film mulching on crop yields, soil water, nitrate, and organic carbon in Northwestern China: A meta-analysis. *Agricultural Water Management, 202*, 166–173.

Malghani, S., Yoo, G.-Y., Giesemann, A., Well, R., & Kang, H. (2020). Combined application of organic manure with urea does not alter the dominant biochemical pathway producing N_2O from urea treated soil. *Biology and Fertility of Soils, 56*, 331–343.

Mosier, A. R., Halvorson, A. D., Reule, C. A., & Liu, X. J. (2006). Net global warming potential and greenhouse gas intensity in irrigated cropping systems in Northeastern Colorado. *Journal of Environmental Quality, 35*, 1584–1598.

Okuda, H., Noda, K., Sawamoto, T., Tsuruta, H., Hirabayashi, T., Yonemoto, J. Y., & Yagi, K. (2007). Emission of N_2O and CO_2 and uptake of CH_4 in soil from a satsuma mandarin orchard under mulching cultivation in central Japan. *Journal of the Japanese Society for Horticultural Science, 76*, 279–287.

Park, S.-I., Yang, H. I., Park, H.-J., Seo, B.-S., Jeong, Y.-J., Lim, S.-S., Kwak, J.-H., Kim, H.-Y., Yoon, K.-S., Lee, S.-M., & Choi, W.-J. (2021). Rice straw cover decreases soil erosion and sediment-bound C, N, and P losses but increases dissolved organic C export from upland maize fields as evidenced by $\delta^{13}C$. *Science of the Total Environment, 753*, 142053.

Peters, S. J. W., Saikawa, E., Markewitz, D., Sutter, L., Avramov, A., Sanders, Z. P., Yosen, B., Wakabayashi, K., Martin, G., Andrews, J. S., & Hill, N. S. (2020). Soil trace gas fluxes in living mulch and conventional agricultural systems. *Journal of Environmental Quality, 49*, 268–280.

Pi, X., Zhang, T., Sun, B., Cui, Q., Guo, Y., Gao, M., Feng, H., & Hopkins, D. W. (2017). Effects of mulching for water conservation on soil carbon, nitrogen and biological properties. *Frontiers of Agricultural Science and Engineering, 4*, 146–154.

Pittelkow, C. M., Liang, X., Linquist, B. A., van Groenigen, K. J., Lee, J., Lundy, M. E., van Gestel, N., Six, J., Venterea, R. T., & van Kessel, C. (2015). Productivity limits and potentials of the principles of conservation agriculture. *Nature, 517*, 365–368.

Pratibha, G., Srinivas, I., Rao, K. V., Shanker, A. K., Raju, B. M. K., Choudhary, D. K., Srinivas Rao, K., Srinivasarao, C., & Maheswari, M. (2016). Net global warming potential and greenhouse gas intensity of conventional and conservation agriculture system in rainfed semi arid tropics of India. *Atmospheric Environment, 145*, 239–250.

Robertson, G. P., & Grace, P. R. (2004). Greenhouse gas fluxes in tropical and temperate agriculture: The need for a full-cost accounting of global warming potentials. *Environment, Development and Sustainability, 6*, 51–63.

Robertson, G. P., Paul, E. A., & Harwood, R. R. (2000). Greenhouse gases in intensive agriculture: Contributions of individual gases to the radiative forcing of the atmosphere. *Science, 289*, 1922–1925.

Rusinamhodzi, L., Corbeels, M., van Wijk, M. T., Rufino, M. C., Nyamangara, J., & Giller, K. E. (2011). A meta-analysis of long-term effects of conservation agriculture on maize grain yield under rain-fed conditions. *Agronomy for Sustainable Development, 31*, 657.

Sainju, U. M. (2016). A global meta-analysis on the impact of management practices on net global warming potential and greenhouse gas Intensity from cropland soils. *PLoS ONE, 11*, e0148527.

Sainju, U. M. (2020). Net global warming potential and greenhouse gas intensity. *Soil Science Society of America Journal, 84*, 1393–1404.

Shah, F., & Wu, W. (2020). Use of plastic mulch in agriculture and strategies to mitigate the associated environmental concerns. In D. L. Sparks (Ed.), *Advances in agronomy* (pp. 231–287). Academic Press.

Smith, P., Martino, D., Cai, Z., Gwary, D., Janzen, H., Kumar, P., McCarl, B., Ogle, S., O'Mara, F., Rice, C., Scholes, B., Sirotenko, O., Howden, M., McAllister, T., Pan, G., Romanenkov, V., Schneider, U., Towprayoon, S., Wattenbach, M., & Smith, J. (2008). Greenhouse gas mitigation in agriculture. *Philosophical Transactions of the Royal Society B: Biological Sciences, 363*, 789–813.

Smith, P., Lanigan, G., Kutsch, W. L., Buchmann, N., Eugster, W., Aubinet, M., Ceschia, E., Béziat, P., Yeluripati, J. B., Osborne, B., Moors, E. J., Brut, A., Wattenbach, M., Saunders, M., & Jones, M. (2010). Measurements necessary for assessing the net ecosystem carbon budget of croplands. *Agriculture, Ecosystems & Environment, 139*, 302–315.

Srinivasarao, C., Lal, R., Kundu, S., & Thakur, P. B. (2015). Conservation agriculture and soil carbon sequestration. In M. Farooq & K. H. M. Siddique (Eds.), *Conservation agriculture* (pp. 479–524). Springer International Publishing.

Steinmetz, Z., Wollmann, C., Schaefer, M., Buchmann, C., David, J., Tröger, J., Muñoz, K., Frör, O., & Schaumann, G. E. (2016). Plastic mulching in agriculture. Trading short-term agronomic benefits for long-term soil degradation? *Science of the Total Environment, 550*, 690–705.

Tubiello, F. N. (2019). Greenhouse gas emissions due to agriculture. In P. Ferranti, E. M. Berry, & J. R. Anderson (Eds.), *Encyclopedia of food security and sustainability* (pp. 196–205). Elsevier.

Varughese, A. M. (2011). Mulching and tillage effects on GHG emissions and properties of an Alfisol in Central Ohio. The Ohio State University.

Wang, J., Rothausen, S. G. S. A., Conway, D., Zhang, L., Xiong, W., Holman, I. P., & Li, Y. (2012). China's water–energy nexus: Greenhouse-gas emissions from groundwater use for agriculture. *Environmental Research Letters, 7*, 014035.

Wang, Z.-B., Chen, J., Mao, S.-C., Han, Y.-C., Chen, F., Zhang, L.-F., Li, Y.-B., & Li, C.-D. (2017). Comparison of greenhouse gas emissions of chemical fertilizer types in China's crop production. *Journal of Cleaner Production, 141*, 1267–1274.

Wang, W., Akhtar, K., Ren, G., Yang, G., Feng, Y., & Yuan, L. (2019). Impact of straw management on seasonal soil carbon dioxide emissions, soil water content, and temperature in a semiarid region of China. *Science of the Total Environment, 652*, 471–482.

Wang, L., Qin, T., Liu, T., Guo, L., Li, C., & Zhai, Z. (2020). Inclusion of microbial inoculants with straw mulch enhances grain yields from rice fields in central China. *Food and Energy Security, 9.*

Wang, L., Li, L., Xie, J., Luo, Z., Zhang, R., Cai, L., Coulter, J. A., & Palta, J. A. (2021). Managing the trade-offs among yield, economic benefits and carbon and nitrogen footprints of wheat cropping in a semiarid region of China. *Science of the Total Environment, 768*, 145280.

West, T. O., & Marland, G. (2002). A synthesis of carbon sequestration, carbon emissions, and net carbon flux in agriculture: Comparing tillage practices in the United States. *Agriculture, Ecosystems & Environment, 91*, 217–232.

Wu, X., Wang, W., Xie, K., Yin, C., Hou, H., & Xie, X. (2019). Combined effects of straw and water management on CH_4 emissions from rice fields. *Journal of Environmental Management, 231*, 1257–1262.

Xia, L., Lam, S. K., Wolf, B., Kiese, R., Chen, D., & Butterbach-Bahl, K. (2018). Trade-offs between soil carbon sequestration and reactive nitrogen losses under straw return in global agroecosystems. *Global Change Biology, 24*, 5919–5932.

Xu, Y. C., Shen, Q. R., Li, M. L., Dittert, K., & Sattelmacher, B. (2004). Effect of soil water status and mulching on N_2O and CH_4 emission from lowland rice field in China. *Biology and Fertility of Soils, 39*, 215–217.

Xu, Y., Wang, Y., Ma, X., Liu, X., Zhang, P., Cai, T., & Jia, Z. (2020). Ridge-furrow mulching system and supplementary irrigation can reduce the greenhouse gas emission intensity. *Science of the Total Environment, 717.*

Yagioka, A., Komatsuzaki, M., Kaneko, N., & Ueno, H. (2015). Effect of no-tillage with weed cover mulching versus conventional tillage on global warming potential and nitrate leaching. *Agriculture, Ecosystems & Environment, 200*, 42–53.

Yan, F., Sun, Y., Hui, X., Jiang, M., Xiang, K., Wu, Y., Zhang, Q., Tang, Y., Yang, Z., Sun, Y., & Jun, M. (2019). The effect of straw mulch on nitrogen, phosphorus and potassium uptake and use in hybrid rice. *Paddy and Water Environment, 17*, 23–33.

Yao, Z., Du, Y., Tao, Y., Zheng, X., Liu, C., Lin, S., & Butterbach-Bahl, K. (2014). Water-saving ground cover rice production system reduces net greenhouse gas fluxes in an annual rice-based cropping system. *Biogeosciences, 11*, 6221–6236.

Yao, Z., Zheng, X., Liu, C., Lin, S., Zuo, Q., & Butterbach-Bahl, K. (2017). Improving rice production sustainability by reducing water demand and greenhouse gas emissions with biodegradable films. *Scientific Reports, 7*, 39855.

Yin, W., Yu, A., Chai, Q., Hu, F., Feng, F., & Gan, Y. (2015). Wheat and maize relay-planting with straw covering increases water use efficiency up to 46%. *Agronomy for Sustainable Development, 35*, 815–825.

Yu, J., Bingcheng, X., Fengmin, L., & Xiaoling, W. (2007). Availability and contributions of soil phosphorus to forage production of seeded alfalfa in semiarid Loess Plateau. *Acta Ecologica Sinica, 27*, 42–47.

Zhang, W.-F., Dou, Z.-X., He, P., Ju, X.-T., Powlson, D., Chadwick, D., Norse, D., Lu, Y.-L., Zhang, Y., Wu, L., Chen, X.-P., Cassman, K. G., & Zhang, F.-S. (2013). New technologies reduce greenhouse gas emissions from nitrogenous fertilizer in China. *Proceedings of the National Academy of Sciences, 110*, 8375–8380.

Zhang, Z. S., Cao, C. G., Guo, L. J., & Li, C. F. (2014). The effects of rape residue mulching on net global warming potential and greenhouse gas intensity from no-tillage paddy fields. *The Scientific World Journal, 2014*, 198231.

Zhang, Z.-S., Cao, C.-G., Guo, L.-J., & Li, C.-F. (2016). Emissions of CH_4 and CO_2 from paddy fields as affected by tillage practices and crop residues in central China. *Paddy and Water Environment, 14*, 85–92.

Zhang, G., Ma, J., Yang, Y., Yu, H., Song, K., Dong, Y., Lv, S., & Xu, H. (2018). Achieving low methane and nitrous oxide emissions with high economic incomes in a rice-based cropping system. *Agricultural and Forest Meteorology, 259*, 95–106.

Zhang, G., Yang, Y., Huang, Q., Ma, J., Yu, H., Song, K., Dong, Y., Lv, S., & Xu, H. (2020). Reducing yield-scaled global warming potential and water use by rice plastic film mulching in a winter flooded paddy field. *European Journal of Agronomy, 114*.

Zheng, J., Fan, J., Zhang, F., Guo, J., Yan, S., Zhuang, Q., Cui, N., & Guo, L. (2021). Interactive effects of mulching practice and nitrogen rate on grain yield, water productivity, fertilizer use efficiency and greenhouse gas emissions of rainfed summer maize in northwest China. *Agricultural Water Management, 248*, 106778.

Zou, J., Huang, Y., Jiang, J., Zheng, X., & Sass, R. L. (2005). A 3-year field measurement of methane and nitrous oxide emissions from rice paddies in China: Effects of water regime, crop residue, and fertilizer application. *Global Biogeochemical Cycles, 19*.

GPSR Compliance
The European Union's (EU) General Product Safety Regulation (GPSR) is a set of rules that requires consumer products to be safe and our obligations to ensure this.

If you have any concerns about our products, you can contact us on

ProductSafety@springernature.com

In case Publisher is established outside the EU, the EU authorized representative is:

Springer Nature Customer Service Center GmbH
Europaplatz 3
69115 Heidelberg, Germany

www.ingramcontent.com/pod-product-compliance
Lightning Source LLC
Chambersburg PA
CBHW071350220926
51613CB00008B/39
9789811964091